SELECTED PAPERS

S. Chandrasekhar

SELECTED PAPERS

S. Chandrasekhar

*

VOLUME 1

Stellar Structure and Stellar Atmospheres

THE UNIVERSITY OF CHICAGO PRESS
CHICAGO AND LONDON

The University of Chicago Press, Chicago 60637
The University of Chicago Press, Ltd., London

Printed in the United States of America

98 97 96 95 94 93 92 91 90 89 5 4 3 2 1

 Acknowledgments appear on page 516.

Library of Congress Cataloging in Publication Data

Chandrasekhar, S. (Subrahmanyan), 1910–
Stellar structure and stellar atmosphere.
p. cm.—(Selected papers of S. Chandrasekhar; v. 1)
1. Stars—Structure. 2. Stars—Atmospheres. I. Title.
II. Series: Chandrasekhar, S. (Subrahmanyan), 1910– Selections.
1989; v. 1.
QB808.C47 1989 88-26110
523.8—dc19 CIP
ISBN 0-226-10089-8 (Vol. 1)
ISBN 0-226-10090-1 (Vol. 1, pbk.)

The paper used in this publication meets the minimum requirements of the American National Standard for Information Sciences—Permanence of Paper for Printed Library Materials, ANSI Z39.48-1984.

S. Chandrasekhar is the Morton D. Hull Distinguished Service Professor Emeritus in the Department of Astronomy and Astrophysics, the Department of Physics, and the Enrico Fermi Institute at the University of Chicago.

Contents

PART TWO

The Equilibrium of Distorted Polytropes

PART THREE

Stellar Evolution

PART FOUR

Integral Theorems on the Equilibrium of a Star

PART FIVE

Theory of Stellar Atmospheres

A complete list of publications by S. Chandrasekhar will appear at the end of the final volume

Preface

The suggestion that I might consider publishing a collection of my scientific papers (or a selection of them) was first made to me by my wife in the early sixties. While I was flattered by the suggestion—the idea had not occurred to me before—I did not take it seriously. The same suggestion was made a few years later by the science editors of the University of Chicago Press. By that time I had been prepared to be more receptive, and a contract was signed in July 1970.

Once I took the suggestion seriously, I saw a major obstacle in proceeding with the project. I had periodically assembled in a monograph my matured outlook on a particular area in which I had concentrated for some years and incorporated, in a coherent framework, much of my published work on the subject. On this account, it seemed to me that the publication of a collection of my papers would be superfluous. Nevertheless, I suggested to the editors of the Press that they seek the opinions of a selected group of scientists familiar with my work in different areas. While the responses of these scientists were uniformly favorable, they did not address themselves to my particular problem, and the matter was left in abeyance.

Again, at the persuasion of the science editors of the Press, in September 1984 I wrote to Martin Schwarzschild, Robert Mullikin, H. C. van de Hulst, Norman Lebovitz, Kip Thorne, and Basilis Xanthopoulos (in view of their familiarity with one or another of the major areas in which I had worked), wondering if they would assist me in the selection process, keeping especially in mind the overlapping contents of my papers and my books. They cordially agreed, and during the course of the following year they did provide me with lists of their choices. But various commitments, including the pressure of my own continuing scientific research, prevented my devoting any time to the project. Now at long last I have agreed to follow a schedule provided by the University of Chicago Press.

The present volume, including papers published during the years 1929–1945, is the first of a projected series of five or six volumes, all hopefully to be published by 1990. In selecting the papers for this volume, I have benefitted from the advice of my long-time friend and colleague Martin Schwarzschild. The criteria for selection are essentially two: first, the papers have not been included in any of my published books; and second, the papers treat matters of possible historical interest not treated in sufficient detail elsewhere.

The papers included in this volume are grouped into five sections. The first section consists of papers on the theory of white dwarfs. The implications of these papers for the broader problems of stellar evolution, and the controversy with Eddington which ensued, have been discussed and written about so extensively that there is very little that I can add to the published accounts. But I may state that my interest in the possible role of electron degeneracy to the structure of white dwarf stars was stimulated by my meeting Arnold Sommerfeld during his visit to Madras, India, in the late fall of 1928. On that occasion Sommerfeld presented me with copies of his papers on the electron theory of metals from which even an undergraduate student (as I was then) could learn.

Perhaps an additional historical remark may be made regarding the concluding sentence of paper 8, which has been quoted in various accounts. This paper was written in 1931 very soon after the publication of paper 6, but I withheld publication because of E. A. Milne's dissent. A year later, when I joined the Institut For Teoretisk Fysik in Copenhagen, I discussed the paper with Leon Rosenfeld, who strongly urged me to publish it. To avoid additional discussions with Milne, I sent it to the *Zeitschrift für Astrophysik*, with editorial offices at the Astrophysikalisches Observatorium in Potsdam. But Professor Milne happened to be visiting Potsdam at that time and he was asked to referee the paper; and in a letter directly written to me, he stated, "Unfortunately I have been unable to recommend acceptance, as the paper contains a mistake in principle, and in any case it would only do your reputation harm if it were published." Again I discussed the matter with Rosenfeld, and at his urging I returned the paper to the editor (W. Grotrian) with some pro forma changes.

This section also contains the discussion by Eddington and Milne (paper 13) that followed the reading of my papers 11 and 12 at the January 1935 meeting of the Royal Astronomical Society. The section ends with my obituary notice for R. H. Fowler.

The second group of papers, on the equilibrium of distorted polytropes, represents the substance of my thesis for the Ph.D. degree at Cambridge (and also for a thesis later submitted for a Fellowship at Trinity College). The original idea for these investigations came from an early paper of E. A. Milne. The papers were written from December 1932 to May 1933 in Copenhagen at the Institute for Theoretical Physics, where I was spending my third year as a Cambridge research student. The subject matter of the thesis was quite outside the interests of Niels Bohr and his associates, who at that time included, among others,

Leon Rosenfeld, Victor Weisskopf, George Placzek, and Christian Møller—all destined to become very distinguished physicists. To have been outside the circumference of this distinguished group was frustrating then and disappointing now: I was simply too immature to take advantage of the company I was in.

When I gave an account of my papers on distorted polytropes at the June 1933 meeting of the Royal Astronomical Society, I was naturally pleased by the reactions of Russell, Eddington, and Milne, which are contained in paper 24.

A parallel line of investigation which bore fruit during the years that followed was that of T. G. Cowling on the apsidal motion in binary stars.

The papers on stellar evolution were written after my *Introduction to the Study of Stellar Structure* had been published. Paper 27 (written with Mario Schönberg) was later to provide the stimulus for fundamental investigations on stellar evolution which followed in the forties with the pioneering investigations of Martin Schwarzschild. These last papers were written during the years of my waning interest in stellar structure.

The group of papers on integral theorems was stimulated by some correspondence with Milne. Perhaps I may take this occasion to expand on the circumstances which led to these papers, since they shed some light on the conflicting personalities of Eddington and Milne.

Eddington had shown that the pressure at the center of a star must exceed the value it would have were the star homogeneous. The converse of this theorem (that the central pressure in a star, under the same premises, must be less than that in a homogeneous configuration of the same mass with a density equal to that at its center) leads to some far-reaching conclusions: in particular, that massive stars with radiation pressures exceeding 10 percent of the total pressure cannot have degenerate cores (paper 30)—a conclusion contrary to Milne's premise of those years that "all stars must have degenerate cores." Milne therefore concluded that the converse of Eddington's theorem must be false and he set about disproving it. When I sent to Milne a proof of the converse theorem (which was vitiated by an error), Milne, pointing out the error, wrote as follows: "After wasting as much time as I thought profitable to get inequalities the other way around . . . I abandoned the matter. If the inequality is invalid it ought to be possible by a simple calculation to construct an example contradicting it. I once thought I had done so but the thing was not so. . . . The resulting inequality looks most doubtful or did when I last wrote it down. . . . The ground is thorny and I advise you not to rush in too rapidly."

The day after I received this letter from Milne, I was in fact able to prove the inequality in question and sent the revised proof (paper 31) to Milne. That evening, Eddington was also dining in Hall; and meeting him after dinner, I told him of my correspondence with Milne on his (Eddington's) proof of his theorem and of Milne's opinion that Eddington's proof was invalid. I then showed

Eddington my own proof, and although he agreed with it, he was convinced that his own proof was equally valid. I said that there was one difference: "Milne would approve of my proof but he does not approve of yours." I further told Eddington that I expected Milne's reaction to my proof the following day. Eddington expressed doubt that Milne would in fact agree with my proof. The following evening after dinner, Eddington came up to me and asked, "Have you received Milne's reply?" I was able to show him a postcard from Milne on which he had written, "I congratulate you on this very elegant solution which makes the matter look very simple but it is really quite deep. At least it is very easily missed—I failed to get it myself."

I have described this incident in detail because the question that was involved *was* in fact a very simple one and not "really deep." It was "deep" only because Milne set out to disprove it on the grounds that it must be wrong: "It contradicted very obvious, much more immediate considerations."

With regard to the last group of papers, on the theory of stellar atmosphere, I need only say that my interest in this area was largely a result of my association with E. A. Milne during those years. Papers 42, 43, and 44, however, originated in some conversations I had with Viktor Ambartsumian, whom I met in the Soviet Union in August 1934. Also, the basic idea in my paper on the blanketing effect of the reversing layers of stars (paper 41) was independently explored by Ambartsumian along parallel lines in a paper published at about the same time.

S. CHANDRASEKHAR

PART ONE

The Theory of White Dwarf Stars

Original page numbers appear in brackets at the bottoms of pages

The Compton Scattering and the New Statistics.

By S. Chandrasekhar, The Presidency College, Madras.

(Communicated by R. H. Fowler, F.R.S.—Received June 20, 1929.)

1. *Introduction.*

Great success has been achieved by Sommerfeld in the electron theory of metals by assuming that there are free electrons in them which obey the Fermi-Dirac statistics. It has been assumed in the case of univalent metals that on the average one electron per atom is free. In general, however, the valency electrons can be considered as free.* These free electrons will take part in the Compton scattering. The analysis of such a Compton effect reduces to the analysis of the collisions between radiation quanta and an electron gas. The general features of such a scattering was first considered by Dirac.† But he has assumed a Maxwellian distribution for the electrons which will not be applicable to the case under consideration, because the electrons in a conductor being degenerate do not obey the Maxwell's law, but the Fermian distribution.

In considering such a process we take it that the conservation of momentum and energy principles are satisfied for each particular collision just as in Compton's theory—only we are here dealing with moving electrons instead of stationary electrons which Compton considers. Thus electrons of different momenta components will produce different Compton shifts, and the intensity of any particular shift will depend on the number of electrons in that state. Thus we have to average for the radiation falling on an assembly of electrons whose momenta are distributed according to the Fermi-Dirac law.

The above is just a natural extension of Compton's theory. In this connection mention should be made of Jauncey's‡ theory of bound electrons whose arguments are essentially what we have put forward in the previous paragraph. But his theory has not been quite satisfactory because he has not assumed any definite distribution of the electrons.

* Rosenfeld, 'Naturw.,' p. 49 (1929).
† 'M.N.R.A.S.,' vol. 85, p. 825 (1925).
‡ 'Phys. Rev.,' vol. 25, p. 723 (1925).

2. *Compton scattering with Moving Electrons.*

Let m_x, m_y, m_z be the momentum of the scattering electron and g_x, g_y, g_z those of the quantum, g_t, m_t represent the masses of the electron and the quantum multiplied by the velocity of light c. If we take polar co-ordinates

$$g_x = h\nu \cos \theta/c \; ; \; g_y = h\nu \sin \theta \cos \phi/c \; ; \; g_z = \sin \theta \sin \phi \;\; g_t = h\nu/c. \tag{1}$$

Then the conservation of momentum and energy gives

$$(m_u, g_u) - (m_u, g_u') = (g_u, g_u'). \tag{2}$$

The above equation gives the frequency of the scattered quantum in terms of the initial momentum of the electron and the incident quantum, and the directions of the incident and scattered quanta.

Equation (2) reduces to

$$m_t - m_x \cos \theta' - m_y \sin \theta' = \frac{\nu}{\nu'}(m_t - m_x) - \frac{h\nu}{c}(1 - \cos \theta')\,, \tag{3}$$

if we assume that the directions of the incident quantum is along the x axis and that of the scattered quantum in the xy plane. Here θ' is simply the angle of scattering.

3. *The Spectral-intensity Distribution Function.*

Before considering the case of scattering of monochromatic X-radiation, we will consider first the more general case when the incident radiation is continuous. Suppose we have such a pencil of radiation confined to a small solid angle $d\omega$ and let I_ν be the intensity per unit frequency range. Let this radiation be incident on an assembly of dn electrons of momentum m_x, m_y, m_z. Let the intensity of radiation scattered in the solid angle $d\omega'$ and frequency range ν' and $\nu' + d\nu'$ be given by

$$R\,(\nu')\, d\nu'\, d\omega'. \tag{4}$$

Then it has been shown by Dirac (*loc. cit.*, equation (8)) that

$$R\,(\nu') = \frac{h^2}{m^2c^3} \cdot dn \,.\, I_\nu\, d\omega \frac{\nu' F\,(a, b)}{\nu m_t}\,. \tag{5}$$

Here ν' is to be regarded as a function of g_x', g_y', g_z' and m_x, m_y, m_z, being that frequency of the incident quantum which will be scattered by an (m_x, m_y, m_z) electron into the frequency range ν' to $\nu' + d\nu'$.

In the above equation $F\,(a, b)$ is a function which depends on the scattering law adopted and a and b the two invariants connected with the scattering process which as well as the initial momentum m_x, m_y, m_z of the electron and g_x, g_y, g_z of the quantum specify the collision.

Now for dn in equation (5), we have to put the Fermi-expression

$$dn = \frac{V}{h^3} \cdot G \cdot \frac{dm_x\, dm_y\, dm_z}{\exp\left(\Sigma m_x^2/2mkT\right)/A + 1}, \tag{6}$$

and integrate with respect to m_x, m_y, m_z. In the above equation A is the constant appearing in the Fermi-Dirac statistics. It has different values according as we consider a degenerate or a non-degenerate gas. When the system is non-degenerate A is a small positive quantity and then has the value

$$A = nh^3 \,.\, (2\pi mkT)^{-3/2}/G. \tag{7}$$

A degenerate system corresponds to A being a large quantity and in that case

$$\log A = \left(\frac{3n}{4\pi G}\right)^{2/3} \cdot \frac{h^2}{2mkT}. \tag{8}$$

Then by equation (6)

$$R(\nu') = \frac{h^2}{m^2c^3} \cdot \frac{V}{h^3} \cdot G \iiint_{-\infty}^{\infty} \frac{I_\nu d\omega \nu' F}{\nu m_t} \cdot \frac{dm_x\, dm_y\, dm_z}{\exp\left(\Sigma m_x^2/2mkT\right)/A + 1} \tag{9}$$

$$= \frac{h^2}{m^2c^3} \cdot \frac{V}{h^3} \cdot G \int_0^{\infty} I_\nu\, d\omega\, \psi(\nu, \nu')\, d\nu. \tag{10}$$

Where

$$\psi(\nu, \nu') = \iint_{-\infty}^{\infty} \frac{\nu' F}{\nu m_t} \cdot \frac{dm_y\, dm_z}{\exp\left(\Sigma m_x^2/2mkT\right)/A + 1} \Big/ \frac{\partial \nu}{\partial m_x}. \tag{11}$$

Where m_x and $\partial\nu/\partial m_x$ are to be evaluated in terms of m_y, m_z and ν by means of equation (3)

$$(6) \quad \frac{\partial m_x}{\partial \nu} = \frac{mc}{\nu'(1 - \cos\theta')} - \frac{h}{c}, \tag{12}$$

and

$$m_x^2 + m_y^2 = \frac{\beta}{\gamma^2}\left[m_y - \frac{K\sin\theta'}{\beta}\right]^2 + \frac{K^2}{\gamma^2}\left(1 - \frac{\sin^2\theta'}{\beta}\right), \tag{13}$$

where

$$\beta = 1 - 2\nu\cos\theta'/\nu' + (\nu/\nu')^2, \quad \gamma = \nu/\nu' - \cos\theta',$$
$$K = -mc(\nu/\nu' - 1) + h\nu(1 - \cos\theta')/c. \tag{14}$$

Then

$$\psi(\nu, \nu') = \iint_{-\infty}^{\infty} \frac{F\left[\dfrac{1}{\nu(1-\cos\theta')} - \dfrac{h\nu'}{mc^2\nu}\right] dm_y\, dm_z}{\dfrac{1}{A}\exp\left\{\dfrac{\dfrac{\beta}{\gamma^2}\left[m_y - \dfrac{K\sin\theta'}{\beta}\right]^2 + \dfrac{K^2}{\gamma^2}\cdot\left[1 - \dfrac{\sin^2\theta'}{\beta}\right] + m_z^2}{2mkT}\right\} + 1}. \tag{15}$$

Suppose now that the radiation is monochromatic, then the number of quanta scattered between the frequency range ν' and $\nu' + d\nu'$ into the solid angle $d\omega'$ is given by

$$R(\nu') = \frac{h^2}{m^2c^3} \cdot \frac{V}{h^3} \cdot G \cdot I_\nu \, d\nu \, dw \, \psi(\nu, \nu'). \tag{16}$$

So that the (spectral) distribution of intensity about the primary frequency ν is given by $\psi(\nu, \nu')$, which we shall now evaluate.

$$\psi(\nu, \nu') = 4F_0\left[\frac{1}{\nu(1 - \cos\theta')} - \frac{h\nu'}{mc^2\nu}\right]\iint_0^\infty \frac{dm_y \, dm_z}{\frac{1}{B}\exp\left(\frac{\beta}{\gamma^2} \cdot \frac{m'^2_y + m_z^2}{2mkT}\right) + 1}, \tag{17}$$

where

$$\left.\begin{aligned} m'_y &= m_y - \frac{K\sin\theta'}{\beta} \\ \frac{1}{B} &= \frac{1}{A} \cdot \exp\left\{\frac{K^2}{\gamma^2 \cdot 2mkT} \cdot \left(1 - \frac{\sin^2\theta'}{\beta}\right)\right\} \end{aligned}\right\}. \tag{18}$$

If we introduce the new variables

$$y = m'^2_y \, \beta/\gamma^2 \, . \, 2mkT, \quad z = m_z^2/2mkT.$$

Then

$$\psi(\nu, \nu') = 2F_0\left[\frac{1}{\nu(1 - \cos\theta')} - \frac{h\nu'}{mc^2\nu}\right]\frac{\gamma \, . \, mkT}{\beta^{\frac{1}{2}}} \cdot \iint_0^\infty \frac{y^{-\frac{1}{2}} \, . \, z^{-\frac{1}{2}} . \, dy \, dz}{e^{y+z}/B + 1}$$

$$= 2F_0\left[\frac{1}{\nu(1 - \cos\theta')} - \frac{h\nu'}{mc^2\nu}\right] \cdot \frac{\gamma \, . \, mkT}{\beta^{\frac{1}{2}}} \, . \, U_0, \tag{19}$$

where U_0 is the special case of the general Sommerfeld integral

$$U_\rho = \frac{1}{\Gamma(\rho + 1)} \cdot \int_0^\infty \frac{u^\rho \, du}{e^u/B + 1}, \tag{20}$$

which gives for $\rho = 0$*

$$U_0 = \pi \log(B + 1). \tag{21}$$

Hence we get our intensity distribution function

$$\psi(\nu, \nu') = 2F_0\left[\frac{1}{\nu(1 - \cos\theta')} - \frac{h\nu'}{mc^2\nu}\right] \cdot \frac{\gamma \, . \, mkT\pi}{\beta^{\frac{1}{2}}} \, . \log(B + 1). \tag{22}$$

Case I.—If B is large and positive we get

$$\psi(\nu, \nu') = 2F_0\left[\frac{1}{\nu(1 - \cos\theta')} \cdot \frac{h\nu'}{mc^2\nu}\right] \cdot \frac{\gamma \, . \, mkT\pi}{\beta^{\frac{1}{2}}}$$

$$\times\left[\log A - \frac{K^2}{\gamma^2 \, . \, 2mkT}\left(1 - \frac{\sin^2\theta'}{\beta}\right)\right], \tag{23}$$

the value of log A being given by (8).

* Sommerfeld, 'Z. Physik,' vol. 47, p. 1 (1928), equation (31A).

An approximation of the above equation to an order of accuracy where the Compton-shift is neglected is

$$\psi(\nu, \nu') = \frac{\sqrt{2}\, F_0 \pi m k T}{\nu (1 - \cos \theta')^{\frac{1}{2}}} \cdot \left[\log A - \frac{(\nu' - \nu)^2}{a \nu^2}\right]. \tag{24}$$

where $a = 4kT(1 - \cos \theta')/mc^2$.

Case II.—If B is small due to the smallness of A we get, to the same order of accuracy as (24), the equation

$$\psi(\nu, \nu') = \frac{n}{h^3 . G} \cdot (2\pi m k T)^{-3/2} . \frac{\sqrt{2F_0 \pi m k T}}{\nu (1 - \cos \theta')^{\frac{1}{2}}} \cdot e^{-(\nu' - \nu)^2 / a\nu^2}, \tag{25}$$

the one given by Dirac (*loc. cit.*, equation (13)).

Equation (25) gives an exponential distribution of intensity about the primary frequency for the scattered radiation. But equations (23) and (24) indicate that the distribution of intensity of the radiation scattered by a degenerate electron gas does not follow an exponential law but gives a parabolic distribution. This perhaps explains the rather broad structure of the Compton modified radiation.*

4. *The Compton effect.*

It is natural that the distribution of intensity predicted by equation (23) places the maximum peak of intensity at a place where the Compton's theory for a free-stationary electron predicts a line. Remembering that in any case $\nu/\nu' = 1$ the maximum frequency will be at a modified frequency where $K = 0$ where

$$K = -mc(\nu/\nu' - 1) + h\nu(1 - \cos \theta')/c = 0,$$

i.e., where

$$\lambda' - \lambda = h(1 - \cos \theta')/mc, \tag{26}$$

i.e., on an intensity-frequency graph the maximum occurs at a place corresponding to the Compton shift.

5. *The Effect of Temperature.*

If we consider the Compton scattering by an electron-gas, the distribution function of which depends on temperature, it would naturally be expected

* Mr. J. W. Du Mond in a private communication to the author from the California Institute of Technology, Pasadena, has kindly pointed out that the above is the characteristic of the Compton-radiation from conductors. His paper in the May issue of the 'Physical Review' (vol. 33, p. 643) gives experimental details and theoretical calculations as well. He has independently derived the parabolic structure.

that the spectral intensity distribution function in the Compton scattering would also depend on temperature and Dirac's classical expression (25) does indicate this by the explicit appearance of the temperature factor in $\psi(\nu, \nu')$. But if we substitute the value of log A given by (8) in (23) we get

$$\psi(\nu, \nu') = 2F_0\left[\frac{1}{\nu(1-\cos\theta')} - \frac{h\nu'}{mc^2\nu}\right] \cdot \frac{\gamma \cdot \pi}{\beta^{\frac{1}{2}}}$$
$$\times\left[\tfrac{1}{8}\left(\frac{6}{\pi}n\right)^{2/3}h^2 - \frac{K^2}{\gamma^2}\cdot\left(1-\frac{\sin^2\theta'}{\beta}\right)\right], \qquad (27)$$

where all the temperature factors have cancelled out. Thus Compton-scattering by an *electron-gas* will not be influenced by temperature. Further the Compton scattering by the bound electrons will also not be influenced by the ranges of temperature available in the laboratory. Thus it appears that the *total* Compton scattering will not be affected by temperature.*

6. *The Effect of a Magnetic-field.*

We will consider the scattering by the conduction electrons only. When the scatterer is placed in a magnetic-field the distribution function for the electrons changes, and in that case the number of electrons in the momentum range m_x, m_y, m_z and $m_x + dm_x$, $m_y + dm_y$, $m_z + dm_z$ is given by the Pauli's expression†

$$dn = \frac{V}{h^3} \cdot \frac{dm_x \cdot dm_y \cdot dm_z}{\exp\left(\dfrac{\varepsilon_m}{kT} + \dfrac{\Sigma m_x^2}{2m_0kT}\right)\Big/A + 1}, \qquad (28)$$

where $\varepsilon_m = mg\,\mu_0 H$; where $\mu_0 = -eh/4\pi m_0 c$ = a Bohr magneton, g = the Lande factor, H = the field strength.

For the summation over all the values of the quantum number m from $-j$ to $+j$ we have the relations

$$\left.\begin{aligned}\sum_{m=-j}^{+j} \varepsilon_m &= 0 \\ \sum_{m=-j}^{+j} \varepsilon_m^2 &= \tfrac{1}{3}G\mu^2H^2\end{aligned}\right\}. \qquad (29)$$

To derive the spectral-intensity distribution function we have to substitute

* Since the writing of the above a report by Jauncey and Bowers has appeared (' Bull. Amer. Phys. Soc.,' vol. 4, p. 26 (1929)) giving experimental observations which support the above conclusions.

† ' Z. Physik,' vol. 41, p. 81 (1927).

(28) instead of (6) in equation (5) and carry out the integration as before. The final result as one can easily see is

$$\psi(\nu, \nu') = 2F_0 \left| \frac{1}{\nu(1 - \cos\theta')} - \frac{h\nu'}{mc^2\nu} \right| \cdot \frac{\gamma \cdot mkT \cdot \pi}{\beta^{\frac{1}{2}}}$$

$$\times \left[\log A + \sum_{m=-j}^{+j} \frac{-\varepsilon_m}{kT} - \frac{K^2}{2\gamma^2 \cdot mkT}\left(1 - \frac{\sin^2\theta'}{\beta}\right)\right], \quad (30)$$

which on account of (29) becomes identical with (23). Thus it appears on Pauli's theory of the paramagnetism of an electron-gas that the scattering of such an assembly should not be influenced by the presence of a magnetic field. In this connection mention should be made of an experimental observation of Bothe* where he tried the influence of a magnetic field. The scatterer he used was paraffin, and he tried up the field strengths of the order of 20,000 Γ. But he could detect no influence.

Summary.

In this paper the Compton scattering by an electron-gas on the Fermi-Dirac statistics is considered. The theory predicts a distribution of spectral intensity not exponentially falling off about the maximum but *parabolically.* It places the peak of maximum intensity at a place where the Compton relation $\lambda' - \lambda = h(1 - \cos\theta')/mc$ is satisfied. Further, the theory indicates that there should be no influence of temperature or magnetic field in Compton scattering.

In conclusion the author wishes to express his thanks to Dr. R. H. Fowler, F.R.S., and Mr. N. F. Mott for kindly going through the manuscript and suggesting improvements.

* 'Z. Physik,' vol. 41, p. 872 (1927).

The Ionization-Formula and the New Statistics.

By S. CHANDRASEKHAR.

(1) *Introduction.*

THE derivation of the formula

$$\frac{n_+ \, . \, n_-}{n} = \frac{1}{h^3} (2\pi mkT)^{3/2} \, . \, e^{-\chi/kt} \, . \, \sigma \, .$$

(n_+ = the number of ionized atoms,
n_- = the number of electrons,
n = the number of neutral atoms)

has attracted much attention after its first derivation by Saha. The rigorous way of deriving the above equation has been successfully attempted by Fowler †; but no attention has so far been given to modifying the above formula on the new statistics of Fermi-Dirac. Obviously it will be of no practical importance if in the application of the new statistics we treat all the constituents—the atoms of both kinds and

† R. H. Fowler, Phil. Mag. xlv. p. 1 (1923); also Fowler and Milne, 'Monthly Notices,' R. A. S. lxxxiii. p. 407 (1923). A simple way of deriving the equation (1) has been given by W. F. G. Swann, Journ. of Frank. Inst. cc. p. 591 (1925).

the electrons—as degenerate, because the ionization formula, being most applied to the thermodynamics of a star, would not justify our assumption as to the degeneracy of the atoms themselves. But the electrons, even at that high temperature, if the pressure is sufficiently high, will be a degenerate system *. Thus Sommerfeld's † condition of degeneracy,

$$\frac{nh^3}{2} \cdot \frac{1}{(2\pi mkT)^{3/2}} > 1 \quad . \quad . \quad . \quad . \quad . \quad (2)$$

gives, if the number of electrons per c.c. is 10^{30}, the system to be degenerate up to temperatures below $3{\cdot}5^{\circ} \times 10^{9}$, and if $n_{-} = 10^{27}$ the system will be degenerate even up to a temperature of $3^{\circ} \times 10^{7}$.

The object of this paper is to modify the ionization formula on the Fermi-Dirac statistics for the degeneracy of the electrons. But, before dealing with that problem, the case of complete degeneracy will be discussed. The modified formula when the degeneracy of the electrons alone is considered will follow naturally from the case of complete degeneracy.

(2) *General Relation between n_{+}, n, and n .*

According to the new statistics of Fermi-Dirac, the entropy of a system is given by ‡

$$S = nk\left[\frac{5}{2}\frac{U_{3/2}}{U_{1/2}} - \log A\right], \quad . \quad . \quad . \quad . \quad (3)$$

where $U_{3/2}$ and $U_{1/2}$ are the special cases of the general Sommerfeld § integral

$$U_\rho = \frac{1}{\Gamma(\rho+1)}\int_0^\infty \frac{u^\rho du}{\frac{1}{A}e^u + 1} \quad . \quad . \quad . \quad . \quad . \quad (4)$$

For a mixture of gases the entropy is given by

$$S = \Sigma\, kn_\nu\left[\frac{5}{2}\frac{U_{3/2}}{U_{1/2}} - \log A_\nu\right]. \quad . \quad . \quad . \quad (5)$$

In the case under consideration, that of a monatomic gas dissociating into positive ions and electrons, the summation comprises these terms.

* In this connexion, see an interesting contribution by Dr. E. C. Stoner, Phil. Mag. vii. p. 63 (1929).

† Sommerfeld, *Zeits. f. Phys.* xlvii. p. 1 (1928).

‡ Fermi, *Zeits. f. Phys.* xxxvi. p. 902 (1926).

§ Sommerfeld, *Zeits. f. Phys.* xlvii. p. 1 (1928).

If δn represents an arbitrary change in the number of neutral atoms, δn_+ and δn_- the corresponding changes for the positive ions and the electrons, the corresponding change in the entropy is given by

$$\delta S = \Sigma \frac{\partial}{\partial n}\left[nk\left\{\frac{5}{2}\frac{U_{3/2}}{U_{1/2}} - \log A\right\}\right]$$

$$= \Sigma\left\{k\left[\frac{5}{2}\frac{U_{3/2}}{U_{1/2}} - \log A\right]\right.$$

$$\left. + nk\frac{\partial}{\partial A}\left[\frac{5}{2}\frac{U_{3/2}}{U_{1/2}} - \log A\right] \times \frac{\partial A}{\partial n}\right\}\delta n. \quad . \quad (6)$$

Now

$$\frac{\partial}{\partial A}\left[\frac{5}{2}\frac{U_{3/2}}{U_{1/2}} - \log A\right] = \left\{\frac{5}{2}\left[\frac{U'_{3/2}}{U_{1/2}} - \frac{U_{3/2} . U'_{1/2}}{U^2_{1/2}}\right] - \frac{1}{A}\right\}. \quad (7)$$

Now

$$U_\rho' = \frac{1}{\Gamma(\rho+1)}\int_0^\infty \frac{1}{A^2}.\frac{e^u u^\rho du}{\left(\frac{1}{A}e^u+1\right)^2}$$

$$= -\frac{1}{\Gamma(\rho+1)}\int_0^\infty \frac{1}{A}u^\rho . \frac{\partial}{\partial u}\left(\frac{1}{\frac{1}{A}e^u+1}\right)du$$

$$= \frac{1}{A}U_{\rho-1}. \quad . \quad . \quad . \quad , \quad . \quad . \quad . \quad . \quad . \quad . \quad . \quad . \quad (8)$$

Using relation (8), we get for

$$\delta S = \Sigma\left\{k\left[\frac{5}{2}\frac{U_{3/2}}{U_{1/2}} - \log A\right]\right.$$

$$\left. + \frac{nk}{A}\left[\frac{3}{2} - \frac{5}{2}\frac{U_{3/2} . U_{-1/2}}{U^2_{1/2}}\right]\frac{\partial A}{\partial n}\right\}\delta n. \quad . \quad . \quad (9)$$

If Q is the heat absorbed due to the ionization of δn atoms,

$$\delta S = \frac{Q}{T} = \frac{-\chi\delta n + \Sigma\frac{\partial E}{\partial n}.\delta n}{T}. \quad . \quad . \quad . \quad (10)$$

Now E is given in the new statistics by the expression

$$E = \frac{3}{2}\frac{VGkT}{h^3}.(2\pi mkT)^{3/2}.U_{3/2}. \quad . \quad . \quad . \quad (11)$$

Then

$$\frac{\partial E}{\partial n} = \frac{3}{2}\frac{VGkT}{h^3}.(2\pi mkT)^{3/2}\frac{1}{A}U_{1/2}.\frac{\partial A}{\partial n}. \quad . \quad . \quad (12)$$

Hence we get, using relation (12),

$$\delta S = \frac{\chi}{T}\delta n + \Sigma \frac{3}{2} V k \beta_\nu . \frac{1}{A} U_{1/2} \frac{\partial A}{\partial n} \delta n, \quad . \quad . \quad (10\,a)$$

where we put

$$\beta_\nu = \frac{(2\pi m_\nu k T)^{3/2}}{h^3} . G. \quad . \quad . \quad . \quad . \quad (10\,b)$$

Equating (9) and (10 *a*), we get the general relation applicable to the new as well as the classical statistics,

$$\Sigma k \left[\frac{5}{2}\frac{U_{3/2}}{U_{1/2}} - \log A\right] + \Sigma \frac{nk}{A}\left[\frac{3}{2} - \frac{5}{2} . \frac{U_{3/2} . U_{-1/2}}{U^2_{1/2}}\right]\frac{\partial A}{\partial n}$$

$$= -\frac{\chi}{T} + \frac{3}{2}\Sigma V k \beta . \frac{1}{A}\frac{\partial A}{\partial n} . U_{1/2}. \quad . \quad . \quad (13)$$

(3) *The Classical Ionization Formula.*

With the help of relation (13) we will arrive at the classical result (1) already quoted. The classical statistics correspond to a non-degenerate system, where in the distribution function A << 1.

In that case

$$U_\rho = U_{\rho-1} = \ldots . = U_{3/2} = U_{1/2} = U_{-1/2} = A \quad . \quad (14)$$

and $$A = \frac{nh^3}{VG} \cdot (2\pi m k T)^{-3/2}$$

$$= \frac{n}{V\beta} \cdot \quad . \quad . \quad . \quad . \quad . \quad . \quad . \quad . \quad . \quad (15)$$

Substituting relations (14) and (15) in (13), we get

$$-k\Sigma^\nu \log A_\nu = -\frac{\chi}{T}, \quad . \quad . \quad . \quad . \quad . \quad (16)$$

$$-\log\frac{n}{V\beta} + \log\frac{n_+}{V\beta_+} + \log\frac{n_-}{V\beta} = -\frac{\chi^*}{kT}. \quad (16\,a)$$

Substituting values for β, we get at once the classical dissociation formula

$$\frac{n_+ n_-}{n} = \frac{V}{h^3} . (2\pi m k T)^{3/2} . G e^{-\chi/kT}, \quad . \quad . \quad . \quad (17)$$

which is identical with equation (1). In the application of

* The positive sign is given to $\log n_+$ and $\log n_-$ because we have taken δn as an increase in the neutral atoms, and therefore δn_+ and δn_- are negative. Also

$$\delta n = \delta n_+ = \delta n_-.$$

this formula G can be omitted (G = 2, corresponding to spin statistical weight).

(4) *A Completely Degenerate System.*

For this we start again with the fundamental equation (13). In this case the values for U_ρ and A are completely different, and, as it will be seen, this complicates the ionization formula. Now

$$U_\rho = \frac{u_0^{\rho+1}}{\Gamma(\rho+2)} \times \left[1+2\left\{\frac{(\rho+1)\rho . c_2}{u_0^2} + \frac{(\rho+1)\rho . (\rho-1)(\rho-2)\, c_4}{u_0^4}, \ldots\right\}\right], \quad (18)$$

where

$$u_0 = \log A, \quad . \quad . \quad . \quad . \quad . \quad . \quad (18\,a)$$

and

$$c_\nu = 1 - \frac{1}{2^\nu} + \frac{1}{3^\nu} - \frac{1}{4^\nu}, \quad . \quad . \quad . \quad . \quad (19)$$

giving in the special case for $\nu = 2$ the value $\pi^2/12$. The value for A in the case of degeneracy is also given by

$$\log A_\nu = \frac{h^2}{2m_\nu kT}\left(\frac{3n_\nu}{4\pi G}\right)^{2/3} \quad . \quad . \quad . \quad . \quad (20)$$

$$= \frac{\alpha n^{2/3}}{mkT} . \quad . \quad . \quad . \quad . \quad . \quad . \quad . \quad . \quad (21)$$

Equation (13) modifies by the substitution of the values for $U_{3/2}$, $U_{1/2}$, and $U_{-1/2}$ in the following manner:

$$\Sigma \frac{k\pi^2}{2}(\log A_\nu)^{-1} - \Sigma \frac{nk\pi^2}{2}(\log A_\nu)^{-2} . \frac{1}{A_\nu}\frac{\partial A_\nu}{\partial n}$$

$$= -\frac{\chi}{T} + \frac{3}{2}\Sigma V k\beta_\nu . \frac{4}{3}\sqrt{\pi}\,(\log A_\nu)^{3/2}$$

$$\times \left[1 + \frac{\pi^2}{8\,(\log A_\nu)^2}\ldots\right]\frac{1}{A_\nu}\frac{\partial A_\nu}{\partial n} . \quad . \quad . \quad (22)$$

Relation (21) gives

$$\frac{1}{A}\frac{\partial A}{\partial n} = \frac{2}{3}\frac{\alpha n^{-1/3}}{mkT} . \quad . \quad . \quad . \quad . \quad . \quad (23)$$

Making all the necessary simplifications, we get

$$\Sigma \frac{k\pi^2}{6}\frac{mkT}{\alpha n^{2/3}} - \Sigma \frac{2^{7/2} . \pi^2 V \alpha^{5/2} G}{3h^3 mT} . n^{2/}$$

$$- \Sigma \frac{\pi^4 V}{6h^3} . \frac{2^{3/2} k^2 \alpha^{1/2} m T G}{n^{2/3}} = -\frac{\chi}{T} . \quad . \quad . \quad (24)$$

Taking the volume as unity, we get our final ionization formula,

$$\frac{k^2\pi^2T}{6}\left[\frac{1}{\alpha}-\frac{G\pi^2 2^{3/2}\alpha^{1/2}}{h^3}\right]\left[m\left(\frac{1}{n^{2/3}}-\frac{1}{n_+^{2/3}}\right)-\frac{m_-}{n_-^{2/3}}\right]$$
$$-\frac{2^{7/2}\pi^2\alpha^{5/2}G}{3h^3T}\left[\frac{1}{m}(n^{2/3}-n_+^{2/3})-\frac{n_-^{2/3}}{m_-}\right]=-\frac{\chi}{T}. \quad . \quad (25)$$

(5) *Ionization Formula for the Degeneracy of the Electrons only.*

As has been pointed out, a modification of the ionization formula will be useful only *if the electrons alone are treated as degenerate.* This can easily be done, if in summing up the entropy of the system in equation (5), we use for the atoms the classical values for $U_{3/2}$, $U_{1/2}$, $U_{-1/2}$, and A, and the corresponding values in the case for degeneracy for the electrons. Using relations (16) and (24), we get our ionization formula, taking into consideration the degeneracy of the electrons alone :

$$k\log\frac{n_+}{n}-\frac{k^2\pi^2T}{6}\left[\frac{1}{\alpha}-\frac{G\pi^2 2^{3/2}\alpha^{1/2}}{h^3}\right]m_-n_-^{-2/3}$$
$$+\frac{2^{7/2}\pi^2\alpha^{5/2}}{3h^3T}\cdot\frac{n_-^{2/3}}{m_-}=-\frac{\chi}{T}. \quad . \quad . \quad (26)$$

Equation (26) can be put in the more convenient form

$$\log\frac{n_+}{n}=-\frac{\chi}{kT}+\Phi n^{-2/3}T-\Theta n^{2/3}T^{-1}, \quad . \quad . \quad (27)$$

where Φ and Θ have the obvious equivalents, which can be seen by comparison with equation (26). Finally, we get

$$\frac{n_+}{n}=e^{\Phi n^{-2/3}T-\Theta n^{2/3}T^{-1}-\chi/kT}. \quad . \quad . \quad . \quad (28)$$

(6) *Numerical Calculations and Discussion with Reference to the Theory of Dwarf Stars.*

Formulæ (17) and (28) are the classical and the modified ionization formulæ connecting the ionization potential, electron-density, and the temperature. If x is the degree of ionization, we get, on substituting the known numerical values for Φ and Θ, the final numerical formula in the two cases as

$$\log\frac{1-x}{x}=\frac{\chi}{kT}+\frac{3{\cdot}34\times10^{10}\times T}{n^{2/3}}+\frac{2{\cdot}1\times n^{2/3}}{10^{10}\times T}, \quad . \quad (29)$$

and

$$\frac{1-x}{x}=\frac{n\times4{\cdot}1\times10^{-16}}{T^{3/2}}e^{\chi/kT}. \quad . \quad . \quad . \quad (30)$$

The condition of degeneracy gives that the electrons are degenerate at a temperature of 10^9 Å. when the density is 10^{30}. Now we will calculate the degree of dissociation for various temperatures on both the formulæ for an ionization potential corresponding to $\lambda = 1{\cdot}22$ Å. The results are tabulated below :—

Electron density = 10^{30} per c.c., $\lambda = 1{\cdot}22$ Å.

Temperature.	x by the classical formula (30).	x by the modified formula (29).
A.	%.	%.
10^9	6·4	0
2×10^9	17	0
3×10^9	28	0
4×10^9	38	0
5×10^9	45	0

It is very interesting to note that our formula, including the degeneracy of the electrons, gives uniformly 0 per cent. ionization, while the classical formula gives varying degrees. As a matter of fact, it gives 0 per cent. up to a temperature above which the application of the formula itself becomes invalid, due to the Sommerfeld degeneracy condition not being satisfied. The degree of dissociation given by the classical formula has no significance, because the conditions on which the formula has been derived do not hold good at such high pressures.

Our theory predicting zero per cent. ionization goes against the fundamental assumption of Fowler, Stoner, and others in the theory of dwarf stars, where the first postulate is to take for granted the complete ionization in it. The temperature and density we have taken for our calculation of the table roughly correspond to the centre of the Companion of Sirius, and there our formula predicts no ionization.

If that is the case, one may be tempted to question the validity of the fundamental assumption in the theory of dwarf stars. But for all purposes of calculation we can take the electrons as free, if we define a free electron as one which is not under the influence of *one and the same* nucleus for any finite time. Then obviously the electrons at that pressure are free electrons. The "bound electrons" are changing partners continually, and go about knocking, thus really travelling the whole of the phase-space, just what a "free electron" is expected to do. But, none the less, the statement that *every electron* is under the influence of one or other of the nuclei at any arbitrarily chosen time remains valid. In other words, the ionization is zero, a result predicted by our theory. Thus the paradoxical statement results that the electrons, though all of them bound, are free.

(7) *Summary.*

In this paper the ionization formula is considered afresh on the Fermi-Dirac statistics. The modified formula where the degeneracy of the electrons is considered gives, under the conditions existing in the interior of a dwarf star, a degree of ionization = 0 per cent. This result is discussed with reference to the theory of dwarf stars, where complete ionization is usually assumed.

The Presidency College,
Madras, India,
23rd May, 1929.

The Density of White Dwarf Stars.

By S. CHANDRASEKHAR *.

1. THE first application of the Fermi-Dirac statistics to stellar problems was by Fowler † in connexion with the well-known problem of the companion of Sirius. This idea has lately been taken up by Stoner ‡ and others to calculate the limiting density of white dwarf stars. In this paper another way of arriving at the order of magnitude of the density of white dwarfs from different considerations is given.

2. Let p_r denote the radiation pressure and p_G the gas pressure, and the total pressure P is then given by

$$P = p_r + p_G. \qquad (1)$$

We introduce the constant β, such that

$$\left. \begin{aligned} p_r &= (1-\beta)P, \\ p_G &= \beta P. \end{aligned} \right\} \qquad (2)$$

We will make the assumption that $\beta = 1$ approximately, *i.e.*, we leave the radiation pressure out of account. We are dealing therefore with *ideal* conditions which can perhaps exist only in stars which are much higher in the white dwarf stage than even O_2, Eridani B.

* Communicated by R. H. Fowler, F.R.S.
† R. H. Fowler, Month. Not. Roy. A. S. lxxxvii. p. 114 (1926).
‡ E. C. Stoner, Phil. Mag. vii. p. 63 (1929); ix. p. 944 (1930).

Now for a fully degenerate electron gas (in the Sommerfeld sense) the pressure is given by

$$p_e = \frac{\pi}{60}\frac{h^2}{m}\left(\frac{3n}{\pi}\right)^{5/3} . \quad . \quad . \quad . \quad . \quad . \quad . \quad (3)$$

We assume that it is this electron pressure which is by far the greatest contribution to the gas pressure, and therefore to the total pressure. Further, if ρ is the density of the stellar material, the number of electrons is given by

$$n = \frac{\rho}{\mu \mathrm{H}(1+f)}, \quad . \quad . \quad . \quad . \quad . \quad . \quad . \quad (4)$$

where f is the ratio of the number of ions to the number of electrons (we can in practice neglect f), H the mass of the hydrogen atom, and μ the molecular weight. For a fully ionized material of the type we are dealing with $\mu = 2\cdot5$ nearly. We will use this value later. We have therefore

$$p_{\mathrm{G}} = \frac{\pi}{60}\frac{h^2}{m}\left(\frac{3}{\pi \mathrm{H}}\right)^{5/3} \frac{\rho^{5/3}}{\mu^{5/3}(1+f)^{5/3}}$$

$$= 9\cdot845 \times 10^{12}\left[\frac{\rho}{\mu(1+f)}\right]^{5/3}, \quad . \quad . \quad . \quad (5)$$

(The values used for h, m, etc. are those given in A. S. Eddington's 'Internal Constitution of Stars,' Appendix (1).)

Putting

$$\mathrm{K} = \frac{9\cdot845 \times 10^{12}}{\mu^{5/3}(1+f)^{5/3}}, \quad . \quad . \quad . \quad . \quad . \quad . \quad (6)$$

we have for the total pressure

$$\mathrm{P} = \mathrm{K}\rho^{5/3}. \quad . \quad . \quad . \quad . \quad . \quad . \quad . \quad (7)$$

We can now straightway apply the theory of the polytropic gas spheres, where for the exponent γ we have

$$\gamma = 5/3 \text{ or } 1 + \frac{1}{n} = 5/3,$$

giving $$n = 3/2. \quad . \quad . \quad . \quad . \quad . \quad . \quad . \quad . \quad (8)$$

We have therefore *

$$\left(\frac{\mathrm{GM}}{\mathrm{M}'}\right)^{+1/2}\left(\frac{\mathrm{R}'}{\mathrm{R}}\right)^{-3/2} = \frac{[5/2\mathrm{K}]^{3/2}}{4\pi \mathrm{G}}, \quad . \quad . \quad . \quad (9)$$

* A. S. Eddington, 'Internal Constitution of Stars,' p. 83 *et seq.* The notation is the same as that used in his book and now generally adopted. The particular equation (9) follows from the second of the equations (57.3).

$$\text{or} \qquad \frac{GM}{M'} = \frac{125 \times 9{\cdot}845^3 \times 10^{36}}{128\pi^2 G^2} \cdot \frac{1}{\mu^5 (1+f)^5} \left(\frac{R'}{R}\right)^3 . \quad . \quad (10)$$

The values of R′ and M′ can be obtained from the extensive tables given by Emden in his 'Gas-Kugeln,' and are (page 79, tabbelle 4)

$$\left.\begin{aligned} R' &= 3{\cdot}6571, \\ M' &= 2{\cdot}7176. \end{aligned}\right\} \quad . \quad . \quad . \quad . \quad . \quad . \quad (11)$$

Using these values in (10), and expressing the mass in terms of that of the Sun ($= 1{\cdot}985 \times 10^{33}$ grams), we get the result

$$(M/\odot)R^3 = \frac{2{\cdot}14 \times 10^{28}}{\mu^5} (= 2{\cdot}192 \times 10^{26}). \quad . \quad (12)$$

The second value for $(M/\odot)R^3$, given in brackets, we get by using the value 2·5 for μ. We can express (12) differently, as follows:

$$R^6 \rho = \frac{1{\cdot}014 \times 10^{61}}{\mu^5} (= 1{\cdot}039 \times 10^{59}), \quad . \quad . \quad (13)$$

$$\rho = 2{\cdot}162 \times 10^6 (M/\odot)^2. \quad . \quad . \quad . \quad . \quad . \quad . \quad (14)$$

We will apply the above equations to the case of the companion of Sirius. The mass of it, as determined from the double star orbit, is trustworthy, and equals $\cdot 85\odot$. The computed radius $= 1{\cdot}8 \times 10^{8}$. (But we cannot use this value 9 in (13) to calculate the density, as it is based on formulæ which may not be applicable to this case.) From the mass we can derive the radius and equals $6{\cdot}361 \times 10^8$ ~~(about thirty times the accepted value)~~. For the density of the companion of Sirius we get from (14), *provided* it were completely degenerate (which, however, is extremely unlikely),

$$\rho_{\text{C. Sirius}} = 1{\cdot}562 \times 10^6 \text{ grams per cm.}^3 \quad . \quad (15)$$

The mean density assumed is $\cdot 5 \times 10^5$, being thus thirty times smaller than that given by (15). We can, however, take the value given by (15) as indicating the *maximum* density which a stellar material having a mass equal to that of the companion of Sirius can have. A similar calculation can be made for O_2 Eridani B and Procyon B, and the calculated values are collected in a table below. The calculations for the *limiting density* on Stoner's theory give different values, and they are also given for comparison. We discuss the cause of the difference below.

We further note (i.) that the radius of a white dwarf is inversely proportional to the cube root of the mass, (ii.) the density is proportional to the square of the mass, (iii.) the central density would be six times the mean density ρ.

Handwritten corrections on this page are the author's.

3. Stoner (*loc. cit.*) arrives at a formula for the *limiting* density for a material composed of completely ionized atoms on the following argument:—

The density increases as the sphere shrinks, and the limit is reached when the gravitational energy released just supplies the "energy required to squeeze the electrons closer together." The limiting condition would then be given by

$$\frac{d}{dn}(E_G + E_K) = 0, \quad . \quad . \quad . \quad . \quad . \quad (16)$$

E_G being the gravitational energy and E_K the kinetic energy, for which, of course, the Fermi formula is used. The formula he gets is (without his latter relativity-mass correction)

$$\rho_{max.} = 3{\cdot}977 \times 10^6 (M/\odot)^2, \quad . \quad . \quad . \quad . \quad (17)$$

which is exactly the same as our (14) with a difference in the

Star.	Mass.	Radius.	Density. As calc. by (14).	Density. Accepted value.	Density. By Stoner's formula (17).
O_2 Eridani B.	$\cdot 44\ \odot$	$7{\cdot}927 \times 10^8$	$4{\cdot}186 \times 10^5$	$\cdot 98 \times 10^5$	$7{\cdot}8 \times 10^5$
Procyon B.	$\cdot 37\ \odot$	$8{\cdot}399 \times 10^8$	$2{\cdot}960 \times 10^5$	—	$5{\cdot}445 \times 10^5$
Companion of Sirius.	$\cdot 85\ \odot$	$6{\cdot}361 \times 10^8$	$1{\cdot}562 \times 10^6$	$\cdot 5 \times 10^5$	$2{\cdot}872 \times 10^6$

numerical factor only, the discrepancy being about 1 : 2. The difference in the two is obviously due to the fact that our value for ρ is not the "limiting density" in the sense in which Stoner uses the term ; but our calculation gives us a much nearer approximation to the conditions actually existent in white dwarfs than Stoner's calculation does. At any rate, it brings out clearly that the *order of magnitude* of the density which one can on purely theoretical considerations attribute to a white dwarf is the same.

Our results (see table) agree with Stoner's in showing that O_2 Eridani B is much nearer the ideal dwarf-star stage than the companion of Sirius, but indicate also that neither of them is so far from the ideal stage as Stoner's calculation would seem to indicate.

Summary.

The density of the white dwarf stars is reconsidered from the point of view of the theory of the polytropic gas spheres, and gives for the *mean density* of a white dwarf (under ideal conditions) the formula

$$\rho = 2{\cdot}162 \times 10^6 \times (M/\odot)^2.$$

The above formula is derived on considerations which are a much nearer approximation to the conditions *actually existent* in a white dwarf than the previous calculations of Stoner based on uniform density distribution in the star and which gave for the limiting density the formula

$$\rho = 3{\cdot}977 \times 10^6 \times (M/\odot)^2.$$

THE MAXIMUM MASS OF IDEAL WHITE DWARFS

By S. CHANDRASEKHAR

ABSTRACT

The theory of the *polytropic gas spheres* in conjunction with the equation of state of a *relativistically degenerate electron-gas* leads to a *unique value for the mass of a star* built on this model. This mass (=0.91⊙) is interpreted as representing the upper limit to the mass of an ideal white dwarf.

In a paper appearing in the *Philosophical Magazine*,[1] the author has considered the density of white dwarfs from the point of view of the theory of the polytropic gas spheres, in conjunction with the degenerate non-relativistic form of the Fermi-Dirac statistics. The expression obtained for the density was

$$\rho = 2.162 \times 10^6 \times \left(\frac{M}{\odot}\right)^2 , \tag{1}$$

where $M/\odot$ equals the mass of the star in units of the sun. This formula was found to give a much better agreement with facts than the theory of E. C. Stoner,[2] based also on Fermi-Dirac statistics but on uniform distribution of density in the star which is not quite justifiable.

In this note it is proposed to inquire as to what we are able to get when we use the relativistic form of the Fermi-Dirac statistics for the degenerate case (an approximation applicable if the number of electrons per cubic centimeter is $> 6 \times 10^{29}$). The pressure of such a gas is given by (which can be shown to be rigorously true)

$$P = \tfrac{1}{8}\left(\frac{3}{\pi}\right)^{\frac{1}{3}} \cdot hc \cdot n^{4/3} , \tag{2}$$

where h equals Planck's constant, c equals velocity of light; and as

$$n = \frac{\rho}{\mu H(1+f)} , \tag{3}$$

[1] 11, No. 70, 592, 1931.

[2] *Philosophical Magazine*, 7, 63, 1929.

μ equals the molecular weight, 2.5, for a fully ionized material, H equals the mass of hydrogen atom, and f equals the ratio of number of ions to number of electrons, a factor usually negligible. Or, putting in the numerical values,

$$P=K\rho^{4/3}\ , \tag{4}$$

where K equals 3.619×10^{14}. We can now immediately apply the theory of polytropic gas spheres for the equation of state given by (4), where for the exponent γ we have

$$\gamma=\frac{4}{3} \text{ or } 1+\frac{1}{n}=\frac{4}{3} \text{ or } n=3\ .$$

We have therefore the relation[1]

$$\left(\frac{GM}{M'}\right)^2=\frac{(4K)^3}{4\pi G}\ ,$$

or

$$\begin{aligned} M&=1.822\times 10^{33}\ ,\\ &=.91\odot\text{(nearly)}\ . \end{aligned} \tag{5}$$

As we have derived this mass for the star under ideal conditions of extreme degeneracy, we can regard 1.822×10^{33} as the maximum mass of an ideal white dwarf. This can be compared with the earlier estimate of Stoner[2]

$$M_{\text{max}}=2.2\times 10^{33}\ , \tag{6}$$

based again on uniform density distribution. The "agreement" between the accurate working out, based on the theory of the polytropes, and the cruder form of the theory is rather surprising in view of the fact that in the corresponding non-relativistic case the deviations were rather serious.

Trinity College
Cambridge
November 12, 1930

[1] A. S. Eddington, *Internal Constitution of Stars*, p. 83, eq. (57.3.)

[2] *Philosophical Magazine*, **9**, 944, 1930.

The Dissociation Formula according to the Relativistic Statistics.
By S. Chandrasekhar.

(*Communicated by R. H. Fowler and E. A. Milne.*)

1. The temperatures conceived by Professor Milne * for the interiors of stars quite transcend all previous expectations and are precisely those temperatures at which the relativistic effect becomes important in statistical theory—at least for the electron-assembly, if not for the atomic assembly. It was thought worth while therefore to work out the dissociation formulæ on the relativistic statistics. The method followed in this paper will be the very elegant method developed by Milne in his recent paper.† Indeed the following will have to be considered as merely supplementing his results.

* *The Observatory*, **53**, 238, 1930.
† *M.N.*, **90**, 769, 1930.

2. *The Enunciation of the Dissociation Formulæ for Particles obeying Different Statistics.*—If we consider the case of a neutral atom dissociating into an ion and an electron, we know for certain that the electrons obey the Fermi-Dirac statistics, but the neutral atoms in general obey only the Einstein-Bose statistics. We have in fact the empirical rule that particles carrying a charge of an even or an odd multiple of the electronic charge obey the Einstein-Bose or the Fermi-Dirac statistics respectively. Hence in considering the dissociative equilibrium of different types of particles " chemically " reacting with each other, in the light of the new statistics we must, in general, consider the case of the different particles obeying different statistics.*

Let us for simplicity consider the case of an equilibrium characterised by the chemical equation

$$aA + bB = A_aB_b + \chi \quad . \quad . \quad . \quad (1)$$

representing the reaction in which a particles of type A react reversibly with b particles of type B to produce one particle of type A_aB_b with an evolution of energy χ. Let the particles A and B obey the Fermi-Dirac statistics, and A_aB_b the Einstein-Bose statistics,† which must happen when $a + b$ is even. Suppose that the total number of A's present in volume V be X, that of the B's Y. Let these occur in the assembly (in equilibrium) as L *free* particles of type A, and M *free* particles of type B. Then we have obviously

$$X = L + aN \quad . \quad . \quad . \quad . \quad (2)$$

$$Y = M + bN \quad . \quad . \quad . \quad . \quad (3)$$

If, further, T_A, T_B, and T_{AB} are kinetic energies of the respective particles, the total energy U is given by

$$U = T_A + T_B + T_{AB} - N\chi \quad . \quad . \quad . \quad (4)$$

Let x_s, y_s, z_s represent the numbers of cells associated with the energy-range E_s to $E_s + dE_s$ for the particles A, B, and A_aB_b respectively. The entropy of the whole assembly by the Boltzmann principle is

$$S = k \log \prod_s \frac{x_s\,!}{(x_s - p_s')\,!\,p_s'\,!} \prod_s \frac{y_s\,!}{(y_s - q_s')\,!\,q_s'\,!} \prod_s \frac{(z_s + r_s')\,!}{z_s\,!\,r_s'\,!} \quad . \quad (5)$$

where each separate factor corresponds to the total number of " complexions " obtained by arranging p_s' particles of type A into x_s cells, q_s' particles of type B into y_s cells, and r_s' particles of type A_aB_b into z_s cells with the conditions

$$\sum p_s' = L, \quad \sum q_s' = M, \quad \sum r_s' = M \quad . \quad . \quad (6)$$

and that the particles A_aB_b obey the Einstein-Bose statistics, while the

* I am indebted to Mr. Fowler for a full appreciation of this point.

† If $a=b=1$, this would correspond to a neutral atom dissociating into an ion and an electron.

other two types of particles obey the Fermi-Dirac statistics. It is convenient to introduce

$$\left.\begin{aligned} p_s^0 &= x_s - p_s' \\ q_s^0 &= y_s - q_s' \\ r_s^0 &= z_s + r_s' \end{aligned}\right\} \quad . \quad . \quad . \quad . \quad (7)$$

The expression for the entropy (5) then reduces to

$$\mathrm{S} = k \log \prod_s \frac{x_s\,!}{p_s^0\,!\,p_s'\,!} \prod_s \frac{y_s\,!}{q_s^0\,!\,q_s'\,!} \prod_s \frac{r_s^0\,!}{z_s\,!\,r_s'\,!} \quad . \quad . \quad (5')$$

Now the entropy will have to be a maximum subject to the conditions (2), (3), (6), (7), and the further condition

$$\mathrm{U} = \sum_s p_s'\mathrm{E}_s^{\mathrm{A}} + \sum_s q_s'\mathrm{E}_s^{\mathrm{B}} + \sum_s r_s'(\mathrm{E}_s^{\mathrm{AB}} - \chi) \quad . \quad (8)$$

We apply, as usual in such problems, the Lagrange method.

Applying Stirling's theorem in the form

$$\log n\,! \sim n \log n - n,$$

and effecting the variation in the p_s', etc. (with x_s, y_s, z_s constant), we find

$$0 = -\sum \delta p_s^0(1 + \log p_s^0) - \sum \delta p_s'(1 + \log p_s') - (\text{similar terms in } q)$$

$$+ \sum \delta r_s^0(1 + \log r_s^0) - \sum \delta r_s'(1 + \log r_s') \quad . \quad . \quad . \quad (9)$$

$$0 = \delta p_s^0 + \delta p_s' \quad . \quad . \quad . \quad . \quad (10)$$

$$0 = \delta q_s^0 + \delta q_s' \quad . \quad . \quad . \quad . \quad (10')$$

$$0 = \delta r_s^0 - \delta r_s' \quad . \quad . \quad . \quad . \quad (10'')$$

$$0 = \sum \delta p_s' + a \sum \delta r_s' \quad . \quad . \quad . \quad (11)$$

$$0 = \sum \delta q_s' + b \sum \delta r_s' \quad . \quad . \quad . \quad (11')$$

$$0 = \sum_s \mathrm{E}_s^{\mathrm{A}} \delta p_s' + \sum_s \mathrm{E}_s^{\mathrm{B}} \delta q_s' + \sum_s (\mathrm{E}_s^{\mathrm{AB}} - \chi)\delta r_s' \quad . \quad (12)$$

In applying the Lagrange method multiply the equations (10), (10′), (10″), (11), (11′), and (12) by $\mathrm{F}_s + 1$, $\mathrm{G}_s + 1$, $\mathrm{H}_s - 1$, α, β, and γ respectively. Adding the complete set of equations and equating to zero the coefficients of $\delta p_s^0 \ldots \delta r_s'$, we have easily

$$-\log p_s^0 + \mathrm{F}_s = 0, \qquad -\log q_s^0 + \mathrm{G}_s = 0, \qquad \log r_s^0 + \mathrm{H}_s = 0.$$

$$-\log p_s' + \mathrm{F}_s - \alpha - \gamma \mathrm{E}_s^{\mathrm{A}} = 0,$$

$$-\log q_s' + \mathrm{G}_s - \beta - \gamma \mathrm{E}_s^{\mathrm{B}} = 0,$$

$$-\log r_s' - \mathrm{H}_s - a\alpha - b\beta - \gamma(\mathrm{E}_s^{\mathrm{AB}} - \chi) = 0.$$

Altering the constants F_s, G_s, H_s to new constants $\mathrm{C}_s^{\mathrm{AB}}$, etc., such that we have $r_s^0 = \mathrm{C}_s^{\mathrm{AB}}$, etc.,

$$r_s' = \mathrm{C}_s^{\mathrm{AB}} e^{-a\alpha - b\beta - \gamma(\mathrm{E}_s^{\mathrm{AB}} - \chi)},$$

or $$z_s = r_s^0 - r_s' = C_s^{AB}(1 - e^{-aa - b\beta - \gamma(E_s^{AB} - \chi)}),$$

giving us

$$r_s' = \frac{z_s}{e^{aa+b\beta+\gamma(E_s^{AB}-\chi)} - 1}.$$

Similarly

$$p_s' = \frac{x_s}{e^{a+\gamma E_s^A} + 1}, \qquad q_s' = \frac{y_s}{e^{\beta+\gamma E_s^B} + 1}.$$

γ is easily identified to be $\frac{1}{kT}$. The values of L, M, and N follow from (6). Replacing the sums by integrals and writing $kT\theta$ for E, we have

$$L = kT\int_0^\infty \frac{x_s d\theta}{e^{a+\theta} + 1},$$

$$M = kT\int_0^\infty \frac{y_s d\theta}{e^{\beta+\theta} + 1},$$

$$N = kT\int_0^\infty \frac{z_s d\theta}{e^{\left(aa+b\beta-\frac{\chi}{kT}\right)+\theta} - 1}.$$

If we write in the integral for N

$$a' = aa + b\beta - \frac{\chi}{kT}$$

we have its equivalent

$$aa + b\beta = \frac{\chi}{kT} + a'. \qquad . \quad . \quad . \quad . \quad (13)$$

This is for the simple reaction (1). This result is easily generalised for more complicated reactions, and we can enunciate the following generalisation of the theorem proved by Milne (*loc. cit.*) :—

Let us consider a gaseous mixture in which the constituents C_1, C_2 . . . C_r . . . react reversibly according to the chemical equation

$$\sum_r n_r C_r = \sum_{r'} n_{r'} C_{r'} + \chi$$

where χ is energy (expressed in ergs) evolved when n_r molecules of C_r, etc. react to form $n_{r'}$ molecules of $C_{r'}$, etc. Let the gaseous mixture be contained in a volume V at temperature T and consist of N_r molecules of C_r, etc., $N_{r'}$ molecules of $C_{r'}$, etc. The condition for dissociative equilibrium is obtained by eliminating a_1, . . . a_r ; $a_{1'}$ . . . $a_{r'}$ from the set of equations

$$N_r = kT\int_0^\infty \frac{x_r d\theta}{e^{a_r+\theta} + \epsilon_r}; \text{ etc.} \qquad . \quad . \quad . \quad . \quad (14)$$

$$N_{r'} = kT\int_0^\infty \frac{x_{r'} d\theta}{e^{a_{r'}+\theta} + \epsilon_{r'}}; \text{ etc.} \qquad . \quad . \quad . \quad . \quad (15)$$

and $$\sum n_r a_r = \sum_{r'} n_{r'} a_{r'} + \frac{\chi}{kT} \qquad . \quad . \quad . \quad . \quad (16)$$

where x_r, etc., $x_{r'}$, etc., are the numbers of cells in a given energy range E to E + dE for the particles of the type C_r, etc., $C_{r'}$, etc., respectively, and ϵ_r equals + 1 or − 1 according as the particles "C_r" obey the Einstein-Bose or the Fermi-Dirac statistics.

3. *The Equations of Relativistic Statistics.*—When the relativistic effect is taken into account, the usual formula for the number of cells associated with a given energy range E_s to $E_s + dE_s$ (*e.g.*)

$$Z_s dE_s = \frac{2\pi Vq}{h^3}(2m)^{\frac{3}{2}} E_s^{\frac{1}{2}} dE_s \quad . \quad . \quad . \quad (17)$$

(q = the statistical weight) has to be replaced by

$$Z_s dE_s = \frac{2\pi Vq}{h^3}(2m)^{\frac{3}{2}} E_s^{\frac{1}{2}}\left(1 + \frac{E_s}{2mc^2}\right)^{\frac{1}{2}}\left(1 + \frac{E_s}{mc^2}\right) dE_s \quad . \quad (17')$$

If the relativistic effect becomes predominant (we state the physical conditions for this later), we shall have to work with

$$Z_s dE_s = \frac{4\pi Vq}{h^3 c^3} E_s^2 dE_s \quad . \quad . \quad . \quad (18)$$

To work out the dissociation formula when certain types of particles have a predominant relativistic effect while others have not, we have simply to use for the corresponding x_r or $x_{r'}$ the expression (17) or (18) in equation (14) or (15). But we take this opportunity to write down the equations of the relativistic statistics for the *Fermi-Dirac* case in a form convenient for application. Working with (18), we can derive

$$\left.\begin{aligned} N &= 8\pi Vq\left(\frac{kT}{hc}\right)^3 U_2 \\ E &= 24\pi VqkT\left(\frac{kT}{hc}\right)^3 U_3 \\ pV &= \tfrac{1}{3}E \\ S &= nk\left[4\frac{U_3}{U_2} - \log A\right] \end{aligned}\right\} \quad . \quad . \quad . \quad (18')$$

where
$$U_\rho = \frac{1}{\Gamma(\rho + 1)}\int_0^\infty \frac{\theta^\rho d\theta}{\frac{1}{A}e^\theta + 1}.$$

In making the transition from the degenerate to the non-degenerate case we simply pass from one limiting value for U_ρ characterised by large A to the other limiting value characterised by small A. It can easily be shown that the condition for degeneracy in this case is different from the Sommerfeld criterion for the non-relativistic case. If we should regard an assembly as degenerate when the relativistic effect is predominant, we must have the inequality

$$n >> \left(\frac{kT}{hc}\right)^3 \frac{4\pi q}{3} \left(= 2{\cdot}86 T^3\right)$$

satisfied. The assembly must satisfy also another condition, namely, that which secures that the relativistic effect is predominant.

For convenience of reference we collect below the conditions which an assembly must satisfy in order that we can regard a system as degenerate or non-degenerate in the two distinct cases—where the relativistic effect is negligible and where it is highly predominant:

Magnitude of Relativistic Effect.	State of the Assembly.	Conditions. (The Numerical Values are those for an *Electron*-Assembly.)
Second order effect	(i) Degenerate	$n << \frac{8\pi m^3c^3}{3h^3} = 5 \cdot 88 \times 10^{29}$
		$n >> \frac{q}{h^3}(2\pi mkT)^{\frac{3}{2}} = 4 \cdot 87 \times 10^{15} \times T^{\frac{3}{2}}$
	(ii) Non-degenerate	$n << \frac{q}{h^3}(2\pi mkT)^{\frac{3}{2}} = 4 \cdot 87 \times 10^{15} \times T^{\frac{3}{2}}$
		$T << \frac{mc^2}{k} = 5 \cdot 9 \times 10^9.$
Predominant	(iii) Degenerate	$n >> \left(\frac{kT}{hc}\right)^3 \frac{4\pi q}{3} = 2 \cdot 86 T^3$
		$n >> \frac{8\pi m^3c^3}{3h^3} = 5 \cdot 88 \times 10^{29}$
	(iv) Non-degenerate	$n << \left(\frac{kT}{hc}\right)^3 \frac{4\pi q}{3} = 2 \cdot 86 T^3$
		$T >> \frac{mc^2}{k} = 5 \cdot 9 \times 10^9.$

The conditions (ii) and (iv) for the case of non-degeneracy hold good for the Einstein-Bose case also. This is clear when we note that the condition for the rapid convergence of the series expansion of (14) for large α_r, when ϵ_r equals either $+1$ or -1, is just the same (and the non-degeneracy condition is merely the analytical expression of this). In the sequel we do *not* consider the case of degeneracy for the atomic assembly (some constituents of which will obey the Einstein-Bose statistics), for, first, such cases are of no practical value, and secondly, the question of degeneracy for the particles obeying the Einstein-Bose statistics is closely bound up with the difficulty pointed out by Uhlenbeck.* In fact, in the degenerate case the particles can *never* have a predominant relativistic effect because all the particles go down to the zero state. (In the Fermi-Dirac statistics this is prevented by Pauli's principle.)

4. *The Dissociation Formulæ for the Relativistic Case.*—(*a*) Let all the constituents be non-degenerate. We consider the case of the ionisation of an atom A capable of being in a state of lowest energy content $A^{(1)}$, its ionised form being $A_+^{(1)}$. Let χ be the energy of ionisation. If $C^{(1)}$ and $C_+^{(1)}$ are the atomic concentrations per unit

* G. E. Uhlenbeck, *Over Statistische Methoden in die Theorie der Quanta*, Thesis, Leiden (1927), p. 70.

volume we have equations corresponding to (14) and (15) reducing, on using for the number of cells, the value given by (18), to *

$$C^{(1)} = 8\pi q^{(1)}\left(\frac{kT}{hc}\right)^3 e^{-\alpha^{(1)}} \quad . \quad . \quad . \quad . \quad (19)$$

$$C_+^{(1)} = 8\pi q_+^{(1)}\left(\frac{kT}{hc}\right)^3 e^{-\alpha_+^{(1)}} \quad . \quad . \quad . \quad (20)$$

$q^{(1)}$ and $q_+^{(1)}$ being the statistical weights. For C_e, the electron-concentration, we have

$$C_e = 8\pi q_e\left(\frac{kT}{hc}\right)^3 e^{-\alpha_e} \quad . \quad . \quad . \quad (21)$$

and as the chemical equation in this case is

$$A_+^{(1)} + e = A^{(1)} + \chi,$$

we have for our fourth equation, in addition to (19), (20), and (21), according to (16),

$$\alpha_+^{(1)} + \alpha^{(e)} = \alpha^{(1)} + \frac{\chi}{kT} \quad . \quad . \quad . \quad (22)$$

Hence our dissociation formula in this case is

$$\frac{C_+^{(1)} \,.\, C_e}{C^{(1)}} = \frac{q_+^{(1)}q^{(e)}}{q^{(1)}} 8\pi\left(\frac{kT}{hc}\right)^3 e^{-\frac{\chi}{kT}} \quad . \quad . \quad (23)$$

The above equation is analogous to the Saha equation for the non-relativistic case.

(*b*) Let us now consider the case where the electrons are degenerate but the atoms and the ionised atoms are non-degenerate. Then we have in this case for the electron-concentration (because it obeys the Fermi-Dirac statistics)

$$C_e = \frac{4\pi}{3} q_e\left(\frac{kT}{hc}\right)^3 (-\alpha_e)^3 \text{ approximately} \quad . \quad . \quad (24)$$

giving

$$-\alpha_e = \left(\frac{3C_e}{4\pi q_e}\right)^{\frac{1}{3}} \frac{hc}{kT} \quad . \quad . \quad . \quad . \quad (24')$$

We have therefore for our eliminant

$$\log\left(\frac{C^{(1)}}{C_+^{(1)}} \frac{q_+^{(1)}}{q^{(1)}}\right) = \frac{\chi}{kT} + \frac{hc}{kT}\left(\frac{3C_e}{4\pi q_e}\right)^{\frac{1}{3}} \quad . \quad . \quad (25)$$

* (19) and (20) are true, independent of what statistics they obey, but (19) would in general correspond to the Einstein-Bose case, and (20) to the Fermi-Dirac case.

We give below the corresponding equation for the non-relativistic case for comparison : *

$$\log\left(\frac{C^{(1)}}{C_+^{(1)}}\frac{q_+^{(1)}}{q^{(1)}}\right) = \frac{\chi}{kT} + \frac{h^2}{2m_e kT}\left(\frac{3C_e}{4\pi q_e}\right)^{\frac{2}{3}} \quad . \quad . \quad (26)$$

5. *Other Possible Cases.*—We now pass on to the consideration of cases intermediate to the relativistic and the non-relativistic cases. As our table shows, it is quite possible for the electron-assembly to be governed by the relativistic equations, while the atomic assembly obeys reasonably well the non-relativistic equations. We assume that the atomic system is non-degenerate. We have then two cases corresponding to the two possible behaviours of the electron-assembly.

(i) *Non-degenerate Electrons.*—In the equations corresponding to (14) and (15) for $C^{(1)}$ and $C_+^{(1)}$ we have to use for the x_r's the expression (17) and for C_e the expression (18). We have then

$$C^{(1)} = \frac{q^{(1)}(2\pi m^{(1)}kT)^{\frac{3}{2}}}{h^3}e^{-\alpha^{(1)}} \quad . \quad . \quad . \quad (27)$$

$$C_+^{(1)} = \frac{q_+^{(1)}(2\pi m_+^{(1)}kT)^{\frac{3}{2}}}{h^3}e^{-\alpha_+^{(1)}} . \quad . \quad . \quad (28)$$

But we have for the electron-concentration

$$C_e = \frac{8\pi q_e}{h^3c^3}(kT)^3e^{-\alpha_e} . \quad . \quad . \quad . \quad (29)$$

We have therefore for our dissociation formula, remembering (22),

$$\frac{C_+^{(1)}C_e}{C^{(1)}} = \frac{q_+^{(1)} . q_e}{q^{(1)}}\left(\frac{m_+^{(1)}}{m^{(1)}}\right)^{\frac{3}{2}}8\pi\left(\frac{kT}{hc}\right)^3e^{-\chi/kT} \quad . \quad . \quad (30)$$

which, neglecting the trivial factor $m_+^{(1)}/m^{(1)}$ is exactly the same as (23) derived for the case when the relativistic-non-degenerate equations were used for the atomic assembly.

(ii) *Degenerate Electrons.*—In this case, instead of (29) we have to use (24), and our dissociation formula is again

$$\log\left[\frac{C^{(1)}}{C_+^{(1)}}\frac{q_+^{(1)}}{q^{(1)}}\left(\frac{m_+^{(1)}}{m^{(1)}}\right)^{\frac{3}{2}}\right] = \frac{\chi}{kT} + \frac{hc}{kT}\left(\frac{3C_e}{4\pi q_e}\right)^{\frac{1}{3}} . \quad . \quad (31)$$

which is exactly the same as (25).

Thus we have the result that the dissociation formula is independent of whether the atomic assembly is governed by the relativistic or the non-relativistic equations in so far as it is postulated that it is in any case non-degenerate. It is the behaviour of the electron-assembly (*i.e.* whether it is degenerate or non-degenerate) that matters. Since this is so in the two extreme cases, we can reasonably generalise and assume that the equations (23) and (25) hold for all intermediate

* General formulæ for dissociative equilibrium were first given by R. H. Fowler, *Proc. Roy. Soc.*, A, **113**, 432, 1926. Explicit forms were given by S. Chandrasekhar, *Phil. Mag.*, **9**, 292, 1930, and also by E. A. Milne (*loc. cit.*).

behaviour of the atomic assembly. It is unlikely that the atomic assembly can be degenerate at those high temperatures (of the order of 10^{10}) when the concentration would have to be greater than 5×10^{38}. (The electron-assembly is degenerate at very much higher temperatures and much lower concentration obviously because of its very small mass.) So far as the atomic assembly is concerned, therefore, the only question is whether the relativistic effect is large or small for the non-degenerate case. This question, as we have already seen, does not affect our dissociation formulæ. If we could be sure about the behaviour of the electron-assembly, we could immediately choose our dissociation formula independent of the behaviour of the atomic assembly.

What has been said regarding the dissociation formula with the relativistic equations is equally true for the electron-assembly with the non-relativistic equations, as can easily be verified. But a state of affairs where the electron-assembly is governed by non-relativistic equations while the atomic assembly obeys the relativistic equations is physically inconceivable. We have therefore to deal with four kinds of dissociation formulæ—namely, for the two possible behaviours of the electron-assembly for both the relativistic and the non-relativistic cases.

6. *Numerical Calculations.*—(i) According to Milne,* the central density, and the central and the effective temperatures of his highly collapsed configurations (which he identifies with the white dwarfs), are respectively for $M = \cdot 85 \odot$:

$$\left.\begin{aligned} \rho_c &= 2\cdot 87 \times 10^6 \beta^3 \\ T_c &= 2\cdot 92 \times 10^9 \beta(1-\beta)^{\frac{1}{4}} \\ T_e &= 11{,}300 \beta^{\frac{1}{2}} \end{aligned}\right\} \quad . \quad . \quad . \quad (32)$$

where

$$\beta = 1 - \frac{\kappa L}{4\pi c G M} \qquad (\kappa = \text{opacity}).$$

If the effective temperature should correspond to, say, 10000°, β must be about 0·9. This gives us immediately

$$\left.\begin{aligned} C_e &= 7 \times 10^{29} \text{ per cm.}^3 \\ T_c &= 1\cdot 4 \times 10^9 \text{ degrees} \end{aligned}\right\}.$$

The electron-assembly therefore just satisfies those conditions necessary for the system to be degenerate and at the same time have the relativistic effect predominant. Hence the dissociation formula we ought to use is (25), which reduces on using numerical values (x = degree of dissociation) to

$$\log_e \frac{1-x}{x} = \frac{\chi}{kT} + \cdot 7045 \frac{C_e^{\frac{1}{3}}}{T}$$

$$= \frac{\chi}{kT} + 4\cdot 483 \quad . \quad . \quad . \quad (33)$$

* E. A. Milne, *M.N.*, **91**, 4, 1930.

For temperatures of the order of 10^9 the first term in (33) is negligible compared with the second, even when χ corresponds to the K-ionisation potentials for the heaviest elements. The degree of dissociation given by (34) is therefore practically zero, and independent of the ionisation potential. This at first sight would appear to contradict the common-sense view that at these extreme conditions of temperature and pressure matter must necessarily be ionised down to the bare nuclei. At the same time we have to recognise that at such high concentrations every electron must be under the influence of one atomic nucleus or another all the time. The whole system is, in fact, "a gigantic molecule in the lowest quantum state." * But "statistically" they must all be free in the sense that they travel the whole of the phase space. Our analysis giving 0 per cent. ionisation will have to be interpreted this way. Thus, whenever we get the calculated degree of dissociation as being practically zero and independent to a large extent of our choice of the ionisation potential, we can then safely conclude that matter has reached such extreme conditions that we must regard (statistically) all the atoms as being "ionised" down to their bare nuclei, and, further, the electron-assembly as being highly degenerate. We can use this result as a "degeneracy-condition." †

(ii) We now use the *non*-relativistic dissociation formula (26). At the interface of the "composite configurations," where we pass from a solution of Emden's equation "$n = 3$" to another of Emden's, "$n = \frac{3}{2}$," the pressure and temperature are (Milne, *loc. cit.*, equations (82) and (83))

$$\rho' = 3{\cdot}66 \times 10^7 \frac{1-\beta}{\beta}\,; \quad T' = {\cdot}843 \times 10^{10}\left(\frac{1-\beta}{\beta}\right)^{\frac{2}{3}}$$

or

$$\frac{C_e'^{\frac{2}{3}}}{T'} = \frac{(\rho'/2m_H)^{\frac{2}{3}}}{T'} = 5{\cdot}9 \times 10^{10} \quad . \quad . \quad . \quad (34)$$

Introducing (34) in (26) we have simply

$$\log_e \frac{1-x}{x} = \frac{\chi}{kT} + 2{\cdot}482 \quad . \quad . \quad . \quad (35)$$

Comparing (35) with (33) we can deduce that matter must from now onwards be treated as "degenerate"—justifying therefore the transition to the degenerate Fermi-Dirac equation of state for the electrons.

In conclusion I wish to express my thanks to Mr. R. H. Fowler and Professor E. A. Milne for their advice and encouragement during the course of the work.

* R. H. Fowler, "On Dense Matter," *M.N.*, **87**, 114, 1926.

† In this connection see W. Anderson, *Physik. Zeits.*, **30**, 360, 1929, also S. Chandrasekhar (*loc. cit.*), § 6.

6

THE HIGHLY COLLAPSED CONFIGURATIONS OF A STELLAR MASS*

S. CHANDRASEKHAR

1

Professor Milne in his recent paper** on 'The Analysis of Stellar Structure' has put forward some essentially new considerations on the possible steady-state configurations of stellar aggregates of varying mass, luminosity, and opacity. One of the main consequences of the analysis is the explanation not only of the existence of white dwarfs – his collapsed configurations – but also of the principal physical characteristics of these configurations. The following is devoted to the development of Milne's theory of these collapsed configurations a stage further.

2

Milne's estimates for the central density and temperature of these collapsed configurations indicate that in some cases we pass beyond the range of validity of the degenerate form of the Fermi-Dirac equation of state ($p=K\varrho^{5/3}$). It can be shown that the pressure of an electron gas which is highly degenerate and which has a very highly predominant relativistic-mass variation effect, takes the limiting form †

$$p = \frac{n^{4/3}hc}{8}\left(\frac{3}{\pi}\right)^{1/3} \tag{1}$$

(c = velocity of light, h = Planck's constant) if the following two conditions are satisfied

$$n > \left(\frac{kT}{hc}\right)^3 \frac{4\pi G}{3} \equiv 2.86T^3 \quad (G=2) \tag{i}$$

$$n > \frac{8\pi m^3c^3}{3h^3} \equiv 5.88 \times 10^{29} \tag{ii}‡$$

(i) replaces Sommerfeld's degeneracy criterion, and (ii) ensures the predominance of the relativistic effect. As the condition that configurations are highly collapsed is precisely equivalent to the condition that in the central regions the electron assembly

* Reprinted from *Monthly Notices of the Royal Astronomical Society* **91**.

** *M.N.R.A.S.*, **91** (1930) 4–55, referred to hereafter as *loc. cit.* For a general exposition of his main ideas see *Nature* (1931 January 3) 'Stellar Structure and the Origin of Stellar Energy.'

† The corresponding expression for the energy E was obtained by E. C. Stoner, *Phil. Mag.* **9** (1930), 944. That $p=\frac{1}{3}E$ is generally true when the relativistic effect is highly predominant can easily be proved.

‡ (ii) is given in Stoner's paper; (i) now replaces Sommerfeld's criterion. These and other inequalities are briefly discussed in my paper on 'The Dissociation Formula according to the Relativistic Statistics', *M.N.R.A.S.* **91** (1931), 446. (in course of publication).

This version is reprinted from Herbert Gursky and Remo Ruffini, eds., *Neutron Stars, Black Holes, and Binary X-ray Sources*, vol. 48 of the Astrophysics and Space Science Library (Dordrecht, The Netherlands: Kluwer, 1975), in which some minor misprints found in the original publication were corrected.

is degenerate,* it would be sufficient to consider whether (ii) is satisfied in order to count the relativistic effect as highly predominant. Further, as L for these configurations is very small, T_c is normally of the order of 10^9,** and (ii) then would automatically provide for (i). (See Equation (19′)).

Now the central density of a highly, collapsed configuration considered as an Emden polytrope '$n=\frac{3}{2}$' is given by †

$$\varrho_c = \frac{32\pi G^3 M^2 \beta^3}{125 K_1^3 \times 7.385} \tag{2}$$

where K_1 is the degenerate-gas constant and

$$\beta = 1 - \frac{\kappa L}{4\pi c G M}. \tag{3}$$

It is clear then that the relativistic effect will be predominant in the central regions of those collapsed configurations whose masses satisfy the inequality

$$\frac{32\pi G^3 M^2 \beta^3}{125 K_1^3 \times 7.385 \times \mu} > \frac{8\pi m^3 c^3}{3h^3}$$

or

$$M\beta^{3/2} > 5\left[\frac{5\mu \times 7.385}{12}\right]^{1/2}\left(\frac{mcK_1}{Gh}\right)^{3/2} = 0.434 \odot \quad (\text{if } \mu = 2.5 m_H) \tag{4}$$

where μ is the mean molecular weight and $\odot$ denotes the mass of the Sun ($=1.985 \times 10^{33}$ gms).

The purpose of this paper is to find out the consequences of introducing the equation of state $p = K_2 \varrho^{4/3}$. It will be shown that we can enumerate the complete linear sequence of steady-state configurations for an *assigned* small luminosity as the mass varies, the opacity and source-strength being constant and uniform (standard model).

3. The Equations of the Problem

We base our subsequent discussion exclusively on the standard model as it is considerably easier to work with.

We have the following set of equations for the standard model independent of the equation of state we may adopt. (The notation is identical with that used in Milne's paper.)

$$\frac{\mathrm{d}p}{\mathrm{d}r} + \frac{\mathrm{d}p'}{\mathrm{d}r} = -\frac{GM(r)}{r^2}\varrho \tag{5}$$

* Milne, *loc. cit.*, § 22.
** Milne, *loc. cit.*, p. 39.
† Equation (55), *loc. cit.*

$$\frac{\mathrm{d}p'}{\mathrm{d}r} = -\frac{\kappa L(r)}{4\pi c r^2}\varrho \tag{6}$$

$$\frac{L(r)}{M(r)} = \varepsilon = \frac{L}{M}.$$

We have also the following relation between the gas kinetic-pressure (p) and the radiation-pressure p':

$$p = p' \frac{\beta}{1-\beta} \tag{7}$$

where β is defined by (3).

CASE I. (The Relativistic-degenerate Case). – We have for the gas kinetic-pressure, taking into account only the electronic contribution, which is certainly by far the most important

$$p = \frac{n^{4/3}hc}{8}\left(\frac{3}{\pi}\right)^{1/3}. \tag{1'}$$

If we assume the molecular weight $\mu = 2.5\ m_H$, then

$$p = K_2 \varrho^{4/3} \tag{1''}$$

where

$$K_2 = \frac{hc}{8(2.5 m_H)^{4/3}}\left(\frac{3}{\pi}\right)^{1/3} = 3.619 \times 10^{14}. \tag{1'''}$$

With the equation of state given by (1″), the equation of mechanical equilibrium reduces to

$$\frac{4K_2}{3G\beta} r^2 \varrho^{-2/3} \frac{\mathrm{d}\varrho}{\mathrm{d}r} = -M(r). \tag{8}$$

Remembering that

$$\frac{\mathrm{d}M(r)}{\mathrm{d}r} = 4\pi r^2 \varrho,$$

we have on differentiating (8)

$$\frac{4K_2}{3G\beta}\frac{\mathrm{d}}{\mathrm{d}r}\left(r^2 \varrho^{-2/3} \frac{\mathrm{d}\varrho}{\mathrm{d}r}\right) = -4\pi r^2 \varrho. \tag{9}$$

Putting

$$\varrho = \lambda_3 \chi^3 \tag{10}$$

(9) reduces to

$$\frac{K_2 \lambda_3^{-2/3}}{\pi G \beta} \cdot \frac{1}{r^2} \frac{\mathrm{d}}{\mathrm{d}r}\left(r^2 \frac{\mathrm{d}\chi}{\mathrm{d}r}\right) = -\chi^3. \tag{11}$$

Changing r to the variable ζ given by

$$r = \zeta \left[\frac{K_2 \lambda_3^{-2/3}}{\pi G \beta}\right]^{1/2} \tag{12}$$

we have finally

$$\frac{1}{\zeta^2} \frac{\mathrm{d}}{\mathrm{d}\zeta}\left(\zeta^2 \frac{\mathrm{d}\chi}{\mathrm{d}\zeta}\right) = -\chi^3 \tag{13}$$

which is Emden's polytropic equation with index 3.

From (8), using the values of ϱ and r given by (10) and (12), we have

$$M(r) = -\frac{4}{\pi^{1/2}}\left(\frac{K_2}{G\beta}\right)^{3/2} \zeta^2 \frac{\mathrm{d}\chi}{\mathrm{d}\zeta}. \tag{14}$$

It may be noted that, unlike the degenerate non-relativistic case, λ_3 has disappeared from (14).

Case II. (The Non-relativistic-degenerate Case). – We can use the equations given by Milne, *loc. cit.*, Equations (49), (48), (50), and (47) respectively:

$$\frac{1}{\eta^2} \frac{\mathrm{d}}{\mathrm{d}\eta}\left(\eta^2 \frac{\mathrm{d}\psi}{\mathrm{d}\eta}\right) = -\psi^{3/2} \tag{15}$$

where

$$\varrho = \lambda_2 \psi^{3/2} \tag{16}$$

$$r = \eta \left[\frac{5K_1}{8\pi G\beta} \lambda_2^{-1/3}\right]^{1/2}. \tag{17}$$

We have also

$$M(r) = -\frac{1}{4(2\pi)^{1/2}}\left(\frac{5K_1}{G\beta}\right)^{3/2} \lambda_2^{1/2} \eta^2 \frac{\mathrm{d}\psi}{\mathrm{d}\eta}. \tag{18}$$

K_1 in (17) and (18) is the degenerate-gas constant.*

4. The Equations of Fit

Let $r = r'$ be the radius of the surface of demarcation, *i.e.* outside $r = r'$ the distribution of density is given by a solution of Emden's equation '$n = \frac{3}{2}$', and inside $r = r'$ by a solution of Emden's equation '$n = 3$'.

* $K_1 = 2.138 \times 10^{12}$ or 3.17×10^{12}, according as $\mu = 2.5 m_H$ or $2 m_H$. The former value will be used in the numerical work of this paper.

Outside $r=r'$ the equation of state is $p=K_1\varrho^{5/3}$ and inside $r=r'$ we have $p=K_2\varrho^{4/3}$. Let ϱ' and T' be the values of ϱ and T at the interface, so that we have

$$K_1\varrho'^{\,5/3} = K_2\varrho'^{\,4/3} = \tfrac{1}{3}aT'^{\,4}\frac{\beta}{1-\beta}$$

or

$$\varrho' = \left(\frac{K_2}{K_1}\right)^3; \qquad T' = \frac{K_2}{K_1}\left[\frac{K_2}{\frac{1}{3}a}\frac{1-\beta}{\beta}\right]^{1/4} \tag{19}$$

or, using numerical values, *

$$\left.\begin{aligned}\varrho' &= 4.84\times 10^6\\ \text{or}\qquad n' &= 1.165\times 10^{30}\end{aligned}\right\}; \qquad T' = 3.29\times 10^9\left(\frac{1-\beta}{\beta}\right)^{1/4}. \tag{19'}$$

Let $\phi(\eta)$ be any solution of

$$\frac{1}{\eta^2}\frac{\mathrm{d}}{\mathrm{d}\eta}\left(\eta^2\frac{\mathrm{d}\psi}{\mathrm{d}\eta}\right) = -\,\psi^{3/2} \tag{20}$$

and suppose that it vanishes at $\eta=\eta_0$. Then $B^4\phi(B\eta)$ is also a solution of (20) and it vanishes at $\eta=\eta_1$, where $B\eta_1=\eta_0$. Write $B\eta'=b_1$. By (19) and (16) we have

$$\varrho' = \left(\frac{K_2}{K_1}\right)^3 = \lambda_2\psi'^{\,3/2} = \lambda_2 B^6\left[\phi(b_1)\right]^{3/2} \tag{21}$$

and

$$M(r') = -\frac{1}{4(2\pi)^{1/2}}\left(\frac{5K_1}{G\beta}\right)^{3/2}\lambda_2^{1/2}B^3 b_1^2\phi'(b_1) \tag{22}$$

or, eliminating $\lambda_2^{1/2}B^3$ by (21), we have

$$M(r') = -\frac{1}{4(2\pi)^{1/2}}\left(\frac{5K_2}{G\beta}\right)^{3/2}\frac{b_1^2\phi'(b_1)}{[\phi(b_1)]^{3/4}}. \tag{23}$$

Also by (17)

$$r' = \frac{b_1}{B}\left[\frac{5K_1}{8\pi G\beta}\right]^{1/2}\lambda_2^{-1/6} \tag{24}$$

or, eliminating $B\lambda_2^{1/6}$ by (21), we have

$$r' = \left[\frac{5}{8\pi G\beta K_2}\right]^{1/2} K_1 b_1\left[\phi(b_1)\right]^{1/4}. \tag{25}$$

* Remembering that for collapsed configurations $\beta\sim 1$ we see that $n' = 1.165\times 10^{30}$ provides for the condition (i) (Section 2) also.

Now let $g(\zeta)$ be any solution of the equation

$$\frac{1}{\zeta^2}\frac{\mathrm{d}}{\mathrm{d}\zeta}\left(\zeta^2\frac{\mathrm{d}\chi}{\mathrm{d}\zeta}\right) = -\chi^3 \tag{26}$$

which has the first zero at $\zeta=\zeta_0$. Then $Ag(A\zeta)$ is also a solution of (26) which vanishes at $\zeta=\zeta_1$, where $A\zeta_1=\zeta_0$. Write $c_1=A\zeta'$. We have

$$\varrho' = \left(\frac{K_2}{K_1}\right)^3 = \lambda_3\chi'^3 = \lambda_3 A^3 g(c_1)^3 . \tag{27}$$

By (12)

$$r' = \frac{c_1}{A}\left(\frac{K_2}{\pi G\beta}\right)^{1/2} \lambda_3^{-1/3}$$

or, eliminating $A\lambda_3^{1/3}$ by (27), we have

$$r' = \left[\frac{1}{\pi G\beta K_2}\right]^{1/2} K_1 c_1 g(c_1) . \tag{28}$$

By (14) we have also

$$M(r') = -\frac{4}{\pi^{1/2}}\left(\frac{K_2}{G\beta}\right)^{3/2} c_1^2 g'(c_1) . \tag{29}$$

Now our conditions of the fitting of our two configurations are that r' given by (25) and (28) as also $M(r')$ given by (23) and (29) are identical. Equating the respective sides we find we are simply left with

$$\left(\frac{8}{5}\right)^{1/2} c_1 g(c_1) = b_1[\phi(b_1)]^{1/4} \tag{30}$$

$$\left(\frac{8}{5}\right)^{3/2} c_1^2 g'(c_1) = \frac{b_1^2\phi'(b_1)}{[\phi(b_1)]^{3/4}} \tag{31}$$

which are just Milne's equations of fit (100) and (101),* for the transition from a gaseous (Maxwellian) envelope to a degenerate core. We see therefore that when the conditions do not become so drastic as to necessitate the introduction of a homogeneous core, the analytical and the computational difficulties are reduced, as we have to solve the same set of equations for the two transitions – namely, that from a gaseous to a degenerate atmosphere, and then that from the degenerate to the relativistically degenerate atmosphere.

* Professor Milne has since drawn my attention to the fact that this is just what we ought to expect, and that the ($\frac{8}{5}$) occurring in (30) and (31) is just $(n_1+1)/(n_2+1)$, where n_1 and n_2 are the indices of the two Emden equations.

5. A Completely-Relativistically Degenerate Configuration

We consider now a configuration built *entirely* on the relativistic-degenerate equation of state $p=K_2\varrho^{4/3}$. This is therefore an Emden polytrope '$n=3$' and is similar to the Emden-Eddington diffuse configurations. But since we assume the validity of the equation of state $p=K_2\varrho^{4/3}$ right from the boundary it is clear that if this configuration is to approximate to anything practically realisable, the central density must be sufficiently high to make the correction due to the degenerate fringe negligible. We show later that this Emden polytrope has a $\varrho_c=\varrho_{\max}$ (the maximum density matter is capable of), in which case the correction due to the degenerate 'fringe' does become negligible.

We choose λ_3 such that the value of ζ at which χ vanishes is unity, *i.e.* by (12) we choose λ_3 such that

$$r_1^2 = \frac{K_2\lambda_3^{-2/3}}{\pi G\beta} \tag{32}$$

where r_1 is the radius of the star. Hence

$$\lambda_3 = \left(\frac{K_2}{\pi G\beta}\right)^{3/2} \cdot \frac{1}{r_1^3}. \tag{33}$$

The central density is given by

$$\varrho_c = \lambda_3 (\chi)_0^3 = \lambda_3 \zeta_0^3 \tag{34'}$$

or by (33)

$$\varrho_c = \left(\frac{K_2}{\pi G\beta}\right)^{3/2} \frac{\zeta_0^3}{r_1^3}. \tag{34''}$$

The central temperature would be given by

$$\tfrac{1}{3}aT_c^4 \frac{\beta}{1-\beta} = K_2\varrho_c^{4/3}$$

or

$$T_c = \left(\frac{K_2}{\tfrac{1}{3}a}\right)^{1/4} \left(\frac{1-\beta}{\beta}\right)^{1/4} \left(\frac{K_2}{\pi G\beta}\right)^{1/2} \cdot \frac{\zeta_0}{r_1}. \tag{35}$$

As is well known, for the Emden polytrope '$n=3$' the central density ϱ_c and the mean density ϱ_m are related by

$$\frac{\varrho_c}{\varrho_m} = -\frac{\zeta_0}{3\chi'(\zeta_0)} = 54.36\,. \tag{35'}$$

Finally we have a relation connecting the mass and luminosity, which is merely the condition that the whole mass shall be representable as a relativistic configuration of Emden type

$$M = -\frac{4}{\pi^{1/2}}\left(\frac{K_2}{G\beta}\right)^{3/2} \zeta_0^2 \left(\frac{d\chi}{d\zeta}\right)_{\zeta=\zeta_0} \tag{36}$$

where, since for Emden's solution $\zeta_0^2(d\chi/d\zeta)_{\zeta=\zeta_0} = -2.015$, we have, introducing numerical values in (36),

$$M = 0.9177 \odot \beta^{-2/3} \equiv M_3 \tag{36'}$$

we have also the limiting relation

$$\underset{L\to 0}{M} \to \frac{2.015 \times 4}{\pi^{1/2}}\left(\frac{K_2}{G}\right)^{3/2} = 0.92 \odot . \tag{36''}$$

6

If the white dwarf under consideration could legitimately be considered as obeying down to its central regions the Emden equation '$n=\frac{3}{2}$', it is clear that ϱ_c so calculated should not exceed $\varrho' [= (K_2/K_1^3)]$. The central density of a highly collapsed configuration which is a complete Emden polytrope '$n=\frac{3}{2}$' is

$$\varrho_c = \frac{32\pi G^3 M^2 \beta^3}{125 K_1^3 \times 7.385}. \tag{2}$$

We must have therefore

$$\frac{32\pi G^3 M^2 \beta^3}{125 K_1^3 \times 7.385} \leqslant \left(\frac{K_2}{K_1}\right)^3.$$

Hence for considerations based on Emden's '$n=\frac{3}{2}$' alone to be valid we must have the inequality

$$M \leqslant 1.214 \times 10^{33}\, \beta^{-3/2}\ \text{gm} = 0.6115 \odot \beta^{-3/2} \quad (= M_{3/2},\ \text{say})$$

satisfied. Hence collapsed configurations of mass *less than* $M_{3/2}$ are Emden polytropes '$n=\frac{3}{2}$'. For $M=M_{3/2}$, ϱ_c is just equal to our 'interfacial density'. If M becomes greater than $M_{3/2}$ the relativistic core spreads and the configurations become composite. We proceed now to the study of these composite-configurations.

7. The Composite Series I

In working out the composite series we again consider only Emden's solution for the relativistic core, *i.e.* we exclude for the present the possibility of the conditions becoming so drastic as to necessitate the changing over from $p=K_2\varrho^{4/3}$ equation of state. (It will be seen that when ϱ_c becomes equal to ϱ_{max} – the maximum density matter is capable of – the degenerate fringe becomes negligible, and so we are not required to introduce the relativistic and the homogeneous core simultaneously.)

We have, since $\chi(0)=1$, if χ is Emden's solution for '$n=3$',

$$\varrho_c = \lambda_3 A^3 \tag{37}$$

or by (27)

$$\varrho_c = \left(\frac{K_2}{K_1}\right)^3 \frac{1}{[g(c_1)]^3}. \tag{38}$$

For the central temperature we have, since

$$\tfrac{1}{3} a T_c^4 = K_2 \varrho_c^{4/3} \frac{1-\beta}{\beta},$$

$$T_c = \left(\frac{3K_2}{a}\right)^{1/4} \frac{K_2}{K_1} \left(\frac{1-\beta}{\beta}\right)^{1/4} \cdot \frac{1}{g(c_1)}. \tag{39}$$

We have also

$$\frac{\varrho'}{\varrho_c} = [g(c_1)]^3; \qquad \frac{T'}{T_c} = g(c_1). \tag{40}$$

The radius r_1 of the whole configuration is given by

$$\frac{r'}{r_1} = \frac{\eta'}{\eta_1} = \frac{B\eta'}{B\eta_1} = \frac{b_1}{\eta_0} \tag{41}$$

or by (25)

$$r_1 = \left[\frac{5}{8\pi G\beta K_2}\right]^{1/2} K_1 \eta_0 [\phi(b_1)]^{1/4}. \tag{42}$$

The effective temperature and the mean density are easily found to be

$$T_e = \left(\frac{L}{ac}\right)^{1/4} \left(\frac{8G\beta K_2}{5}\right)^{1/4} \cdot \frac{1}{K_1^{1/2}} \cdot \frac{1}{\eta_0^{1/2} [\phi(b_1)]^{1/8}} \tag{43}$$

$$\varrho_m = \tfrac{3}{4} \left[\frac{8\pi G\beta K_2}{5}\right]^{3/2} \frac{M}{\pi K_1^3} \frac{1}{\eta_0^3 [\phi(b_1)]^{3/4}}. \tag{44}$$

It is not difficult to put the above equations in a form which makes it clear that as $b_1 \to \eta_0$ and $c_1 \to \zeta_0$ these composite configurations *continuously* pass over into the complete relativistic Emden polytrope '$n=3$' with $M=0.92\odot\beta^{-3/2}$. In making the reduction we make free use of (30), (31), and (42):

$$\varrho_c = \left(\frac{K_2}{\pi G\beta}\right)^{3/2} \frac{c_1^3}{r_1^3} \cdot \frac{\eta_0^3}{b_1^3} \tag{45}$$

$$\frac{\varrho_c}{\varrho_m} = -\tfrac{1}{3}\frac{c_1}{g'(c_1)}\frac{M(r')}{M(r_1)}\cdot\frac{\eta_0^3}{b_1^3} \tag{45'}$$

$$M(r') = -\frac{4}{\pi^{1/2}}\left(\frac{K_2}{G\beta}\right)^{3/2}\zeta_0^2\left(\frac{\mathrm{d}\chi}{\mathrm{d}\zeta}\right)_{\zeta=\zeta_0}\left[\frac{c_1^2 g'(c_1)}{\zeta_0^2 g'(\zeta_0)}\right] \tag{46}$$

$$T_c = \left(\frac{K_2}{\tfrac{1}{3}a}\right)^{1/4}\left(\frac{1-\beta}{\beta}\right)^{1/4}\left(\frac{K_2}{\pi G\beta}\right)^{1/2}\frac{c_1}{r_1}\frac{\eta_0}{b_1^3}. \tag{46'}$$

But when $b_1 \to \eta_0$ and $c_1 \to \zeta_0$ it is clear from (42) and (38) that simultaneously $r_1 \to 0$ and $\varrho_c \to \infty$.* Thus the completely relativistic model considered as the limit of the composite series is a point-mass with $\varrho_c = \infty$! The theory gives this result because $p = K_2\varrho^{4/3}$ allows any density provided the pressure be sufficiently high. We are bound to assume therefore that a stage must come beyond which the equation of state $p = K_2\varrho^{4/3}$ is not valid, for otherwise we are led to the physically inconceivable result that for $M = 0.92 \odot \beta^{-3/2}$, $r_1 = 0$, and $\varrho = \infty$. As we do not know physically what the next equation of state is that we are to take, we assume for definiteness the equation for the homogeneous incompressible material $\varrho = \varrho_{max}$, where ϱ_{max} is the maximum density of which matter is capable. The preceding analysis would then break down when ϱ_c given by (38) exceeds ϱ_{max}. Now ϱ_{max} must at the lowest estimate be of the oider of 10^{12} gm cm^{-3}, for if the "maximum density of matter is limited only by the sizes of the electrons and nuclei, densities of the order 10^{14} should not be impossible." ** Our interfacial density is only 4.84×10^6 and the ratio of this to the central density $\varrho_c = \varrho_{max}$ is of the order of 10^{-6}, and a reference to the Emden tables shows that if for the moment we assume the star as *completely* relativistically degenerate, before we have proceeded $\frac{1}{1000}$th of the radius of the star into the interior, our assumption becomes valid. Hence the correction due to the degenerate 'fringe' is negligible. We can therefore to a high degree of approximation consider, as the limit of these composite series, the Emden polytrope '$n=3$' with $\varrho_c = \varrho_{max}$ and $M = 0.92 \odot \beta^{-3/2}$, it being understood that ϱ_{max} is sufficiently high to make the correction due to the degenerate fringe negligible. (Hence for highly collapsed configurations of mass greater than $0.92 \odot \beta^{-3/2}$ we can neglect the degenerate fringe.) Thus the highly collapsed configurations for which $0.612 \odot \beta^{-3/2} < M < 0.92 \odot \beta'^{-3/2}$ are composite, consisting of a degenerate envelope surrounding a relativistic core. These composite configurations are bordered on either side by two completely determinable configurations – on the one side by an Emden polytrope '$n = \frac{3}{2}$,' and on the other by an Emden polytrope '$n = 3$' with $\varrho_c = \varrho_{max}$.

8. Composite Series II

So far we have considered only the Emden solutions for the relativistic core, and we have been able to specify the steady-state configurations only for $M \leqslant M_3$. Further, $M = M_3$ corresponds to a high degree of approximation to an Emden-polytrope

* This suggests that when g is Emden's solution, φ is of the centrally condensed type.

** R. H. Fowler, *M. N.* **87** (1926), 114, 'Dense Matter'.

'$n=3$' with $\varrho_c=\varrho_{\max}$. Hence for $M>M_3$ we should expect the homogeneous core to have spread out. We should therefore consider the fit of a relativistic envelope surrounding a homogeneous core with $\varrho=\varrho_{\max}$, the core and the envelope being continuous at the interface. If $r=r''$ (where $\zeta=\zeta''$ and $\chi=\chi''$) is the radius of the surface of demarcation, the equation of fit is found to reduce to

$$\tfrac{1}{3}\zeta''^3\chi''^3=-\zeta''^2\left(\frac{d\chi}{d\zeta}\right)_{\zeta=\zeta''}. \tag{47}$$

It can easily be shown that if χ be an Emden or a centrally condensed type of solution, then (47) has no roots, except that in the former case we have the trivial solution $\zeta''=0$. Hence the introduction of the homogeneous core compels us to a consideration of only the collapsed-type solutions for χ.

Now by (36)

$$-\zeta_0^2\left(\frac{d\chi}{d\zeta}\right)_{\zeta=\zeta_0}=\frac{M}{\dfrac{4}{\pi^{1/2}}\left(\dfrac{K_2}{G\beta}\right)^{3/2}}=\frac{1}{C_2^{1/2}} \tag{48}$$

where C_2 can now be called the 'discriminant' of the relativistic standard model for $M>M_3$. It may be noted here that C_2 is primarily a function *only* of M, since the hypothesis that the configurations are highly collapsed provides us with $\beta\sim 1$ in any case. We can write (48) differently as

$$-\zeta_0^2\left(\frac{d\chi}{d\zeta}\right)_{\zeta=\zeta_0}=\frac{M}{M_3}\times 2.015 \tag{49}$$

where 2.015 is the corresponding boundary value for the Emden solution. Hence for $M>M_3$ we clearly see from (49) that the solutions for χ now belong to the collapsed family. Thus the condition that $M>M_3$ is precisely equivalent to the condition that there is a homogeneous core. By methods similar to Section 4 we easily obtain

$$M(r'')=\tfrac{4}{3}\pi r''^3\varrho_{\max}=-\frac{4}{\pi^{1/2}}\left(\frac{K_2}{G\beta}\right)^{3/2}c_1^2 g'(c_1) \tag{50}$$

$$r_1=\frac{r''\zeta_0}{c_1} \tag{51}$$

where $c_1=A\zeta''$ and r_1 is the radius of the whole configuration. Comparing (50) with (49) we see that as M increases beyond M_3 the collapse proceeds further and further till finally, when $M\to\infty$, $c_1\to\zeta_0$, and $r_1\to r''$, and the whole configuration has completely 'collapsed' into one mass of incompressible matter at the highest density matter is capable of. We have in the limit, so to say, a '*solid star*'.

Further, if $g(c_1)$ corresponds to the Emden solution there is only one 'trivial' root for (47), namely, $c_1=0$, *i.e.* the central density of this completely relativistic Emden polytrope is just equal to ϱ_{max} and the radius of the star is then obviously given by

$$\frac{\frac{4}{3}\pi r_1^3 \varrho_{max}}{54.36} = 0.92 \odot \beta^{-3/2}. \tag{52}$$

Thus this Composite Series II joins continuously the Composite Series I (Section 7) and the Emden polytrope '$n=3$' with $\varrho_c = \varrho_{max}$, and $M = .92 \odot \beta^{-3/2}$ is the *common limit* of both the series. We have therefore the following complete classification of the highly collapsed configurations ($L \ll L_0$, $\beta \sim 1$) for *M considered as a variable taking the whole range of values.*

Mass	Description
Class I – $M < 0.61 \odot \beta^{-3/2}$	Emden polytropes '$n=\frac{3}{2}$'
$M_{3/2} = M = .61 \odot \beta^{-3/2}$	An Emden polytrope $n=\frac{3}{2}$ with $\varrho_c = (K_2/K_1)^3$.
Class II – $0.61 \odot \beta^{-3/2} < M < 0.92 \odot \beta'^{-3/2}$	Composite I – Degenerate envelope surrounding a relativistic core.
$M_3 = M = 0.92 \odot \beta'^{-3/2}$	Approximately an Emden polytrope '$n=3$' with $\varrho_c = \varrho_{max}$.
Class III – $M > .92 \odot \beta'^{-3/2}$	Composite II – relativistic envelope and homogeneous core.
$M \to \infty$	Completely homogeneous ($\varrho = \varrho_{max}$).

That we are thus able to enumerate definitely the steady-state configurations for the whole range of M appears to be in complete conformity with the general scheme of Milne's ideas.

To apply the above classification to the known white dwarfs – o_2 Eridani B, Procyon B, and Van Maanen's star possibly belong to Class I. That the companion of Sirius is in Class II is also likely.But it appears that no white dwarf has yet been discovered which has a homogeneous core at the centre. This classification is made with caution, since, though they are certainly of the collapsed type, they are by no means '*highly*' collapsed, for then L and T_e would be so small that we could not see the stars.

9. Summary

In this paper Milne's theory of collapsed configurations is developed a stage further, the essential refinement being in the introduction of a relativistically degenerate core with the equation of state $p = K_2 \varrho^{4/3}$. This enables an enumeration to be made of the steady-state configurations for the whole range of M considered as a variable with $L \sim 0$. The classification arrived at is shown in the table in Section 8.

In conclusion, I wish to record my best thanks to Professor Milne for much valuable advice and criticism during the course of the work.

The Stellar Coefficients of Absorption and Opacity.

By S. Chandrasekhar.

(Communicated by P. A. M. Dirac, F.R.S.—Received May 14, 1931.)

1. The evaluation of the atomic absorption and opacity coefficients* due to atomic nucleii in an enclosure containing free electrons distributed according to the Fermi-Dirac statistics, apart from its theoretical interest, has of late gained some physical importance since Professor Milne's recent analysis of stellar structure has disclosed the possibility of *all* the stars possessing at least a zone of complete degeneracy. It is proposed, therefore, to examine this question in this paper. The formulæ here arrived at for the degenerate case are true only for high degeneracy. It has not been found possible to evaluate them for " incipient " degeneracy.

2. *Absorption by an Atomic Nucleus in an Electron-gas obeying the Fermi-Dirac Statistics.*—We shall define $a_0(\mathrm{E}, \nu)$ to be the rate of absorption of energy from radiation of frequency ν and unit intensity, by electrons of energy E, and unit mean density. Then it is known from the work of Kramers and Gaunt† that

$$a_0(\mathrm{E}, \nu) = 4\pi \mathrm{Z}^2 e^6 / 3\sqrt{3} hcm^2 \nu^3 v, \tag{1}$$

where Z*e* is the charge of the atomic nucleus, and v the velocity of the electrons, the other symbols having their usual significance. Let this nucleus be in an enclosure of unit volume containing N_e free electrons distributed according to the Fermi-Dirac statistics. We want then the total absorption of energy Q_v from the incident radiation by all the electrons with velocity v, in the enclosure, in unit time. On the classical theory we should have clearly

$$\mathrm{Q}_v \, dv = a_0(\mathrm{E}, \nu) \, \mathrm{N}_v \, dv. \tag{2}$$

This, however, is not true on the Fermi-Dirac statistics. On this theory the probability of an encounter depends also on the density of the phase space in the region in which the electron finds itself after the collision. If, for instance, this region is fully packed—*i.e.*, all the phase cells are occupied—then the collision simply cannot occur. This is a consequence of Pauli's exclusion

* In conformity with the definitions due to A. S. Eddington, " The Internal Constitution of Stars," p. 119. For a very neat exposition of the general theory of the stellar absorption coefficient see S. Rosseland, 'Handbuch der Astrophysik,' Band III/1, p. 412, *et seq.*

† 'Phil. Trans.,' A, vol. 229, p. 163 (1930), equation (6.10).

b

principle (as applied to the statistical mechanics) which prevents more than two electrons occupying the same phase cell. (There must, therefore, be in (2) an extra factor depending on the energy $E + h\nu$ of the electron *after* the absorption of the quantum $h\nu$). The extra factor (in general), as has been shown by Pauli and Nordheim,* is

$$(1 - N_{E'}/P_{E'}), \tag{3}$$

where $N_{E'}$ and $P_{E'}$ are the number of electrons and phase cells corresponding to the energy E' of the electron after the encounter. We have, therefore, for the total absorption due to all the electrons with velocity v

$$Q_v\, dv = a_0\,(E, \nu)\, N_E\, (1 - N_{E+h\nu}/P_{E+h\nu})\, dv. \tag{4}$$

Now, on the Fermi-Dirac statistics

$$N_E\, dv = \frac{4\pi G m^3}{h^3} \cdot \frac{v^2\, dv}{\frac{1}{A} e^{E/kT} + 1}, \tag{5}$$

where G is the "spin-weight" which for the case of an electron-gas equals two, and A is determined by the "normalisation condition"

$$N_e = \frac{G\,(2\pi m k T)^{3/2}}{h^3} \cdot \frac{2}{\sqrt{\pi}} \int_0^\infty \frac{u^{1/2}\, du}{\frac{1}{A} e^u + 1}. \tag{6}$$

The precise value of A will depend on the state of the gas—degenerate or otherwise—but we will leave it for the present unspecified. Also, it is easily seen that

$$1 - \frac{N_{E+h\nu}}{P_{E+h\nu}} = \frac{\frac{1}{A} e^{(E+h\nu)/kT}}{\frac{1}{A} e^{(E+h\nu)/kT} + 1}. \tag{7}$$

Hence by (4), (5), (6) and (7) we have for the total absorption α_ν due to *all* the electrons in the enclosure,

$$\alpha_\nu = \int_0^\infty Q_v\, dv = \frac{16\pi^2 Z^2 e^6 m G}{3\sqrt{3}\, c h^4 \nu^3} \int_0^\infty \frac{\frac{1}{A} e^{(E+h\nu)/kT}\, v\, dv}{\left(\frac{1}{A} e^{E/kT} + 1\right)\left(\frac{1}{A} e^{(E+h\nu)/kT} + 1\right)}$$

$$= \frac{16\pi^2 Z^2 e^6 G k T}{3\sqrt{3}\, c h^4 \nu^3} \frac{1}{A} e^{h\nu/kT} \int_1^\infty \frac{dx}{\left(\frac{1}{A} x + 1\right)\left(\frac{1}{A} e^{h\nu/kT} x + 1\right)} \tag{8}$$

* Nordheim, 'Proc. Roy. Soc.,' A, vol. 119, p. 689 (1928). Also R. H. Fowler, "Statistical Mechanics," pp. 556 *et seq.*

by an obvious change in the variable. We have therefore

$$\alpha_\nu = \frac{16\pi^2 Z^2 e^6 G kT}{3\sqrt{3}\,ch^4\nu^3} \cdot \frac{e^{h\nu/kT}}{e^{h\nu/kT}-1} \log\left[\frac{e^{h\nu/kT}(A+1)}{e^{h\nu/kT}+A}\right]. \tag{9}$$

3. *Absorption in a Non-degenerate Gas.*—A non-degenerate gas corresponds to the " A " in (6) being a small constant. The value of A is then given by

$$A \sim \frac{N_e h^3}{G}(2\pi mkT)^{-3/2} \ll 1. \tag{10}$$

Also,

$$\log\left[\frac{e^{h\nu/kT}(A+1)}{e^{h\nu/kT}+A}\right] = (1-e^{-h\nu/kT})\log(A+1) - (1-e^{-h\nu/kT})\frac{A^2}{2}e^{-h\nu/kT} + \dots$$

$$\sim (1-e^{-h\nu/kT})\,A, \tag{11}$$

using (10) and (11) in our equation (9) for the absorption coefficient α_ν, we have

$$_M\alpha_\nu = N_e \frac{16\pi^2 Z^2 e^6}{3\sqrt{3}\,hc\nu^3} \cdot \frac{1}{(2\pi m)^{3/2}(kT)^{1/2}}. \tag{12}$$

Milne* gives, working from the beginning on the Maxwell-distribution law,

$$_M\alpha_\nu = N_e \frac{16\pi^2 Z^2 e^6}{3\sqrt{3}\,hc\nu^3 (2\pi m)^{3/2}(kT)^{1/2}}(1-e^{-h\nu/kT}), \tag{13}$$

which differs from (12) by an extra-factor $(1-e^{-h\nu/kT})$. As Gaunt† has already pointed out, this extra-factor in Milne's expression is not justifiable. It would be correct if Kramer's expression for the emission included the stimulated emission also. This, however, is not the case.‡

* 'M.N.R.A.S.,' vol. 85, p. 750 (1925), equation (2).

† *Loc. cit.*, p. 197.

‡ S. Rosseland (*loc. cit.*) also evaluates $_M\alpha_\nu$. His equation (377) is exactly the same as (12) but with an extra-factor $(1+h\nu/kT)$. This is clearly an oversight. Again, his equation (389) contains an extra-factor $e^{h\nu/kT}$. This extra-factor arises because, working on Kramer's theory of emission, to calculate the total *emission* in the frequency range ν and $\nu + d\nu$ he integrates over all the *initial* velocities of the electrons from 0 to ∞. This he does because, according to him, this takes into account the contribution to the absorption by the " bound-free " transitions also. Rosseland, however, recognises that this is a " gross-exaggeration." *If* Rosseland is right in including the " bound-free " transitions in this way, a comparison of (23) and (24) shows that in the Maxwellian case the greater part of the total absorption is due to the " bound-free " transitions. This result is known to be true from the researches of Professor Milne. But it will be seen that in the degenerate case the " bound-free " transitions can be neglected (see § 8).

b 2

4. *The Coefficients of Absorption and Opacity in a Non-degenerate Electron-gas.*—As all the existing treatments for the evaluation of the atomic-opacity coefficient contain some error or other, it seems worth while to give a derivation of it free from such errors.* As is well known, there are two quantities of interest in astrophysical calculations: (i) the *atomic absorption coefficient* defined as the "straight mean," *i.e.*,

$$\alpha_a = \int_0^\infty \alpha'_\nu I_\nu \, d\nu \Big/ \int_0^\infty I_\nu \, d\nu, \tag{14}$$

and (ii) the *atomic opacity coefficient* defined as the Rosseland mean, *i.e.*,

$$\alpha_0^{-1} = \int_0^\infty \frac{1}{\alpha'_\nu} \frac{\partial I_\nu}{\partial T} d\nu \Big/ \int_0^\infty \frac{\partial I_\nu}{\partial T} d\nu. \tag{15}$$

Further, as has been explained by Rosseland,† it is not the pure absorption coefficient α_ν that has to be introduced in (15). It is the absorption coefficient α_ν multiplied by $(1 - e^{-h\nu/kT})$, *i.e.*,

$$\alpha'_\nu = \alpha_\nu (1 - e^{-h\nu/kT}). \tag{16}$$

This is necessary to take into account properly the stimulated emission process. "This result is brought about by the fact that the stimulated emission process will show exactly the same asymmetry as the field of radiation, and this fact entails the consequence that the outward flow of radiation will be reinforced by a stimulated emission in its own direction." Though Rosseland makes this remark in connection with the evaluation of the coefficient of opacity, it is clear, however, that the same arguments will also apply to the "straight-mean" giving the atomic absorption coefficient.‡ For, an incident radiation in a given direction will stimulate emission in a direction exactly parallel to itself, and what is essential in the evaluation of the absorption is the random (and *therefore* the unstimulated) emission. Hence, even in (14) we use α_ν' defined by (16), and not merely α_ν. Thus we are effectively working with Milne's expression (13) though the extra-factor now owes its origin to an entirely different reason and is not due to the assumption that the Kramers' expression for the emission includes the stimulated emission also.

* The treatment in Eddington's book is not without errors. In his equation (156.3 for the emission in the frequency range ν and $\nu + d\nu$, he replaces $1/v$ by the mean harmonic-velocity. This clearly should not be done since $1/v$ should be averaged only from ∞ to v_0 where $\frac{1}{2} mv_0^2 = h\nu$, and not from ∞ to 0 as Eddington does. Neither does Eddington make clear that this effectively means (according to Rosseland) taking into account the "bound-free" transitions also, and is a "gross-exaggeration."

† *Loc. cit.*, pp. 456 and 469.

‡ I owe this remark to Dr. Dirac. See Rosseland, *loc. cit.*, equation (352).

Case I. The Rosseland Mean.—We have, therefore, by (12), (15) and (16)

$$_{\mathrm{M}}\alpha_o = \mathrm{N}_e \frac{16\pi^2 \mathrm{Z}^2 e^6}{3\sqrt{3}\, hc\,(2\pi m)^{3/2}\,(k\mathrm{T})^{1/2}} \cdot \frac{\partial}{\partial \mathrm{T}}\int_0^\infty \mathrm{I}_\nu\, d\nu \Bigg/ \frac{\mathrm{B}h}{k\mathrm{T}^2}\int_0^\infty \frac{\nu^3 e^{h\nu/k\mathrm{T}}}{e^{h\nu/k\mathrm{T}}-1} \cdot \frac{\nu^4 e^{h\nu/k\mathrm{T}}\, d\nu}{(e^{h\nu/k\mathrm{T}}-1)^2}, \tag{17}$$

where I_ν is the usual Planck function

$$\mathrm{I}_\nu = \mathrm{B}\nu^3/(e^{h\nu/k\mathrm{T}}-1). \tag{18}$$

Now

$$\frac{\mathrm{B}h}{k\mathrm{T}^2}\int_0^\infty \frac{\nu^7 e^{2h\nu/k\mathrm{T}}}{(e^{h\nu/k\mathrm{T}}-1)^3} = \frac{\mathrm{B}}{\mathrm{T}}\left(\frac{k\mathrm{T}}{h}\right)^7 7\,!\ \vartheta_1 \tag{19}$$

where

$$\vartheta_1 = \tfrac{1}{2}\left\{\sum_{n=1}^{\infty}\frac{1}{n^6} + \sum_{n=1}^{\infty}\frac{1}{n^7}\right\} = 1{\cdot}0128. \tag{20}$$

Also

$$\frac{\partial}{\partial \mathrm{T}}\int_0^\infty \mathrm{I}_\nu\, d\nu = \frac{\mathrm{B}}{\mathrm{T}}\left(\frac{k\mathrm{T}}{h}\right)^4 4\,!\ \vartheta_2, \tag{21}$$

where

$$\vartheta_2 = \sum_{n=1}^{\infty}\frac{1}{n^4} = \frac{\pi^4}{90} = 1{\cdot}0823. \tag{22}$$

Hence we have finally

$$_{\mathrm{M}}\alpha_o = \frac{8\pi^2\vartheta_2}{315\sqrt{3}\vartheta_1} \cdot \frac{\mathrm{Z}^2 e^6 h^2}{c\,(2\pi m)^{3/2}} \cdot \frac{\mathrm{N}_e}{(k\mathrm{T})^{7/2}}. \tag{23}$$

Case II. The Straight Mean.—We calculate now the straight mean giving the atomic absorption coefficient. By (12), (14) and (16) we have

$$_{\mathrm{M}}\alpha_a = \mathrm{N}_e \cdot \frac{16\pi^2 \mathrm{Z}^2 e^6}{3\sqrt{3}\, hc\,(2\pi m)^{3/2}\,(k\mathrm{T})^{7/2}} \cdot \frac{\mathrm{B}\displaystyle\int_0^\infty e^{-h\nu/k\mathrm{T}}\, d\nu}{\mathrm{B}\displaystyle\int_0^\infty \frac{\nu^3\, d\nu}{e^{h\nu/k\mathrm{T}}-1}}$$

$$= \frac{80}{\pi^2\sqrt{3}} \cdot \frac{\mathrm{Z}^2 e^6 h^2}{c\,(2\pi m)^{3/2}} \cdot \frac{\mathrm{N}_e}{(k\mathrm{T})^{7/2}}. \tag{24}$$

which agrees with Milne's formula (*loc. cit.*, equation (3)). A comparison of (23) and (24) shows the effect of averaging by the two different methods. The difference in the numerical factors in (23) and (24) (which amounts to a factor 30), is almost incredible, but I have been unable to detect any mistake in my algebra.

5. *The Coefficients of Absorption and Opacity in a Degenerate Gas.*—We proceed now to the evaluation of the atomic absorption and opacity coefficients in a degenerate gas, *i.e.*, corresponding to " A " in (6) being a large positive

constant. The first step is to obtain a suitable approximation to the log-factor in (9), consistent with our assumption that the gas is degenerate. Now

$$\log \frac{e^{h\nu/kT}(A+1)}{e^{h\nu/kT}+A} = \log\left(1+\frac{1}{A}\right) + \frac{h\nu}{kT} - \log\left(1+\frac{1}{A}e^{h\nu/kT}\right). \tag{25}$$

The following argument shows that

$$\frac{1}{A}e^{h\nu/kT} \ll 1$$

for the range of frequencies, about which the greater part of the radiant energy is distributed according to the Planck's law, provided the electron-gas is sufficiently degenerate.

Let λ_m be the wave-length at which there is the maximum of energy in the black-body spectrum corresponding to the temperature T. Then we have

$$\lambda_m T = \frac{ch}{4{\cdot}965k} \text{ (“ Wien's displacement law ”) or } \frac{h\nu_m}{kT} = 4{\cdot}965. \tag{26}$$

Also if

$$\frac{1}{A}e^{h\nu/kT} \ll 1$$

then

$$\log A \gg h\nu/kT. \tag{27}$$

Also the degeneracy condition is

$$\log A = \frac{h^2}{2mkT}\left(\frac{3N_e}{4\pi G}\right)^{2/3} \gg 1. \tag{28}$$

Now log A is for a highly degenerate gas of the order 20 or 30.* Hence comparing (26), (27) and (28), it can be concluded that for the main part of the Planck's curve where most of the radiant energy is concentrated, we can when working to the first order neglect not only $1/A$ but also $\frac{1}{A}e^{h\nu/kT}$. Hence by (25) we have with this understanding

$$\log \frac{e^{h\nu/kT}(A+1)}{e^{h\nu/kT}+A} \sim \frac{h\nu}{kT} \tag{29}$$

Using (29) in (9) we have simply,

$$_D\alpha_\nu = \frac{16\pi^2 Z^2 e^6 G}{3\sqrt{3}\,ch^3} \cdot \frac{e^{h\nu/kT}}{\nu^2(e^{h\nu/kT}-1)}. \tag{30}$$

* For $n = 10^{30}$, $T = 10^8$, $\log A = 4{\cdot}2 \times 10^{-11}$. $N^{2/3}/T = 42$.

Case I. The Rosseland Mean.—By (30), (15) and (16) we have for the atomic-opacity coefficient in a highly degenerate electron-gas

$$
{}_{\mathrm{D}}\alpha_0 = \frac{32\pi^2 Z^2 e^6}{3\sqrt{3}\, ch^3} \frac{\dfrac{\partial}{\partial T}\displaystyle\int_0^\infty I_\nu \, d\nu}{\dfrac{Bh}{kT^2}\displaystyle\int_0^\infty \frac{\nu^6 e^{h\nu/kT}\, d\nu}{(e^{h\nu/kT}-1)^2}}. \tag{31}
$$

Since

$$
\frac{Bh}{kT^2}\int_0^\infty \frac{\nu^6 e^{h\nu/kT}\, d\nu}{(e^{h\nu/kT}-1)^2} = \frac{B}{T}\left(\frac{kT}{h}\right)^6 6!\,\frac{\pi^6}{945} \tag{32}
$$

We have by (22), (31) and (32)

$$
{}_{\mathrm{D}}\alpha_0 = \frac{56}{15\sqrt{3}} \frac{Z^2 e^6}{ch\,(kT)^2}. \tag{33}
$$

Case II. "The Straight Mean."—We have for the atomic-absorption coefficient in a degenerate gas by (30), (14) and (16)

$$
\begin{aligned}
{}_{\mathrm{D}}\alpha_a &= \frac{32\pi^2 Z^2 e^6}{3\sqrt{3}\, ch^3} \cdot \frac{B\displaystyle\int_0^\infty \frac{\nu d\nu}{e^{h\nu/kT}-1}}{B\displaystyle\int_0^\infty \frac{\nu^3 d\nu}{e^{h\nu/kT}-1}} \\
&= \frac{80}{3\sqrt{3}} \frac{Z^2 e^6}{ch\,(kT)^2}.
\end{aligned} \tag{34}
$$

Here again we find a large difference between the straight and the Rosseland means.

6. *The Degenerate-atomic-absorption Coefficient Formula.*—(34) then is our atomic-absorption-coefficient for a highly degenerate gas (on the assumption of the validity of the equation (1) for a_0 (E, ν)). The most striking point about (33) and (34) is that they are independent of the electron-concentration and therefore of the density. But the following shows that this result is a natural consequence of the properties of a highly degenerate gas and the *form* of our expression for a_0(E, ν).

An idealisation is necessary to explain the result simply. In a highly degenerate gas the distribution of energy is as shown in the figure.

Or algebraically

$$
\left.
\begin{aligned}
N(E) &= \frac{4\pi (2m)^{3/2}}{h^3} E^{1/2} = KE^{1/2} \ \text{(say)} \quad (E \le E^*) \\
&= 0 \qquad\qquad\qquad\qquad\qquad\qquad (E > E^*)
\end{aligned}
\right\}, \tag{35}
$$

where

$$E^* = \frac{h^2}{8m}\left(\frac{3N_e}{\pi}\right)^{2/3}. \tag{36}$$

Now, an electron with energy E' can absorb a quantum $h\nu$ *only when*

$$E' + h\nu > E^* \tag{37}$$

for otherwise the energy of the electron after the absorption of the quantum $h\nu$ is less than E^*, and the phase-cells here are all occupied. But it can absorb quanta of energy-magnitudes greater than $h\nu_0 (= E^* - E')$ since it then goes

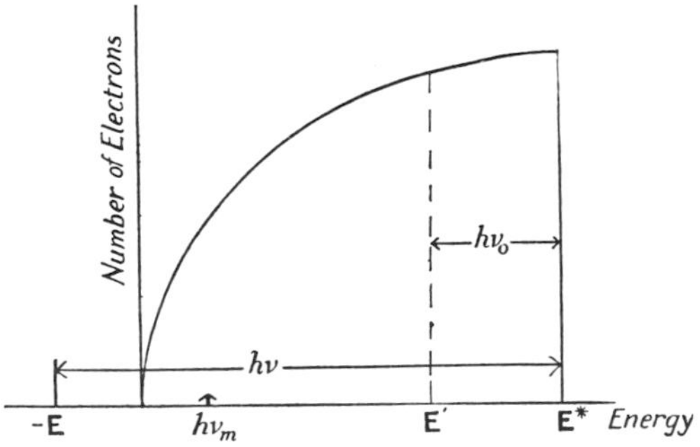

into a region of phase-space which is completely vacant.† If we now increase the electron-concentration we merely increase E^*, and no alteration is made in the distribution of the electrons with energy less than E^*. We merely add a few more electrons, *all* of which can have energies only greater than the original E^*. From (35) we have obviously

$$N_{E^*} = KE^{*1/2}. \tag{38}$$

Also from (1) we have

$$a_0(E^*, \nu) \propto N_{E^*}/E^{*1/2} = K.$$

Hence

$$\delta\alpha_\nu = \delta\,(a_0(E,^*\,\nu)) = 0 \qquad \text{for } all\ \nu. \tag{39}$$

Hence our result that the atomic absorption (or opacity) coefficient is independent of density, is understood as a very natural consequence of the properties of a highly degenerate-gas and the form for our expression for the rate of absorption of energy.

† Similar reasoning (but not exactly the same) has been employed by R. H. Fowler ('Proc. Roy. Soc.,' A, vol. 118, p. 229 (1928)) to explain the sharp photoelectric threshold frequency.

In this connection we can on this highly idealised picture of a degenerate-gas obtain the atomic absorption coefficient also. Now by (1) we have

$$a_0\,(\mathrm{E},\,\nu) = \frac{4\pi Z^2 e^6}{3\sqrt{3}hcm^{3/2}\nu^3\sqrt{2\mathrm{E}}} \tag{40}$$

To find the total absorption α_ν due to all the electrons we must multiply (40) by $\mathrm{N_E}d\mathrm{E}$ and integrate only from E^* to $(\mathrm{E}^* - h\nu)$ as is clear from equation (37) and the diagram.

We find using (35) that

$$\alpha_\nu = \frac{32\pi^2 Z^2 e^6}{3\sqrt{3}\,h^4 c\nu^3}\int_{\mathrm{E}^*-h\nu}^{\mathrm{E}^*} d\mathrm{E}$$

$$= \frac{32\pi^2 Z^2 e^6}{3\sqrt{3}\,h^3 c\nu^2}, \tag{41}$$

which is exactly the same as (30) but with the factor $(1 - e^{-h\nu/k\mathrm{T}})^{-1}$ missing.† That this is not inconsistent with our assumption of degeneracy in § 5 can be seen by using (41) to find the absorption coefficient. We find readily (using the "Rosseland correction factor" also)

$$_{\mathrm{D}}\alpha_a = \frac{32\pi^2 Z^2 e^6}{3\sqrt{3}\,h^3 c}\;\frac{\mathrm{B}\displaystyle\int_0^\infty \nu e^{-h\nu/k\mathrm{T}} d\nu}{\mathrm{B}\displaystyle\int_0^\infty \frac{\nu^3 d\nu}{e^{h\nu/k\mathrm{T}} - 1}}$$

$$= \frac{160}{\pi^2\sqrt{3}}\cdot\frac{Z^2 e^6}{ch\,(k\mathrm{T})^2}. \tag{42}$$

Similarly for α_ν given by (42) the atomic-opacity coefficient is easily found to be

$$_{\mathrm{D}}\alpha_0 = \frac{16\pi^2\vartheta_2}{45\sqrt{3}\vartheta_3}\cdot\frac{Z^2 e^6}{ch\,(k\mathrm{T})^2} \tag{42'}$$

where ϑ_2 is defined in (22) and

$$\vartheta_3 = \tfrac{1}{2}\left\{\sum_{n=1}^{\infty}\frac{1}{n^6} + \sum_{n=1}^{\infty}\frac{1}{n^5}\right\} = 1{\cdot}0271.$$

The atomic absorption coefficients given by (42) and (34) are exactly the same, but for a difference in the numerical factors amounting to $\pi^2/6$ or 1·6. The difference, however, between the Rosseland means (42′) and (33) is almost

† The approximation (35) is the one corresponding to "$\mathrm{T} \to 0$." It is clear that (30) → (41) as $\mathrm{T} \to 0$.

negligible, the ratio of the two numerical factors in the two equations being $1:1{\cdot}01$.†

Presumably (34) and (33) are the more reliable ones, because the approximations used there are derived from a less idealised picture and take account of the partial vacancy in the region below E^* and the non-vacancy beyond E^* as well.

7. *Application of one of Gaunt's Formulæ to the case of Absorption by a Degenerate Gas.*—So far we have based our discussion exclusively on the "Kramers' approximation" for the continuous absorption. This, however, is not likely to be valid for *extreme* degeneracy, for then it is only the electrons of very high velocity that take part in the absorption-process, and since the condition that the electron-gas is very highly degenerate provides for the maximum of the Planck-curve falling well within the "threshold" energy, a better approximation (than Kramers') would appear to be the one corresponding to $E \gg h\nu$. Now, Gaunt‡ gives for this case

$$a_0\,(E,\,\nu) = \frac{4Z^2e^6}{3hcm^2v\nu^3}\log\frac{E}{h\nu}. \tag{43}$$

Since

$$\alpha_\nu = \int_{E^*-h\nu}^{E^*} a_0\,(E,\,\nu)\,N_E\,dE$$

we have by (35) and (43)

$$\begin{aligned}\alpha_\nu &= \frac{32\pi Z^2e^6}{3h^4c\nu^3}\int_{E^*-h\nu}^{E^*}\log\frac{E}{h\nu}\,dE \\ &= \frac{32\pi Z^2e^6}{3h^4c\nu^3}\left[h\nu\log\left(\frac{E^*}{h\nu}-1\right) - E^*\log\left(1-\frac{h\nu}{E^*}\right) - h\nu\right],\end{aligned} \tag{44}$$

or to the same order of accuracy as (43)

$$\alpha_\nu \sim \frac{32\pi Z^2e^6}{3h^3c\nu^2}\log\frac{E^*}{h\nu}, \tag{45}$$

† In deriving (42) and (42′) we have not averaged α_ν given by (41) but

$$\alpha_\nu' = \alpha_\nu\,(1 - e^{-h\nu/kT}).$$

One cannot definitely say whether the use of α_ν or α_ν' is more consistent when working with the highly idealised picture of a degenerate gas. For (41) is true only when $h\nu \gg kT$, in which case we cannot differentiate between α_ν and α_ν'. If however, we had used α_ν (given by (41)) as such, to obtain the "means," the atomic-absorption and opacity coefficients so derived would have been identical with (33) and (34).

‡ Gaunt, *loc. cit.*, equation (5.39).

(45) immediately gives a value for the atomic absorption coefficient. We have by (14), (16) and (45)

$$_{D'}\alpha_a = \frac{32\pi Z^2 e^6}{3h^3c} \frac{B\int_0^\infty \nu \log \frac{E^*}{h\nu} e^{-h\nu/kT} d\nu}{B\int_0^\infty \frac{\nu^3 d\nu}{e^{h\nu/kT}-1}}. \tag{46}$$

Now

$$\int_0^\infty \nu \log \frac{E^*}{h\nu} e^{-h\nu/kT} d\nu = \left(\frac{kT}{h}\right)^2 \int_0^\infty x \log \frac{E^*}{kTx} e^{-x} dx$$
$$= \left(\frac{kT}{h}\right)^2 \left[\log \frac{E^*}{kT} + \gamma - 1\right], \tag{47}$$

where γ is the Euler-Mascheroni's constant $= 0{\cdot}5772$. Hence we have by (46) and (47)

$$_{D'}\alpha_a = \frac{160}{\pi^3} \cdot \frac{Z^2 e^6}{ch\,(kT)^2}\left[\log \frac{E^*}{kT} - 0{\cdot}4228\right]. \tag{48}†$$

which differs from the formula (42) derived on the Kramers' approximation by a factor

$$g = \frac{\sqrt{3}}{\pi}\left[\log \frac{E^*}{kT} + \gamma - 1\right], \tag{49}$$

so that $_{D'}\alpha_a = g\ _D\alpha_a$. g can appropriately be called the *Gaunt-factor*. Also

$$\frac{E^*}{kT} = \frac{h^2}{8mkT}\left(\frac{3Ne}{\pi}\right)^{2/3} = \log A \tag{50}$$

in our previous notation. Thus the Gaunt-factor for the atomic absorption coefficient is

$$g = \sqrt{3}\,[\log \log A + \gamma - 1]/\pi. \tag{51}‡$$

† In deriving (48) we have averaged α_ν' ($= \alpha_\nu (1 - e^{-h\nu/kT})$). As was pointed out earlier, when working with the ideal picture of a degenerate gas, we cannot be certain as to which is more correct to use α_ν or α_ν'. If we had used α_ν given by (45) as such, then we would have obtained

$$_{D'}\alpha_a = \frac{80}{\pi\,.\,3} \cdot \frac{Z^2 e^6}{ch\,(kT)^2}\left[\log \frac{E^*}{kT} + \gamma - 1\right], \tag{48'}$$

which differs from (48) by a factor $\pi^2/6$. Since, however, (43) itself is valid only for extreme degeneracy, it is likely that (48) is the more reliable one.

‡ In deriving (51) we have compared (48) with (42), since both these are derived from the same ideal picture of a degenerate gas. But, as was pointed out earlier, (34) is the more reliable one. The Gaunt-factor is then given by

$$g' = 6\sqrt{3}\,[\log \log A + \gamma - 1]/\pi^3 = 0{\cdot}6081g. \tag{51'}$$

If, however, we had used (48′) instead of (48), g would then be given by (51).

Now, for $T = 10^8$ and $N_e = 10^{30}$ we have $\log A = 42$. The Gaunt-factor then amounts to $1\cdot87$. Also $\log_e \log A = 3\cdot78$. Thus it appears that even for this not too high a degeneracy, it would have been a good enough approximation if instead of (45) we had merely used

$$\alpha_\nu = \frac{32\pi Z^2 e^6}{3h^3 c\nu^2} \log \frac{E^*}{kT} \tag{45'}$$

in which case the Gaunt-factor would have been simply

$$g \sim \frac{\sqrt{3}}{\pi} \log \log A, \tag{51''}$$

and would mean the neglect of $-0\cdot4228$ in (51). Now, the previous sections fully illustrate the extreme sensitiveness of the Rosseland mean to the structure of α_ν, and we cannot rely on (45′) as a valid approximation to evaluate the atomic-opacity coefficient, though we could depend on it (as it now appears) to evaluate the atomic absorption coefficient. If, however, (45′) is a valid approximation,† then it would follow that the Gaunt-factor for the atomic-absorption as well as the opacity coefficients is given by (51″).

From the nature of the approximations which Gaunt makes in deriving his formula (43), it appears *not* quite safe to use (48) even for the degeneracy contemplated by $T = 10^8$ and $N_e = 10^{30}$. It is, however, satisfactory that the Gaunt-factor is not very large and that $\log \log A$ is a slowly increasing function of N_e. Though it was shown in § 6 that the atomic absorption and opacity coefficients (derived on the basis of the Kramers' approximation) being independent of density could be inferred from first principles, yet it is physically rather hard to conceive of how this could be so. The introduction of the Gaunt-factor removes incidentally this difficulty.

8. *General Discussion and possible Applications.*—In the evaluation of the opacity and absorption coefficients we have considered only the "free-free" transitions, *i.e.*, the contribution to the total absorption due to the electron transitions from one occupied phase-cell of positive energy to another unoccupied phase-cell of positive energy. Now it is known‡ that in the classical Maxwell case most of the absorption is photoelectric, even in a highly ionised gas. It arises from the absorption by the very few electrons bound to atoms

† I am unable to see any *a priori* justification for (45′), and it is not an easy matter in any case to evaluate the Rosseland mean corresponding to the correct form for the α_ν given by (45).

‡ Milne, 'M.N.R.A.S.,' vol. 85, p. 750 (1925), also Rosseland, 'Astrophys. J.,' vol. 61, p. 424 (1925).

at any one instant of time. However high the degree of ionisation, there is a large photoelectric absorption from just those very few electrons. The question which now arises is as to how important the "bound-free" transitions are in the degenerate case. The following arguments show that the contribution to the total absorption by the "bound-free" transitions must be negligible.

Firstly, it is to be noted that even the "free-electrons" can in a sense be regarded as "bound." For, an electron with energy E′ can absorb a quantum $h\nu$ only if

$$h\nu > E^* - E', \tag{37}$$

which is the same thing as saying that the E′ electrons are "bound electrons" with an "ionisation energy" $\chi = E^* - E'$. In this sense, it is only the E* electrons that deserve to be called free!

Now, an electron in an orbit with a *negative* energy E, if it has to become a "free electron," must "switch off" only to states of positive energy greater than E*, *i.e.*, quanta of magnitude $h\nu > E^* + E$, alone are effective in causing transitions. Also, even towards the contribution to the absorption by the "free-free" transitions, it is only the electrons with energies near the threshold energy that are operative. Further (as has been pointed out earlier) the condition that the gas is degenerate provides for the maximum of the Planck curve falling fairly well inside the threshold energy (see figure). All these considerations make clear that the "bound-free" transitions contribute very little, if at all, to the total absorption.

Now, we will briefly discuss some possible applications of the degenerate opacity formula.

Neglecting the Gaunt-factor, we have for the coefficient of opacity per unit mass κ for a degenerate gas,

$$\kappa \propto T^{-2}. \tag{52}$$

Now in a collapsed star the central regions are degenerate, and the temperature law is†

$$T \propto \rho^{5/12}$$

or by (52)

$$\kappa \propto \rho^{-5/6}. \tag{53}$$

On account of the fair degree of homogeneity of the Emden polytrope "$n=3/2$," and since we can also expect the small decrease in the opacity towards the central regions to be compensated by a corresponding increase in the average rate of liberation of energy ε_r, we can reasonably be sure that $\kappa_r\varepsilon_r$ is fairly

† Milne, 'M.N.R.A.S.,' vol. 91, p. 4 (1930) (especially § 24).

constant. Hence we can feel confident that the standard model equations for the degenerate core of a collapsed star are likely to be a very good approximation.

We cannot, however, apply our formulæ immediately to the case of the white dwarfs. The " astronomical value " for $\kappa_c \varepsilon_c$ is extremely sensitive to what we assume the effective temperature to be. Thus in the case of O_2 Eridani B the assumed effective temperature is 11,200 degrees. Also a Te $\sim$ 11,750 degrees would make the star *completely* of the collapsed type, *i.e.*,

$$1 - \beta = \kappa L / 4\pi c G M = 0.$$

This, combined with the fact that Professor Milne has raised strong doubts against the derived effective temperatures of these stars,* makes it clear that the white dwarfs in any case cannot form a test of the opacity question.

Summary.

(1) A general expression for the atomic absorption coefficient α_ν, due to atomic nuclei in an enclosure containing free electrons distributed according to the Fermi-Dirac statistics, is derived (equation (9)).

(2) The atomic absorption coefficient (the " straight mean ") and the atomic opacity coefficient (the " Rosseland mean ") are evaluated both for the cases of degeneracy and non-degeneracy. The two means are found to differ more considerably than what has hitherto been thought they would.

(3) The " degenerate " opacity and absorption coefficients based on the Kramers' approximation are found to be independent of density and inversely proportional to the square of the temperature. A simple explanation for these results is given based on an idealised picture of a degenerate gas.

(4) Certain correction factors called " Gaunt-factors " are defined, the introduction of which makes the absorption and the opacity coefficients of a degenerate gas slowly increasing functions of the electron concentration.

(5) The bearing of these results to Milne's theory of stellar structure is indicated.

In conclusion, I wish to express my thanks to Dr. P. A. M. Dirac, F.R.S., for many valuable suggestions and helpful criticisms.

* 'M.N.R.A.S.,' vol. 91, p. 39, footnote (1930).

Some Remarks on the State of Matter in the Interior of Stars.

By **S. Chandrasekhar** (Copenhagen).

With 3 figures. (Received September 28, 1932.)

It is shown that for *all* stars for which the radiation-pressure is greater than a tenth of the total pressure, an appeal to the Fermi-Dirac statistics to avoid the central singularity which arises in the discussions of the centrally condensed and the collapsed stars cannot be made. The bearing of this result on the possible state of matter in the interior of stars is indicated.

Since the publication of Milne's memoir on the "Analysis of Stellar Structure" in the Monthly Notices for November 1930[1]) a great deal of work has been done to consider "composite" stellar models. But the following simple considerations seem to have escaped notice and it seems worth while to state them explicitly.

§ 1. The Surfaces of Demarcation. As we approach the centre of a centrally-condensed or a collapsed star, we change over to the equation of state $p = K_1 \varrho^{5/3}$ if the perfect gas law breaks down. If the perfect gas-law breaks down at all, the actual transition from the perfect-gas envelope to the degenerate core must occupy a certain zone, but we could for the sake of convenience consider a definite surface of demarcation defined as the surface at which the two equations of state give the same gas-pressure.

Now in the perfect gas envelope the *total* pressure is given by

$$P = \left[\left(\frac{k}{\mu}\right)^4 \frac{3}{a}\left(\frac{1-\beta}{\beta^4}\right)\right]^{1/3} \varrho^{4/3}, \tag{1}$$

where

$$\beta = 1 - \frac{\varkappa L}{4\pi c G M}, \tag{2}$$

$\varkappa$ = the opacity coefficient, L = luminosity in ergs cm^{-3}, M = mass in grams, k = Boltzmann's Constant, μ = molecular weight $= \alpha m_H$ (say), m_H = mass of the hydrogen atom. Since for the standard model the gas pressure p is given by

$$p = \beta P, \tag{3}$$

we have

$$p = C \varrho^{4/3}, \tag{4}$$

where

$$C = \left[\left(\frac{k}{\mu}\right)^4 \frac{3}{a} \frac{1-\beta}{\beta}\right]^{1/3} = \frac{2.632 \cdot 10^{15}}{\alpha^{4/3}} \left[\frac{1-\beta}{\beta}\right]^{1/3}. \tag{4'}$$

[1]) Refered to as l. c.

The equation of state in the degenerate zone is

$$p = K_1 \varrho^{5/3}, \tag{5}$$

where

$$K_1 = \frac{1}{20}\left(\frac{3}{\pi}\right)^{2/3} \frac{h^2}{m\,\mu^{5/3}} = \frac{9.890 \cdot 10^{12}}{\alpha^{5/3}}. \tag{6}$$

At the first surface of demarcation which we will call S_1, the density ϱ_1 is given by

$$C\,\varrho_1^{4/3} = K_1\,\varrho_1^{5/3},$$

or

$$\varrho_1 = \left(\frac{C}{K_1}\right)^3. \tag{7}$$

Now, it is well known that the equation of state (5) changes over again into the relativistic-degenerate-equation of state

$$p = K_2 \varrho^{4/3}, \tag{8}$$

where

$$K_2 = \frac{h\,c}{8\,\mu^{4/3}}\left(\frac{3}{\pi}\right)^{1/3} = \frac{1.228 \cdot 10^{15}}{\alpha^{4/3}}. \tag{8'}$$

Hence if *circumstances permit* we have to consider a second surface of demarcation, S_2, where the density ϱ_2 is given by

$$\varrho_2 = \left(\frac{K_2}{K_1}\right)^3. \tag{9}$$

Hence we have *two* surfaces of demarcation if and only if

$$\varrho_2 > \varrho_1,$$

or

$$\left(\frac{K_2}{K_1}\right)^3 > \left(\frac{C}{K_1}\right)^3, \tag{10}$$

i. e. only when [cf. equations (4′), (6), (8′)]

$$\frac{1-\beta}{\beta} < \frac{h^3\,c^3\,a}{512\,\pi\,k^4} = 0.1015,$$

or

$$\boldsymbol{\beta > 0,9079.} \tag{11}$$

It may be remarked in passing that the above value for β is independent of the assumed molecular weight. *It depends only on the mass, luminosity and opacity in the gaseous envelope.* It is also independent of whether we consider the same opacity for the degenerate zone and the gaseous envelope, or different opacities in the two regions.

§ 2. The meaning of the fundamental inequality (11) is made clear by the following.

In the following graph I plot $\log p$ against $\log \varrho$.

For numerical calculations I use $\alpha = 2$. The straight line ABK represents the equation of state $p = K_1 \varrho^{5/3}$ and BC the equation of state $p = K_2 \varrho^{4/3}$. These two intersect at B where the density is that which corresponds to the *second* surface of demarcation, namely ϱ_2. ABC gives roughly the equation of state of a degenerate gas.

Let us consider a star for which $\beta = 0.98$. By (4) we get

$$\log p = 14.455 + {}^4/_3 \log \varrho. \tag{12}$$

DE represents this equation. It intersects the degenerate equation of state AB, C at E. The point E corresponds to the first surface of demarcation S_1. Hence for all stars for which $\beta = 0.98$, we first traverse a perfect gas envelope with an equation of state represented by DE. Then we traverse a degenerate zone corresponding to EB and finally (if we have not yet reached the centre) a relativistically degenerate zone.

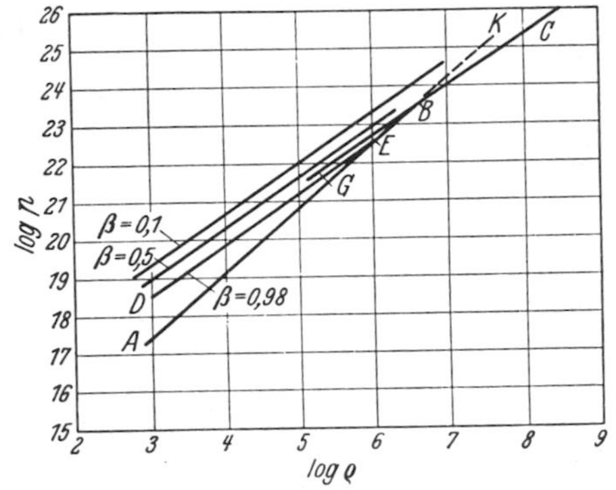

Fig. 1.

Now, if $\beta = 0.9079$, then GB represents the perfect gas equation of state and the degenerate zone reduces to a single layer, and the relativistically degenerate zone is described equally well by the perfect gas equation.

Now if $\beta < 0.9079$[1]) the perfect gas equation of state has *no* intersections with ABC and this means that however high the density may become the temperature rises sufficiently rapidly to prevent the matter from becoming degenerate.

In this connection it will have to be remembered that considerations of relativity do not affect the equation of state of a perfect gas. $p = NkT$, *is true independent of relativity.*

§ 3. *Centrally-Condensed Stars.* Now, for each mass M there is a unique luminosity L_0 — the "Eddington luminosity" which makes the star

[1]) The radiation pressure is *greater* than a tenth of the total pressure if $\beta < 0.9079$.

a perfect gas sphere, with a polytropic index 3. This L_0 characteristizes a unique β_0 which is in fact related to M by means of EDDINGTON's quartic equation:

$$1 - \beta = 0.003\,09 \left(\frac{M}{\odot}\right)^2 \alpha^4 \beta^4. \tag{13}$$

Now from the definition of a centrally-condensed and a collapsed star, it is clear that

$$\left.\begin{aligned} \beta_{c \cdot c} &< \beta_0 \\ \beta_{\text{col.}} &> \beta_0. \end{aligned}\right\} \tag{14}$$

Consider first the mass $\mathfrak{M}$ for which $\beta_0 = 0.9079$. By (13) we have

$$\mathfrak{M}/\odot = 6{,}623\,\alpha^{-2}. \tag{15}$$

If we assume $\alpha = 2$,

$$\mathfrak{M}/\odot = 1.656. \tag{15'}$$

Now consider a centrally-condensed star of mass M *greater* than (or equal to) $\mathfrak{M}$. Then we obviously have

$${}_M\beta_0 < {}_{\mathfrak{M}}\beta_0 = 0.908.$$

$${}_M\beta_{c \cdot c} < {}_{\mathfrak{M}}\beta_0 < 0{,}908. \tag{16}$$

Hence, we have the result that for *all centrally condensed stars of mass greater tham $\mathfrak{M}$, the perfect gas equation of state does not break down, however high the density may become, and the matter does not become degenerate. An appeal to the Fermi-Dirac statistics to avoid the central singularity cannot be made.*

Since however we cannot allow the infinite density which the centrally condensed solution of EMDEN's differential equation — index 3 — allows at the centre and in the absence of our knowledge of any equation of state governing the perfect gas other than that of degenerate matter, our only way out of the singularity is to assume that there exists a maximum density $\varrho_{\max}$ which matter is capable of. We have therefore to consider the "fit" of a gaseous envelope of the centrally condensed type on to a homogeneous core at the maximum density of matter. *If we insist on the density to be continuous at the interface* the equation of "fit" is found to be[1])

$$\frac{1}{3}\,\xi'\,\Theta'^3 = -\left(\frac{d\Theta}{d\xi}\right)_{\xi=\xi'}, \tag{17}$$

[1]) S. CHANDRASEKHAR, M. N. **91**, 456, 1931, equation (47).

where the polytropic equation describing the gaseous part of the star is

$$\frac{1}{\xi^2}\frac{d}{d\xi}\left(\xi^2\frac{d\Theta}{d\xi}\right) = -\Theta^3, \tag{17'}$$

where ξ' is the value of ξ at which $\varrho_{\max}$ begins. In (17′) the meaning of Θ and ξ are the following:

$$\varrho = \lambda_3\Theta^3, \quad r = \xi\left[\frac{C}{\pi G\beta}\right]^{-1/3}\lambda_3^{-1/3}. \tag{17''}$$

(λ_3 is a *homology constant*). But (17) has *no* solutions if Θ is of the EMDEN's or of the centrally-condensed type. Hence the acceptance of a $\varrho_{\max}$ does not

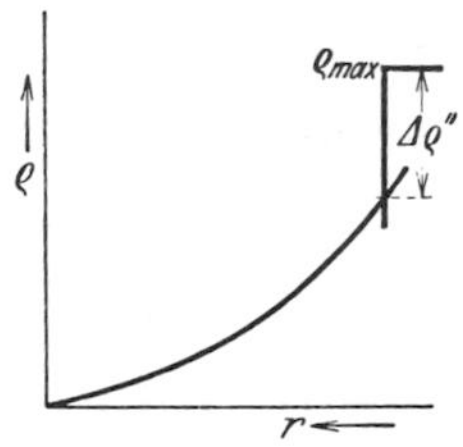

Fig. 2.

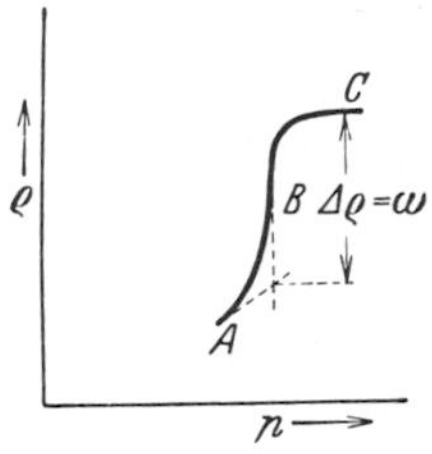

Fig. 3.

help us out of the difficulty if we insist on the density to be continuous at the interface. The procedure then to construct an equilibrium configuration would be to proceed along the centrally condensed solution until the mean density $\varrho_m(r)$ of the *surviving mass* $M(r)$ equals $\varrho_{\max}$ which will occur at a determinate $r = r''$ (say) where

$$M(r'') = {}^4/_3\,\pi r''^3\,\varrho_{\max}; \tag{18}$$

we then replace the material inside $r = r''$ by a sphere of incompressible matter at the density $\varrho_{\max}$. At r'' there will be a discontinuity of density (see Fig. 2).

Now the form of Θ as $\xi \to 0$ for a centrally condensed solution is (MILNE, l. c.):

$$\Theta \sim \frac{1/\sqrt{2}}{\xi\,[\log(D/\xi)]^{1/2}}, \tag{19}$$

where D is a constant. D is *fixed* by the condition that the analytic continuation of (19) passes through $\xi = 1$ and $\Theta = 0$ and satisfies here the requisite boundary condition, namely

$$M = -\frac{4}{\pi^{1/2}}\left(\frac{C}{G\beta}\right)^{3/2}\left(\xi^2\frac{d\Theta}{d\xi}\right)_0. \tag{20 [1]}$$

[1]) C is given by equation (4′).

Hence we get the result that D is a function of L, M and $\varkappa$ only and hence fixed. Since D is fixed by the boundary condition, it follows that the value of ξ'' at which $\Theta(\xi'')$ becomes equal to Θ_{max} (where cf. equation (17''))

$$\Theta_{\max} = \varrho_{\max}^{1/3} \lambda_3^{-1/3}, \tag{21}$$

is fixed as a function of λ_3. In other words the discontinuity in Θ, $\Delta\Theta''$ at the interface ξ'' is a single-valued function of L, M, $\varkappa$ and λ or

$$\Delta\Theta'' = F(L, M, \varkappa; \lambda_3) \tag{21'}$$

or by (21)

$$\Delta\varrho'' = f(L, M, \varkappa; \lambda_3) \tag{22}$$

where $\Delta\varrho''$ is the discontinuity of density at the interface.

But it has been suggested by LANDAU[1]) (among others) that the maximum density of matter will arise *after* some kind of *overcompressibility, the incompressibility setting in later* (see Fig. 3).

Further it has been suggested that 1) the pressure at which the overcompressibility sets in must be a physical property of the atomic nuclei and the electrons in the enclosure, and 2) the form of the curve ABC is again an intrinsic physical property of matter. If we idealise the situation of Fig. 3, we see that *$\Delta\varrho$ ought to be a physical property of matter.* Let this $\Delta\varrho$ be ω. Then by (22) we have to so choose the homology constant λ_3, that $\Delta\varrho''$ equals ω:

$$f(L, M, \varkappa; \lambda_3) = \omega. \tag{23}$$

This fixes λ_3 and hence by (17'') fixes r_0 — the radius of the configuration. Hence *we are able to obtain equilibrium configurations for arbitrary mass, and arbitrary luminosity, the radius however bieng determinate in each case.*

§ *4.* In the above section we have tried to construct the equilibrium configurations for all centrally-condensed stars of mass greater than $\mathfrak{M}$[2]), and found that the introduction of a homogeneous core at the maximum density of matter (ϱ_{max}) with a discontinuity of density at the interface was necessary. We may now ask about the equilibrium configurations for centrally condensed stars with $\beta > 0.908$. Now the star has clearly a degenerate zone (see Fig. 1). A little consideration shows that if we come along a centrally-condensed solution in the perfect gas part of the star then at the interface S_1 (cf. § 1) we are compelled to choose a centrally-condensed solution for the polytropic equation of index "$^3/_2$"

1) I am indebted to Dr. STRÖMGREN for advice on these matters.

2) Or more generally, centrally-condensed stars with $\beta < 0.908$.

to describe the non-relativistic degenerate part of the star[1]); also at the second surface demarcation S_2 we are again forced to choose a centrally-condensed solution for the potytropic equation of index "3". Hence in this case also we are unable to avoid the central singularity by appealing to the FERMI-DIRAC statistics alone. The star must have a homogeneous core with a discontinuity of density at the interface. The considerations of the previous section apply and we see that the centrally-condensed stars $\beta > 0.908$ differ from the centrally-condensed stars with $\beta < 0.908$ only in this, that while in the former type of stars we have to traverse a degenerate zone before reaching the homogeneous core, in the latter type, the stellar material continues to be a perfect gas till we reach the homogeneous core. Thus we find that *all centrally-condensed stars (on the standard model) must have a homogeneous core at the centre with a discontinuity of density at the interface.*

§ 5. Collapsed-Stars. Just a few remarks about collapsed stars may be permitted. A detailed analysis of *highly-collapsed* stars has been given elsewhere (CHANDRASEKHAR, l. c.).

Consider a collapsed star of mass greater than $\mathfrak{M}$ and let further $\beta_0 < \beta_{\text{col.}} < 0.9078$. In other words the "collapse" has not proceeded sufficiently far to increase β beyond 0.9078. In such a case the collapse can occur only on a homogeneous core. But if the collapse proceeds sufficiently far, such that $\beta_{\text{col.}} > 0.9078$ in spite of β_0 being less than 0,9078, the star will then possess a degenerate zone as well.

Conclusion: We may conclude that great progress in the analysis of stellar structure is not possible before we can answer the following fundamental question:

Given an enclosure containing electrons and atomic nuclei, (total charge zero) what happens if we go on compressing the material indefinitely?

[1]) This is also true if we ascribe different opacities to the gaseous and the degenerate part of the star.

9

The Physical State of Matter in the Interior of Stars.

In 1924 Sir Arthur Eddington established his mass-luminosity relation on the basis of the perfect gas hypothesis. During the ten years that have elapsed since, our knowledge of the possible states of an ionized gas has advanced considerably, and the purpose of this review is to show how this new knowledge has gone far towards a clarification of ideas regarding the physical conditions prevailing in the interior of stars. It is necessary, however, to state explicitly that lack of complete information regarding the internal distribution of energy sources is not very serious when we are primarily concerned with the hydrostatic equilibrium of the star. The various numerical integrations that have been carried out with widely different laws for opacity and source distribution have shown that the qualitative nature of the information that can in principle be derived from steady state considerations alone is, within limits, independent of the particular type of laws we choose to discuss ; when it is therefore found necessary to particularize the situation, we shall work with the *Standard Model.*

Now the possible equations of state of a more or less completely ionized gas are :—

$$p=\left(\frac{k}{m_{\mathrm{H}}\mu}\right)\rho \mathrm{T}, \quad \ldots \quad (1)$$

$$p=\frac{1}{20}\left(\frac{3}{\pi}\right)^{2/3}\frac{h^2}{m(m_{\mathrm{H}}\mu)^{5/3}}\rho^{5/3}=\frac{9{\cdot}890\times 10^{12}}{\mu^{5/3}}\rho^{5/3}$$
$$=\mathrm{K}_1\rho^{5/3}\ (\text{say}), \quad \ldots \quad (2)$$

$$p=\left(\frac{3}{\pi}\right)^{1/3}\frac{hc}{8(m_{\mathrm{H}}\mu)^{4/3}}\rho^{4/3}=\frac{1{\cdot}228\times 10^{15}}{\mu^{4/3}}\rho^{4/3}$$
$$=\mathrm{K}_2\rho^{4/3}\ (\text{say}), \quad \ldots \quad (3)$$

where p=pressure, ρ=density, T=temperature, k= Boltzmann's constant, m_{H}=mass of the hydrogen atom, μ=mean molecular weight, h=Planck's constant, c= velocity of light, m=mass of the electron. Equations (1), (2), and (3) correspond to the perfect *, degenerate and relativistically degenerate gases respectively †. Equation (2) will be valid if the pressure given by this formula is, firstly, much greater than that given by (1), and, secondly, much less than that given by (3). These two conditions give the criteria for ordinary degeneracy. If the first criterion is satisfied and the second violated, then the material will be relativistically degenerate. On the other hand, if the pressure given by (1) is greater than that given by either (2) or (3) then the gas remains an ideal gas in the classical sense. Bearing these considerations in mind we shall examine the state of matter in the interior of stars.

(A) *Massive Stars.*—We shall first attempt to give precision as to what is meant by " massive " in this connection. To do that we assume that in the perfect gas regions of the star the equations of the standard

* We shall use the term "perfect gas" to denote an ideal gas in the classical sense (*i. e.* (1) will be valid).

† It may be remarked here that equation (1) is true, independent of relativity.

model hold, *i. e.*, the gas pressure p is a constant fraction * β of the total pressure (=the sum of the gas pressure and the radiation pressure which is equal to $aT^4/3$). Then one easily finds by eliminating the temperature in (1) that in the perfect gas regions

$$p = C\rho^{4/3}, \quad \ldots \ldots \quad (4)$$

where

$$C = \left[\left(\frac{k}{m_H\mu}\right)^4 \frac{3}{a}\frac{1-\beta}{\beta}\right]^{1/3} = \frac{2{\cdot}632\times10^{15}}{\mu^{4/3}}\left[\frac{1-\beta}{\beta}\right]^{1/3}, \quad \ldots \quad (5)$$

where, in a well-known notation, $\beta = 1 - \kappa\eta L/4\pi cGM$ (G is the constant of gravitation).

It is now immediately obvious that if C as defined by (5) be greater than the relativistic-degenerate constant K_2 then the perfect gas equation (1) cannot break down in any part of the internal regions of the star.

One finds by (3) and (4) that if $K_2 < C$ then

$$\frac{1-\beta}{\beta} > \frac{\pi^4}{960} = 0{\cdot}1015 \quad \text{or} \quad \beta < 0{\cdot}9079. \quad \ldots \quad (6)$$

The meaning of (6) is simply this. There is normally a certain temperature gradient in the star, and if the radiation pressure is greater than a tenth of the total pressure then the temperature increases sufficiently rapidly to prevent the matter from becoming degenerate †. We can in fact say much more than this. If for a moment we now consider that the configuration is a perfect gas sphere, then, on the standard model, β is a function of the mass M and μ only, and is given by Eddington's quartic equation :

$$1-\beta = 0{\cdot}00309\,(M/\odot)^2\mu^4\beta^4, \quad \ldots \ldots \quad (7)$$

where $\odot$ refers to the mass of the Sun. For perfect gas configurations, the inequality (6) is formally equivalent to the inequality

$$M > 6{\cdot}623\mu^{-2}\times\odot. \quad \ldots \ldots \quad (8)$$

Hence for all stars for which β is less than or equal to

* That β should be an absolute constant is not essential for the argument (see the following footnote).

† This result is actually much more general than the derivation would suggest. Defining β abstractly as the ratio, gas pressure/total pressure, then it is clear that even if β were not a constant, the stellar material would continue to be a perfect gas provided only $\beta < 0{\cdot}91$ throughout the perfect gas parts of the star.

that given by Eddington's quartic equation * and mass M greater than $6{\cdot}623\mu^{-2}\times\odot$ the perfect gas equation of state cannot pass over into the degenerate equations of state. The conclusion is that *if equations* (1), (2), *and* (3) *represent the only possible states of stellar material then all stars with* $M>6{\cdot}623\mu^{-2}\times\odot$ *and which are not white dwarfs are necessarily wholly perfect gas configurations.* But the perfect gas law (1) can conceivably deviate in ways yet unknown to us. We shall come back to this point towards the end.

(B) *Stars of Small Mass.*—Some general conclusions regarding the physical conditions in the interior of stars of small mass can be deduced by an application of a theorem due to Eddington which states that the total pressure cannot anywhere exceed

$$P_{max.}=\frac{1}{2}\left(\frac{4}{3}\pi\right)^{1/3}GM^{2/3}\rho_0^{4/3}=B\rho_0^{4/3}\text{ (say)}, \quad (9)$$

where ρ_0 is the greatest density inside the star. P_{max} given by (9) is just equal to the central pressure in a configuration of mass M with a uniform density ρ_0.

From (9) we can at once set a lower limit to the mass of stars for which zones of relativistic degeneracy can possibly occur. For if there are regions in the star which are relativistically degenerate then clearly by the above theorem

$$K_2<B \quad \text{or} \quad M>\sqrt{\left(\frac{K_2}{G}\right)^3\frac{6}{\pi}}, \quad (10)$$

or by (3)

$$M>1{\cdot}743\mu^{-2}\times\odot. \quad (11)$$

Hence *for stars of mass less than* $1{\cdot}743\mu^{-2}\times\odot$ *there can be no regions in which matter is relativistically degenerate.* For stars of mass less than this limit matter could be incipiently relativistically degenerate, but if equation (2) describes the state of affairs sufficiently well then by (9) we can now set an upper limit to the density (as was first shown by Eddington). Thus one finds that

$$\rho<\left(\frac{B}{K_1}\right)^3=6{\cdot}301\times10^5\,.\,\mu^5(M/\odot)^2\text{ gms. cm.}^{-3}. \quad (12)$$

We can also formally set an upper limit to the tempera-

* The former case corresponds to "centrally condensed" configurations in the sense defined by Milne.

ture using equation (9), but, since the matter is assumed to be degenerate, it is clear that physical considerations alone require T to be such that the pressure given by (1) is much less than that given by (2). This yields an inequality for T.

$$T << \left(\frac{\mu m_H}{k}\right)\frac{B^2}{K_1} = 8{\cdot}808 \times 10^8 \,.\, \mu^{8/3}(M/\odot)^{4/3}. \qquad (13)$$

We see therefore that for stars of small mass (*i. e.*, $< 1{\cdot}743\mu^{-2} \times \odot$) the physical conditions cannot be more extreme than the limits set above.

(C) *White Dwarfs.*—The degenerate equations of state have essentially clarified our views regarding the constitution of white dwarfs. Since these stars are of very small luminosity, radiation pressure must play a minor role, and the equilibrium can be studied more or less thoroughly *. The analysis yields in addition a confirmation of the inequalities obtained in (B). One finds that when $M < M_{3/2} = 3{\cdot}822\mu^{-2}\beta^{-3/2} \times \odot$ (β has the usual meaning, but by hypothesis ~ 1) relativistic degeneracy does not appreciably set in, and the central density is given by

$$\rho_c = 1{\cdot}310 \times 10^5 \,.\, \mu^5(M/\odot)^2\beta^3 \text{ gms. cm.}^{-3}. \quad . \quad (14)$$

We see that $M_{3/2} > 1{\cdot}743\mu^{-2} \times \odot$ and that $\rho_c < \rho_{\max.}$ given by (12). But when M becomes greater than $M_{3/2}$, relativistic degeneracy sets in very rapidly, and in fact when $M \rightarrow 5{\cdot}736\mu^{-2} \times \odot\beta^{-3/2}$ the star tends to contract to a point. Hence, by taking the mass sufficiently near this limit we can obtain arbitrarily high values for the central density ; but it is very doubtful if this result has any particular significance.

(D) $1{\cdot}743\mu^{-2} \times \odot < M < 6{\cdot}623\mu^{-2} \times \odot$.—We now come to discuss these stars of intermediate mass. The situation is rather complicated, because the star can have relativistically degenerate zones †. The problem now is : Can these stars have cores of high density consistent with the equations of state (1), (2), and (3) ? Now Milne has shown (*M. N.* **92.** 610, 1932) that if $\beta < 4/5$, then it is not possible to have centrally condensed stars. (This result was obtained on the assumption that the degenerate

* The results quoted in this section are taken from the author's paper (*M. N.* **91.** 456, 1931).

† If the star has no regions where the relativistic degeneracy has appreciably set in then the upper limits (12) and (13) continue to hold.

parts of the star are characterized by opacity which is negligible in comparison with that of the outer gaseous envelope. Actually Milne considered only two phase configurations, but the result quoted appears to be true even if we consider three phase configurations.) The stars in the range above specified necessarily satisfy this condition, and hence, when $M < 6{\cdot}623\mu^{-2} \times \odot$, "centrally-condensed" stars are not possible. By this statement one merely means that it is not possible to reach the centre with finite density provided only with the equations of state (1), (2), and (3), if the perfect gas part of the star is described by centrally-condensed solutions of the Emden's equation of index 3.

One can therefore summarize the present situation as follows :—

Given that (1), (2), *and* (3) *are the only possible equations of state for an ionized gas, then for a star (which is not a white dwarf) it is possible to have finite physical conditions at the centre if, and only if, the star is wholly a perfect gas configuration in the sense of equation* (1).

In this connection it is necessary to draw attention to another point. Strömgren's investigations on the (minimum) hydrogen content of stars indicate that the massive B-type stars should be practically wholly composed of hydrogen if there is to be no opacity-discrepancy for these stars on Eddington's model. If now these stars had dense central regions (governed by equations of state yet unknown to us) then to predict the correct luminosity we should have to increase the hydrogen content over the minimum value, and this would not be possible as the limit has already been reached.

The general evidence then is in favour of Eddington's perfect gas hypothesis for ordinary stars, and it would follow that the physical conditions in the interior of stars derived by him * should be near the truth. But it is well to emphasise here that one cannot be too cautious in making this statement. One has to bear in mind that if, as is likely, transmutations of elements are an important source of stellar energy, the steady state of a star is consistent only with the equilibrium of the transmutations occurring in it. The reason for this is that if the

* Modified to take account of the known abundance of hydrogen (such calculations have been made by Eddington and Strömgren).

energy be liberated by non-equilibrium processes of transmutations then a star would be overstable because of the very high power of the temperature dependence of this mode of energy liberation. This conclusion has been reached by Steensholt and Sterne. If the transmutations then are to occur at equilibrium rates, the central temperatures must indeed be very much higher than is provided by the perfect gas hypothesis. But our earlier discussion now shows that if such high temperatures do at all exist in the interior of stars then it must be due to deviations from the known ideal gas laws in ways about which we have at present no precise information. It is conceivable, for instance, that at a very high critical density the atomic nuclei come so near one another that the nature of the interaction might suddenly change and be followed subsequently by a sharp alteration in the equation of state in the sense of giving a maximum density of which matter is capable. However, we are now entering a region of pure speculation, and it is best to conclude the discussion at this stage.

Trinity College, Cambridge, S. CHANDRASEKHAR.
1934 February 15.

Stellar Configurations with degenerate Cores.

In an article published in the March number of *The Observatory* the new orientation towards the general problem of Stellar Structure resulting from the use of degenerate statistics was discussed. A result obtained in that article and which is of importance in our present discussion can be recalled in the following terms:— For stellar material at a specified temperature T and density ρ we can define abstractly a quantity β denoting the ratio between the gas pressure p and the total pressure P (which is the sum of the gas kinetic and radiation pressure). Then if

$$\frac{960}{\pi^4}\frac{1-\beta}{\beta}>1, \quad . \quad . \quad . \quad . \quad . \quad . \quad (1)$$

the material at density ρ and temperature T will be a perfect gas in the classical sense. If β_ω be such that relation (1) is an *equality* then in stellar configurations in which $(1-\beta)$ is always greater than $(1-\beta_\omega)$ the stellar material continues to be a perfect gas however high the density may become. On the standard model this means that if the mass be greater than a certain critical mass (say $\mathfrak{M}$) then finiteness of central density restricts us to consider only the non-singular solutions of Emden's equation with index 3. This means that if we plot $(1-\beta)$ against the radius R then the curves of constant mass $(M>\mathfrak{M})$ are lines parallel to the R-axis. For $(1-\beta)>(1-\beta_\omega)$ these lines (which I shall refer to as "Eddington lines") are not distorted by the introduction of degenerate states. The question of using degenerate states for these configurations does not arise at all. If, however $(1-\beta)<(1-\beta_\omega)$, the curves of constant mass are no longer fully represented by the Eddington lines, and in this region we have a non-trivial solution to Milne's fundamental problem, which for our purposes can be formulated as follows :—" For a star to be wholly gaseous the mass is a function of β only. Call the appropriate β, β_M. Has the star equilibrium configurations for $\beta\neq\beta_M$?" This problem is important, for it is precisely by formulating the problem of stellar structure in this way that we can fully analyse the structure of stars of mass less than $\mathfrak{M}$ $(=6{\cdot}623\,\mu^{-2}\odot)$ *.

* *Cf.* S. Chandrasekhar, *Zs. f. Astrophysik*, **5**, 321 (1932), equation (15).

To answer Milne's problem for $(1-\beta)<(1-\beta_\omega)$ it is essential to take the equation of state for the degenerate matter in the exact form. We cannot neglect relativistic degeneracy since we have seen already that precisely because of the relativistic effects the Eddington lines are undistorted in the greater part of the $(1-\beta, R)$ diagram. I shall refer to such a plot as a *Milne diagram.*

Now the equation of state for the degenerate state can be written parametrically as follows :—

$$p=\frac{\pi m^4c^5}{3h^3}f(x)\ ;\ \rho=\frac{8\pi m^3c^3\mu m_H}{3h^3}x^3=Bx^3\ (\text{say}), \quad . \quad (2)$$

where

$$f(x)=[x(x^2+1)^{1/2}(2x^2-3)+3\sinh^{-1}x], \quad . \quad . \quad (3)$$

the other symbols having their usual meaning. It may be noticed here that with the same definition for ρ as in (2) the pressure for a classical gas can be written as

$$p=\frac{\pi m^4c^5}{3h^3}\left(\frac{960}{\pi^4}\frac{1-\beta_1}{\beta_1}\right)^{1/3}.\,2x^4. \quad . \quad . \quad . \quad (4)$$

As a preliminary to the study of composite configurations with degenerate cores we firstly consider the structure of *completely* collapsed configurations with $\beta=1$. In this case the radiation pressure p' is zero and the total pressure p is given by (2). If one introduces the function ϕ defined as

$$\rho=\frac{\rho_c}{\left(1-\frac{1}{y_0^2}\right)^{3/2}}\left(\phi^2-\frac{1}{y_0^2}\right)^{3/2}, \quad . \quad . \quad . \quad (5)$$

where

$$y_0^2=x_0^2+1\ ;\ \rho_c=\rho_{\text{central}}=Bx_0^3, \quad . \quad . \quad . \quad (5')$$

then one can prove that the structure of the configurations is completely specified by the solution of the differential equation

$$\frac{1}{\eta^2}\frac{d}{d\eta}\left(\eta^2\frac{d\phi}{d\eta}\right)=-\left(\phi^2-\frac{1}{y_0^2}\right)^{3/2}, \quad . \quad . \quad (6)$$

with

$$\phi=1 \text{ at } \eta=0\ ;\ \phi(\eta_1)=\frac{1}{y_0},\ \eta_1 \text{ referring to the boundary}, \quad . \quad . \quad . \quad (7)$$

where η measures the radius vector in a suitable scale. (6) is an *exact* equation, and it is surprising that it has not been isolated before. The derivation of this exact equation has led to a considerable simplification in the analysis of the problem of stellar structure. For a specified y_0, *i.e.* for a specified central density, the structure is completely determined and in particular its mass. We see from (6) that as $y_0 \to \infty$, $\phi \to$ the Emden-function with index 3. The mass of these configurations therefore tends to a unique limit as $y_0 \to \infty$. This mass is naturally M_3 (which was first obtained by the writer (*M.N.* **91.** p. 456, 1931)). Configurations with mass less than M_3 have finite radii. On the Milne diagram we can therefore plot on the radius-axis a series of points corresponding to the radii of different masses of these configurations. M_3 in particular is at the origin of the two axes $(1-\beta, R)$.

With this necessary preliminary analysis of these configurations with $\beta=1$ we can now see how the Eddington lines should be distorted in the region of the Milne diagram $(1-\beta)<(1-\beta_\omega)$. If we consider a star of mass M less than $\mathfrak{M}$ and contract it from infinite extension the star continues to be wholly gaseous till the central density is such that (*cf.* equations (2) and (4))

$$\left(\frac{960}{\pi^4}\frac{1-\beta_M}{\beta_M}\right)^{1/3}=\frac{f(x_0)}{2x_0^4}\ ;\ \rho_c=Bx_0^3 . . \quad . \quad . \quad (8)$$

The radius R_0 of this configuration (with $\rho_c=Bx_0^3$) can now be determined. We can therefore draw the curve $(R_0, 1-\beta_M)$ in the Milne diagram. This curve naturally intersects the $(1-\beta)$ axis, where $\beta_M=\beta_\omega$ corresponding to $M=\mathfrak{M}$. Hence the curves of constant mass are vertical lines parallel to the R-axis until they intersect the $(R_0, 1-\beta_M)$ curve. Below this curve the Eddington lines are distorted, and to study the curves of constant mass inside this region we have to consider composite configurations where the structure of the degenerate core is governed by the differential equation (6) and the outer envelope by Emden's equation with index 3. In considering these configurations it would be natural to work the *generalized*-standard-model in which "$\kappa\eta$" takes different values in the envelope and in the core. We will, however, first consider the usual standard model where "$\kappa\eta$" has the same constant value throughout the star.

The composite configurations can now be studied

by writing down the "equations of fit" and solving them. It may be stated that for solving the equations of fit one can with some modifications adopt here the methods developed by Milne in a rather different connection. One can first prove that when β_1 has the same value in the envelope as in the core then *only collapsed configurations are possible*, i.e., a composite configuration has a "$1-\beta$" which is always less than the value $(1-\beta_M)$ which it has in the wholly gaseous state. The nature of the curves of constant mass can at once be predicted. If the mass is less than M_3 then in the completely collapsed state $(\beta_1=1)$ it has a unique radius already determined from our analysis of these configurations. For each mass M′ we can calculate $\beta_{M'}$. The vertical line through $(1-\beta_{M'})$ cuts the $(1-\beta_M, R_0)$ curve at the point $(1-\beta_{M'}, R_0(M'))$. When the star contracts further it goes along some smooth curve joining the point $(1-\beta_{M'}, R_0(M'))$ with a point on the R-axis corresponding to the radius which this M′ has in the completely collapsed state. In particular, the curve of constant mass for M_3 passes through the origin. One finds that β_{M_3} is specified by

$$\frac{960}{\pi^4}\frac{1-\beta_{M_3}}{\beta^4_{M_3}}=1, \quad . \quad . \quad . \quad . \quad . \quad . \qquad (9)$$

Let $\beta_{M_3}=\beta_0$. Clearly $(1-\beta_\omega)>(1-\beta_0)$.

The question arises what happens for stars with $M_3<M\leq\mathfrak{M}$. Now when $(1-\beta)<(1-\beta_\omega)$ then the configuration has a mass $M_3\beta^{-3/2}$ as $y_0\to\infty$. (This result was obtained in my paper in *M. N.* already referred to.) Hence when $M_3<M\leq\mathfrak{M}$ the curves of constant mass intersect the $(1-\beta)$ axis at a point β^* such that

$$M=M_3\beta^{*-3/2}. \quad . \quad . \quad . \quad . \quad . \quad . \qquad (10)$$

One can show that β^* is related to $\beta\dagger$—the value it has in the wholly gaseous state—by the relation

$$\beta^*=\left(\frac{\pi^4}{960}\frac{\beta\dagger^4}{1-\beta\dagger}\right)^{1/3} \quad . \quad . \quad . \quad . \quad . \quad . \qquad (11)$$

We notice that $\beta^*=\beta\dagger=\beta_\omega$ is a solution. Also $\beta^*=1$ when $\beta\dagger=\beta_0$. These results are of course necessary for consistency. We should further have

$$\mathfrak{M}=M_3\beta_\omega^{-3/2}. \quad . \quad . \quad . \quad . \quad . \quad . \qquad (12)$$

Relation (12) can in fact be shown to be true. Hence Milne's problem admits of a solution (consistent with our present knowledge of the equations of state for ionized material) for $(1-\beta)<(1-\beta_\omega)$, and in this region only collapsed configurations are possible on the standard model. We have also seen how the curves of constant mass run in this region.

The treatment of the generalized standard model ("β_2" of the core different from "β_1" of the envelope) can be carried out in a similar way, though the analysis is very much more complicated. If we consider $\beta_2=1$ as an extreme case, then one can prove for instance that *composite configurations with* $M>M_3$ *are necessarily centrally condensed.* When $M\leq M_3$, but greater than another critical mass, "quasi-diffuse" and centrally-condensed configurations make their appearance in addition to the usual collapsed configurations. It is clearly impossible to describe these results in this short communication, which is intended primarily as a preliminary statement of some of the results of the author's recent studies. The detailed investigations with full tables of solutions will be published elsewhere, but the purpose of writing this article was to show how the setting up of an exact differential equation to describe the degenerate state has led to an almost complete solution of the general problem of stellar structure along the lines indicated above.

Finally, it is necessary to emphasize one major result of the whole investigation, namely, that it must be taken as well established that the life-history of a star of small mass must be essentially different from the life-history of a star of large mass. For a star of small mass the natural white-dwarf stage is an initial step towards complete extinction. A star of large mass ($>\mathfrak{M}$) cannot pass into the white-dwarf stage, and one is left speculating on other possibilities.

S. Chandrasekhar.

Trinity College, Cambridge.
1934 October 24.

11

THE HIGHLY COLLAPSED CONFIGURATIONS OF A STELLAR MASS. (SECOND PAPER.)

S. Chandrasekhar, Ph.D.

1. A study of the equilibrium of degenerate gas spheres has a twofold significance in the analysis of stellar structure, namely, in providing an approach to a proper theory of white dwarfs, and also, we shall see, in providing a certain limiting sequence of configurations to which all stars must tend eventually. A beginning in the study of these configurations was made by the author in a previous communication,* where for convenience the equation of state of degenerate matter was taken to correspond to one or other of the two limiting forms $p = K_1\rho^{5/3}$ or $p = K_2\rho^{4/3}$ according as the density was less than or greater than a certain density ρ' where

$$\rho' = (K_2/K_1)^3,$$

ρ' itself being such that both the equations of state yield the same calculated value for the pressure. Actually in the analysis a certain small temperature gradient was allowed for. Working on the standard model it was assumed that the ratio β of the gas pressure to the total pressure was a constant, but by hypothesis ("highly collapsed") β was taken to be very nearly unity. On these assumptions it followed that stars of mass less than a certain specified $M_{3/2}$ (see I, § 6, page 462) were complete Emden polytropes with index $n = 3/2$, and further that configurations of greater mass must be *composite*, *i.e.* must have inner regions where degeneracy is predominantly relativistic. Lastly, and this was the most important conclusion reached, these composite configurations have a ***natural limit***: On the standard model a completely relativistically degenerate configuration has a mass given by (*cf.* I, equation (36))

$$M = -\frac{4}{\pi^{1/2}}\left(\frac{K_2}{G}\right)^{3/2}\left(\xi^2\frac{d\theta_3}{d\xi}\right)_1 \cdot \beta^{-3/2} = M_3\beta^{-3/2} \text{ (say)}, \tag{1}†$$

where θ_3 is the Emden function with index $n = 3$. These configurations have zero radius (*cf.* the remarks in I following the equations (45), (46), page 463).‡

* *M.N.*, **91**, 456, 1931 (referred to as I). See also the earlier papers of the author in *Phil. Mag.*, **11**, 592, 1931, and *Astrophysical Journal*, **64**, 92.

† In I we denoted by M_3 what we have now defined as $M_3\beta^{-3/2}$. It is convenient to separate out the term involving β from the purely "mass factor."

‡ In I this "singularity" was formally avoided by introducing a state of "maximum density" for matter, but now we shall not introduce any such hypothetical states, mainly for the reason that it appears from general considerations that when the central density is high enough for marked deviations from the known gas laws (degenerate or otherwise) to occur the configurations then would have such small radii that they would cease to have any practical importance in astrophysics.

Apart from the above results of a general character, the analysis in I did not lead to any further quantitative results. To obtain by the methods of I anything more exact would have meant very considerable numerical work to "fit" an appropriate solution of Emden's equation with index $n=3/2$ (to describe the outer ordinarily degenerate envelope) with an *Emden function* of index 3 (to describe the inner relativistically degenerate core). It would be very much more satisfactory to take the exact equation describing the degenerate state and treat the whole degenerate parts of a star on the same footing instead of as in I, further subdividing it to correspond to one or other of the two limiting forms of the equation describing the degenerate state. By a very remarkable coincidence the differential equation (governing the structure of a degenerate gas sphere in hydrostatic equilibrium) based on the exact equation of state takes an extremely simple form. We show, in fact, that the structure of the configuration is governed by a solution of the differential equation,

$$\frac{1}{\eta^2}\frac{d}{d\eta}\left(\eta^2\frac{d\phi}{d\eta}\right)=-\left(\phi^2-\frac{1}{y_0^{\,2}}\right)^{3/2}. \tag{2}*$$

It is to be noticed that there is only one parameter occurring in the equation, and a single system of integrations should suffice to obtain a clear insight into these configurations. Equation (2) has a formal similarity with Emden's equation. Indeed, we shall show that under certain circumstances ϕ can be expressed in terms of the Emden functions with appropriate indices. It is the derivation of the above equation that has led to the developments summarised in this and the following paper. In this paper we shall establish this equation and provide tables of solutions. In the analysis we shall omit all references to radiation pressure, *i.e.* this paper strictly deals with configurations having $\beta=1$. The introduction of radiation in these configurations involves quite delicate considerations, and all these find a proper treatment in the paper following this one.

2. *The Differential Equation governing the Structure of Degenerate Matter in Gravitational Equilibrium.*—The pressure-density relation for a degenerate gas can be written parametrically as follows:—

$$\left.\begin{aligned} p&=\frac{\pi m^4c^5}{3h^3}[x(2x^2-3)(x^2+1)^{1/2}+3\sinh^{-1}x],\\ \rho&=\frac{8\pi m^3c^3\mu H}{3h^3}x^3, \end{aligned}\right\} \tag{3}$$

where m = mass of the electron, c = velocity of light, h = Planck's constant, H = mass of the proton, μ = molecular weight. Equation (3) is established in Appendix I to this paper, where also $f(x)$ is tabulated. We rewrite (3) as

$$p=A_2f(x)\,; \qquad \rho=Bx^3, \tag{4}$$

* This equation was given without proof in the author's preliminary note in the *Observatory*, **57**, 373, 1934.

where
$$\left.\begin{aligned} A_2 &= \frac{\pi m^4 c^5}{3h^3}; \qquad B = \frac{8\pi m^3 c^3 \mu H}{3h^3}, \\ f(x) &= x(2x^2-3)(x^2+1)^{1/2} + 3\sinh^{-1} x. \end{aligned}\right\} \tag{5}$$

The equations of equilibrium are, as usual,
$$\left.\begin{aligned} \frac{dp}{dr} &= -\frac{GM(r)}{r^2}\rho, \\ \frac{dM(r)}{dr} &= 4\pi\rho r^2. \end{aligned}\right\} \tag{6}$$

From (6) we have
$$\frac{1}{r^2}\frac{d}{dr}\left(\frac{r^2}{\rho}\frac{dp}{dr}\right) = -4\pi G\rho. \tag{7}$$

Substitute for p and ρ from (4). We have
$$\frac{A_2}{B}\frac{1}{r^2}\frac{d}{dr}\left(\frac{r^2}{x^3}\frac{df(x)}{dr}\right) = -4\pi G B x^3. \tag{8}$$

From the definition of $f(x)$ in (5) we easily verify that
$$\frac{df(x)}{dr} = \frac{8x^4}{(x^2+1)^{1/2}}\frac{dx}{dr}, \tag{9}$$

or
$$\frac{1}{x^3}\frac{df(x)}{dr} = \frac{8x}{(x^2+1)^{1/2}}\frac{dx}{dr} = 8\frac{d\sqrt{x^2+1}}{dr}. \tag{10}$$

Hence (8) can be rewritten as
$$\frac{1}{r^2}\frac{d}{dr}\left(r^2\frac{d\sqrt{x^2+1}}{dr}\right) = -\frac{\pi G B^2}{2A_2}x^3. \tag{11}$$

Put
$$y^2 = x^2 + 1. \tag{12}$$

Then
$$\frac{1}{r^2}\frac{d}{dr}\left(r^2\frac{dy}{dr}\right) = -\frac{\pi G B^2}{2A_2}(y^2-1)^{3/2}. \tag{13}$$

Let x take the value x_0 at the centre.

Further, let y_0 be the corresponding value of y at the centre. Introduce the new variables η and ϕ defined as follows :—
$$r = \alpha\eta; \qquad y = y_0\phi, \tag{14}$$

where
$$\left.\begin{aligned} \alpha &= \left(\frac{2A_2}{\pi G}\right)^{1/2}\frac{1}{By_0}, \\ y_0^2 &= x_0^2 + 1. \end{aligned}\right\} \tag{15}$$

Our differential equation finally takes the form
$$\frac{1}{\eta^2}\frac{d}{d\eta}\left(\eta^2\frac{d\phi}{d\eta}\right) = -\left(\phi^2 - \frac{1}{y_0^2}\right)^{3/2}. \tag{16}$$

By (14) we have to seek a solution of (16) such that ϕ takes the value unity at the origin. Further, from symmetry the derivative of ϕ must

14

vanish at the origin. The *boundary* is defined at the point where the density vanishes, and this by (12) means that if η_1 specifies the boundary

$$\phi(\eta_1) = \frac{1}{y_0}. \tag{17}$$

3. From our definitions of the various quantities we find that

$$\rho = \rho_0 \frac{y_0^3}{(y_0^2 - 1)^{3/2}} \left(\phi^2 - \frac{1}{y_0^2} \right)^{3/2}, \tag{18}$$

where

$$\rho_0 = Bx_0^3 = B(y_0^2 - 1)^{3/2} \tag{18'}$$

specifies the central density. Also we may notice that the scale of length a introduced in (15) has in terms of the physical quantities the form

$$a = \frac{1}{4\pi m\mu H y_0} \left(\frac{3h^3}{2cG} \right)^{1/2}, \tag{19}$$

or putting in numerical values

$$a = \frac{7{\cdot}720 \times 10^8}{\mu y_0} = l_1 y_0^{-1} \text{ cm. (say).} \tag{20}$$

4. *The Potential.*—The function ϕ itself has a physical meaning. If V is the inner gravitational potential, then from general theory we have

$$\frac{dV}{dr} = \frac{1}{\rho} \frac{dP}{dr}. \tag{21}$$

From (5) and (10) we see that

$$\frac{dV}{dr} = \frac{8A_2}{B} y_0 \frac{d\phi}{dr}, \tag{22}$$

or integrating

$$V = \frac{8A_2}{B} y_0 \phi + \text{constant.} \tag{23}$$

If we choose the arbitrary zero of the potential on the boundary of the configuration we have by (17) that the "constant" in (23) is $(8A_2/B)$. Hence finally

$$V = \frac{8A_2}{B} y_0 \left(\phi - \frac{1}{y_0} \right). \tag{24}$$

5. *The Mass Relation.*—The mass of the material enclosed up to a point η is clearly

$$M(\eta) = 4\pi \int_0^\eta \rho r^2 dr = 4\pi a^3 \int_0^\eta \rho \eta^2 d\eta. \tag{25}$$

By (18),

$$M(\eta) = 4\pi\rho_0 \frac{a^3 y_0^3}{(y_0^2 - 1)^{3/2}} \int_0^\eta \left(\phi^2 - \frac{1}{y_0^2} \right)^{3/2} \eta^2 d\eta, \tag{26}$$

or using our differential equation (16)

$$M(\eta) = -4\pi\rho_0 \frac{a^3 y_0^3}{(y_0^2 - 1)^{3/2}} \int_0^\eta \frac{d}{d\eta} \left(\eta^2 \frac{d\phi}{d\eta} \right) d\eta. \tag{27}$$

Remembering that ρ_0 is given by (18) we have explicitly

$$M(\eta) = -4\pi\left(\frac{2A_2}{\pi G}\right)^{3/2}\frac{1}{B^2}\eta^2\frac{d\phi}{d\eta}. \tag{28}$$

The mass of the whole configuration is therefore

$$M = -4\pi\left(\frac{2A_2}{\pi G}\right)^{3/2}\frac{1}{B^2}\left(\eta^2\frac{d\phi}{d\eta}\right)_{\eta=\eta_1}. \tag{29}$$

We notice that in (28) and (29) y_0 does not *explicitly* occur. It is of course implicitly present inasmuch as in the differential equation defining ϕ, y_0 occurs.

6. *The Relation between the Mean and the Central Density.*—Let $\bar{\rho}(\eta)$ be the mean density of the material inside η. Then

$$M(\eta) = \tfrac{4}{3}\pi a^3\eta^3\bar{\rho}(\eta). \tag{30}$$

Comparing (28) and (30), we have

$$\tfrac{1}{3}\eta^3\bar{\rho}(\eta) = -\rho_0\frac{y_0^3}{(y_0^2-1)^{3/2}}\eta^2\frac{d\phi}{d\eta}, \tag{31}$$

or

$$\frac{\bar{\rho}(\eta)}{\rho_0} = -3\frac{y_0^3}{(y_0^2-1)^{3/2}}\frac{1}{\eta}\frac{d\phi}{d\eta}. \tag{32}$$

From (32) we deduce that *the relation between the mean and the central density of the whole configuration is*

$$\rho_0 = -\bar{\rho}\left(1-\frac{1}{y_0^2}\right)^{3/2}\frac{\eta_1}{3\phi'(\eta_1)} \tag{33}$$

(ϕ' denoting the derivative)—a relation analogous to the corresponding relation in the theory of polytropes.

7. *An Approximation for Configurations with Small Central Densities.*—When the central density is small we should have the law $p = K_1{}^{5/3}$ holding approximately, and the corresponding configurations must have structures which can approximately be represented by an Emden polytrope with index $n = 3/2$. We establish this on our differential equation in the following way:—

Now by definition $y_0^2 = x_0^2 + 1$, and we need a first-order approximation when x_0^2 is small. *We shall neglect all quantities of order x_0^4 or higher.* Then

$$y_0 = 1 + \tfrac{1}{2}x_0^2. \tag{34}$$

Put

$$\phi^2 - \frac{1}{y_0^2} = \theta. \tag{35}$$

In our approximation we have

$$\phi = 1 - \tfrac{1}{2}(x_0^2 - \theta). \tag{36}$$

At the origin ϕ takes the value unity. Hence

$$\theta(0) = x_0^2. \tag{37}$$

From (16) we derive the following differential equation for θ :—

$$\frac{1}{2}\frac{d^2\theta}{d\eta^2}+\frac{1}{\eta}\frac{d\theta}{d\eta}=-\theta^{3/2}. \tag{38}$$

Put

$$\xi=2^{1/2}\eta. \tag{39}$$

Then

$$\frac{1}{\xi^2}\frac{d}{d\xi}\left(\xi^2\frac{d\theta}{d\xi}\right)=-\theta^{3/2}, \tag{40}$$

which is Emden's equation with index $n=3/2$, but *the solution we need is not the Emden function in the usual normalisation* * *with* $\theta=1$ *at* $\xi=0$. By (37) our θ takes the value x_0^2 at the origin. Denote by $\theta_{3/2}$ the Emden function. Now it is a property of the differential equation (40) that if θ is any solution then $C^4\theta(C\xi)$ is also a solution where C is any arbitrary constant. Hence if we put

$$C=x_0^{1/2}, \tag{41}$$

and take for θ, $\theta_{3/2}$, we would obtain the solution we need. Hence

$$\theta=x_0^2\theta_{3/2}(x_0^{1/2}\xi)=x_0^2\theta_{3/2}(\sqrt{2x_0}\eta). \tag{42}$$

By (37) then

$$\phi=1-\tfrac{1}{2}x_0^2\{1-\theta_{3/2}(\sqrt{2x_0}\eta)\}+O(x_0^4), \tag{43}$$

which relates ϕ with $\theta_{3/2}$. From (43) we see that for these configurations the boundary η_1 must be such that

$$(\theta_{3/2}\sqrt{2x_0}\eta_1)=0. \tag{44}$$

Let $\xi_1(\theta_{3/2})$ be the boundary of the *Emden function*. Then from (44) we deduce that

$$\eta_1=\frac{\xi_1(\theta_{3/2})}{\sqrt{2x_0}}. \tag{45}$$

From (45) we see that as $y_0\to 1$, $x_0\to 0$, $\eta_1\to\infty$. The radius tends to infinity with the same singularity.

Again from (43) we have

$$\frac{d\phi}{d\eta}=\tfrac{1}{2}x_0^2\sqrt{2x_0}\frac{d\theta_{3/2}(\xi)}{d\xi}. \tag{46}$$

Combining (45) and (46) we have a relation we shall need later:

$$\left(\eta^2\frac{d\phi}{d\eta}\right)_1=\left(\frac{x_0}{2}\right)^{3/2}\left(\xi^2\frac{d\theta_{3/2}}{d\xi}\right)_1. \tag{47}$$

Further,

$$\left(\frac{1}{\eta}\frac{d\phi}{d\eta}\right)_1=x_0^3\left(\frac{1}{\xi}\frac{d\theta_{3/2}}{d\xi}\right)_1. \tag{48}$$

We shall find the above expressions useful when we come to discuss "highly"

* In the sequel by "*Emden function*" *we shall always mean the one which takes the value unity at the origin.* We shall denote the *Emden function* with index n by θ_n.

collapsed configurations ($(1-\beta)$ finite but small), but now we verify that the scheme is consistent. From (48) and (33) we have

$$\rho_0 = -\bar{\rho}\left(\frac{\xi}{3\theta'_{3/2}}\right)_1, \tag{49}$$

which is precisely the formula for an Emden polytrope with index $n = 3/2$. Again from (29) and (47)

$$M = -4\pi\left(\frac{2A_2}{\pi G}\right)^{3/2}\frac{1}{B^2}\left(\frac{x_0}{2}\right)^{3/2}\left(\xi^2\frac{d\theta_{3/2}}{d\xi}\right)_1. \tag{50}$$

To compare the above with the formula derived on the law $p = K_1\rho^{5/3}$ we note that the degenerate constant K_1, given by

$$K_1 = \frac{1}{20}\left(\frac{3}{\pi}\right)^{2/3}\frac{h^2}{m(\mu H)^{5/3}}, \tag{51}$$

is related to our A_1 and B by the relation

$$K_1 = \frac{8}{5}\frac{A_2}{B^{5/3}}. \tag{52}$$

Combining (50) and (52) and setting λ_2 to denote the central density ($= Bx_0^3$) we find that

$$M = -4\pi\left(\frac{5K_1}{8\pi G}\right)^{3/2}\lambda_2^{1/2}\left(\xi^2\frac{d\theta_{3/2}}{d\xi}\right)_1, \tag{53}$$

which is the usual formula since on the law $p = K\rho^{1+\frac{1}{n}}$ the polytropic relation is

$$M = -4\pi\left(\frac{(n+1)K}{4\pi G}\right)^{3/2}\lambda_2^{\frac{3-n}{2n}}\left(\xi^2\frac{d\theta_n}{d\xi}\right)_1. \tag{53'}$$

8. *The Limiting Mass.*—From our differential equation (16) we see that

$$\phi \to \theta_3 \quad \text{as} \quad y_0 \to \infty. \tag{54}$$

But from (20) we see that at the same time the radius tends to zero. From (28) then

$$M \to -4\pi\left(\frac{2A_2}{\pi G}\right)^{3/2}\frac{1}{B^2}\left(\xi^2\frac{d\theta_3}{d\xi}\right)_1. \tag{55}$$

To see that we have now simply recovered our earlier result in I (equation (36)) we have only to notice that the relativistic degenerate constant K_2, defined by

$$K_2 = \left(\frac{3}{\pi}\right)^{1/3}\frac{hc}{8(\mu H)^{4/3}}, \tag{56}$$

is related to our A_2 and B by the relation

$$K_2 = \frac{2A_2}{B^{4/3}}. \tag{57}$$

9. As mentioned in § 1, we shall denote by M_3 the mass

$$M_3 = 4\pi\left(\frac{2A_2}{\pi G}\right)^{3/2}\frac{1}{B^2}\omega_3^0, \tag{58}$$

where following Milne we have introduced the quantity $\omega_3{}^0$ defined by

$$\omega_3{}^0 = -\left(\xi^2\frac{d\theta_3}{d\xi}\right)_1. \tag{59}$$

If we define correspondingly that

$$\Omega(y_0) = -\left(\eta^2\frac{d\phi}{d\eta}\right)_{\eta=\eta_1} \tag{60}$$

for our function ϕ, then the mass relation can be written as

$$M(y_0)\omega_3{}^0 = M_3\Omega(y_0). \tag{61}$$

As the mass of the configuration increases monotonically with increasing y_0, we have the useful inequality

$$\Omega(y_0) > \omega_3{}^0 \qquad (y_0 \text{ finite}). \tag{62}$$

Finally we may note that the insertion of numerical values in our formula for M_3 yields

$$M_3 = 5{\cdot}728\mu^{-2} \times \odot, \tag{63}$$

where $\odot$ represents the mass of the Sun.

10. *The General Results.*—In the previous sections, §§ 7, 8, 9, we have merely related our present treatment with the results obtained in I on the basis of the polytropic theory. Those results appear as simple limiting cases. However, the exact treatment on the basis of our differential equation

$$\frac{1}{\eta^2}\frac{d}{d\eta}\left(\eta^2\frac{d\phi}{d\eta}\right) = -\left(\phi^2 - \frac{1}{y_0{}^2}\right)^{3/2} \tag{64}$$

at the same time provides much more quantitative information. The boundary conditions

$$\phi = 1, \qquad \frac{d\phi}{d\eta} = 0 \quad \text{at} \quad \eta = 0, \tag{65}$$

combined with a particular value for y_0, would determine ϕ completely, and therefore the mass of the configuration as well. The equation (64) does not admit of a "homology constant," and hence *each mass has a density distribution characteristic of itself which cannot be inferred from the density distribution in a configuration of a different mass.* This difference between our configurations governed by (64) and polytropes has, as we shall see, an important bearing in the theory of general stellar models considered in the following paper.

Each specified value for y_0 determines uniquely the mass M, the radius R_1 and the ratio of the mean to the central density. We have (collecting together our earlier results):

$$M/M_3 = \Omega(y_0)/\omega_3{}^0, \tag{66}$$

$$R_1/l_1 = \eta_1/y_0, \tag{67}$$

$$\rho_0/B = (y_0{}^2 - 1)^{3/2}, \tag{68}$$

$$\bar{\rho}/\rho_0 = -\frac{1}{\left(1-\frac{1}{y_0^2}\right)^{3/2}} \frac{3}{\eta_1}\left(\frac{d\phi}{d\eta}\right)_1. \tag{69}$$

In (67) we have introduced a new unit of length ($l_1 = ay_0$),

$$l_1 = \frac{1}{4\pi m\mu H}\left(\frac{3h^3}{2cG}\right) = 7{\cdot}720\mu^{-1} \times 10^8 \text{ cm.,} \tag{67'}$$

and which therefore does not involve factors in y_0. Further, the physical variables determining the structure of the configuration are:

$$\rho = \rho_0 \frac{1}{\left(1-\frac{1}{y_0^2}\right)^{3/2}}\left(\phi^2 - \frac{1}{y_0^2}\right)^{3/2}, \tag{70}$$

$$\bar{\rho} = -\rho_0 \frac{1}{\left(1-\frac{1}{y_0^2}\right)^{3/2}} \frac{3}{\eta}\frac{d\phi}{d\eta}, \tag{71}$$

$$M(\eta) \propto -\eta^2 \frac{d\phi}{d\eta}. \tag{72}$$

11. In § 10 we have reduced the problem of the structure of degenerate gas spheres to a study of our functions ϕ for different initially prescribed values for the parameter y_0. The integration has been numerically effected for the following ten different values of the parameter :—

$$1/y_0^2 = 0{\cdot}8,\ 0{\cdot}6,\ 0{\cdot}5,\ 0{\cdot}4,\ 0{\cdot}3,\ 0{\cdot}2,\ 0{\cdot}1,\ 0{\cdot}05,\ 0{\cdot}02,\ 0{\cdot}01. \tag{73}$$

The following expansion for ϕ near the origin may be noted here for further reference :—

$$\phi = 1 - \frac{q^3}{6}\xi^2 + \frac{q^4}{40}\xi^4 - \frac{q^5(5q^2+14)}{7\,!}\xi^6 + \frac{q^6(339q^2+280)}{3\times 9\,!}\xi^8 - \frac{q^7(1425q^4+11436q^2+4256)}{5\times 11\,!}\xi^{10} + \cdots \tag{74}$$

where

$$q^2 = 1 - \frac{1}{y_0^2}. \tag{75}^*$$

The important quantities of interest are the boundary quantities occurring in equations (66), (67), (69). These are tabulated in Table I for the different values of y_0.

12. From the figures of Table I it is easy to calculate the mass in units of M_3, the radius in units of l_1 and the central density ($=x_0^3$) in units of B

* When $y_0 \to \infty$, $q \to 1$ and the series (74) goes over into the expansion for Emden θ_3 near the origin (cf. *British Association Tables*, **2**, Introduction, equation on top of page v).

($=9{\cdot}8848\times 10^5\ \mu$ grams cm.$^{-3}$). These express the chief physical characteristics of these configurations in the "natural system" of units occurring in the theory of these configurations. In Table III they are converted into the more conventional system of units expressing the radius and the density in C.G.S. units and the mass in units of the Sun. The actual figures tabulated are for $\mu=1$. The figures for other values of μ can be obtained by multiplying M by μ^{-2}, R_1 by μ^{-1} and ρ by μ. To see the order of magnitudes involved here it is of interest to point out that the mass $4{\cdot}852\odot\mu^{-2}$ has a radius only slightly over the radius of the Earth (radius of the Earth 6×10^8 cm. compared to $7{\cdot}7\times 10^8$ cm. for the radius of $4{\cdot}852\odot$). The mass $0{\cdot}957 M_3$ has a radius considerably less than the radius of the Earth.

TABLE I

$\frac{1}{y_0^2}$	η_1	$-\eta_1^2\phi'(\eta_1)$	$\rho_0/\bar{\rho}$
0	6·8968	2·0182	54·182
·01	5·3571	1·9321	26·203
·02	4·9857	1·8652	21·486
·05	4·4601	1·7096	16·018
·1	4·0690	1·5186	12·626
·2	3·7271	1·2430	9·9348
·3	3·5803	1·0337	8·6673
·4	3·5245	0·8598	7·8886
·5	3·5330	0·7070	7·3505
·6	3·6038	0·5679	6·9504
·8	4·0446	0·3091	6·3814
1	∞	0	5·9907

TABLE II

The Physical Characteristics of Degenerate Spheres in the "Natural" Units

$\frac{1}{y_0^2}$	M/M_3	R_1/l_1	ρ_0/B
0	1	0	∞
·01	0·95733	0·53571	985·038
·02	0·92419	0·70508	343
·05	0·84709	0·99732	82·8191
·1	0·75243	1·28674	27
·2	0·61589	1·66682	8
·3	0·51218	1·96102	3·56423
·4	0·42600	2·22908	1·83711
·5	0·35033	2·49818	1
·6	0·28137	2·79148	0·54433
·8	0·15316	3·61760	0·125
1·0	0	∞	0

TABLE III

The Physical Characteristics of Degenerate Spheres in the Usual Units

(Calculations are for $\mu = 1$. For other values μ, M should be multiplied by μ^{-2}, R_1 by μ^{-1}, ρ_c by μ)

$\frac{1}{y_0^2}$	$M/\odot$	ρ_0 in grm./cm.$^{-3}$	ρ_{mean} in grm./cm.$^{-3}$	Radius in cm.
0	5·728	∞	∞	0
·01	5·484	$9{\cdot}737 \times 10^8$	$4{\cdot}716 \times 10^7$	$4{\cdot}136 \times 10^8$
·02	5·294	$3{\cdot}391 \times 10^8$	$1{\cdot}578 \times 10^7$	$5{\cdot}443 \times 10^8$
·05	4·852	$8{\cdot}187 \times 10^7$	$5{\cdot}111 \times 10^6$	$7{\cdot}699 \times 10^8$
·1	4·310	$2{\cdot}669 \times 10^7$	$2{\cdot}114 \times 10^6$	$9{\cdot}936 \times 10^8$
·2	3·528	$7{\cdot}908 \times 10^6$	$7{\cdot}960 \times 10^5$	$1{\cdot}287 \times 10^9$
·3	2·934	$3{\cdot}523 \times 10^6$	$4{\cdot}065 \times 10^5$	$1{\cdot}514 \times 10^9$
·4	2·440	$1{\cdot}816 \times 10^6$	$2{\cdot}302 \times 10^5$	$1{\cdot}721 \times 10^9$
·5	2·007	$9{\cdot}885 \times 10^5$	$1{\cdot}345 \times 10^5$	$1{\cdot}929 \times 10^9$
·6	1·612	$5{\cdot}381 \times 10^5$	$7{\cdot}741 \times 10^4$	$2{\cdot}155 \times 10^9$
·8	0·877	$1{\cdot}236 \times 10^5$	$1{\cdot}936 \times 10^4$	$2{\cdot}793 \times 10^9$
1·0	0	0	0	∞

Now if we define that matter is "relativistically degenerate" for densities greater than $\rho'(=(K_2/K_1)^3)$, then we can from our results easily find the masses which are characterised by central regions of "relativistic degeneracy." The value of x corresponding to ρ' is readily seen to be 1·25. Hence

$$\frac{1}{y_0'^2} = \frac{1}{x'^2+1} = 0{\cdot}39024. \tag{76}$$

From fig. 1 we now see that for $M \leqslant 0{\cdot}43M_3$ there are no regions which are "relativistically degenerate" on this convention. For $M > 0{\cdot}43M_3$ there are regions in which $x > x'(=1{\cdot}25)$, and the fraction of the whole radius inside which $x > x'$ rapidly increases to unity. In the mass-radius curve we can therefore draw circles about each point with radii proportional to the actual radii of the corresponding configurations, and draw inside each a concentric circle to represent the "relativistic" region. This has been done in fig. 2 at a few points. We see that even for $M = 0{\cdot}75M_3$ there is barely a "fringe" of ordinarily degenerate regions. This diagram clearly illustrates a general principle that degeneracy never usually sets in without being relativistic.

13. *Comparison with the Results on Emden Polytrope $n = 3/2$.*—It is of interest to see in how far the results of the above exact treatment differ from what one would obtain on the law $p = K_1\rho^{5/3}$. We have already shown in § 7 that one gets these Emden configurations as limiting cases for zero density and therefore for small masses (expressed in units of M_3). Our comparison here therefore amounts to a comparison of the results based on an exact treatment of the equation (64) with the limiting form for $y_0 \to 1$ extrapolated for all masses. For this purpose it is convenient to rewrite the formulæ for the case of the polytrope $n = 3/2$ in the following way.

From (45) and (50) we have now

$$R_1 = \frac{l_1 \xi_1(\theta_{3/2})}{\sqrt{2x_0}}, \tag{77}$$

$$M/M_3 = \left(\frac{x_0}{2}\right)^{3/2} \frac{1}{\omega_3{}^0}\left(\xi^2 \frac{d\theta_{3/2}}{d\xi}\right)_1. \tag{77'}$$

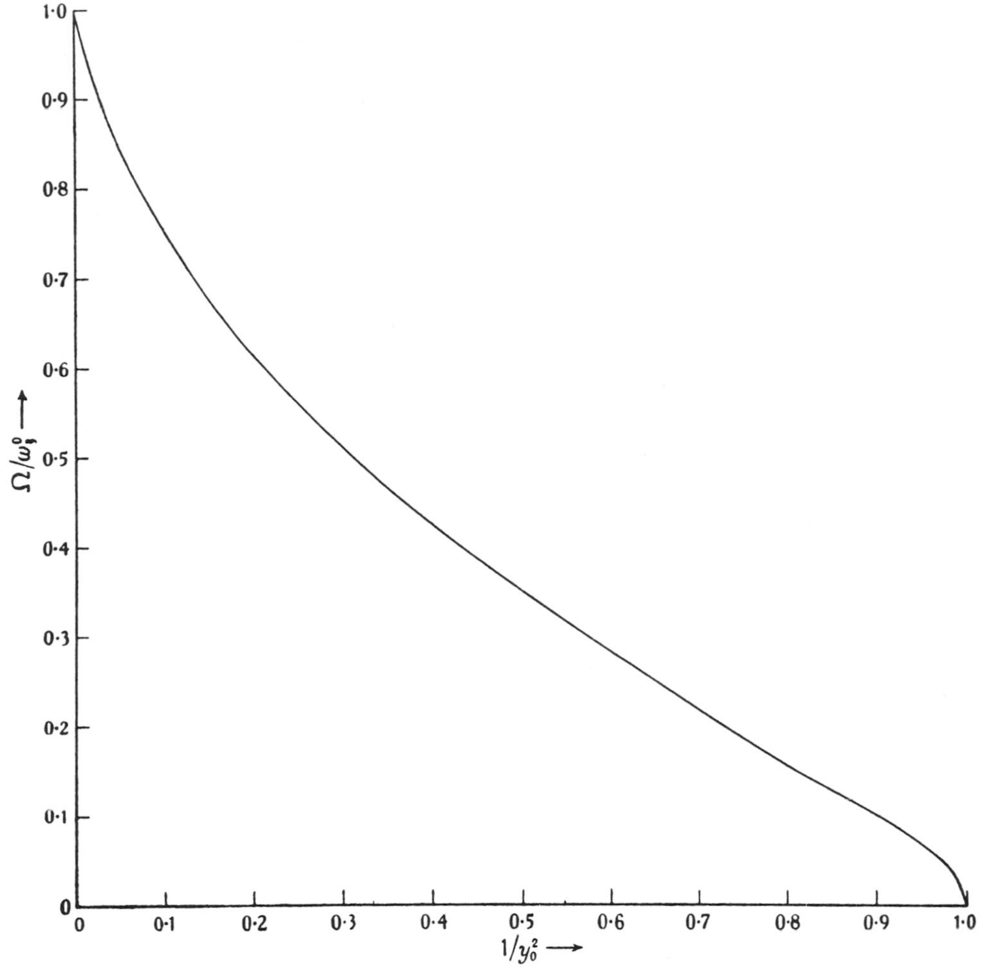

FIG. 1.—$\{(\Omega/\omega_3{}^0),\ 1/y_0{}^2\}$-*relation.*

From (77) and (77′) we have on eliminating x_0

$$2R_1 = \left(\frac{\omega_{3/2}{}^0 M_3}{\omega_3{}^0 M}\right)^{1/3} . \, l_1, \tag{78}$$

where following Milne we have introduced the "invariant" $\omega_{3/2}{}^0$ defined by

$$\omega_{3/2}{}^0 = -\left(\xi^5 \frac{d\theta_{3/2}}{d\xi}\right)_1 = 132{\cdot}3843. \tag{79}$$

It is of interest to notice that the two invariants ω_3^0 and $\omega_{3/2}^0$ of the Emden equation with the indices $n=3$ and $3/2$ occur in (78) in a "symmetrical way." Numerically (78) is found to be

$$R_1 = 2{\cdot}01647\left(\frac{M_3}{M}\right)^{1/3} . \, l_1. \tag{80}$$

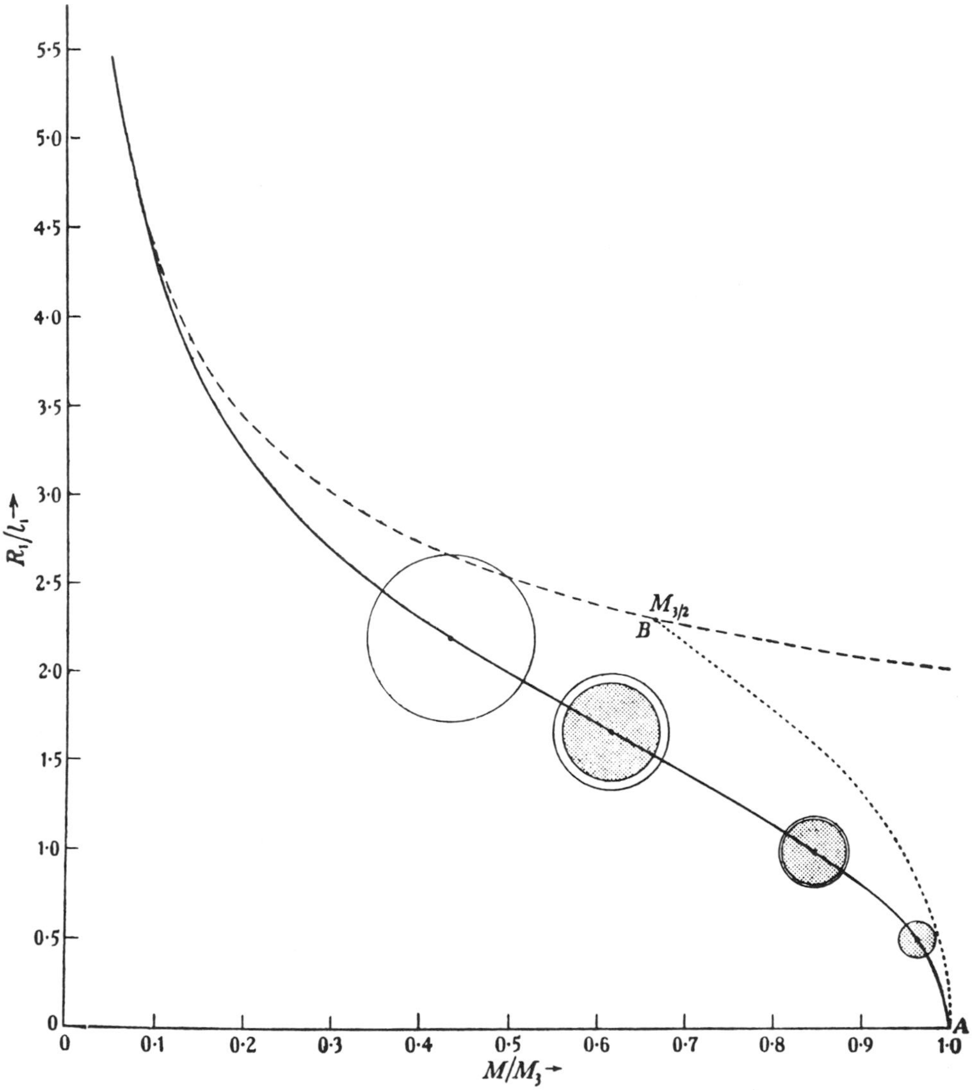

FIG. 2.—*The full line curve represents the exact (mass-radius)-relation for the highly collapsed configurations. This curve tends asymptotically to the - - - - curve as $M \to 0$.*

(80) expresses the *mass-radius* relation for the polytropic limit, the radius and the mass expressed in the same units as the quantities in Table II. Similarly the mass-central density relation now reads

$$x_0^3 = 4{\cdot}42381(M/M_3)^2. \tag{81}$$

The results calculated on the basis of (80) and (81) for the same masses as in Table II are summarised in Table IV. The corresponding curves are shown dotted in figs. 2 and 3.

TABLE IV

M/M_3	R_1/l_1	x_0^3
1	2·0165	4·4238
0·9573	2·0459	4·0538
0·9242	2·0700	3·7780
0·8471	2·1311	3·1739
0·7524	2·2174	2·5042
0·6159	2·3701	1·6778
0·5122	2·5203	1·1603
0·4260	2·6801	0·8027
0·3503	2·8606	0·5429
0·2814	3·0772	0·3502
0·1532	3·7691	0·1038

One notices clearly from these two curves how marked the deviations from the limiting curves become even for quite small masses. Thus for $M = 0{\cdot}15 M_3$ the central density predicted by our exact treatment is about 25 per cent. greater and the radius about 5 per cent. smaller. The relativistic effects are therefore quite significant even for small masses. They certainly cannot be ignored for masses greater than $0{\cdot}2 M_3$. Of course the extrapolation of the $n = 3/2$ configurations for masses (in units of M_3) approaching unity is quite misleading. These completely collapsed configurations have a natural limit, and our exact treatment now shows how this limit is reached.

It is of interest to compare the full-line curve in fig. 2 representing our exact (mass-radius) curve with what one would obtain by the methods of I, where the degenerate spheres of mass greater than a certain limit $M_{3/2}$ were considered as "composite configurations." The mass $M_{3/2}$ was defined as one in which the Emden polytrope with $n = 3/2$ * would have a central density $\rho' (= (K_2/K_1)^3)$. In our present notation we have by (81)

$$M_{3/2} = \sqrt{\frac{(1{\cdot}25)^3}{4{\cdot}42381}} \cdot M_3 = 0{\cdot}66446 M_3. \tag{82}$$

This particular point is marked as B in fig. 2 on the - - - - curve. A treatment of the composite configurations by the methods of I would have led to some kind of curve like the dotted one in fig. 2 conjecturally drawn. But fortunately it is now not necessary to go into the very elaborate numerical work that would have been involved to fix the part BA by the methods of I. By a single system of integrations we have now fixed the exact nature of the (mass-radius) curve for these completely collapsed configurations.

* The equation of state being $p = K_1 \rho^{5/3}$.

14. *The Relative Density Distributions in the Different Configurations.*—Our main diagram (fig. 4) now illustrates the relative density distributions in the configurations studied. Here we have plotted (ρ/ρ_0) against (η/η_1) for the different masses for which we have numerical results. The two limiting density distributions specified by Emden, θ_3 and $\theta_{3/2}$, are also shown (dotted) in the same figure. Fig. 4, which is *the* principal outcome of our studies, presents a set of ten out of a continuous family of density distributions

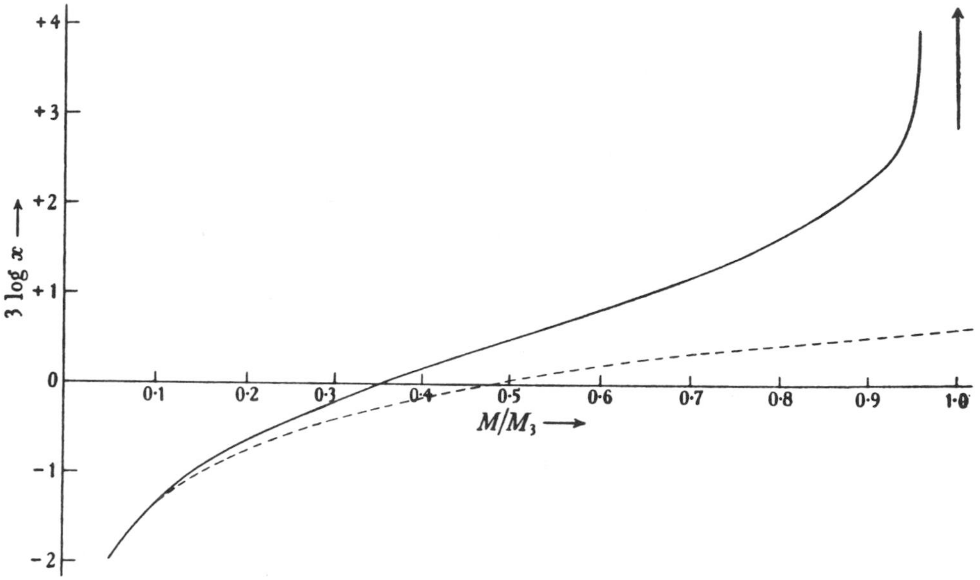

FIG. 3.—*The full line curve represents the exact (mass, log ρ_0)-relation for the highly collapsed configurations. This curve tends asymptotically to the dotted curve as $M \to 0$.*

covering the range specified by the polytropic distributions of indices 3/2 and 3.

15. *Concluding Remarks.*—In this paper we have strictly confined ourselves to the case "$\beta=1$." But in stellar problem the radiation pressure (even if small) necessarily plays a deciding rôle, and the question as to in what sense we have to understand the completely degenerate spheres studied here as representing "the limiting sequence of configurations to which all stars must tend eventually" can be answered only by introducing radiation in these configurations. To do this properly we have first to develop adequate methods to treat composite configurations consisting of degenerate cores (of the structures studied here) surrounded by gaseous envelopes. These and related problems are studied in the following paper (p. 226).

16. *Manuscript Copy of Tables.*—The functions ϕ and their derivatives ϕ' (to six and five significant figures respectively) have been computed by the author for the values of $1/y_0^2$ specified in (73). In addition to ϕ and ϕ' the auxiliary functions ρ/ρ_0, $\rho_0/\bar{\rho}$, $-\eta^2\phi'$ and two other functions U and V (defined in equation (91) of the following paper) have also been tabulated. The auxiliary functions were calculated correct to five significant figures. All

the functions were tabulated for steps of 0·1 for the argument η. A manuscript copy of these tables has been deposited in the Library of the Society.*

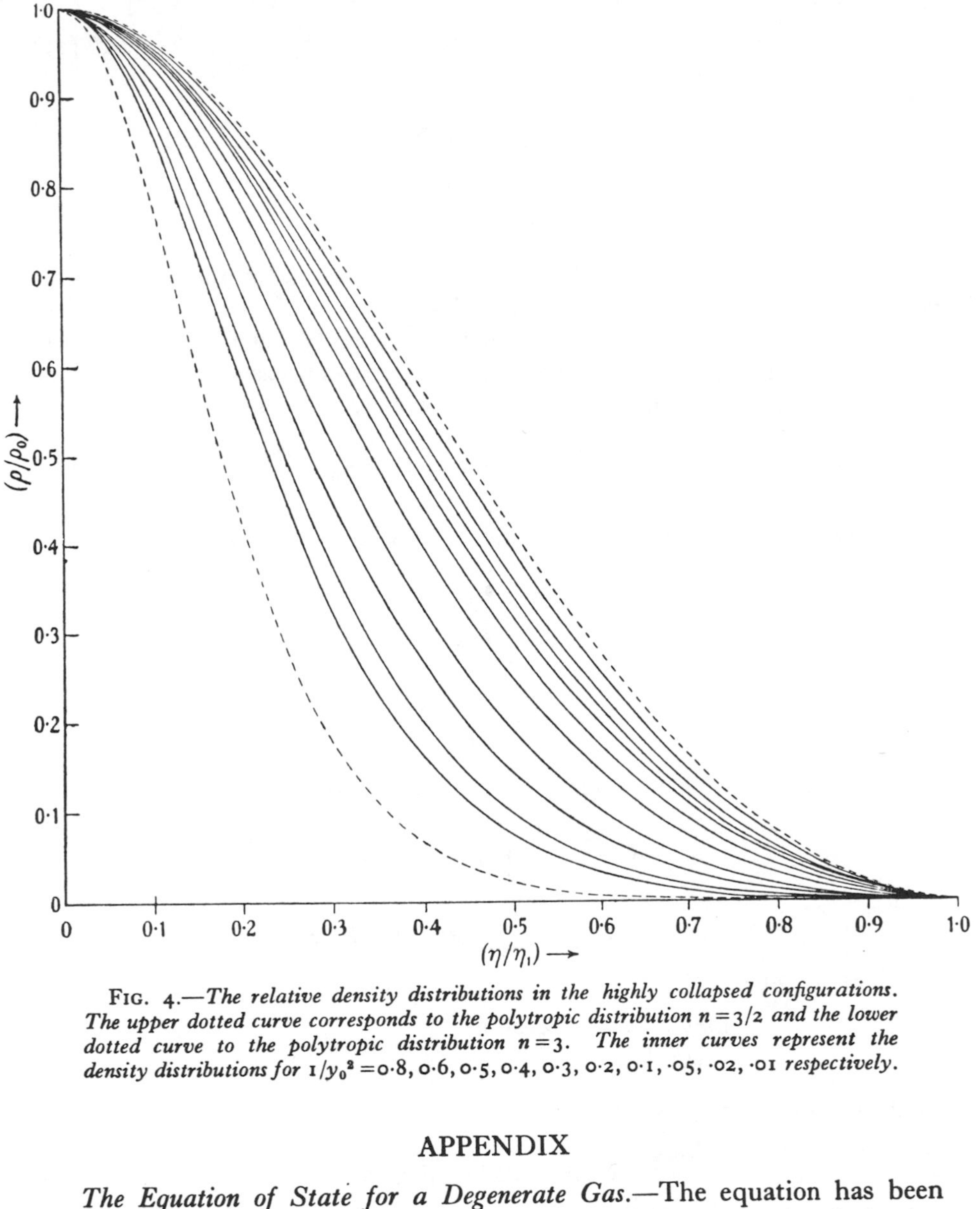

FIG. 4.—*The relative density distributions in the highly collapsed configurations. The upper dotted curve corresponds to the polytropic distribution* $n = 3/2$ *and the lower dotted curve to the polytropic distribution* $n = 3$. *The inner curves represent the density distributions for* $1/y_0^2 = 0·8, 0·6, 0·5, 0·4, 0·3, 0·2, 0·1, ·05, ·02, ·01$ *respectively.*

APPENDIX

The Equation of State for a Degenerate Gas.—The equation has been derived by Stoner (among others),† but we shall give a simpler derivation of the same.

In a completely degenerate electron assembly all the electrons have momenta less than a certain "threshold" value p_0, and in the region of the

* **Dr. Chandrasekhar's Tables can be consulted by Fellows on application to the Assistant Secretary (Editors).**

† *M.N.*, **92**, 444, 1931.

available phase space of volume $\frac{4}{3}\pi p_0^3 V$ every cell of volume h^3 contains just two electrons. Clearly then we have

$$n=\frac{8\pi}{h^3}\int_0^{p_0} p^2 dp, \tag{1}*$$

$$\mathfrak{E}=\frac{8\pi V}{h^3}\int_0^{p_0} E p^2 dp, \tag{2}$$

$$P=\frac{8\pi}{3h^3}\int_0^{p_0} p^3 \frac{dE}{dp} dp, \tag{3}$$

where n is the number of electrons per unit volume in the assembly of volume V, $\mathfrak{E}$ the total energy and E the kinetic energy of a free electron. We have now denoted the pressure by P instead of by "p" as in the text of the paper to avoid confusion with the momentum, which has to be denoted by "p." From (1) and from (2) and (3) we have respectively

$$p_0^3=\frac{3h^3 n}{8\pi}; \qquad P=\frac{8\pi}{3h^3}E(p_0)p_0^3-\frac{\mathfrak{E}}{V}. \tag{4}$$

Equations (1) to (4) are quite general. Now in the relativistic mechanics we have

$$E=mc^2\left\{\left(1+\frac{p^2}{m^2c^2}\right)^{1/2}-1\right\}, \tag{5}$$

or

$$p^2=\frac{E(E+2mc^2)}{c^2}. \tag{5'}$$

Using (5′) in (3) we have, after some minor transformations, that

$$P=\frac{8\pi m^4c^5}{3h^3}\int_0^{\theta_0} \sinh^4 \theta d\theta, \tag{6}$$

where

$$\sinh\theta=p/mc; \qquad \sinh\theta_0=p_0/mc. \tag{7}†$$

(7) yields at once that

$$P=\frac{8\pi m^4c^5}{3h^3}\left[\frac{\sinh^3\theta\cosh\theta}{4}-\frac{3}{16}\sinh 2\theta+\frac{3}{8}\theta\right]_{\theta=\theta_0}. \tag{8}$$

Writing x for (p_0/mc) we have

$$P=\frac{\pi m^4c^5}{3h^3}\left[x(2x^2-3)(x^2+1)^{1/2}+3\sinh^{-1}x\right], \tag{9}$$

* This equation follows directly from the expression for the number of waves associated with electrons whose energies lie between E and $E+dE$ given by Dirac (*P.R.S.*, **112**, 660, 1926, his unnumbered equation on p. 671). Actually Dirac obtains this result using the Klein-Gordon relativistic wave equation. That the same result would follow from Dirac's relativistic wave equation (on neglecting the states of kinetic energy—which is permissible when no external perturbations are present) is clear from J. von Neumann, *Z. f. Physik*, **48**, 868, 1928.

† θ here introduced will not be confused with the Emden function.

$$\rho = n\mu H = \frac{8\pi m^3 c^3 \mu H}{3h^3} x^3, \tag{10}$$

which are the equations quoted in the text. Our derivation now shows "why" we are able to reduce the differential equation for degenerate gas spheres in gravitational equilibrium to such a simple form. The "reason" is that we have such an elementary integral for P as in (6).

The function $f(x)$ on the right-hand side of (9) has the following asymptotic forms:—

$$f(x) \sim \tfrac{8}{5}x^5 - \tfrac{4}{7}x^7 + \tfrac{1}{3}x^9 - \tfrac{5}{22}x^{11} + \ldots \quad x \to 0, \tag{11}$$

$$f(x) \sim 2x^4 - 3x^2 + \ldots \quad x \to \infty. \tag{12}$$

Finally we notice that

$$\frac{f(x)}{2x^4} < 1 \quad \text{for all finite } x. \tag{13}$$

The inequality in (13) is a *strict* one. If only the first terms in the expansions (11) and (12) are retained, we can easily eliminate x from (9) and (10) for these limiting cases and obtain, as we should expect, that

$$P = K_1 \rho^{5/3} \quad (x \to 0); \qquad P = K_2 \rho^{4/3} \quad (x \to \infty), \tag{14}$$

with

$$K_1 = \frac{1}{20}\left(\frac{3}{\pi}\right)^{2/3} \frac{h^2}{m(\mu H)^{5/3}}; \qquad K_2 = \left(\frac{3}{\pi}\right)^{1/3} \frac{hc}{8(\mu H)^{4/3}}. \tag{15}^*$$

If we write our "equation of state" (9) and (10) parametrically as (changing to "p" to denote pressure),

$$p = A_2 f(x); \qquad \rho = Bx^3, \tag{16}$$

we find, on putting in the numerical values for the constants, that (in C.G.S. units)

$$A_2 = 6{\cdot}0406 \times 10^{22}; \qquad B = 9{\cdot}8848 \times 10^5 \mu, \tag{17}$$

or

$$\left.\begin{aligned} \log \rho &= 5{\cdot}9950 + 3 \log x + \log \mu, \\ \log p &= 22{\cdot}7811 + \log f(x) \end{aligned}\right\}. \tag{18}$$

Stoner has previously made some calculations concerning the (p, ρ) relation for a degenerate gas, but for the study in the following paper more accurate tables for $f(x)$ were needed. Accordingly the whole computation was re-

* The law $P = K_2 \rho^{4/3}$ was first used by the author in his paper on "Highly Collapsed Configurations," etc. (*M.N.*, **91**, 456, 1931). This law has also been derived by E. C. Stoner (*M.N.*, **92**, 444, 1932), T. E. Sterne (*M.N.*, **93**, 764, 1933), and is also implicitly contained in J. Frenkel (*Z. f. Physik*, **50**, 234, 1928). The law has also been used by L. Landau (*Physik. Zeits. d. Soviet Union*, **1**, 285, 1932). It may also be pointed out that the law $P = K_2 \rho^{4/3}$ is implicit in certain equations in a paper by F. Juttner (*Z. f. Physik*, **47**, 542, 1928, equations in §§ 13, 17; our equation (6) above is a limiting form of Juttner's integral $Q(\alpha, \gamma; +1)$). This last work of Juttner is related to his earlier work on the relativistic theory of an ideal classical gas, for a convenient summary of which see W. Pauli, *Relativitätstheorie* (Leipzig, Teubner), § 49.

done and the results are tabulated in Table V. I am indebted to Dr. Comrie and Mr. Sadler for the loan of a manuscript copy of a seven-figure table for $\sinh^{-1} x$, which was valuable in the computations of $f(x)$.

TABLE V

x	$f(x)$	$f(x)/2x^4$
0	0	0
0·2	0·000505	0·15785
0·4	0·015527	·30325
0·6	0·111126	·42873
0·8	0·435865	·53206
1·0	1·229907	·61495
1·2	2·82298	·68070
1·4	5·62991	·73276
1·6	10·14696	·77415
1·8	16·94969	·80731
2·0	26·69159	·83411
2·2	40·10347	·85598
2·4	57·99311	·87398
2·6	81·24509	·88894
2·8	110·8207	·90149
3·0	147·7578	·91209
3·5	279·8113	·93232
4·0	484·5644	·94641
4·5	784·5271	·95659
5·0	1205·2069	·96417
6·0	2525·739	·97444
7·0	4710·192	·98088
8·0	8070·587	·98518
9·0	$1·296694 \times 10^4$	·98818
10·0	$1·980725 \times 10^4$	·99036
20·0	$3·192093 \times 10^5$	·99753
30·0	$1·618212 \times 10^6$	·99890
40·0	$5·116812 \times 10^6$	·99938
50·0	$1·249501 \times 10^7$	·99960
60·0	$2·591280 \times 10^7$	·99972
70·0	$4·801018 \times 10^7$	·99980
80·0	$8·190727 \times 10^7$	·99984
90·0	$13·12039 \times 10^7$	·99988
100·0	$19·9980 \times 10^7$	·99990

Trinity College, Cambridge :
1935 *January* 1.

STELLAR CONFIGURATIONS WITH DEGENERATE CORES.

S. Chandrasekhar, Ph.D.

1. When Professor Milne began his investigations on stellar structure the following problem in specific relation to the standard model was in the forefront of his studies. On the hypothesis of a perfect gas * the mass M of the configuration is a single-valued function of β, defining the constant ratio of the gas to the total pressure in the configuration. The relation in question is of course Eddington's quartic equation. Call the appropriate β, β_M. *Has the star equilibrium configurations when* $\beta \neq \beta_M$*?*

Professor Milne himself supplied the first part of the answer. If, for a prescribed mass M, $(1-\beta)$ were greater than $(1-\beta_M)$, then the outer parts of the configuration must be described by *centrally condensed* † singularity possessing solutions of Emden's equation with index 3; on the other hand, if $(1-\beta)$ were less than $(1-\beta_M)$, then the outer parts must be described by *collapsed solutions* of Emden's equation.

It is of course clear that if we agree to describe the outer parts of a configuration by singularity possessing solutions of the differential equations involved, then we must assume that somewhere in the inner regions the perfect gas laws break down. Hence the possibility of the usefulness or otherwise of these centrally condensed and collapsed solutions of Emden's equation ($n=3$) depends essentially upon whether the physics of an ionised gas predicts marked deviations from the perfect gas law under any circumstances. Professor Milne therefore drew attention to the fact that such deviations were predicted by atomic physics, and further pointed out that according to a suggestion originally due to R. H. Fowler such deviations were realised in Nature under the observed white-dwarf conditions.

The above restatement of Milne's problem is of importance in our present discussion, and emphasises that the physics of the situation is the most important consideration. What is meant can be exemplified as follows. Suppose, for instance, that for some prescribed values of β the physical conditions are always of a character that the ideal gas laws do not break down. Then of course the question of using the singular solutions for such configurations does not arise. The general point to realise in this context is that we cannot infer from the existence of singular solutions of the differential equations involved that the gas laws must break down in the inner regions. It is thus more important to examine whether by following these singular solutions we do ever reach physical circumstances where consistent with our knowledge of the equations of state of an ionised gas we have, in fact,

* By *perfect gas* we shall always mean a gas ideal in the classical sense; *i.e.* the corresponding equation of state is $p=(k/\mu H)\rho T$.

† We assume that the reader is generally familiar with the arrangement of the solutions of Emden's differential equation. Otherwise reference should be made to the researches of R. H. Fowler and others.

marked deviations from the ideal gas laws. If there are no deviations at densities, however high (along these singular solutions), then we have simply to abandon the use of them for those particular configurations.

2. The above remarks elaborating Milne's original point of view are necessary, because there exists a large class of stellar configurations for which the use of the singular solutions has to be abandoned precisely for the reasons explained towards the end of the last paragraph. Thus it has already been shown by the author that for stellar configurations in which the radiation pressure continues to be always greater than about a tenth of the total pressure, degeneracy does not set in anywhere.* On the standard model this result has the consequence that for all stars of mass greater than a certain mass $\mathfrak{M}$, Milne's problem has the trivial answer that *there exists no equilibrium configuration which is characterised by a* $\beta \neq \beta_M$ *for* $M \geqslant \mathfrak{M}$, *since consistent with the physics of degenerate matter stars of mass greater than or equal to* $\mathfrak{M}$ *are necessarily wholly gaseous.* This circumstance, however, does not minimise the importance of Milne's problem, for it *has* a non-trivial solution when the configuration has a mass less than $\mathfrak{M}$, and the value of Milne's fundamental problem lies in this, that without his general formulation it could hardly have been possible to analyse the structure of stellar configurations of mass less than $\mathfrak{M}$.

In this paper a first attempt is made to solve Milne's problem consistent with the exact equation of state for degenerate matter. The analysis has been made possible by the derivation in the previous paper (referred to as II)† of the exact differential equation to describe degenerate matter in gravitational equilibrium. The present paper falls into two distinct parts. In the first part we develop certain consequences based on general principles. A systematic treatment of the composite configurations consisting of a gaseous envelope surrounding a degenerate core is undertaken in the second part of the paper. Towards the end the bearing of the results of these studies on the wider problems of stellar evolution is briefly commented upon.

I. *General Considerations*

3. *Equation of State of a Perfect Gas and Degeneracy Conditions.*— Consider material at density ρ and temperature T. The gas pressure according to the perfect gas law is

$$p = \left(\frac{k}{\mu H}\right)\rho T, \tag{1}$$

where k is the Boltzmann constant, μ the molecular weight and H the mass of the proton. At temperature T the radiation pressure p' is, by Stefan's law,

$$p' = \tfrac{1}{3}aT^4. \tag{2}$$

* *Zeit. für Astrophysik*, **5**, 321, 1932. The arguments in this paper do not make it sufficiently clear that the result is of a very general character. They are better stated in the author's article in the *Observatory*, **67**, 93, 1934—especially the remarks in the footnotes of p. 95.

† The earlier paper (*M.N.*, **91**, 456, 1931) will be referred to as I.

Let P denote the total pressure, and let the gas pressure p be a fraction β of the total pressure.* Then

$$P = p + p' = \frac{1}{\beta} p = \frac{1}{1-\beta} p'. \tag{3}$$

Eliminating T in (1) we have

$$p = \left[\left(\frac{k}{\mu H} \right)^4 \frac{3}{a} \frac{1-\beta}{\beta} \right]^{1/3} \rho^{4/3}. \tag{4}$$

Instead of (4) we shall introduce a parametric representation as in II, equations (4), (5). We set

$$\rho = Bx^3; \qquad B = \frac{8\pi m^3 c^3 \mu H}{3h^3}. \tag{5}$$

Substituting (5) in (4) we have

$$p = A_2 \left(\frac{512\pi k^4}{h^3 c^3 a} \frac{1-\beta}{\beta} \right)^{1/3} . \, 2x^4, \tag{6}$$

where A_2 as defined in II is given by

$$A_2 = \frac{\pi m^4 c^5}{3h^3}. \tag{7}$$

In (6) we substitute for a the theoretical expression

$$a = \frac{8}{15} \frac{\pi^5 k^4}{h^3 c^3}, \tag{8}$$

and obtain the simple expression

$$p = A_2 \left(\frac{960}{\pi^4} \frac{1-\beta}{\beta} \right)^{1/3} . \, 2x^4 = 2A_1 x^4 \quad \text{(say)}. \tag{9}$$

It has of course to be understood that equation (9) is merely another form for (1).

Now for material at density ρ and temperature T we can also formally calculate the pressure which would be given by the degenerate formula. With the same definition for ρ as in (6) we have

$$p_{\text{deg}} = A_2 f(x), \tag{10}$$

where as in II, equation (5),

$$f(x) = x(2x^2 - 3)(x^2 + 1)^{1/2} + 3 \sinh^{-1} x. \tag{11}$$

It is clear that degeneracy for a system with a specified β would set in if for some finite value for x the pressures given by the two formulæ (9) and (10) were equal, *i.e.* if the equation

$$\left(\frac{960}{\pi^4} \frac{1-\beta}{\beta} \right)^{1/3} . \, 2x^4 = f(x) \tag{12}$$

* No implication of β being a *constant* is made here. We are simply defining a parameter to describe material at ρ and T.

is soluble in x. If a solution for (12) exists, then for values of x much smaller than the root of (12) the pressure would be given by (9), and for values of x much larger than the root of (12) the pressure would be given by (10). The criterion for degeneracy then is the following: For a given ρ and T *formally* calculate on the perfect gas law the value for β (as we have done in equation (3)), and with this value for β seek a solution for (12); if a solution exists, and if the value of x at the specified density ρ is much greater than the root, then the system is degenerate; if a solution exists, and if the value of x at the density ρ is much smaller than the root, then the system is a perfect gas in the classical sense. If (12) has no solution for a prescribed β, then the system is *a fortiori* not degenerate. Now we know that (see Appendix to II)

$$\frac{f(x)}{2x^4} < 1 \quad \text{for all finite } x, \tag{13}$$

and

$$\frac{f(x)}{2x^4} \to 1 \quad \text{as} \quad x \to \infty \quad (i.e.\ \rho \to \infty). \tag{13'}$$

Hence if

$$\left(\frac{960}{\pi^4}\frac{1-\beta}{\beta}\right) \geqslant 1, \tag{14}$$

then the system at temperature T and density ρ is necessarily not degenerate. Let β_ω be such that the relation (14) is an equality, *i.e.*

$$\frac{1-\beta_\omega}{\beta_\omega} = \frac{\pi^4}{960} = 0{\cdot}1014678\ \ldots, \tag{15}$$

or

$$1-\beta_\omega = 0{\cdot}09212\ \ldots; \qquad \beta_\omega = 0{\cdot}90788.\ \ldots \tag{16}$$

If for material at density ρ and temperature T the fraction $(1-\beta)$ calculated by means of the equations (1), (2) and (3) is greater than $1-\beta_\omega$, then the system is definitely not degenerate.

For values of β greater than β_ω (12) admits of a solution in finite x, and this value of x would define a convenient measure as to at what densities degeneracy would set in for a system with this prescribed β. We shall adopt in these circumstances the following *approximation for the real equation of state of an ionized gas*:—

$$\left.\begin{aligned} p &= A_2 f(x), \qquad x \geqslant x', \\ \text{and} \qquad p &= 2A_1 x^4, \qquad x \leqslant x', \end{aligned}\right\} \tag{17}$$

x' being such that

$$\left(\frac{960}{\pi^4}\frac{1-\beta}{\beta}\right)^{1/3} = \frac{f(x')}{2x'^4}. \tag{18}$$

When the density is exactly Bx'^3 we shall say that degeneracy is "just beginning to develop."

In Table I the solutions of (18) for different x's are tabulated and graphically illustrated in fig. 1. From this graph we can directly read off the values of x below which we can regard the material as a perfect gas for a calculated β.

4. *The Equations for the Gaseous Envelope.*—These have been set up by various authors from Eddington onwards, but we shall briefly give the

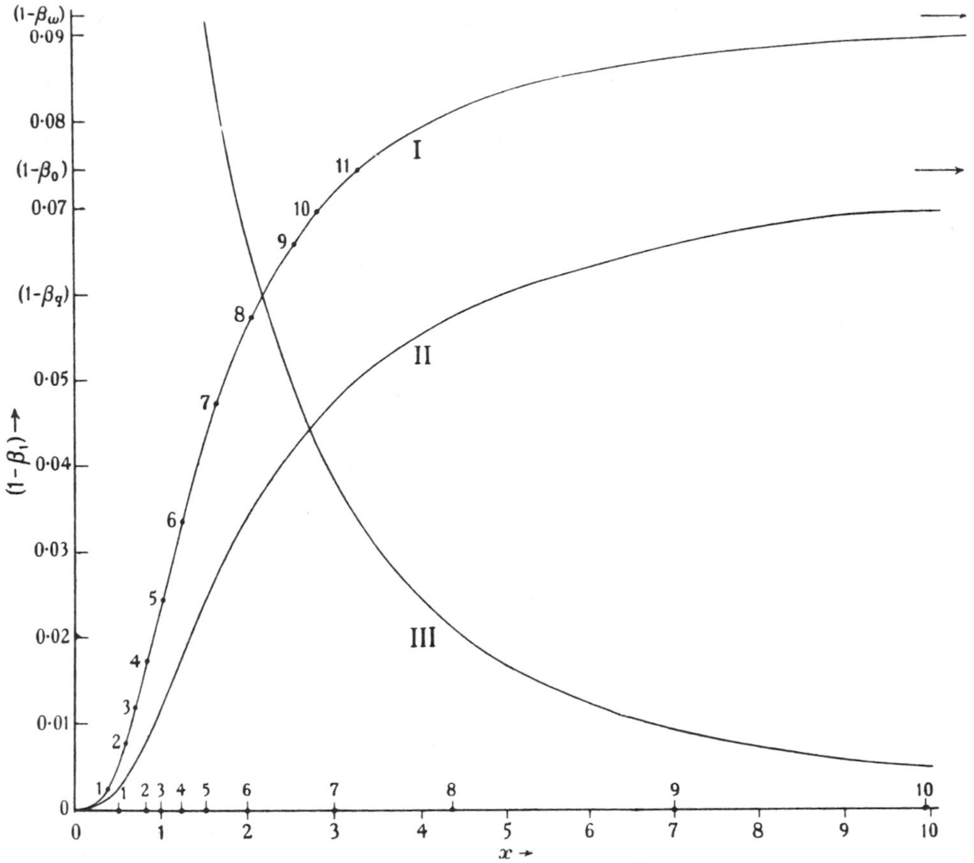

FIG. 1.—I. $(x_0, 1-\beta_1)$*-curve (see equations* (18) *and* (29)). *The points marked* (1 . . . 10) *on this curve and the x-axis are the end-points of the curves of constant mass in the domain of degeneracy for values of M in Table II.*

II. $(x_0, 1-\beta^\dagger)$*-curve (see equation* (44)).

III. $(x_0, 1-\beta_c)$*-curve (see equation* (150)).

The intersection of curves I and III defines $(1-\beta_q)$. *For* $(1-\beta_1) > (1-\beta_q)$ *we have centrally condensed configurations on the generalized standard model.*

main formulæ, as the particular form (9) for the equation of state of a perfect gas we are now using has not been used before. We shall confine ourselves to the standard model. In this model (as also in the "generalised model") for the gaseous portions the ratio $\beta_1 : (1-\beta_1)$ between the gas and the radiation pressure is a constant. Also in a well-known notation

$$\beta_1 = 1 - \frac{(\kappa\eta)_1 L}{4\pi c G M}. \tag{19}$$

The parametric representation for the equation of state in equations (5) and (9) can be used, since "β" is now a constant. The reduction to Emden's equation

$$\frac{1}{\xi^2}\frac{d}{d\xi}\left(\xi^2\frac{d\theta}{d\xi}\right) = -\theta^3, \tag{20}$$

proceeds in the usual way. In our notation the reduction is effected by the following substitutions :—

$$\left.\begin{aligned} x = \lambda_1\theta\,; \qquad r = \alpha_1\xi, \\ \alpha_1 = \left(\frac{2A_1}{\pi G\beta_1}\right)^{1/2}\frac{1}{B\lambda_1}. \end{aligned}\right\} \tag{21}$$

We find also that

$$M(\xi) = -4\pi\left(\frac{2A_1}{\pi G\beta_1}\right)^{3/2}\frac{1}{B^2}\xi^2\frac{d\theta}{d\xi}. \tag{22}$$

Finally it might be recalled that

$$A_1 = A_2\left(\frac{960}{\pi^4}\frac{1-\beta_1}{\beta_1}\right)^{1/3}. \tag{23}$$

5. *The Specification of the "Domain of Degeneracy" in the Milne Diagram.*—Equation (22) applied to the boundary gives

$$M = -4\pi\left(\frac{2A_1}{\pi G\beta_1}\right)^{3/2}\frac{1}{B^2}\left(\xi^2\frac{d\theta}{d\xi}\right)_1. \tag{24}$$

If the configuration is wholly gaseous, then we have

$$M = 4\pi\left(\frac{2A_2}{\pi G}\right)^{3/2}\frac{1}{B^2}\left(\frac{960}{\pi^4}\frac{1-\beta_1}{\beta_1{}^4}\right)^{1/2}\omega_3{}^0, \tag{25}^*$$

where

$$\omega_3{}^0 = -\left(\xi^2\frac{d\theta_3}{d\xi}\right)_1, \tag{26}$$

and θ_3 is now the *Emden function.* Equation (25) is of course Eddington's quartic equation (in a very different notation), and makes M a function of β_1 only and independent of the central density and hence the radius. In a diagram, therefore, in which we plot the radius R of the configuration against $(1-\beta_1)$ the curves of constant mass are lines parallel to the R axis. Milne's problem formulated in § 1 is to ask *whether the curves of constant mass in any part of this plane are distorted by the physical possibility of degeneracy at the centre.* As Milne was the first to use such a plot, I shall refer to any system of curves of constant mass in the $(R, 1-\beta_1)$ plane as a *Milne diagram.*†

Now for a given mass M equation (25) determines a β_1. Start with this mass having an infinite radius and imagine it being slowly contracted. At

* A simpler form for (25) is given in equation (55).

† In his paper in *M.N.*, **91**, 4, 1931, Professor Milne has used such a plot in his fig. 4 (p. 47) for the first time. However, his conjectural drawings here (and elsewhere) have to be very considerably modified on the basis of our present analysis.

first the configuration will be so rarefied that it will be wholly gaseous, and the path of the representative point in the $(R, 1-\beta_1)$ plane will be along the line parallel to the R axis through $\beta=\beta_1$. How far is this process of contraction possible? From our arguments in § 3 this is theoretically possible to an unlimited extent if $(1-\beta_1)>(1-\beta_\omega)$. Denote by $\mathfrak{M}$ the mass of the configuration which has a $\beta_1=\beta_\omega$. By (25)

$$\mathfrak{M}=4\pi\left(\frac{2A_2}{\pi G}\right)^{3/2}\frac{1}{B^2}\left(\frac{960}{\pi^4}\frac{1-\beta_\omega}{\beta_\omega^{\,4}}\right)^{1/2}\omega_3^{\,0}. \tag{27}$$

For configurations with mass greater than $\mathfrak{M}$ the appropriate $(1-\beta_1)$ is greater than $(1-\beta_\omega)$, and hence the "representative point" will travel down an "Eddington line" (by which we mean the line parallel to the R axis through $\beta_1(M)$ on the $(1-\beta_1)$ axis), however far the contraction may proceed. But the situation is different when the mass of the configuration is less than $\mathfrak{M}$. These have a "$(1-\beta_1)$" $<(1-\beta_\omega)$, and hence a stage must come when the configuration should begin to develop central regions of degeneracy. On the scheme of approximation (17) we can now easily see how far we can continue the contraction before degeneracy sets in.

Let the central density be ρ_0. Then

$$\rho_0=Bx_0^{\,3}. \tag{28}$$

Degeneracy would "just begin to develop" at the centre for a value of $x=x_0$ such that

$$\frac{f(x_0)}{2x_0^{\,4}}=\left(\frac{960}{\pi^4}\frac{1-\beta_1}{\beta_1}\right)^{1/3}. \tag{29}$$

For this configuration the mean density $\bar{\rho}$ is simply

$$\bar{\rho}=-3\left(\frac{1}{\xi}\frac{d\theta_3}{d\xi}\right)_1 Bx_0^{\,3}. \tag{30}$$

((30) is just the usual formula expressing the relation between the mean and the central density of a polytrope.) The radius R_0 of the configuration is therefore given by

$$\tfrac{4}{3}\pi R_0^{\,3}=\frac{\text{Mass}}{\text{Mean density}}. \tag{31}$$

Substituting in the above the expressions (25) and (29) we obtain

$$R_0=\left(\frac{2A_2}{\pi G}\right)^{1/2}\left(\frac{960}{\pi^4}\frac{1-\beta_1}{\beta_1^{\,4}}\right)^{1/6}\frac{1}{Bx_0}\xi_1(\theta_3), \tag{32}$$

where $\xi_1(=6{\cdot}897\ldots)$ defines the boundary of the Emden function θ_3.

Define a unit of length l by (see equation (19), II)

$$l=\left(\frac{2A_2}{\pi G}\right)^{1/2}\frac{\xi_1}{B}=\frac{7{\cdot}720\times 6{\cdot}897\times 10^8}{\mu},$$

or

$$l=5{\cdot}324\times 10^9\mu^{-1}. \tag{33}$$

From (32) then

$$\frac{R_0}{l} = \left(\frac{960}{\pi^4}\frac{1-\beta_1}{\beta_1^4}\right)^{1/6}\frac{1}{x_0}, \tag{34}$$

where x_0 is again determined from (29), using which we can rewrite (34) more conveniently as

$$\frac{R_0}{l} = \left(\frac{f(x_0)}{2x_0^4}\frac{1}{\beta_1}\right)^{1/2}\frac{1}{x_0}. \tag{35}$$

It is a fairly simple matter to calculate from (29) and (35) corresponding pairs of values for R_0 and β_1. These are tabulated in Table I. This $(R_0, 1-\beta_1)$ curve can therefore be drawn (see fig. 3). The region bounded by this curve and the two axes then defines the *domain of degeneracy*, meaning that it is only in this region that the Eddington lines are distorted.

TABLE I

x	$1-\beta_1$	R_0/l	$M/\mathfrak{M}$	x	$1-\beta_1$	R_0/l	$M/\mathfrak{M}$
0	0	∞	0	2·8	0·06919	0·3515	0·8245
0·2	0·00040	1·9868	0·0543	3·0	·07149	·3304	·8422
0·4	·00282	1·3787	·1451	3·5	·07598	·2870	·8767
0·6	·00793	1·0956	·2458	4·0	·07920	·2535	·9014
0·8	·01505	0·9187	·3435	4·5	·08158	·2268	·9195
1·0	·02305	·7934	·4320	5·0	·08337	·2051	·9332
1·2	·03101	·6985	·5093	6·0	·08583	·1721	·9520
1·4	·03839	·6235	·5754	7·0	·08739	·1481	·9639
1·6	·04495	·5627	·6313	8·0	·08844	·1299	·9719
1·8	·05068	·5123	·6784	9·0	·08918	·1157	·9776
2·0	·05561	·4699	·7180	10·0	·08972	·1043	·9817
2·2	·05983	·4337	·7515	20·0	·09150	·0524	·9953
2·4	·06344	·4025	·7798	30·0	·09185	·0350	·9979
2·6	·06653	·3753	·8039	∞	·09212	0	1

From (29) and (35) we have that, as $\beta_1 \to \beta_\omega$,

$$x_0 \to \infty, \qquad R_0 \to 0. \tag{36}$$

Hence, as we should expect, the $(R_0, 1-\beta_1)$ curve intersects the $(1-\beta_1)$ axis, where $\beta_1 = \beta_\omega$. One can further prove that the curve in fact cuts the $(1-\beta_1)$ axis *vertically*, but we omit the proof.

Again from (29) we see that as $\beta_1 \to 1$, $x_0 \to 0$. For small x_0 we have in fact (see equation (11) in the Appendix of the previous paper)

$$\frac{f(x_0)}{2x_0^4} \sim \frac{4}{5}x_0, \quad (x_0 \to 0). \tag{37}$$

Hence from (35) as $\beta_1 \to 1$,

$$\frac{R_0}{l} \sim \left(\frac{4}{5}\right)^{1/2}\frac{1}{x_0^{1/2}}. \tag{38}$$

6. Finally we notice here a "mass relation" which we subsequently need. Required the mass $M(x_0)$ of the configuration which intersects the $(R_0, 1-\beta_1)$ curve when the central density corresponds to some specified x_0. From (25) and (29) we have

$$M(x_0) = -4\pi\left(\frac{2A_2}{\pi G}\right)^{3/2}\frac{1}{B^2}\left(\frac{f(x_0)}{2x_0^4}\frac{1}{\beta_1}\right)^{3/2}\omega_3^0. \tag{39}$$

Comparing this with (27) we have

$$M(x_0) = \mathfrak{M}\left(\frac{f(x_0)}{2x_0^4}\frac{\beta_\omega}{\beta_1}\right)^{3/2}. \tag{40}$$

From (40) we again see that, as $\beta_1 \to \beta_\omega$,

$$x_0 \to \infty, \qquad \frac{f(x_0)}{2x_0^4} \to 1, \qquad M \to \mathfrak{M} \tag{41}$$

—as it should (see fig. 2, where $M(x_0)$ is plotted against x_0, and also the mass of a completely collapsed configuration against x_0, which corresponds to its central density).

7. *The Nature of the Curves of Constant Mass for $M \leqslant M_3$ in the Domain of Degeneracy.*—In § 6 we have shown at what stage a configuration of mass less than $\mathfrak{M}$ (contracting from infinite extension) begins to develop degeneracy at the centre. This happens when the appropriate Eddington line for the specified mass intersects the $(R_0, 1-\beta_1)$ curve. If the contraction continues further the configuration will begin to develop finite degenerate cores, and the major problem is to see how the curves of constant mass run inside the domain of degeneracy. To do this properly we should develop methods to treat composite configurations, but before we proceed to do this we can obtain some features from general principles.

Now in the last paper we have already made an analysis of *completely collapsed configurations*. Each mass (less than M_3) has a certain uniquely determined radius. Thus if the mass under consideration has a central density corresponding to $y = y_0$, then the radius R_1 is given

$$R_1 = \alpha\eta_1 = \left(\frac{2A_2}{\pi G}\right)^{1/2}\frac{\eta_1}{By_0}, \tag{42}$$

where η_1 is the boundary of the corresponding function ϕ satisfying the differential equation (16) II. In terms of the unit of length l (equation (33)) we have

$$\frac{R_1}{l} = \frac{1}{y_0}\frac{\eta_1(\phi(y_0))}{\xi_1(\theta_3)}. \tag{43}$$

These completely collapsed configurations correspond to $\beta_1 = 1$. Hence we know from (43) the point at which the curves of constant mass for $M \leqslant M_3$ must intersect the R axis. Also for any mass we can calculate the value β_1 has in the wholly gaseous state. From equation (29) II and equation (25) we have the relation

$$\left(\frac{960}{\pi^4}\frac{1-\beta^\dagger}{\beta^{\dagger 4}}\right)^{1/2} = \frac{\Omega(y_0)}{\omega_3^0} = \frac{M}{M_3}, \tag{44}$$

where as in II, equation (60),

$$\Omega(y_0) = -\eta^2\left(\frac{d\phi}{d\eta}\right)_1, \tag{45}$$

and $\beta^\dagger$ is the value of β_1 for a wholly gaseous configuration, which in its completely collapsed state has a central density corresponding to $y = y_0$.

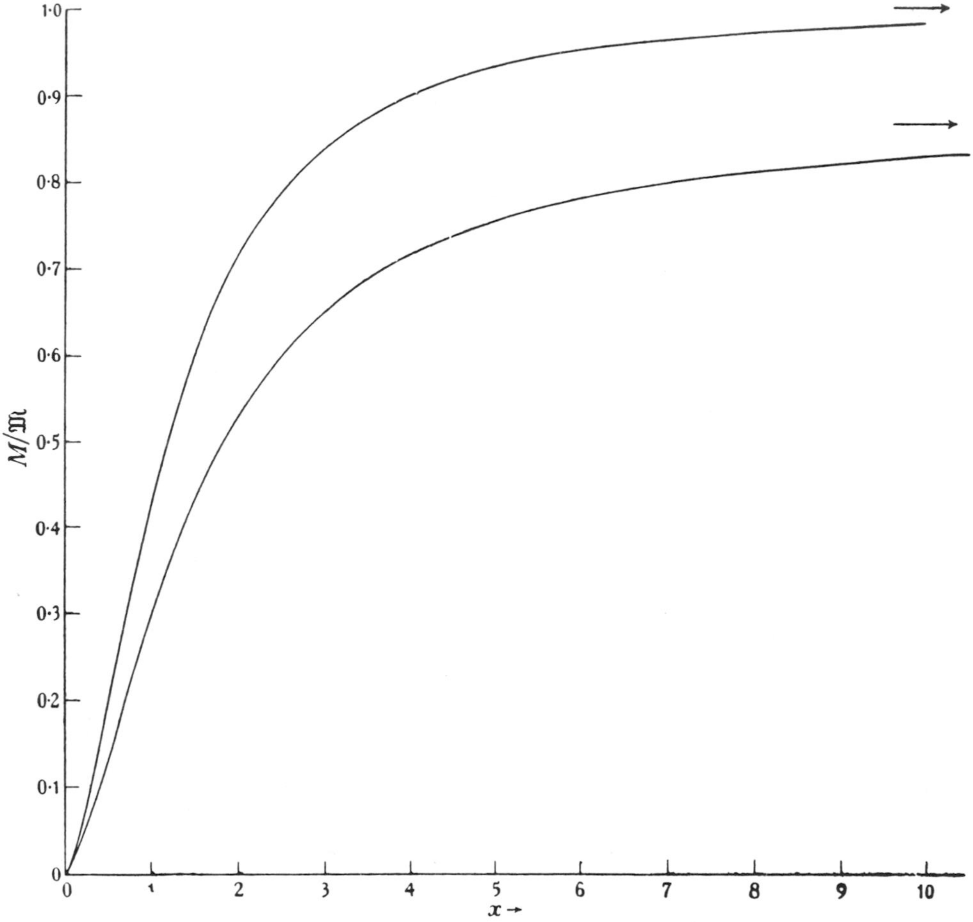

FIG. 2.—*The upper curve illustrates the* (M, x_0)*-relation for configurations in which degeneracy is just developing at the centre. The lower curve illustrates the* (M, x_0)*-relation for completely collapsed configurations.*

Now the line through $\beta^\dagger$ parallel to the R axis will intersect the $(R_0, 1-\beta_1)$ curve at $(R_0(M(y_0)), 1-\beta^\dagger)$. In the domain of degeneracy the continuation of the curve must in some way connect the point $(R_0(M(y_0)), 1-\beta^\dagger)$ and the point R_1 on the R axis, where

$$R_1 = \frac{1}{y_0(M)} \cdot \frac{\eta_1(\phi(y_0(M)))}{\xi_1(\theta_3)}. \tag{46}$$

In the previous paper we have obtained the numerical values for η_1, Ω, etc., for ten different values of y_0, and for the corresponding configurations the values of R_1 and $(1-\beta^\dagger)$ can be numerically evaluated. The results are given in Table II. In fig. 1 the curve $(x_0, 1-\beta^\dagger)$ is drawn. From this curve we can directly read off the value of the central density $(=Bx_0^3)$ in the completely collapsed state for a configuration which in the wholly gaseous state has a given value for $(1-\beta_1)$.

TABLE II

$\frac{1}{y_0^2}$	M/M_3	$(1-\beta^\dagger)\times 10$	R_1/l
0	1	0·7446	0
·01	0·95733	0·6966	·07767
·02	0·92419	0·6596	·10223
·05	0·84709	0·5746	·14460
·1	0·75243	0·4732	·18657
·2	0·61589	0·3358	·24168
·3	0·51218	0·2414	·28434
·4	0·42600	0·1718	·32320
·5	0·35033	0·1187	·36222
·6	0·28137	0·0779	·40475
·8	0·15316	0·0236	·52453
1·0	0	0	∞

We have thus fixed the "end-points" for the curves of constant mass with $M \leqslant M_3$ in the domain of degeneracy. The corresponding pairs of points on the $(R_0, 1-\beta_1)$ curve and the R axis are shown in figs. 1 and 3 (the points marked 5 to 15 on the R axis and also on the $(R_0, 1-\beta_1)$ curve). The corresponding points are also marked in fig. 1 on the $(x_0, 1-\beta_1)$ curve and the x axis. Finally we may notice that from (44) and (34) we have

$$\frac{R_0}{l}=\left(\frac{\Omega(y_0)}{\omega_3^0}\right)^{1/3}\frac{1}{x_0(\beta^\dagger)}=\left(\frac{M}{M_3}\right)^{1/3}\frac{1}{x_0(\beta^\dagger)}. \tag{47}$$

From (47) and (38) we now have on eliminating x_0 that

$$\frac{R_0}{l}\sim 0{\cdot}8\left(\frac{M_3}{M}\right)^{1/3}, \quad (x_0\to 0,\ \beta_1\to 1). \tag{48}$$

We have already obtained in II, § 13 (equation (80)), a similar relation for the radii of completely collapsed configurations of small mass. In our present units we have

$$\frac{R_1}{l}\sim 0{\cdot}29238\left(\frac{M_3}{M}\right)^{1/3}, \quad (x_0\to 0). \tag{49}$$

For a given $M(<<M_3)$ we have

$$\frac{R_0}{R_1}\to 2{\cdot}7362, \quad (M\to 0). \tag{50}$$

Thus for small masses ("small" when expressed in units of M_3) there is a contraction by a factor of about 3 the configuration in passing from a state in which degeneracy is just beginning to develop at the centre to the completely collapsed state. In the gaseous state the central density is about 54 times the mean density, while in the completely collapsed state the central density is only 6 times the mean density. The net result, however, is an increase in the central density by about a factor 2 (more exactly 2·2650) in passing to the completely collapsed state. These results are only approximately true even for a star of mass $=0\cdot 1 M_3$.

8. *Some Relations for the Mass M_3.*—From the arguments in § 7 it is clear that the curve of constant mass for $M=M_3$ passes through the origin of our system of axes. We shall denote by β_0 the value of β_1 which M_3 has in the wholly gaseous state. From (44) we have

$$\frac{960}{\pi^4}\frac{1-\beta_0}{\beta_0{}^4}=1. \tag{51}$$

The numerical solution of (51) is found to be

$$1-\beta_0=0\cdot07446\,; \qquad \beta_0=0\cdot92554. \tag{51'}$$

The asymptote to the $(x_0, 1-\beta^\dagger)$ curve drawn in fig. 1 is the line parallel to the x axis through $(1-\beta_0)$. From (40) and (51) we have the following relation between M_3 and $\mathfrak{M}$:—

$$M_3=\mathfrak{M}\beta_\omega{}^{3/2}=0\cdot86505\mathfrak{M}. \tag{52)*$$

9. *The Nature of the Curves of Constant Mass for $M>M_3$ in the Domain of Degeneracy.*—In §§ 7, 8 we have fixed the "end-points" for the curves of constant mass for configurations with mass less than or equal to M_3, and saw further that the appropriate curve for M_3 must pass through the origin. The question now arises, What happens for configurations with $\mathfrak{M} \geqslant M > M_3$? The answer to this question can be given if in the composite configurations "$\kappa\eta$" has the same value in the core as in the gaseous envelope outside. For it was proved in I that on the standard model the completely relativistic configuration has a mass

$$M=M_3\beta^{-3/2}, \tag{53}$$

and is of zero radius (*cf.* I, p. 463; the result is restated in II, § 1). Hence *the curves of constant mass for $M>M_3$ must cross the* $(1-\beta_1)$ *axis at a point* $(1-\beta^*)$, *say, such that*

$$M=M_3\beta^{*-3/2}. \tag{54}$$

Let us denote by $\beta^\dagger$ the value of β_1 in the wholly gaseous state. There is a simple relation between $\beta^\dagger$ and β^*. To establish this we first notice that equation (25) can be rewritten in the form

$$M=M_3\left(\frac{960}{\pi^4}\frac{1-\beta^\dagger}{\beta^{\dagger 4}}\right)^{1/2}. \tag{55}$$

* This relation without proof was given in the author's preliminary note in *The Observatory*, 57, 373, 1934.

Equating (54) and (55) we have the relation

$$\beta^* = \left(\frac{\pi^4}{960}\frac{\beta^{\dagger 4}}{1-\beta^\dagger}\right)^{1/3}. \tag{56}$$

We will derive (56) again from our analysis of composite configurations, but we now verify that the relation (56) is precisely what would make our scheme consistent. Thus when $\beta^\dagger = \beta_0$, *i.e.* when $M = M_3$, we have from (50) that $\beta^* = 1$; in other words, the appropriate curve for M_3 must pass through the origin, which in fact it does. Again, when $\beta^\dagger = \beta_\omega$, then (50) yields that $\beta^* = \beta_\omega$; the appropriate curve for $\mathfrak{M}$ is therefore the full line through $(1-\beta_\omega)$ parallel to the R axis, as we should have expected.

The following table gives a set of corresponding pairs of values for β^* and $\beta^\dagger$ (see also fig. 3, where the corresponding pairs of points on the $(R_0, 1-\beta_1)$ curve and the $(1-\beta_1)$ axis are marked 1, 2, 3, 4) :—

TABLE III

$1-\beta^\dagger$	$1-\beta^*$	$M/\mathfrak{M}$.
0·09212	0·09212	1
0·090	0·08220	0·9838
0·085	0·05768	0·9457
0·080	0·03143	0·9075
0·075	0·00319	0·8692
0·07446	0	0·8651

The above results are of course true only on the usual standard model. The question as to what happens on the *generalised standard model* for configurations of mass greater than M_3 cannot be satisfactorily answered without a proper treatment of composite configurations, to which we proceed now.

II. *The Analysis of Composite Configurations*

In the treatment of these composite configurations on the generalised standard model we shall adopt with suitable modifications the methods which have been developed by Milne.* The main results are summarised in § 21.

10. On the generalised standard model we have the pressure integral (Milne, *loc. cit.*, equation (7))

$$p = \tfrac{1}{3}aT^4\left(\frac{4\pi cGM}{\kappa\eta L} - 1\right) + D, \tag{57}$$

in any interval in which "$\kappa\eta$" is constant.† D changes discontinuously

* E. A. Milne, *M.N.*, **92**, 610, 1932 (referred to as *loc. cit.*). See also T. G. Cowling, *M.N.*, **91**, 472, 1931.

† The "η" used here will not be confused with our variable η, defining the radius vector. "$\kappa\eta$" (in this combination) occurs only in §§ 10, 11.

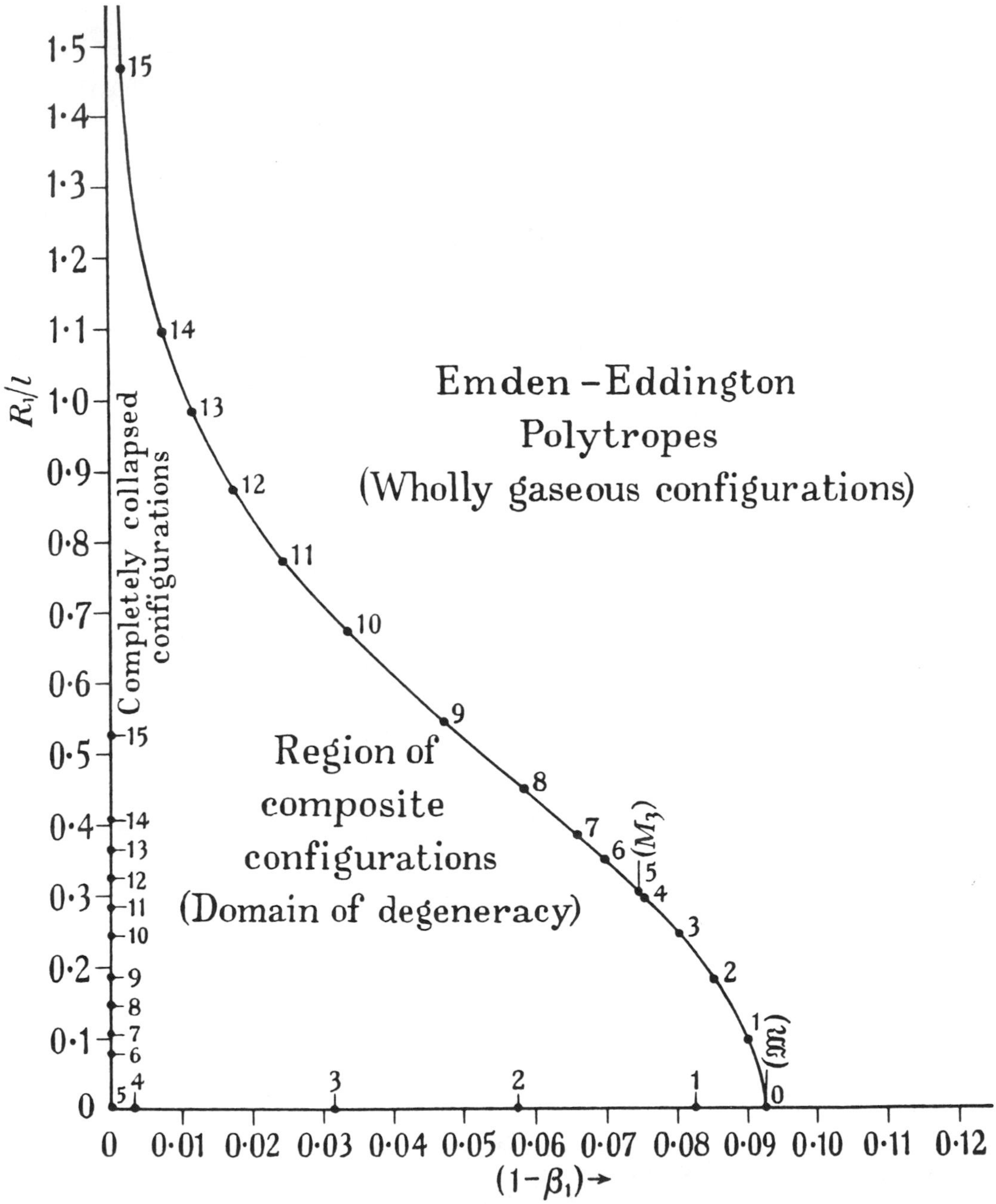

FIG. 3.—*The curve running from* $1-\beta_1 = \cdot 092$ . . . *to infinity along the* R_1 *axis is the* (R_0, $1-\beta_1$)*-curve (see equation* (35)). *The points marked* (5 . . . 15) *on the* (R_0, $1-\beta_1$)*-curve and the* R_1 *axis are the end-points (in the domain of degeneracy) for the curves of constant mass for those values of* M *tabulated in Table II. The points marked* (1 . . . 4) *on the* (R_0, $1-\beta_1$)*-curve and on the* ($1-\beta_1$)*-axis are the corresponding end-points for some curves of constant mass in the domain of degeneracy on the usual standard model* ($\beta_1 = \beta_2$). (*See Table III.*)

from one constant value to another when "$\kappa\eta$" changes. We shall assume that "$\kappa\eta$" assumes different values in the degenerate core and the gaseous envelope.

Let D_1 be the value of D in the gaseous envelope and D_2 in the degenerate core. In the gaseous envelope the usual external boundary condition $T=0$ when $\rho=0$ leads to D_1 being zero. From (57), then in the gaseous envelope we have the constancy of the ratio of the gas to the radiation pressure. The reduction to Emden's equation follows in the usual way. The relevant equations in our notation have already been given in § 4, equations (21) and (22).

11. *Interfacial Conditions.*—Let the interface occur at a density $\rho=\rho'$ and a temperature $T=T'$ such that at this density and temperature the pressures given by the degenerate and the perfect gas formulæ are the same. Hence if $x=x'$ at the interface, then

$$\frac{f(x')}{2x'^4}=\left(\frac{960}{\pi^4}\frac{1-\beta_1}{\beta_1}\right)^{1/3}, \tag{58}$$

where as in § 4

$$\beta_1=\left(1-\frac{(\kappa\eta)_1 L}{4\pi cGM}\right). \tag{59}$$

Again from (57) we have

$$\frac{1-\beta_1}{\beta_1}p'=\frac{1-\beta_2}{\beta_2}p'-D_2\frac{1-\beta_2}{\beta_2}, \tag{60}$$

where

$$\beta_2=\left(1-\frac{(\kappa\eta)_2 L}{4\pi cGM}\right), \tag{61}$$

and D_2 is the value of "D" in the degenerate core, which must be non-zero if $(\kappa\eta)_2\neq(\kappa\eta)_1$. In (60) we have used p' to denote the gas pressure at the interface. Solving (60) we have

$$D_2=\frac{(\beta_1-\beta_2)}{\beta_1(1-\beta_2)}p'. \tag{62}$$

This value of D_2 ensures the continuity of T at the interface and we need no longer consider the temperature.

12. *The Equations for the Degenerate Core.*—From (57) we now have

$$P=p+p_r=\frac{1}{\beta_2}A_2 f(x)-D_2\frac{1-\beta_2}{\beta_2}, \tag{63}$$

where D_2 is defined by (62), A_2 and $f(x)$ having the same meanings as hitherto.

The reduction to our differential equation (16), II, follows at once. We introduce the variables

$$r=\alpha_2\eta\,; \qquad y=(x^2+1)^{1/2}=y_0\phi, \tag{64}$$

where

$$\left.\begin{aligned}\alpha_2&=\left(\frac{2A_2}{\pi G\beta_2}\right)^{1/2}\frac{1}{By_0},\\ y_0^2&=x_0^2+1,\end{aligned}\right\} \tag{65}$$

and find as before that ϕ satisfies the differential equation

$$\frac{1}{\eta^2}\frac{d}{d\eta}\left(\eta^2\frac{d\phi}{d\eta}\right) = -\left(\phi^2 - \frac{1}{y_0^2}\right)^{3/2}. \tag{66}$$

(It has to be especially noticed that if the mass of a composite configuration is specified then y_0 is one of the quantities that has to be determined and has therefore to be regarded as an unknown.) We have the mass relation

$$M(\eta) = -4\pi\left(\frac{2A_2}{\pi G\beta_2}\right)^{3/2}\frac{1}{B^2}\eta^2\frac{d\phi}{d\eta}. \tag{67}$$

13. *The Equations of Fit.*—Let the interface occur at $\xi = \xi'$ and $\eta = \eta'$. We then have two sets of equations for the absolute value of the radius at which degeneracy sets in (*i.e.* at the interface), and also for the mass of the degenerate core :

$$\left.\begin{aligned} x' &= y_0\left(\phi^2(\eta') - \frac{1}{y_0^2}\right)^{1/2}, \\ r' &= \left(\frac{2A_2}{\pi G\beta_2}\right)^{1/2}\frac{1}{By_0}\eta', \\ M(r') &= -4\pi\left(\frac{2A_2}{\pi G\beta_2}\right)^{3/2}\frac{1}{B^2}\left(\eta^2\frac{d\phi}{d\eta}\right)_{\eta=\eta'}, \end{aligned}\right\} \tag{68}$$

and

$$\left.\begin{aligned} x' &= \lambda_1\theta(\xi'), \\ r' &= \left(\frac{2A_1}{\pi G\beta_1}\right)^{1/2}\frac{1}{B\lambda_1}\xi', \\ M(r') &= -4\pi\left(\frac{2A_1}{\pi G\beta_1}\right)^{3/2}\frac{1}{B^2}\left(\xi^2\frac{d\theta}{d\xi}\right)_{\xi=\xi'}, \end{aligned}\right\} \tag{69}$$

where x' is such that

$$\frac{f(x')}{2x'^4} = \left(\frac{960}{\pi^4}\frac{1-\beta_1}{\beta_1}\right)^{1/3} = \frac{A_1}{A_2}. \tag{70}$$

Equating the different expressions for x', r' and $M(r')$ we have

$$\lambda_1\theta(\xi') = y_0\left(\phi^2(\eta') - \frac{1}{y_0^2}\right)^{1/2}, \tag{71}$$

$$\left(\frac{A_1\beta_2}{A_2\beta_1}\right)^{1/2}\frac{\xi'}{\lambda_1} = \frac{\eta'}{y_0}, \tag{72}$$

$$\left(\frac{A_1\beta_2}{A_2\beta_1}\right)^{3/2}\left(\xi^2\frac{d\theta}{d\xi}\right)_{\xi=\xi'} = \left(\eta^2\frac{d\phi}{d\eta}\right)_{\eta=\eta'}. \tag{73}$$

From (71) and (72) we obtain

$$\left(\frac{A_1\beta_2}{A_2\beta_1}\right)^{1/2}\xi'\theta(\xi') = \eta'\left(\phi^2(\eta') - \frac{1}{y_0^2}\right)^{1/2}. \tag{74}$$

We can eliminate the "homology constant" λ_1 of the Emden equation

16

completely from the discussion by employing an argument due to Milne (*loc. cit.*, p. 617):—

If $g(\xi)$ is a solution of Emden's equation ($n=3$) vanishing at $\xi=\xi_0$, then $\theta=Cg(C\xi)$ is also a solution which vanishes at ξ_1, where $C\xi_1=\xi_0$. Put $C\xi'=a$. Then

$$\left(\xi^2\frac{d\theta}{d\xi}\right)_{\xi=\xi'}=a^2g'(a)\,;\qquad \xi'\theta(\xi')=ag(a). \tag{75}$$

(A prime to the g's denotes the derivative.) Substituting these in (73) and (74) we have

$$\left(\frac{A_1\beta_2}{A_2\beta_1}\right)^{1/2}ag(a)=\eta'\left(\phi^2(\eta')-\frac{1}{y_0^{\,2}}\right)^{1/2}, \tag{76}$$

$$\left(\frac{A_1\beta_2}{A_2\beta_1}\right)^{3/2}a^2g'(a)=\left(\eta^2\frac{d\phi}{d\eta}\right)_{\eta=\eta'}. \tag{77}$$

From the above we obtain the more convenient set:

$$\frac{a[g(a)]^3}{g'(a)}=\frac{\eta'\left(\phi^2(\eta')-\dfrac{1}{y_0^{\,2}}\right)^{3/2}}{\phi'(\eta')}, \tag{78}$$

$$\frac{ag'(a)}{g(a)}=\left(\frac{A_2\beta_1}{A_1\beta_2}\right)\frac{\eta'\phi'(\eta')}{\left(\phi^2(\eta')-\dfrac{1}{y_0^{\,2}}\right)^{1/2}}. \tag{79}$$

The above two equations combined with the following two,

$$x'=y_0\left(\phi^2(\eta')-\frac{1}{y_0^{\,2}}\right)^{1/2}, \tag{80}$$

$$\frac{f(x')}{2x'^4}=\left(\frac{960}{\pi^4}\frac{1-\beta_1}{\beta_1}\right)^{1/3}=\frac{A_1}{A_2}, \tag{81}$$

form our complete system of the equations of fit. In his investigations on the "fitting" of one polytropic distribution to another Milne was able to reduce all the equations of fit to just a pair of equations. The circumstance which gave rise to this simplification is the fact that polytropic distributions admit of a constant of "homology" which can always be eliminated (just as we have eliminated λ_1). But our differential equation for ϕ does not admit of a constant of homology,* and we cannot (in principle) simplify the equations (78) to (81) further. We shall presently discuss these equations of fit and outline methods for solving them, but we shall first tabulate the formulæ for the physical characteristics of these configurations.

14. *The Physical Characteristics of the Configurations.*

Radius of the Core—

$$r'=\left(\frac{2A_2}{\pi G}\right)^{1/2}\frac{1}{Bx'}\left(\frac{960}{\pi^4}\frac{1-\beta_1}{\beta_1^{\,4}}\right)^{1/6}ag(a). \tag{82}$$

* We have drawn attention to this circumstance in II, § 10.

Mass of the Core—

$$\frac{M(r')}{M(r)} = \frac{a^2 g'(a)}{\xi_0^2 g'(\xi_0)}. \tag{83}$$

Mean Density of the Core—

$$\rho_m\,(\text{core}) = -3Bx'^3 \frac{a^2 g'(a)}{[ag(a)]^3}. \tag{84}$$

(Mean Density of Core)/(Interfacial Density)—

$$\frac{\rho_m\,(\text{core})}{\rho'} = -3\frac{a^2 g'(a)}{[ag(a)]^3}. \tag{85}$$

Radius of the Configuration—

$$R_1 = \left(\frac{2A_2}{\pi G}\right)^{1/2} \frac{1}{Bx'} \left(\frac{960}{\pi^4} \frac{1-\beta_1}{\beta_1^4}\right)^{1/6} \xi_0 g(a). \tag{86}$$

Mean Density of the Configuration—

$$\rho_m = -3Bx'^3 \frac{\xi_0^2 g'(\xi_0)}{[\xi_0 g(a)]^3}. \tag{87}$$

Effective Temperature—

$$T_e = \left(\frac{L}{ac}\right)^{1/4} \left(\frac{G}{2A_2}\right)^{1/4} \left(\frac{\pi^4}{960} \frac{\beta_1^4}{1-\beta_1}\right)^{1/12} \frac{B^{1/2} x'^{1/2}}{[\xi_0 g(a)]^{1/2}}. \tag{88}$$

Central Density—

$$\rho_0 = B(y_0^2 - 1)^{1/2}. \tag{89}$$

Some of the preceding formulæ become indeterminate when the relative core radius tends to unity (these configurations are not necessarily completely collapsed), but they can be transformed to determinate forms by using the following relation easily obtained from the equations of fit :—

$$g(a) = \left(\frac{A_2 \beta_1}{A_1 \beta_2}\right) \frac{\eta'}{a} \left[\frac{1}{a^2 g'(a)} \left(\eta^2 \frac{d\phi}{d\eta}\right)_{\eta=\eta'}\right]^{1/3} \left(\phi^2(\eta') - \frac{1}{y_0^2}\right)^{1/2}. \tag{79'}$$

The physical variables may then be expressed in the following alternative forms :—

$$r' = \left(\frac{2A_2}{\pi G \beta_2}\right)^{1/2} \frac{1}{By_0} \eta'. \tag{82'}$$

$$R_1 = \left(\frac{2A_2}{\pi G \beta_2}\right)^{1/2} \frac{1}{By_0} \left(\frac{\xi_0}{a}\right) \eta'. \tag{84'}$$

$$\rho_m = 3By_0^3 \frac{M}{M_3} \left(\frac{\beta_2}{\beta_1}\right)^{3/2} \left(\frac{a}{\xi_0}\right)^3 \omega_3^0 \cdot \frac{1}{\eta'^3}. \tag{87'}$$

$$T_e = \left(\frac{L}{ac}\right)^{1/4} \left(\frac{G\beta_2}{2A_2}\right)^{1/4} \frac{B^{1/2} y_0^{1/2}}{\eta'^{1/2}} \left(\frac{a}{\xi_0}\right)^{1/2}. \tag{88'}$$

In all the above formulæ we have (following Milne) used only such functions as ξ/ξ_0, $\xi g(\xi)$, $-\xi^2 g'(\xi)$, so as to be independent of the "normalisation" to the zero ξ_0.

15. *The General Method to Solve the Equations of Fit.*—Following Milne we define the auxiliary functions $u(\xi)$, $v(\xi)$ associated with a solution of Emden's equation ($n=3$) as follows :—

$$u(\xi) = -\frac{\xi\theta^3}{\theta'}; \qquad v(\xi) = -\frac{\xi\theta'}{\theta}. \tag{90}$$

We shall in a similar way define two auxiliary functions $U(\eta)$, $V(\eta)$ associated with our function ϕ as follows :—

$$U(\eta) = -\frac{\eta\left(\phi^2 - \frac{1}{y_0{}^2}\right)^{3/2}}{\phi'}; \qquad V(\eta) = -\frac{\eta\phi'}{\left(\phi^2 - \frac{1}{y_0{}^2}\right)^{1/2}}. \tag{91}$$

The equations of fit then take the form :

$$u(a) = U(\eta'), \tag{92}$$

$$v(a) = \left(\frac{\pi^4}{960}\frac{\beta_1}{1-\beta_1}\right)^{1/3}\left(\frac{\beta_1}{\beta_2}\right)V(\eta'), \tag{92'}$$

$$\frac{f(x')}{2x'^4} = \left(\frac{960}{\pi^4}\frac{1-\beta_1}{\beta_1}\right)^{1/3}, \tag{93}$$

$$x' = y_0\left(\phi^2(\eta') - \frac{1}{y_0{}^2}\right)^{1/2}. \tag{94}$$

The functions U and V have been tabulated by the author, and Fairclough has tabulated u and v for some solutions of Emden's equation. The general procedure, then, to solve the equations of fit would be the following :—

Consider a sequence of configurations having the same initially prescribed central density, and therefore having a given prescribed value for $y_0{}^2$. This means that for this sequence of configurations the degenerate cores are all governed by the same function ϕ. Since the central density is known, and since further the interfacial density must be less than this, it follows that a sequence of configurations having the same central density must be characterised by values of $(1-\beta_1)$ less than a certain critical value $(1-\beta_1(y_0))$, say, such that

$$\frac{f(x_0)}{2x_0{}^4} = \left(\frac{960}{\pi^4}\frac{1-\beta_1(y_0)}{\beta_1(y_0)}\right)^{1/3}, \quad (x_0{}^2 = y_0{}^2 + 1) \tag{95}$$

(we shall sometimes use $(1-\beta_1(x_0))$ instead of $(1-\beta_1(y_0))$).

Now choose a value of β_1 greater than $\beta_1(y_0)$. Equation (93) can be solved for x'. This value of x' would determine the interfacial value of $\eta=\eta'$ from equation (94), and this value of η' would in turn determine $U(\eta')$ and $V(\eta')$. Equations (92) and (92′) finally determine the values of $u(a)$ and $v(a)$. The last stage in the solution of the equations of fit is to determine $\omega_3(=-\xi^2\theta')$ such that along this solution for some value of ξ, u and v have the values already determined. The solution will be *uniquely* determined in this way since u and v are one parameter families of curves (the parameter in fact being ω_3), and the problem simply reduces to one of finding the particular (u, v) curve which passes through a given point in that plane. Once the appropriate (u, v) curve has been determined, the value of "a" is determined at the same time from the "ladder of points" on the (u, v) curve labelling the value of ξ/ξ_0. The value of ω_3 determines also the mass of the configuration, which is given by (*cf.* equation (55))

$$M = M_3\left(\frac{960}{\pi^4}\,\frac{1-\beta_1}{\beta_1{}^4}\right)^{1/2}\left(\frac{\omega_3}{\omega_3{}^0}\right). \tag{96}$$

The physical characteristics now follow from the equations of § 14.

Having constructed in the above way sequences of composite configurations, each sequence being characterised by a different particular value for the central density (but constant along each sequence), we can construct a system of (*mass*, $1-\beta_1$) and (*mass*, *radius*) curves, a pair of curves for each sequence. From these curves (or more accurately by using methods of interpolation) we can then construct the (*mass*, *radius*) curves for different assigned values for $(1-\beta_1)$. From this, the final step of drawing the curves of constant mass in the $(R, 1-\beta)$ diagram follows at once.

The number of solutions of Emden's equation $(n=3)$ that have been integrated so far do not yet provide sufficient material to carry through the above programme with sufficient accuracy, but even a purely analytical discussion of the equations of fit yields information about the different types of composite configurations that exist, and also some general features of the system of curves of constant mass in the $(R, 1-\beta_1)$ diagram.

16. *An Approximate Solution of the Equations of Fit for Highly Collapsed Configurations.*—If the interfacial density is so small that we can regard degeneracy to have set in unrelativistically, then the equations of fit can be solved to a first approximation if in addition the relative core radius is near unity. As the method of obtaining this approximate solution illustrates the general method outlined in § 14, we shall give a short derivation of the same.

Near the zero of ϕ we have the expansion ρ

$$\phi = \frac{1}{y_0} + \frac{\Omega}{\eta_1}(\tau + \tau^2 + \tau^3 + \ldots), \tag{97}$$

where

$$\tau = \frac{\eta_1 - \eta}{\eta_1}. \tag{98}$$

From the above we readily find that

$$\phi^2 - \frac{1}{y_0{}^2} \sim \frac{2\Omega}{y_0\eta_1}\tau, \qquad (\tau \to 0), \tag{99}$$

$$U \sim \left(\frac{2\eta_1}{y_0}\right)^{3/2} \Omega^{1/2}\tau^{3/2}, \qquad (\tau \to 0), \tag{100}$$

$$V \sim \left(\frac{2\eta_1}{y_0}\right)^{-1/2} \Omega^{1/2}\tau^{-1/2}, \qquad (\tau \to 0). \tag{101}$$

The handwritten correction on this page is the author's.

From (100) and (101) we deduce the following asymptotic relation near $\tau=0$:—

$$V \sim \frac{\Omega^{2/3}}{U^{1/3}}. \tag{102}$$

Now if degeneracy sets in unrelativistically, i.e. *if* $(1-\beta_1)$ *is very nearly zero*, equations (93) and (94) can be written as

$$\tfrac{4}{5}x' = \left(\frac{960}{\pi^4}\,\frac{1-\beta_1}{\beta_1}\right)^{1/3}, \tag{103}$$

$$x' = y_0\left(\phi^2(\eta') - \frac{1}{y_0^{\,2}}\right)^{1/2}. \tag{103'}$$

From (99) and (103) we now obtain (neglecting all quantities of order τ^2 and more)

$$\tau' = \left(\frac{5}{8}\right)^2 \frac{2\eta_1}{\Omega y_0}\left(\frac{960}{\pi^4}\,\frac{1-\beta_1}{\beta_1}\right)^{2/3}, \tag{104}$$

thus determining the place where the interface occurs. At $\tau=\tau'$ we have from (100) and (101) that

$$U(\tau') = \left(\frac{5}{8}\right)^3\left(\frac{2\eta_1}{y_0}\right)^3 \frac{1}{\Omega}\left(\frac{960}{\pi^4}\,\frac{1-\beta_1}{\beta_1}\right), \tag{105}$$

$$V(\tau') = \left(\frac{5}{8}\right)^{-1}\left(\frac{2\eta_1}{y_0}\right)^{-1}\Omega\left(\frac{\pi^4}{960}\,\frac{\beta_1}{1-\beta_1}\right)^{1/3}. \tag{106}$$

The equations of fit (92) and (92′) now determine $u(a)$ and $v(a)$. We have

$$u(a) = \left(\frac{5\eta_1}{4y_0}\right)^3 \frac{1}{\Omega}\left(\frac{960}{\pi^4}\,\frac{1-\beta_1}{\beta_1}\right), \tag{107}$$

$$v(a) = \left(\frac{4y_0}{5\eta_1}\right)\Omega\left(\frac{\pi^4}{960}\,\frac{\beta_1}{1-\beta_1}\right)^{2/3}\left(\frac{\beta_1}{\beta_2}\right). \tag{108}$$

Now for a solution of Emden's equation we have (*cf.* Milne, *loc. cit.*, p. 624)

$$u \sim \omega_3^{\,2} t^3\,; \qquad v \sim t^{-1}, \tag{109}$$

where

$$t = \frac{\xi_0 - \xi}{\xi_0}. \tag{110}$$

Equations (107), (108) and (109) determine ω_3, and t' at the interface. We find that

$$t' = \left(\frac{5\eta'}{4y_0}\right)\frac{1}{\Omega}\left(\frac{960}{\pi^4}\,\frac{1-\beta_1}{\beta_1}\right)^{2/3}\left(\frac{\beta_2}{\beta_1}\right), \tag{111}$$

$$\omega_3 = \Omega\left(\frac{\pi^4}{960}\,\frac{\beta_1}{1-\beta_1}\right)^{1/2}\left(\frac{\beta_1}{\beta_2}\right)^{3/2}. \tag{112}$$

If we denote

$$\omega_3^{\,2} = C^{-1} \tag{113}$$

—the "discriminant" in Milne's notation—we can rewrite (104) and (111) in the forms

$$t' = C^{2/3}\left(\frac{5}{8}\frac{\beta_1}{\beta_2}\right)\left(\frac{2\eta_1\Omega^{1/3}}{y_0}\right); \qquad \tau' = C^{2/3}\left(\frac{5}{8}\frac{\beta_1}{\beta_2}\right)^2\left(\frac{2\eta_1\Omega^{1/3}}{y_0}\right) \tag{114}$$

or

$$a = \xi_0\left[1 - C^{2/3}\left(\frac{5}{8}\frac{\beta_1}{\beta_2}\right)\left(\frac{2\eta_1\Omega^{1/3}}{y_0}\right)\right], \tag{115}$$

$$\eta' = \eta_1\left[1 - C^{2/3}\left(\frac{5}{8}\frac{\beta_1}{\beta_2}\right)^2\left(\frac{2\eta_1\Omega^{1/3}}{y_0}\right)\right]. \tag{115'}$$

The above solutions will be useful to evaluate the thicknesses of gaseous envelopes in white dwarfs, but these applications are reserved for future communications.

It is of interest to see that with *further restrictions* our result (115), (115′) goes over into certain of Milne's formulæ.

If the mass of the configuration is small (*expressed in units of* M_3), then we have (II, equations (45) and (48))

$$\eta_1 = \frac{\xi_1(\theta_{3/2})}{\sqrt{2x_0}}, \tag{116}$$

$$\Omega = \left(\frac{x_0}{2}\right)^{3/2}\left(-\xi^2\frac{d\theta_{3/2}}{d\xi}\right)_1. \tag{117}$$

Using these we see that

$$\left(\frac{2\eta_1}{y_0}\Omega^{1/3}\right)^3 \to \left(-\xi^5\frac{d\theta_{3/2}}{d\xi}\right)_1 = \omega_{3/2}, \tag{118}$$

the "invariant" for the Emden equation $n=3/2$. With this substitution in (115), (115′) we have Milne's formulæ (*loc. cit.*, equations (38), (39)). But our derivation of his results shows that his formulæ are valid only under the following conditions: (1) $(1-\beta_1)$ *is very small*, and (2) *the central density of the configuration is also very small*. From our results in II, § 13, it follows then that his formulæ give a fair approximation only for M less than about a tenth of M_3, but even then it can be regarded only as an approximation of an approximation. If one considers formally the fitting of an Emden solution of $n=3$ to an Emden function of index $n=3/2$, then the corresponding formulæ (which are Milne's formulæ (38) and (39)) formally give solutions (a fact pointed out by Milne) also when $\beta_1 \sim 0$. Actually, however, if $\beta_1 \sim 0$ the second terms in the brackets of (115) and (115′) are small, but our earlier approximation (103) which has led to these formulæ will no longer be valid—indeed in our analysis there exist no composite configurations for $\beta_1 < \beta_\omega(=\cdot 908)$, and the formal solutions which Milne's formulæ (38) and (39) predict for $\beta \sim 0$ have clearly no physical meaning.*

17. We can deduce from our approximate formulæ of § 16 some results regarding how the curves of constant mass ($M \leqslant M_3$) intersect the x axis

* See, however, his remarks on top of p. 622 (*loc. cit.*).

in the $(x, 1-\beta_1)$ diagram. From (96) and (112) we have for the mass of the configuration

$$M = M_3\left(\frac{\Omega}{\omega_3{}^0}\right)\beta_2{}^{-3/2}. \qquad (119)$$

But $M_3(\Omega/\omega_3{}^0)$ is the mass of the *completely* collapsed configuration (*cf.* II, equation (61)), which has the same central density as our composite configuration. Hence when the curves of constant mass intersect the x (or the R) axis we have the relation

$$M(0, x_0) = M(1-\beta_1, x_0)\beta_2{}^{-3/2}. \qquad (120)$$

If $\beta_1 = \beta_2$ (*i.e.* in the usual standard model), we have

$$M(0, x_0) = M(1-\beta_1, x_0)\beta_1{}^{-3/2}. \qquad (121)$$

From this one can readily show that *the curves of constant mass intersect the x axis in the $(x, 1-\beta_1)$ diagram with negative slope which tends to zero both when $M \to 0$ and $M \to M_3$. Hence there exists a mass-curve which intersects the x axis with a maximum negative slope.*

On the other hand, if $\beta_2 = 1$ (the extreme case in the generalised standard model), (121) leads to

$$M(0, x_0) = M(1-\beta_1, x_0). \qquad (121')$$

From (121') it follows that *curves of constant mass for all $M \leqslant M_3$ intersect the x axis in the $(x, 1-\beta_1)$ diagram at right angles.* From this it also follows that for $M = M_3$ the appropriate curve passes through the origin, cutting the R axis at right angles. We shall show in § 20 that for $M = M_3$ the whole segment of the $(1-\beta_1)$ axis from zero to $1-\beta_\omega$ is a part of its curve of constant mass.

18. *A General Discussion of the Equations of Fit and the Types of Configurations that Exist.*—The general disposition of the (u, v) curves for the solutions of Emden's equations is known from Milne's work and Fairclough's integrations.* The most important feature is that the Emden $\omega_3{}^0$ curve divides the two families of curves, the collapsed curves $(\omega_3 > \omega_3{}^0)$ all lying entirely above the $\omega_3{}^0$ curve, and the centrally condensed curves all lying entirely below the $\omega_3{}^0$ curve. The $\omega_3{}^0$ curve itself runs from $(u=3, v=0)$ to $(u=0, v=\infty)$. For all types for $u \sim 0$, $v \sim \infty$ the asymptotic form of the curve is (*cf.* equation (109))

$$v \sim \frac{\omega_3{}^{2/3}}{u^{1/3}}. \qquad (122)$$

Further, the initial negative slope for the $\omega_3{}^0$ curve is easily found to be 5/9.

Now the (U, V) curves on the other hand *all* run from $(U=3, V=0)$ to infinity along the V axis, the asymptotic form being (*cf.* equation (102))

$$V \sim \frac{\Omega^{2/3}}{U^{1/3}}. \qquad (123)$$

* *M.N.*, **93**, 40, 1932.

Also the initial negative slope is readily found from the relations (see equation (74), II)

$$U \sim 3\left(1 - \frac{q}{5}\eta^2\right), \tag{124}$$

$$V \sim \tfrac{1}{3}q^2\eta^2, \tag{125}$$

where as in II

$$q = \left(1 - \frac{1}{y_0^2}\right)^{1/2} = \frac{x_0}{(x_0^2 + 1)^{1/2}}. \tag{126}$$

We have then

$$\frac{dV}{dU} \sim -\frac{5}{9}q = -\frac{5}{9}\frac{x_0}{(x_0^2+1)^{1/2}}. \tag{127}$$

From (123) and (127) we see that all the (U, V) curves lie entirely below the Emden ω_3^0 curve.

From the above relations we can infer the different types of configurations (*e.g.* collapsed, centrally condensed, quasi-diffuse) that will occur in our theory. We shall consider only two cases: (*a*) the usual standard model with $\beta_1 = \beta_2$, and (*b*) the extreme case of the generalised standard model with $\beta_2 = 1$. The intermediate cases can be treated similarly but will not be considered here.

19. *The Usual Standard Model* $(\beta_1 = \beta_2)$.—We will denote by $\Gamma(\beta_1, y_0)$ the curve

$$\Gamma(\beta_1, y_0) = \left\{U, \left(\frac{\pi^4}{960}\frac{\beta_1}{1-\beta_1}\right)^{1/3} V\right\}. \tag{128}$$

We shall also denote it by $\Gamma(\beta_1, x_0)$, x_0 being the corresponding value for x when $y = y_0$. For a given y_0 the $\Gamma(\beta_1, y_0)$ curves are defined for only values of $(1 - \beta_1)$ less than a critical value $(1 - \beta_1(y_0))$ depending on y_0 alone. The initial negative slope of a $\Gamma(\beta_1, y_0)$ curve is

$$\frac{5}{9}\frac{2x'^4}{f(x')}\frac{x_0}{(x_0^2+1)^{1/2}}.$$

Now

$$\frac{2x'^4}{f(x')}\frac{x_0}{(x_0^2+1)^{1/2}} \geqslant \frac{2x_0^4}{f(x_0)}\frac{x_0}{(x_0^2+1)^{1/2}} > 1, \tag{129}$$

the second inequality in (129) being a *strict* one for all finite x_0. *Hence all the* $\Gamma(\beta_1, y_0)$ *curves initially start above the Emden curve* ω_3^0.

This has an immediate consequence. Consider a mass M which has in the wholly gaseous state a value for $\beta_1 = \beta^\dagger$. The curve $\Gamma(\beta^\dagger, x_0(\beta^\dagger))$ (where $x_0(\beta^\dagger)$ is the value of x at which degeneracy sets in for $\beta_1 = \beta^\dagger$), like the other Γ curves, starts from the point (3, 0), initially going above the ω_3^0 curve. For this configuration $u = 3$, $v = 0$ and $\omega_3 = \omega_3^0$. When the central density of the configuration slightly increases, the degenerate core is of small but finite dimensions, and the appropriate values for $u(a)$ and $v(a)$ must still be in the neighbourhood of (3, 0), but since the Γ curves all lie initially above the ω_3^0 curve, it follows at once from the disposition of the (u, v) curves that the *configuration is necessarily collapsed.* If centrally condensed configurations exist there must be some with small but finite degenerate cores to ensure the continuity of the curves of constant mass in the $(x, 1-\beta_1)$ diagram, but as we have shown that when degeneracy develops into a core, the formation is one due to the collapse of the configuration. From this it follows that *on the standard model* ($\beta_1 = \beta_2$) *there exist only collapsed configurations.*

Now the $\Gamma(\beta_1, y_0)$ curves do not in general ascend above *all* the collapsed curves, though all the Γ curves start initially below all the collapsed curves. For a $\Gamma(\beta_1, y_0)$ curve goes to infinity with the asymptotic relation

$$\left(\frac{\pi^4}{960}\frac{\beta_1}{1-\beta_1}\right)^{1/3} V \sim \left(\frac{\pi^4}{960}\frac{\beta_1}{1-\beta_1}\right)^{1/3} \frac{\Omega^{2/3}(y_0)}{U^{1/3}}. \tag{130}$$

Hence by (122) these curves ascend above only such (u, v) curves which satisfy the inequality

$$\omega_3^{2/3} \leqslant \left(\frac{\pi^4}{960}\frac{\beta_1}{1-\beta_1}\right)^{1/3} \Omega^{2/3}(y_0), \tag{131}$$

or

$$\omega_3 \leqslant \left(\frac{\pi^4}{960}\frac{\beta_1}{1-\beta_1}\right)^{1/2} \Omega(y_0). \tag{132}$$

When the equality sign in (132) occurs the two curves touch at infinity. From (96) we have

$$\omega_3 = \omega_3^0 \left(\frac{1-\beta^\dagger}{\beta^{\dagger 4}}\frac{\beta_1^4}{1-\beta_1}\right)^{1/2}, \tag{133}$$

where $\beta^\dagger$ has the meaning with which we have used it so far. From (132) and (133) it follows that

$$\omega_3^0 \leqslant \left(\frac{\pi^4}{960}\frac{\beta^{\dagger 4}}{1-\beta^\dagger}\right)^{1/2} \beta_1^{-3/2}\Omega(y_0). \tag{134}$$

Now if $(1-\beta^\dagger) < (1-\beta_0)$, *i.e.* when $M < M_3$,

$$\left(\frac{\pi^4}{960}\frac{\beta^{\dagger 4}}{1-\beta^\dagger}\right) > 1, \tag{135}$$

and there exists a $y_0 = y_0^*$ (say) such that

$$\Omega(y_0^*) = \left(\frac{960}{\pi^4}\frac{1-\beta^{\dagger 4}}{\beta^{\dagger 4}}\right)^{1/2} \omega_3^0. \tag{135'}$$

From (135) and (135′) it follows that for $M < M_3$ there exist collapsed configurations, and in fact also a completely collapsed state for any particular mass $M(< M_3)$ with a central density corresponding to y_0 determined by (135′). When $M = M_3$ (134) can still be satisfied with $\beta_1 = 1$, only when $y_0 = \infty$. The curve for M_3 passes through the origin in the $(R, 1-\beta_1)$ diagram, but in the $(x, 1-\beta_1)$ diagram the appropriate curve goes to infinity with the x axis as the asymptote.

Finally, when $M > M_3$ we have

$$\left(\frac{\pi^4}{960}\frac{\beta^{\dagger 4}}{1-\beta^\dagger}\right) < 1, \tag{136}$$

and since the maximum value of $\Omega(y_0)$ is $\omega_3{}^0$, the inequality (134) can be satisfied only for such values of $\beta_1 \leqslant \beta^*$, where

$$\beta^* = \left(\frac{\pi^4}{960}\frac{\beta^{\dagger 4}}{1-\beta^\dagger}\right)^{1/3}. \tag{137}$$

When $\beta_1 = \beta^*$, we have from (133) that

$$\omega_3 = \omega_3{}^0\left(\frac{\pi^4}{960}\frac{\beta^*}{1-\beta^*}\right)^{1/2}. \tag{138}$$

From (138) it follows that this ω_3 curve touches the $\Gamma(\beta^*, \infty)$ curve at infinity, *i.e.* the configuration is one in which the relative core radius is unity. Hence *when $M > M_3$ configurations collapsed more than to the extent* (138) *do not exist.* The curves of constant mass therefore intersect the $(1-\beta_1)$ axis at the point $(1-\beta^*)$ in the $(R, 1-\beta_1)$ diagram. In the $(x, 1-\beta_1)$ diagram the curves of constant mass for $M > M_3$ run to infinity, having as the asymptote the line through $(1-\beta^*)$ parallel to the x axis. We have thus re-derived some of our earlier results in § 9 from a proper discussion of the equations of fit. Since, however, only collapsed configurations exist, it is now obvious that in the domain of degeneracy the curves of constant mass connect by some "direct" path the point $(R_0(M), 1-\beta^\dagger)$ on the $(R_0, 1-\beta_1)$ curve to a point on the R axis $(M \leqslant M_3)$ or a point on the $(1-\beta_1)$ axis $(M > M_3)$.

20. *The Generalised Standard Model* $(\beta_2 = 1)$.—We shall now define by $\Gamma'(1-\beta_1, y_0)$ the curve

$$\Gamma'(1-\beta_1, y_0) = \left\{U, \left(\frac{\pi^4}{960}\frac{\beta_1{}^4}{1-\beta_1}\right)^{1/3} V\right\}, \tag{139}$$

with $(1-\beta_1)$ of course less than $(1-\beta_1(y_0))$. The initial negative slope of these curves Γ' are

$$\frac{5}{9}\left(\frac{\pi^4}{960}\frac{\beta_1{}^4}{1-\beta_1}\right)^{1/3} q, \qquad \left(q = \left(1 - \frac{1}{y_0{}^2}\right)^{1/2}\right). \tag{140}$$

Further, for $U \sim 0$, $V \sim \infty$, we have the asymptotic relation

$$\left(\frac{\pi^4}{960}\frac{\beta_1{}^4}{1-\beta_1}\right)^{1/3} V \sim \left(\frac{\pi^4}{960}\frac{\beta_1{}^4}{1-\beta_1}\right)^{1/3} \frac{\Omega^{2/3}}{U^{1/3}}. \tag{141}$$

Now if $(1-\beta_1) > (1-\beta_0)$, then

$$\left(\frac{\pi^4}{960}\frac{\beta_1{}^4}{1-\beta_1}\right) < 1, \qquad (1-\beta_1) > (1-\beta_0). \tag{142}$$

Since further $\Omega < \omega_3{}^0$, $q < 1$ (y_0 finite), we have the result that the $\Gamma'(1-\beta_1, y_0)$ *curves for* $(1-\beta_1) > (1-\beta_0)$ *all lie entirely below the* $\omega_3{}^0$ *curve.* Hence

On the generalised standard model with $\beta_2 = 1$ *there exist centrally condensed configurations and only centrally condensed configurations for* $(1-\beta_1) > (1-\beta_0)$.

Consider now the mass M_3. By (133) we have

$$\omega_3 = \omega_3{}^0\left(\frac{\pi^4}{960}\frac{\beta_1{}^4}{1-\beta_1}\right)^{1/2}, \tag{143}$$

since for $M = M_3$, $\beta^\dagger = \beta_0$. Hence the asymptotic forms for the (u, v) curves are

$$v \sim \left(\frac{\pi^4}{960} \frac{\beta_1^4}{1-\beta_1}\right)^{1/3} \frac{\omega_3^{0\,2/3}}{u^{1/3}}. \tag{144}$$

Comparing this with (141), we see that when $\Omega = \omega_3^0$ (in which case $(1-\beta_1)$ need be less than only $(1-\beta_\omega)$) the appropriate (u, v) curves for the different values $0 \leqslant (1-\beta_1) \leqslant (1-\beta_\omega)$ all touch the $\Gamma'(1-\beta_1, \infty)$ at ∞. Hence the segment $0 \leqslant (1-\beta_1) \leqslant (1-\beta_\omega)$ of the $(1-\beta_1)$ axis in the $(R, 1-\beta_1)$ diagram is a part of the curve of constant mass for $M = M_3$. Hence

The curve of constant mass for $M = M_3$ in the domain of degeneracy consists of a centrally condensed branch joining the point $(R_0(M_3), 1-\beta_0)$ (on the $(R_0, 1-\beta_1)$ curve) to the point $(1-\beta_\omega)$ on the $(1-\beta_1)$ axis and the segment $0 \leqslant (1-\beta_1) \leqslant (1-\beta_\omega)$ of the $(1-\beta_1)$ axis.

From the above it follows that:

Composite configurations with $M > M_3$ are all necessarily centrally condensed.

It is easy to see that for these the curves of constant mass in the domain of degeneracy consists of a simple connection between the point $(R_0(M), 1-\beta^\dagger)$ to the point $(0, 1-\beta_\omega)$. But the point configurations with $M_3 \leqslant M \leqslant \mathfrak{M}$ at $(0, 1-\beta_\omega)$ are not all identical when each one of them is considered as the proper limit of an appropriate sequence of configurations (along the respective curves of constant masses). We then find that they all have different relative core radii.

For these point configurations the equations of fit take the limiting forms

$$u(a) = u(b, \omega_3^0), \tag{145}$$

$$v(a) = \beta_\omega v(b, \omega_3^0), \tag{146}$$

i.e. the problem consists in merely fitting two regions, the inner core being governed by the Emden function ($n=3$) and the outer envelope by a centrally condensed solution of Emden's equation ($n=3$). The appropriate (u, v) curve for a mass $M_3 > M \geqslant \mathfrak{M}$ is selected by its value for ω_3, which is readily seen to be

$$\omega_3 = \omega_3^0 \left(\frac{960}{\pi^4} \frac{1-\beta^\dagger}{\beta^{\dagger 4}}\right)^{1/2} \beta_\omega^{3/2} \; (< \omega_3^0). \tag{147}$$

The equations of fit have a unique solution, since for these configurations

$$\left(\frac{960}{\pi^4} \frac{1-\beta^\dagger}{\beta^{\dagger 4}}\right)^{1/3} \beta_\omega > \beta_\omega, \tag{148}$$

and consequently the appropriate centrally condensed solution ultimately ascends above the Emden curve. The two curves must therefore intersect, the values of a and b (a on the centrally condensed curve, b on the Emden curve) at the point of intersection being the required solution of the equations of fit. In particular when $\beta^\dagger = \beta_\omega$ one easily sees that the appropriate point of intersection is $(3, 0)$—in other words, for $M = \mathfrak{M}$ the relative core radius for the point configuration at $(0, 1-\beta_\omega)$ is zero. We have already seen that for $M = M_3$ the point configuration at $(0, 1 - \beta_\omega)$ has a relative core radius of unity. Hence *the limiting relative core radii along the centrally condensed sequences of composite configurations (each sequence being characterised by a mass* $M_3 \leqslant M \leqslant \mathfrak{M}$*) varies from unity for* M_3 *to zero for* $\mathfrak{M}$.

We pass on now to consider the types of configurations that occur for $M < M_3$.

The initial negative slope for $\Gamma'(1 - \beta_1, y_0)$ curves can also be written in the form (*cf.* equation (140)):

$$\tfrac{5}{9}\beta_1 \frac{2x'^4}{f(x')} \frac{x_0}{(x_0^2 + 1)^{1/2}}. \tag{149}$$

Now by the inequality (129) we can always formally calculate a "$\beta_c(x_0)$" such that

$$\beta_c(x_0) = \frac{f(x_0)}{2x_0^4} \frac{(x_0^2 + 1)^{1/2}}{x_0}. \tag{150}$$

If now

$$\beta_1 \geqslant \beta_c(x_0), \tag{151}$$

then

$$\beta_1 \frac{2x'^4}{f(x')} \frac{x_0}{(x_0^2 + 1)^{1/2}} \geqslant \beta_1 \frac{2x_0^4}{f(x_0)} \frac{x_0}{(x_0^2 + 1)^{1/2}} \geqslant 1. \tag{152}$$

On the other hand, if $\beta_1 < \beta_c(x_0)$, then the second inequality in (152) would not necessarily follow. But for this to happen $(1 - \beta_1)$ must satisfy simultaneously two inequalities, since for a given y_0, $(1 - \beta_1)$ must be less than $(1 - \beta_1(y_0))$ determined by (95). The two inequalities then are

$$\left.\begin{aligned} (1 - \beta_1) &> 1 - \beta_c(x_0), \\ (1 - \beta_1) &\leqslant 1 - \beta_1(y_0). \end{aligned}\right\} \tag{153}$$

The two inequalities will not in general be satisfied. In fig. 1 the curve $(1 - \beta_c(x))$ is plotted against x. In the same figure we have also the curve $(1 - \beta_1(y))$. The two curves are seen to intersect. Let $\beta_1 = \beta_q$ and $x = x_q$ at the point of intersection. Then

$$1 - \beta_c(x_q) = 1 - \beta_1(y_q), \qquad (y_q^2 = x_q^2 + 1). \tag{154}$$

Numerically it is found that

$$1 - \beta_q = \cdot 060 \text{ (approximately)}; \qquad x_q = 2 \cdot 2 \text{ (approximately)}. \tag{155}$$

We shall also define a mass M_q by the relation

$$M_q = M_3 \left(\frac{960}{\pi^4} \frac{1 - \beta_q}{\beta_q^4} \right)^{1/2}. \tag{156}$$

i.e. M_q is the mass which has the value β_q for β_1 in the wholly gaseous state. Numerically we find that

$$M_q = 0 \cdot 87 M_3 \text{ (approximately)}. \tag{156'}$$

Now consider a mass less than M_q, then for these configurations $(1 - \beta^\dagger) < (1 - \beta_q)$. Consequently the curve $\Gamma'(1 - \beta^\dagger, x_0(\beta^\dagger))$ starts initially above the $\omega_3{}^0$ curve. From this it follows that the composite configurations of mass $M < M_q$ with small but finite degenerate cores are necessarily of the collapsed type. But a collapsed configuration has a $(1 - \beta_1) < (1 - \beta^\dagger) < (1 - \beta_q)$, and the corresponding Γ' curve starts with a still greater initial negative slope. In this way one sees that *all composite configurations with* $M \leqslant M_q$ *are collapsed configurations.*

When, however, $M > M_q$ the curve $\Gamma'(1 - \beta^\dagger, x_0(\beta^\dagger))$ lies entirely below the $\omega_3{}^0$ curve, and so do the curves for values of x_0 slightly greater. Centrally condensed configurations are possible. *Hence for* $M_q < M \leqslant M_3$, *centrally condensed and quasi-diffuse configurations exist in addition to the usual collapsed configurations.*

The curves of constant mass $(M \leqslant M_3)$ intersect the R axis in the $(R, 1 - \beta_1)$ diagram at the same points as in the usual standard model. Further, as we have already shown in § 17, all these curves intersect the x axis in the $(x, 1 - \beta_1)$ diagram at right angles.

21. *Summary of the Main Results of* §§ 16–20.—In the preceding sections we have obtained some general information regarding the types of composite configurations that exist, and it is convenient to summarise here the main results.

A. There exist no composite configurations for $M \geqslant \mathfrak{M}$, where

$$\mathfrak{M} = M_3 \beta_\omega{}^{-3/2},$$

and β_ω is such that

$$\frac{960}{\pi^4} \frac{1 - \beta_\omega}{\beta_\omega} = 1.$$

B. *The Usual Standard Model* $(\beta_1 = \beta_2)$.—(*a*) All composite configurations $(M < \mathfrak{M})$ are all necessarily collapsed.

(*b*) (i) For a prescribed $M < M_3$ the sequence of equilibrium configurations has as its limit a completely collapsed configuration $(\beta_1 = 1)$ with a central density $Bx_0{}^3$ related to M by

$$M = M_3(\Omega(y_0)/\omega_3{}^0),$$

where

$$\Omega(y_0) = -\left(\eta^2 \frac{d\phi}{d\eta}\right)_1; \qquad \omega_3{}^0 = -\left(\xi^2 \frac{d\theta_3}{d\xi}\right)_1,$$

θ_3 being the Emden function $(n = 3)$ and ϕ satisfying the differential equation

$$\frac{1}{\eta^2}\frac{d}{d\eta}\left(\eta^2 \frac{d\phi}{d\eta}\right) = -\left(\phi^2 - \frac{1}{y_0{}^2}\right)^{3/2}, \qquad (y_0{}^2 = x_0{}^2 + 1).$$

(ii) For $M = M_3$ the limiting completely collapsed configuration is of zero radius, and its value of "β_1" in the wholly gaseous state (denoted by β_0) satisfies the relation

$$\frac{960}{\pi^4} \frac{1 - \beta_0}{\beta_0{}^4} = 1.$$

(iii) For $M_3 < M < \mathfrak{M}$ the sequence of collapsed configurations end with a finite non-zero value for "$1 - \beta_1$," the maximum value β^* of β_1 being related to $\beta^\dagger$ (the value β_1 has for the prescribed mass in its wholly gaseous state) by the equation

$$\beta^* = \left(\frac{\pi^4}{960} \frac{\beta^{\dagger 4}}{1-\beta^\dagger}\right)^{1/3}.$$

C. *The Generalised Standard Model* ($\beta_2 = 1$).—(i) There exists a mass $M_q(\sim 0{\cdot}87 M_3)$ such that all composite configurations with $M \leqslant M_q$ are necessarily collapsed.

(ii) For $M_q < M < M_3$ each mass has a sequence of centrally condensed configurations which passes continuously (through a quasi-diffuse configuration) into a sequence of collapsed configurations, ending in a completely collapsed configuration with a central density related to the mass as in B (i) above.

(iii) For $M = M_3$ the composite configurations of *finite* radii are all centrally condensed. The limiting configuration along the centrally condensed sequence is one of zero radius with $1-\beta_1 = 1-\beta_\omega$. The relative core radius tends to unity along the centrally condensed sequence as the limiting configuration at $(1-\beta_\omega)$ is approached. Finally a whole sequence of point configurations for $0 \leqslant 1-\beta_1 \leqslant 1-\beta_\omega$ exists for this mass.

(iv) For $M_3 < M < \mathfrak{M}$ the composite configurations are *all* centrally condensed. The centrally condensed sequences for the different masses all end with a point configuration at $1-\beta_1 = 1-\beta_\omega$. The relative core radii for the different masses along their respective centrally condensed sequences tend to different limits, decreasing monotonically from unity for $M = M_3$ to zero for $M = \mathfrak{M}$.

The general results on the generalised standard model ($\beta_2 = 1$) are therefore of the character illustrated in fig. 4. To obtain more precise information one would have to solve the equations of fit, a simple and direct method for which has been outlined in § 15. But as mentioned there, the number of quadratures for the singular solutions of Emden's equation ($n = 3$) that have been carried out so far (by Fairclough) do not yet provide sufficient data to solve the equations of fit by the method of § 15 to any reasonable degree of accuracy. There exist numerical quadratures for only four collapsed solutions and three centrally condensed solutions, and our method to solve the equations of fit depends on a process of interpolation among the (u, v) tables for different ω_3. It is, however, possible to avoid the above process of interpolation by adopting an indirect method to solve the equations of fit, but this indirect method requires very much more numerical work to be done on our ϕ functions. These detailed calculations are reserved for future communications, but it is fortunate that the very circumstance that the degenerate core is governed by our differential equation for ϕ has allowed us to infer quite detailed results (summarised above) regarding the types of composite configurations that exist, and conclude therefrom the general character of the results illustrated in figs. 3 and 4.

We pass on now to comment briefly on the bearing of our results on certain aspects of the general problem of stellar evolution and stellar structure.

22. Consider a stellar mass $M < M_3$. In §§ 10–21 we have examined how such a stellar mass tends to the completely collapsed configuration (with the same mass) when the luminosity tends to zero through a sequence

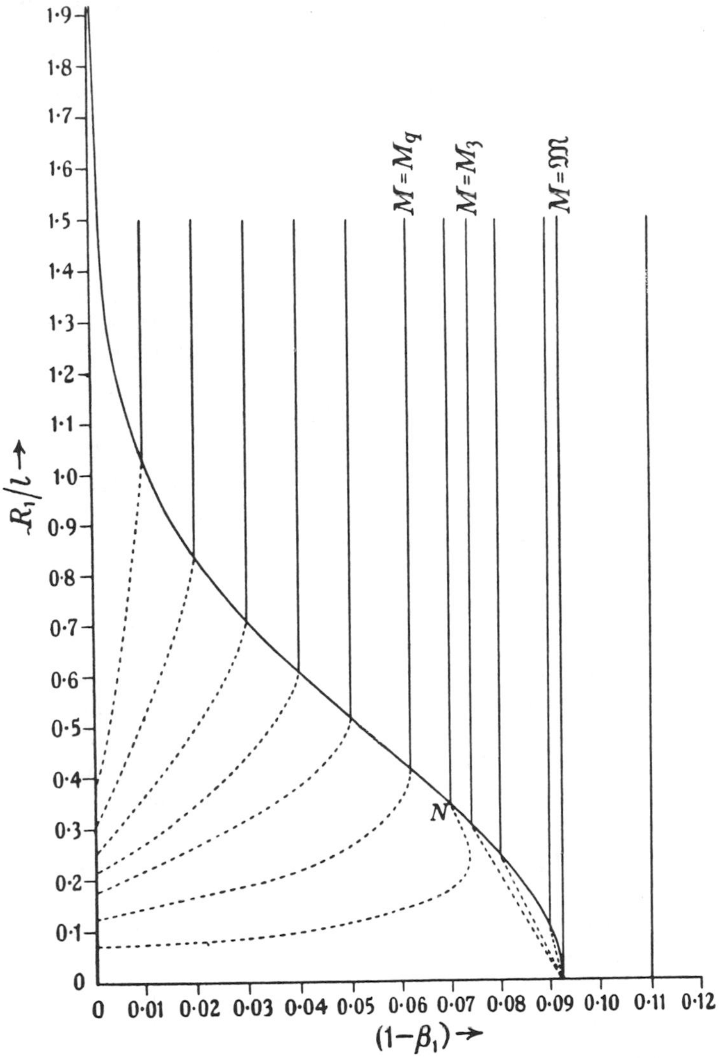

FIG. 4.—*The general nature of the curves of constant mass on the generalized standard model* ($\beta_2=1$). *The end-points of the various curves in the domain of degeneracy have been fixed exactly. The actual curves are shown dotted as the equations of fit have not been solved with any accuracy.*

of equilibrium configurations all conforming to the standard model. As it is clearly immaterial how the limit of zero luminosity is reached, it follows that *any* stellar mass $M < M_3$ must necessarily develop central regions of degeneracy when the luminosity decreases sufficiently, and should eventually become a "black dwarf"—to use a term due to R. H. Fowler—the structure of which will be governed by our differential equation for ϕ. Thus it is seen that all stellar masses less than M_3 will in the course of evolution pass through the white-dwarf stage, which itself is only a preliminary towards a state of complete degeneracy and darkness. It is more than probable that there exists a large number of such faint more or less completely collapsed configurations, though only a few white dwarfs are known. Such faint configurations would be difficult to detect both on account of their faintness and on account of their small dimensions. We have seen in II, § 12 (Table III), that $M = 5{\cdot}48\mu^{-2}\odot$ will in the completely collapsed state have a radius quite significantly smaller than the radius of the Earth.

Again we may expect that the fundamental distinction made on the standard model between masses less than and greater than $\mathfrak{M}$—in that for $M > \mathfrak{M}$ the equilibrium configurations are necessarily wholly gaseous (in the perfect gas sense)—is only a counterpart on the standard model of a more general result that all stars of sufficiently large mass are necessarily perfect gas configurations ; for the increased dominance of radiation pressure for large stellar masses is quite a general result and, as we have already seen, the possibility of degeneracy is entirely excluded if only the radiation pressure is greater than a tenth of the total pressure throughout the entire mass. It should indeed be possible to define a "domain of degeneracy" on the basis of any model (as we have done in § 5 for the standard model) by drawing in the (radius-luminosity) diagram a curve similar to our $(R_0, 1-\beta_1)$ curve.* The result will always be that stars of mass greater than a certain limit will all be perfect gas configurations and will therefore conform to Eddington's mass-luminosity relation however much they may contract. As a result such stars can never pass *directly* into the white-dwarf stage. A possibility is that at some stage in the process of contraction the flux of radiation in the outer layers of the stars will become so large that a profuse ejection of matter will begin to take place. This ejection of atoms will continue till the mass of the star becomes small enough for central degeneracy to be possible. Once degeneracy sets in, the star will evolve along some white-dwarf sequence —possibly at the Kelvin-Helmholtz rate—to end again as a completely collapsed configuration.

One can perhaps look for a confirmation of these ideas in the fact that Wolf-Rayet stars, which are known to eject matter, must be massive if they are to conform to Eddington's mass-luminosity relation.† Thus Kosirev ‡ estimates that an average Wolf-Rayet star has a mass of about $10\odot$, and that further the annual loss of matter due to the ejection of atoms amounts to 10^{-6} of its mass.§ Once the mass becomes less than M_3 the luminosity would begin to decrease relatively more rapidly, the radial ejection would

* The author has already constructed such curves for some more general stellar models, but these extensions of the ideas developed in this paper will not be considered here.

† A massive star will be a perfect gas configuration, and we have therefore no reason to suspect that the Wolf-Rayet stars will not conform to Eddington's mass-luminosity relation.

‡ *M.N.*, 94, 430, 1934.

§ Some earlier estimates by Beals (*Pub. D.A.O.*, 4, No. 17, p. 297) gave rather lower rates, but in the author's opinion Beals's figures are definitely underestimates.

cease and the star collapse into a white dwarf. It is seen that these ideas fit consistently with the general indications that the nuclei of planetary nebula are white dwarfs and that the nebular envelope consists of matter ejected originally from the nuclear star. Our arguments in addition indicate that the masses of the nuclei of planetary nebula must be considerably less than that of an average Wolf-Rayet star. Thus, whether the star is of large or of small mass, the final stage in its evolution is always the white-dwarf stage, the only difference being that a star of large mass must first decrease its mass below M_3 by passing through the Wolf-Rayet stage. But in either case the completely collapsed configurations represent the "limiting sequence of stellar configurations to which all stars must tend eventually."

23. On the usual standard model we have shown that all the composite configurations are of the collapsed type. There is, however, a range of mass between M_3 and $\mathfrak{M}$ for which the configurations tend to zero radius with finite luminosity (which is less than the luminosity in the wholly gaseous state). But before that stage is reached there will occur the loss of mass due to the ejection of atoms, arising from the increased outward flux of radiation, and the final course of evolution for these masses must be more or less similar to that described for the perfect gas configurations in § 22, the only difference being that for the masses in the range $M_3 < M < \mathfrak{M}$ (or its equivalent on more general stellar models) the ejection would begin to take place when the luminosity is less than that predicted by Eddington's mass-luminosity relation.

24. *The Nova Phenomenon.*—We have shown that on the generalised standard model with $\beta_2 = 1$, for $M_3 > M > M_q$, there exist equilibrium configurations which are centrally condensed, and that for a prescribed mass the centrally condensed branch passes continuously into a sequence of collapsed configurations through a quasi-diffuse configuration. The quasi-diffuse configuration has a "$1 - \beta_1$" which is the same as that which characterises the wholly gaseous configuration. A possibility then for a configuration (with $M_q > M > M_3$) which, contracting from infinite extension, has reached a stage when degeneracy is just beginning to develop at the centre (*i.e.* the representative point in the $(R, 1 - \beta_1)$ plane is at a point like N in fig. 4) is that, instead of evolving along the centrally condensed branch, there occurs a discontinuous decrease of the external radius, the configuration passing into the quasi-diffuse state with a *finite* degenerate core with the same luminosity as it had at N. The consequence of such a decrease in radius would be the release of the difference of the total potential energies of the configuration at N and in its quasi-diffuse state vertically below N. It is possible that in this way the nova phenomenon (or more possibly the "super-nova phenomenon") occurs.

Ideas similar to those suggested above have been previously proposed by Milne. But it is perhaps necessary to draw attention to the differences between the suggestion made above and Milne's original ideas. ***Firstly***, he regarded the "nova outburst" as occurring in the passage of a centrally condensed star into the quasi-diffuse state, *i.e.* the configuration is regarded as being initially in the centrally condensed branch, and the "nova outburst" as taking place when on *decreasing* luminosity the radius of the configuration *increases* till a point similar to N is reached. On the other hand, according to the suggestion made above the "nova outburst" is assumed to take place when a wholly gaseous configuration *contracting* from infinite extension passes directly into the collapsed sequence without ever passing into the centrally condensed branch. *Secondly*, on Milne's ideas all stars greater than a certain critical mass * can in principle become novæ, while our analysis shows that stars exhibiting the nova phenomenon must have masses in a comparatively small range.

25. *Have the Centrally Condensed Configurations that Exist for $M_q < M < M_3$ on the Generalised Standard Model any Relation to the Ordinary Stars?*—There is one initial difficulty in making any suggestions as to the relation of these centrally condensed configurations to the ordinary stars. For all these configurations have radii less than $0{\cdot}4l$ (see fig. 4), which is about $2 \times 10^9 \mu^{-1}$ cm. This upper limit is itself very considerably smaller than the radii of ordinary stars. In connection with such a difficulty Milne has suggested † that such a radius discrepancy might be due to the use of the wrong boundary conditions ($T=0$, $\rho=0$), and that the use of the proper boundary conditions ($T=T_0$, $\rho=0$) might remove this discordance. Actually such investigations as have been made by Cowling ‡ and others do not favour this suggestion. But even if it should be possible to remove this discordance in a way which makes ordinary stars have small "cores" of the structure and dimensions similar to our centrally condensed configurations surrounded by extremely tenuous but extended envelopes, there would still be further difficulties to be overcome. The first is the opacity difficulty. In a general investigation Eddington § has shown that theories which postulate such extended envelopes would merely transfer the opacity difficulty from the core to the envelope, with indeed much larger factors in the discrepancy. It does not seem possible either, to remove this increased discordance by the hydrogen-abundance hypothesis. There is another difficulty as well. Milne has suggested in a different connection that if and when centrally condensed configurations are possible, the configuration having the maximum luminosity for the prescribed mass should exhibit the Cepheid phenomenon. If so, then on our analysis the Cepheids must have masses in a rather small range; but this is in very serious discordance with the results of observation. For these centrally condensed configurations, though they have luminosities greater than the corresponding gaseous configurations, are at most only 20 per cent. greater, and the mass-luminosity relation shows that the masses of the Cepheids can be almost anything. ‖ Thus on the whole the author is inclined to the view that the

* His critical mass is defined as one which has a "$1-\beta_1$" $=0{\cdot}2$ in the wholly gaseous state. One finds that this means $M=2{\cdot}1937M_3$, which is *greater* than our $\mathfrak{M}$.

† *Zeit. für Astrophysik*, **4**, 75, 1932.

‡ T. G. Cowling, *Zeit. für Astrophysik*, **4**, 331, 1932.

§ A. S. Eddington, *M.N.*, **91**, 109, 1931. I am indebted to Sir A. S. Eddington for discussions on these and related matters.

‖ See Table xxv in *Internal Constitution of the Stars* (Cambridge).

centrally condensed configurations that occur in the theory of the generalised standard model have no relation to the ordinary stars. But the existence of a centrally condensed branch may have important bearings on other problems—for example in a possible explanation of the nova phenomenon as outlined in § 24.

Concluding Remarks.—In this paper the general problems of stellar structure as they present themselves on the standard model have been rediscussed, using the exact differential equation derived in the preceding paper to describe degenerate matter in gravitational equilibrium. Since we have restricted ourselves exclusively to the standard model, it is clear that only a first preliminary attack has been made on a much wider problem, of how the conclusions regarding stellar constitution and stellar evolution that have been drawn on the perfect gas hypothesis for the stars have to be modified by the physical possibility of degeneracy in stellar interiors. The methods that have been developed in this paper and the results obtained would have to be extended for more general stellar models before any very definite conclusions could be drawn.

The main results of the analysis are summarised in § 21 and figs. 1, 3 and 4; some general conclusions which follow are considered in §§ 22–25.

Trinity College, Cambridge:
1935 *January* 4.

13

Discussion of Papers 11 and 12 by A. S. Eddington and E. A. Milne

Dr. Chandrasekhar read a paper describing the research which he has recently carried out, an account of which has already appeared in *The Observatory*, **57**, 373, 1934, investigating the equilibrium of stellar configurations with degenerate cores. He takes the equation of state for degenerate matter in its exact form, that is to say, taking account of relativistic degeneracy. An important result of the work is that the life history of a star of small mass must be essentially different from that of a star of large mass. There exists a certain critical mass $\mathfrak{M}$. If the star's mass is greater than $\mathfrak{M}$ the star cannot have a degenerate core, but if the star's mass is less then $\mathfrak{M}$ it will tend, at the end of its life history, towards a completely collapsed state.

Prof. Milne. I have had an opportunity of seeing Dr. Chandrasekhar's paper. We have both been working on the same problem. I had intended to present a paper, written around Mr. Fairclough's latest numerical results, to this Meeting of the Society, but it has been unavoidably delayed. In many ways the methods pursued and the results obtained are the same as Dr. Chandrasekhar's. I have pursued a cruder method of analysis, but I believe that my method gives more insight into the fundamental physical postulates underlying the work, takes account of our ignorance of the behaviour of degenerate matter, and gives a more rational picture. A result common to our theory and Dr. Chandrasekhar's is that the more massive a star, the smaller its radius when completely collapsed. This has a bearing on the Russell diagram.

The President. Fellows will wish to return their thanks to Dr. Chandrasekhar. I now invite Sir Arthur Eddington to speak on his paper "Relativistic Degeneracy".

Sir Arthur Eddington. Dr. Chandrasekhar has been referring to degeneracy. There are two expressions commonly used in this connection, "ordinary" degeneracy and "relativistic" degeneracy, and perhaps I had better begin by explaining the difference. They refer to formulæ expressing the electron pressure P in terms of the electron density σ. For ordinary degeneracy $P_e = K\sigma^{5/3}$. But it is generally supposed that this is only the limiting form at low densities of a more complicated relativistic formula, which shows P varying as something between $\sigma^{5/3}$ and $\sigma^{4/3}$, approximating to $\sigma^{4/3}$ at the highest densities. I do not know whether I shall escape from this meeting alive, but the point of my paper is that there is no such thing as relativistic degeneracy !

I would remark first that the relativistic formula has defeated the original intention of Prof. R. H. Fowler, who first applied the theory of degeneracy to astrophysics. When, in 1924, I suggested that owing to ionization we might have to deal with exceedingly dense matter in astronomy, I was troubled by a difficulty that there seemed to be no way in which a dense star could cool down. Apparently it had to go on radiating for ever, getting smaller and smaller. Soon afterwards Fermi-Dirac statistics were discovered, and Prof Fowler applied them to the problem and showed that they solved the difficulty ; but now Dr. Chandrasekhar has revived it again. Fowler used the ordinary formula ; Chandrasekhar, using the relativistic formula which has been accepted for the last five years, shows that a star of mass greater than a certain limit $\mathfrak{M}$ remains a perfect gas and can never cool down. The star has to go on radiating and radiating and contracting and contracting until, I suppose, it gets down to a few km. radius, when gravity becomes strong enough to hold in the radiation, and the star can at last find peace.

Dr. Chandrasekhar had got this result before, but he has rubbed it in in his last paper ; and, when discussing it with him, I felt driven to the conclusion that this was almost a *reductio ad absurdum* of the relativistic degeneracy formula. Various accidents may intervene to save the star, but I want more protection than that. I think there should be a law of Nature to prevent a star from behaving in this absurd way !

If one takes the mathematical derivation of the relativistic degeneracy formula as given in astronomical papers, no fault is to be found. One has to look deeper into its physical foundations, and these are not above suspicion. The formula is based on a combination of relativity mechanics and non-relativity quantum theory, and I do not regard the offspring of such a union as born in lawful wedlock. I feel satisfied myself that the current formula is based on a partial relativity theory, and that if the theory is made complete the relativity corrections are compensated, so that we come back to the " ordinary " formula.

Suppose we are dealing with a cubic centimetre of material in the middle of a star. Ordinarily we analyse this into electrons, protons, etc., travelling about in all directions. In wave mechanics, the electrons are represented by waves. There are two kinds of waves, progressive and standing. In the ordinary analysis of matter into electrons one is dealing with progressive waves ; but in the analysis which leads to the Exclusion Principle (used in deriving the degeneracy formula) the electron is represented by a standing wave. Now an electron represented by a standing wave is a quite different sort of entity from the electron represented by a progressive wave. The former is constantly changing its identity. I might compare the progressive wave with Professor Stratton and the standing wave with the President of the Royal Astronomical Society ; only, to make the analogy a good one, the Society would have to change its President gradually and continuously, instead of suddenly every two years. The formulæ which apply to such a President would be different from the formulæ which apply to an ordinary individual ; and this point has a definite bearing on the question. The electron represented by a progressive wave can be brought to rest by a Lorentz transformation, and it then becomes a standing wave. This transformation introduces a factor into the equation, which is not needed if the waves referred to are standing waves originally. My main point is that the Exclusion Principle presupposes analysis into standing waves, and this has been wrongly combined with formulæ which refer to progressive waves.

The President. The arguments of this paper will need to be very carefully weighed before we can discuss it. I ask you to return thanks to Sir Arthur Eddington.

CORRESPONDENCE.

To the Editors of 'The Observatory'.

The Configuration of Stellar Masses.

GENTLEMEN,—

In view of the fundamental character of the paper read by Sir Arthur Eddington at the meeting of the Royal Astronomical Society on 1935 January 11, perhaps I may be allowed to state that the basis of the calculation just completed by Mr. Norman Fairclough (referred to in my remarks at the meeting) is the equation of state $p=K\rho^{5/3}$ for a degenerate gas. For the sake of simplicity, and to have a well-defined case fully worked out, we had restricted attention to composite configurations for which "relativistic degeneracy", whether it exists or not, was ignored. Sir Arthur Eddington's investigations may now confer on our work a justification to which it is only accidentally entitled. The work consists in the carrying out of the programme sketched in *M.N.R.A.S.*, **92.** 610, 1932, and there left unfinished, namely the enumeration of *radii* for all possible values of L (luminosity) and M (mass) in the form of curves of r_1 (radius) against L for constant M. The work evidently coincides in aim, and partly in results, with the similar work by Dr. S. Chandrasekhar. The differences can more profitably be discussed when our papers are prepared for publication.

I am, Gentlemen,

Yours faithfully,

Oxford, 1935 Jan. 13. E. A. MILNE.

14

STELLAR CONFIGURATIONS WITH DEGENERATE CORES. (SECOND PAPER.)

S. Chandrasekhar, Ph.D.

1. In a previous communication * the general problems of stellar structure as they present themselves on the standard model were rediscussed, using the exact relativistic equation of state to describe degenerate matter.† The method developed in I is, however, quite general and consists essentially in relating the completely degenerate gas spheres governed by the differential equation

$$\frac{1}{\eta^2}\frac{d}{d\eta}\left(\eta^2\frac{d\phi}{d\eta}\right) = -\left(\phi^2 - \frac{1}{y_0^2}\right)^{3/2}, \qquad (1)\ddagger$$

* *M.N.*, **95**, 226–260, 1935. This paper will be referred to as I.

† In a recent paper, *M.N.*, **95**, 297, 1935, Eddington has questioned the validity of the relativistic equation of state for degenerate matter which is still generally accepted. There are, however, grounds for not abandoning the accepted form of the equation of state—the arguments are presented in the preceding paper by Dr. Christian Møller and the writer.

‡ This equation was established in the author's paper, *M.N.*, **95**, 207–226, 1935. This paper will be referred to as H.C. II. The earlier paper, *M.N.*, **91**, 456, 1931, will be referred to as H.C. I.

with the wholly gaseous configurations. Since on the standard model approximation for the gaseous configurations the ratio $(1-\beta_1)$ of the radiation pressure to the total pressure is a function of the mass only, the study of the curves of constant mass in the $(R,\ 1-\beta_1)$-diagram allows a convenient approach to the problem. In this diagram wholly gaseous configurations are represented by lines parallel to the R-axis, while the completely degenerate configurations are represented by points on the R-axis. The relation between these two sets of configurations was obtained by starting with a wholly gaseous configuration of prescribed mass and infinite extension and slowly contracting it and considering whether deviations from perfect gas laws towards degeneracy set in at all and if so when. In this way a *domain of degeneracy* in the $(R,\ 1-\beta_1)$-diagram was defined in which the configurations must be composite. To fix the precise nature of the curves of constant mass in the domain of degeneracy one requires further assumptions regarding the opacity of the degenerate core, but the problem of relating the degenerate spheres with the gaseous configurations was in principle solved.

2. But the discussion in I was incomplete in so far as the explicit appearance of the physically important parameter, namely, the luminosity L was suppressed by the use of $(1-\beta_1)$ as the main variable. To gain further physical insight it is necessary therefore to transform the discussion of the curves of constant mass in the $(R,\ 1-\beta_1)$-plane to a discussion of the curves of constant mass in the $(\log L,\ \log R)$-plane. This is done in Section I of this paper.

3. To complete the discussion we shall have to verify that the general results are not dependent on the very special nature of the model on which they have been obtained. As Jeans has more than once emphasised,* considerable caution is required in interpreting results based on stellar models which make gaseous configurations Emden polytropes of index 3. The more general analysis in which $(1-\beta_1)$ was allowed to vary through the configuration was provided by Jeans.† It follows from his analysis that *for fairly general stellar models the ratio* $(1-\beta_c)$ *of the radiation pressure to the total pressure at the centre of the configuration is a function of the mass only and is independent of the radius.* It is therefore clear that the whole discussion of I (especially that in Section I of that paper) can now be repeated on this more general analysis by considering the curves of constant mass in the $(R,\ 1-\beta_c)$-plane. In this plane the gaseous configurations are represented by lines parallel to the R-axis, and the relation between these "Emden-Jeans" polytropes to the completely degenerate configurations can be examined as before. This is done in Section II of this paper.

4. Lastly, in Section III various miscellaneous problems which arise are briefly considered.

* *Astronomy and Cosmogony* (Cambridge), chap. iii. Also *M.N.*, **85**, 201, 1925.
† *Astronomy and Cosmogony* (Cambridge), §§ 78–86, 88–92.

Section I.

5. *The Physical Variables.*—As shown in I, equation (55), Eddington's quartic equation can be written in the form

$$M = M_3\left(\frac{960}{\pi^4}\frac{1-\beta_1}{\beta_1{}^4}\right)^{1/2}, \tag{2}$$

where

$$M_3 = -4\pi\left(\frac{2A_2}{\pi G}\right)^{3/2}\frac{1}{B^2}\left(\xi^2\frac{d\theta_3}{d\xi}\right)_1, \tag{3}$$

where A_2, B and the other symbols have the same meaning as in I. M_3 of course represents the upper limit to the mass of a completely degenerate configuration.

If R is the radius of the configuration, then one easily finds that the central density ρ_0 is given by

$$\rho_0 = B\,.\,\frac{M}{M_3}\left(\frac{l}{R}\right)^3, \tag{4}$$

where l is the unit of length introduced in I, equation (33), namely,

$$l = \left(\frac{2A_2}{\pi G}\right)^{1/2}\frac{\xi_1(\theta_3)}{B}. \tag{5}$$

Again the central temperature T_0 of the configuration can be determined from the equation

$$T_0 = -\frac{\beta_1 \mu H}{4k}\frac{GM}{R(\xi\theta_3{}')_1}. \tag{6}*$$

Using (2) and (3), (6) can be rewritten as

$$\frac{4kT_0}{mc^2} = \frac{Ml}{M_3 R}\beta_1. \tag{7}$$

If M^*, R^*, ρ^*, T^* denote the mass, the radius, the density and the temperature when expressed in units of M_3, l, B and $(mc^2/4k)$ respectively then we have

$$M^* = \left(\frac{960}{\pi^4}\frac{1-\beta_1}{\beta_1{}^4}\right)^{1/2}; \quad \rho_0{}^* = \frac{M^*}{R^{*3}}; \quad T_0{}^* = \frac{M^*}{R^*}\beta_1. \tag{8}$$

Also we notice the relation

$$\frac{T_0{}^{*3}}{\rho_0{}^*} = \frac{960}{\pi^4}\frac{1-\beta_1}{\beta_1}. \tag{9}$$

6. *Luminosity.*—We start with the equation †

$$L = \frac{4\pi cGM(1-\beta_1)}{\alpha\kappa_c}, \tag{10}$$

* See, for instance, Milne, *Handbuch der Astrophysik*, Band III/1, p. 209.

† A. S. Eddington, *Internal Constitution of the Stars* (Cambridge), p. 124 (equation 90.1).

where κ_c is the central opacity. We shall assume that

$$\kappa = \kappa_1 \frac{\rho^*}{T^{*7/2}}, \tag{11}$$

where κ_1 is the opacity at $\rho = B$ and $T = mc^2/4k$.

From (8), (9), (10) and (11) we derive that

$$L = \frac{4\pi c G M_3}{a\kappa_1}\left(\frac{960}{\pi^4}\frac{1-\beta_1}{\beta_1^4}\right)^{7/4} \beta_1^{7/2}(1-\beta_1) R^{*-1/2}. \tag{12}$$

If we now introduce the unit of luminosity L_1 defined by

$$L_1 = \frac{4\pi c G M_3}{a\kappa_1}, \tag{13}$$

then (12) can be rewritten in the form

$$L^* = (M^*\beta_1)^{7/2}(1-\beta_1) R^{*-1/2}, \tag{14}$$

where L^* is used to denote the luminosity expressed in units of L_1. If we further introduce the quantity $L^*(\beta_1)$ defined by

$$L^*(\beta_1) = \left(\frac{960}{\pi^4}\frac{1-\beta_1}{\beta_1^4}\right)^{7/4} \beta_1^{7/2}(1-\beta_1), \tag{15}$$

then we have from (14) that

$$\log L^* = \log L^*(\beta_1) - \tfrac{1}{2}\log R^*. \tag{16}$$

The first term on the right-hand side of (16) is a function of the mass only, and hence the curves of constant mass in the ($\log L^*$, $\log R^*$) diagram are straight lines.

7. *The Domain of Degeneracy in the* ($\log L^*$, $\log R^*$)*-Diagram.*—In I, § 5, we showed at what stage a configuration of a prescribed mass less than $\mathfrak{M}$ (contracting from infinite extension) would "just begin to develop degeneracy" * at the centre. This occurs when the radius R_0^* of the configuration is given by (I, equation (34))

$$R_0^* = \left(\frac{960}{\pi^4}\frac{1-\beta_1}{\beta_1^4}\right)^{1/6} \frac{1}{x_0(\beta_1)}, \tag{17}$$

where $x_0(\beta_1)$ is such that

$$\frac{f(x_0)}{2x_0^4} = \left(\frac{960}{\pi^4}\frac{1-\beta_1}{\beta_1}\right)^{1/3}. \tag{18}$$

This value of R_0^* substituted in (12) defines the corresponding luminosity L_0^*:

$$L_0^* = \left(\frac{960}{\pi^4}\frac{1-\beta_1}{\beta_1^4}\right)^{5/3} \beta_1^{7/2}(1-\beta_1)(x_0(\beta_1))^{-1/2}. \tag{19}$$

(17) and (19) together define a curve in the ($\log L^*$, $\log R^*$)-plane,

* What is here meant by "degeneracy just beginning to develop" is stated on p. 229 of my last paper (I).

corresponding to the $(R_0^*, 1-\beta_1)$-curve in the $(R^*, 1-\beta_1)$-plane. Thus for any given mass less than $\mathfrak{M}$ the intersection of the line

$$\log L^* = \log L^*(\beta^\dagger) - \tfrac{1}{2}\log R^* \tag{20}$$

with the $(\log L_0^*, \log R_0^*)$-curve defines the stage at which degeneracy would just begin to develop at the centre. In (20), $\beta^\dagger$ represents the value β_1 has for the prescribed mass in the wholly gaseous state. In Table I a set of corresponding pairs of values for $\log L_0^*$ and $\log R_0^*$ is given and the corresponding locus is shown in fig. 1. The part of the plane below this curve defines our *domain of degeneracy* in this plane.

TABLE I

x	$1-\beta_1$	$\log L_0^*$	$\log R_0^*$	x	$1-\beta_1$	$\log L_0^*$	$\log R_0^*$
0	0	$-\infty$	$+\infty$	3·0	·07149	$\bar{2}$·9413	$\bar{1}$·5190
0·2	·00039	$\bar{8}$·2429	0·2982	3·5	·07598	$\bar{1}$·0520	$\bar{1}$·4579
0·4	·00282	$\bar{6}$·6625	·1394	4·0	·07920	$\bar{1}$·1339	$\bar{1}$·4039
0·6	·00793	$\bar{5}$·9546	·0397	4·5	·08158	$\bar{1}$·1969	$\bar{1}$·3556
0·8	·01505	$\bar{4}$·7689	$\bar{1}$·9632	5·0	·08337	$\bar{1}$·2479	$\bar{1}$·3120
1·0	·02305	$\bar{3}$·3221	$\bar{1}$·8995	6·0	·08583	$\bar{1}$·3249	$\bar{1}$·2357
1·2	·03101	$\bar{3}$·7164	$\bar{1}$·8441	7·0	·08739	$\bar{1}$·3817	$\bar{1}$·1706
1·4	·03839	$\bar{2}$·0076	$\bar{1}$·7949	8·0	·08844	$\bar{1}$·4261	$\bar{1}$·1138
1·6	·04496	$\bar{2}$·2290	$\bar{1}$·7503	9·0	·08918	$\bar{1}$·4624	$\bar{1}$·0625
1·8	·05068	$\bar{2}$·4016	$\bar{1}$·7095	10·0	·08972	$\bar{1}$·4931	$\bar{1}$·0182
2·0	·05561	$\bar{2}$·5391	$\bar{1}$·6720	20·0	·09150	$\bar{1}$·6691	$\bar{2}$·7193
2·2	·05983	$\bar{2}$·6506	$\bar{1}$·6372	30·0	·09185	$\bar{1}$·7620	$\bar{2}$·5435
2·4	·06344	$\bar{2}$·7427	$\bar{1}$·6048	∞	·09212	∞	$-\infty$
2·6	·06653	$\bar{2}$·8198	$\bar{1}$·5744				
2·8	·06919	$\bar{2}$·8852	$\bar{1}$·5459				

It is of course clear that the $(\log L_0^*, \log R_0^*)$-curve asymptotically approaches the line

$$\log L^* = \log L^*(\beta_\omega) - \tfrac{1}{2}\log R^* = 1\cdot 0378 - \tfrac{1}{2}\log R^*, \tag{21}$$

where, as in I, equation (15), β_ω is such that

$$\frac{960}{\pi^4}\frac{1-\beta_\omega}{\beta_\omega} = 1. \tag{22}$$

That the $(\log L_0^*, \log R_0^*)$-curve asymptotically approaches the line (21) simply corresponds to the fact that $\mathfrak{M}$ represents the upper limit to the masses for which degeneracy can set in on contraction.

When $M \ll M_3$ one easily obtains the asymptotic relation (*cf.* I, equation 48)

$$L_0^* \sim \frac{\pi^4}{960}\left(\frac{4}{5}\right)^{33/2} R_0^{*-17}, \tag{23}$$

or

$$\log L_0^* = \bar{3}\cdot 4073 - 17\log R_0^*, \quad (M^* \to 0). \tag{24}$$

The line (24) is also shown in fig. 1.

8. *The Nature of the Curves of Constant Mass in the Domain of Degeneracy.*—Consider a mass less than M_3. Then for this mass there exists an equili-

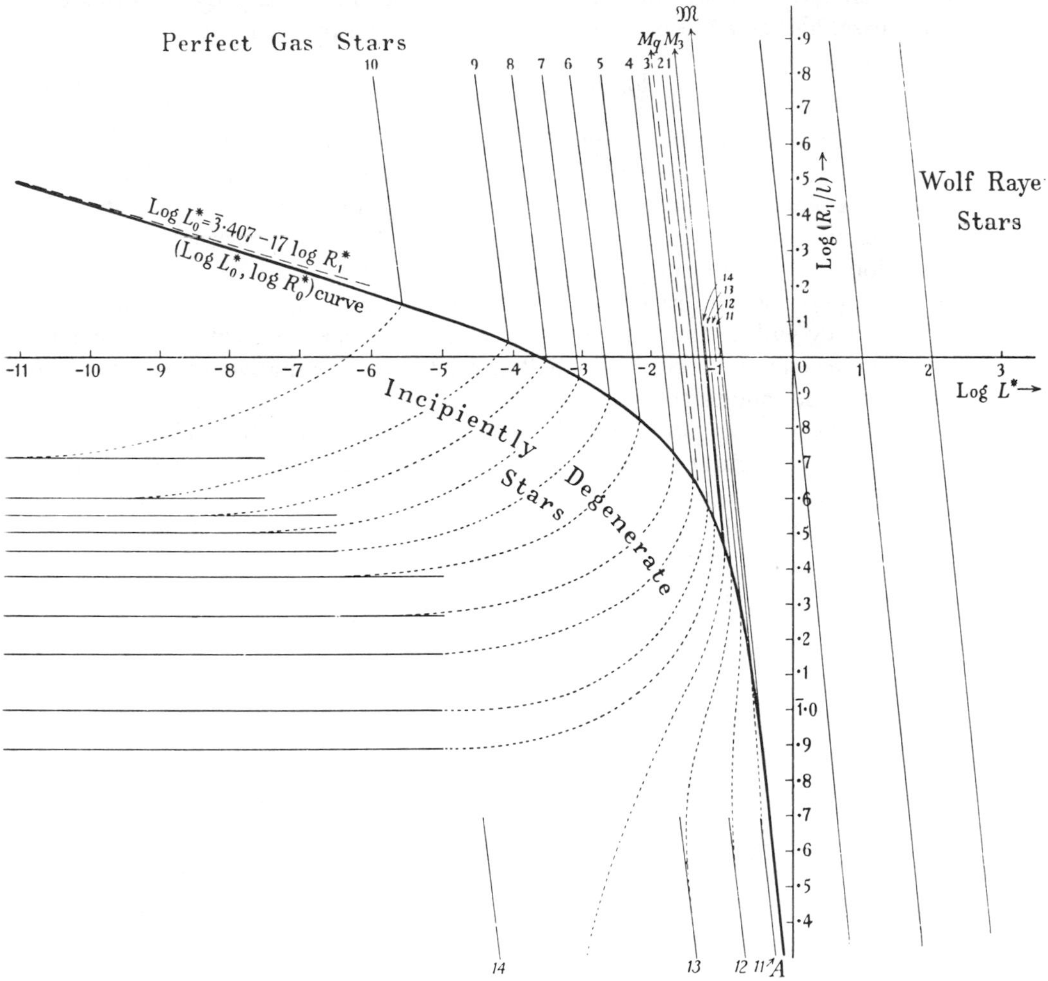

FIG. 1.—*The nature of the curves of constant mass in the (log L*, log R*)-plane on the usual standard model.*

For a general description of the results summarized in the above diagram, see 8 *and also* 23. *On the generalized standard model the system of the curves will look slightly different: thus the continuation of the curves of constant mass marked* (11 . . . 14) *in the perfect gas region will all tend asymptotically to the line* 𝔐*A*. *But the continuation of the curves of constant mass for those marked* (1 . . . 10) *in the perfect gas region must on all models eventually become asymptotic to the lines* (1 . . . 10) *in the domain of degeneracy (which is the region below the (log* L_0^*, *log* R_0^**)-curve).*

brium configuration in which it is completely degenerate and has a radius R_1^* given by (*cf.* I, equation (46))

$$R_1^* = \frac{1}{y_0(M)} \frac{\eta_1(\phi(y_0(M)))}{\xi_1(\theta_3)}. \tag{25}$$

Hence if we start with this mass and contract it from infinite extension, then its luminosity increases according to (20) till this line intersects the ($\log L_0^*$, $\log R_0^*$) locus. On further contraction the configuration develops a degenerate core of finite dimensions and the luminosity must ultimately decrease, and as $L^* \to 0$ the curve must tend asymptotically to

$$\log R^* = \log R_1^* = \text{constant}. \qquad (26)$$

In H.C. II we have obtained the values of η_1, M^*, etc. for ten different values of y_0, and for these configurations the values of $\log R_1^*$ and $\log L^*(\beta^\dagger)$ can be evaluated. The results of such calculations are given in Table II. The corresponding lines are shown in fig. 1 (the lines marked 1 to 10 in the domain of degeneracy and also in the perfect gas region).

The precise nature of the curves of constant mass in the domain of degeneracy will depend on the assumptions one makes regarding the opacity of the degenerate core. We shall indicate the qualitative results for the two extreme models ($\beta_1 = \beta_2$ and $\beta_2 = 1$) discussed in I, §§ 18–21.

(A) *The Usual Standard Model* ($\beta_1 = \beta_2$).—(i) For $M < M_3$, since the composite configurations are all of the collapsed type, it is clear that as soon as the configuration begins to develop degeneracy at the centre the luminosity should begin to increase less rapidly than $R^{-\frac{1}{2}}$. The curves of constant mass must therefore be of the nature shown by the dotted curves in fig. 1.

TABLE II

$\frac{1}{y_0^2}$	M/M_3	$\log L^*(\beta_1)$	$\log R_1^*$
0	1·0	$\bar{2}$·7543	$-\infty$
·01	·95733	$\bar{2}$·6670	$\bar{2}$·8903
·02	·92419	$\bar{2}$·5957	$\bar{1}$·0096
·05	·84709	$\bar{2}$·4172	$\bar{1}$·1602
·1	·75243	$\bar{2}$·1690	$\bar{1}$·2708
·2	·61589	$\bar{3}$·7374	$\bar{1}$·3832
·3	·51218	$\bar{3}$·3286	$\bar{1}$·4538
·4	·42600	$\bar{4}$·9116	$\bar{1}$·5095
·5	·35033	$\bar{4}$·4620	$\bar{1}$·5590
·6	·28137	$\bar{5}$·9521	$\bar{1}$·6072
·8	·15316	$\bar{6}$·5173	$\bar{1}$·7198

(ii) The curve for $M = M_3$ bends inwards as it enters the domain of degeneracy and goes to $(-\infty, -\infty)$.

(iii) For $M_3 < M < \mathfrak{M}$ the luminosity initially increases less rapidly than $R^{-1/2}$ on entering the domain of degeneracy, but should again ultimately increase as $R^{-1/2}$, since for these masses the curves of constant mass should be asymptotic to the line

$$\log L^* = \log L^*(\beta^*) - \tfrac{1}{2} \log R^*, \qquad (27)$$

where β^* is related to $\beta^\dagger$ (the value β_1 has in the wholly gaseous state) by the equation (*cf.* I, equation (56)),

$$\beta^* = \left(\frac{\pi^4}{960}\frac{\beta^{\dagger 4}}{1-\beta^\dagger}\right)^{1/3}. \tag{28}$$

In fig. 1 this feature of the usual standard model is indicated.

(B) *The Generalized Standard Model* ($\beta_2 = 1$).—(i) For $M \leqslant M_q$ ($\sim 0{\cdot}87 M_3$) the composite configurations are all of the collapsed type and the qualitative nature of the curves of constant mass for these masses must therefore be of the same nature as A (i) above.

(ii) For $M_q < M < M_3$, the composite configurations are initially of the centrally condensed type, and hence for these masses the luminosity will begin to increase *more* rapidly than $R^{-1/2}$ on developing degeneracy at the centre. However, the luminosity must begin to decrease after attaining a certain maximum, since eventually the configurations must tend towards the completely degenerate state.

(iii) For $M_3 \leqslant M < \mathfrak{M}$ the luminosity increases more rapidly than $R^{-1/2}$ in the domain of degeneracy and the curves of constant mass for all these masses must asymptotically tend to the line (*cf.* equations (21), (22))

$$\log L^* = \log L^*(\beta_\omega) - \tfrac{1}{2} \log R^*. \tag{29}$$

9. So far we have restricted ourselves to the standard model approximation, and in relating the completely degenerate configurations with $M \leqslant M_3$ with the wholly gaseous configurations we have seen how such a stellar mass tends to the completely degenerate state when the luminosity tends to zero through a sequence of equilibrium configurations all conforming to the standard model. In addition to M_3 there appeared another mass $\mathfrak{M}$ which played an important rôle in the theory. $\mathfrak{M}$ was initially defined as one for which $\beta^\dagger = \beta_\omega$ (*cf.* I, equations (15), (29)), but the relation between M_3 and $\mathfrak{M}$, namely (I, equation (52)),

$$\mathfrak{M} = M_3 \beta_\omega^{-3/2}, \tag{30}$$

merely means that the existence of an upper limit M_3 to the mass of a completely degenerate configuration and the upper limit $\mathfrak{M}$ to the mass of a configuration for which degeneracy can at all set in on contraction are closely related to one another. Thus configurations in the mass range $M_3 \leqslant M \leqslant \mathfrak{M}$ bridge the gap between masses for which we have equilibrium configurations with zero luminosity and those which cannot develop degenerate cores however far the contraction may proceed.

The existence of a mass $\mathfrak{M}$ is by no means surprising, for, as we have already emphasised in I, § 22, the increased dominance of the radiation pressure for large stellar masses is quite a general result,* and the possibility of degeneracy is entirely excluded if only the radiation pressure is greater than a tenth of the total pressure throughout the entire mass. Hence it is clear that the general features that can be inferred from fig. 1 of this paper,

* An elementary proof of this result is given in the author's report in *Nordisk Astronomisk Tidskrift*, **16**, 37, 1935.

for instance, should to a large extent be independent of the model on the basis of which the discussion has been carried out. It is, however, of interest to verify that this is so by discussing the relation between the completely degenerate configurations and the wholly gaseous configurations on the basis of a more general scheme than what the standard model provides. As the analysis required for this verification is given in Section II below, we shall postpone to Section III some general considerations which arise from a closer examination of fig. 1.

Section II

10. The starting-point of our present discussion is provided by Jeans's investigations on gaseous configurations in which "$1-\beta_1$" is allowed to vary. We shall briefly recapitulate Jeans's analysis in our present notation.

It is clear, of course, that $(1-\beta_1)$ must decrease inwards but the precise law of variation will depend on various factors. If one assumes that

$$\frac{M}{L}\frac{L(r)}{M(r)} \propto T^{\delta}, \tag{31}$$

and that further the coefficient of opacity varies according to the law

$$\kappa \propto \frac{\rho}{T^{7/2}}, \tag{32}$$

then one can easily show that to a fair degree of approximation we have

$$\frac{\beta_1}{(1-\beta_1)^2} = \frac{\beta_c}{(1-\beta_c)}\left(\frac{T}{T_0}\right)^{(\frac{1}{2}-\delta)}, \tag{33}^*$$

where β_1 as usual defines the ratio of the gas pressure to the total pressure, and β_c the value of β_1 at the centre of the configuration where the temperature is assumed to be equal to T_0. Equation (33) is valid for layers not immediately near the surface. Further, we have quite generally that

$$\frac{\rho}{\rho_0} = \frac{\beta_1}{1-\beta_1}\frac{1-\beta_c}{\beta_c}\left(\frac{T}{T_0}\right)^3. \tag{34}$$

From (33) and (34) we deduce that

$$\frac{\beta_1}{(1-\beta_1)^2}\left(\frac{\beta_1}{1-\beta_1}\right)^{\frac{1}{3}(\frac{1}{2}-\delta)} = \frac{\beta_c}{(1-\beta_c)^2}\left(\frac{\beta_c}{1-\beta_c}\right)^{\frac{1}{3}(\frac{1}{2}-\delta)}\left(\frac{\rho}{\rho_0}\right)^{\frac{1}{3}(\frac{1}{2}-\delta)}. \tag{35}$$

From (35) we obtain that

$$[(1+\beta_1)+\tfrac{1}{3}(\tfrac{1}{2}-\delta)]\frac{d\beta_1}{\beta_1(1-\beta_1)} = \tfrac{1}{3}(\tfrac{1}{2}-\delta)\frac{d\rho}{\rho}. \tag{36}$$

Since, however, the total pressure P is given,

$$P = \left[\left(\frac{k}{\mu H}\right)^4 \frac{3}{a}\frac{1-\beta_1}{\beta_1{}^4}\right]^{1/3}\rho^{4/3}, \tag{37}$$

* This equation is due to Jeans and Woltjer.

we have (assuming μ constant *)

$$\frac{dP}{P} = -\frac{1}{3}\frac{4-3\beta_1}{\beta_1(1-\beta_1)}d\beta_1 + \frac{4}{3}\frac{d\rho}{\rho}; \tag{38}$$

or using (36),

$$\frac{dP}{P} = \frac{1}{3}\left[4 - \frac{(\frac{1}{2}-\delta)(4-3\beta_1)}{3(1+\beta_1)+(\frac{1}{2}-\delta)}\right]\frac{d\rho}{\rho}. \tag{39}$$

On the other hand, if

$$P = K\rho^{1+\frac{1}{n}}, \tag{40}$$

we should have

$$\frac{dP}{P} = \left(1 + \frac{1}{n}\right)\frac{d\rho}{\rho}. \tag{41}$$

Comparing (39) and (41) we have for the "*effective polytropic index*" n the expression

$$n = 3 + (1-2\delta)\frac{4-3\beta_1}{1+3\beta_1+2\delta(1-\beta_1)}. \tag{42}$$

Equation (42) shows how the effective polytropic index n varies through the star. (With $\delta = \frac{1}{2}$, $n = 3$ = constant, and we go back to the standard model.) However, Jeans considers that a fair approximation is obtained by regarding the whole configuration as a complete Emden polytrope with an index n_{β_c} given by

$$n_{\beta_c} = 3 + (1-2\delta)\frac{4-3\beta_c}{1+3\beta_c+2\delta(1-\beta_c)}. \tag{43}$$

11. If one assumes (43), then (37) can be rewritten as ($n = n_{\beta_c}$),

$$P = \left[\left(\frac{k}{\mu H}\right)^4 \frac{3}{a}\frac{1-\beta_c}{\beta_c^4}\right]^{1/3} \frac{1}{\rho_0^{(3-n)/3n}}\rho^{1+\frac{1}{n}}. \tag{44}$$

Or in terms of our A_2 and B we have

$$P = \frac{2A_2}{B^{4/3}}\left(\frac{960}{\pi^4}\frac{1-\beta_c}{\beta_c^4}\right)^{1/3} \frac{1}{\rho_0^{(3-n)/3n}}\rho^{1+\frac{1}{n}}. \tag{45}$$

The justification for (44) is simply that it gives the same initial variation of P with ρ at the centre as is jointly predicted by (35) and (37) taken together.

With (45) as the "equation of state" the structure of the configuration is completely specified by the Emden function $\theta_{n_{\beta_c}}$ with index n_{β_c}. We easily find that the mass of the configuration is given by

$$M = M_3\left(\frac{960}{\pi^4}\frac{1-\beta_c}{\beta_c^4}\right)^{1/2} \mathfrak{J}_M(n_{\beta_c}), \tag{46}$$

where

$$\mathfrak{J}_M(n) = \left(\frac{n+1}{4}\right)^{3/2}\frac{(\xi^2\theta_n')_1}{(\xi^2\theta_3')_1}. \tag{47}$$

* Variation of μ according to the law $\mu a T^j$ can easily be taken into account, but we shall not consider these refinements here.

If $\delta = 1/2$, $n_{\beta_c} = 3$ and $\mathfrak{J}_M(3) = 1$, and (46) reduces to our earlier equation (2). Hence we can rewrite (46) as

$$M_{(\delta)}(\beta_c) = M_{(1/2)}(\beta_c) \cdot \mathfrak{J}_M(n_{\beta_c}), \tag{48}$$

in an obvious notation.

12. *The Domain of Degeneracy in the* $(R, 1-\beta_c)$*-diagram.*—For a given mass M equations (43), (46) and (47) uniquely determine a β_c. In particular there will be a mass for which $(1-\beta_c) = (1-\beta_\omega)$. We shall denote this mass by $\mathfrak{M}_{(\delta)}$. By (46)

$$\mathfrak{M}_{(\delta)} = M_3 \beta_\omega^{-3/2} \mathfrak{J}_M(n_{\beta_\omega}) = \mathfrak{M}_{(1/2)} \mathfrak{J}_M(n_{\beta_\omega}), \tag{49}^*$$

where

$$n_{\beta_\omega} = 3 + (1-2\delta)\frac{4 - 3\beta_\omega}{1 + 3\beta_\omega + 2\delta(1-\beta_\omega)}. \tag{50}$$

Arguing as in I, § 5, we now see that all configurations with $M > \mathfrak{M}_{(\delta)}$ are necessarily wholly gaseous, and that therefore for these masses the curves of constant mass in the $(R, 1-\beta_c)$-diagram are fully represented by the lines parallel to the R-axis. However, for $M < \mathfrak{M}_{(\delta)}$ degeneracy would begin to develop when the central density is given by

$$\rho_0 = B x_0^3, \tag{51}$$

where x_0 is such that

$$\frac{f(x_0)}{2x_0^4} = \left(\frac{960}{\pi^4}\frac{1-\beta_c}{\beta_c}\right)^{1/3}. \tag{52}$$

The radius $R_0(\beta_c\,;\,\delta)$ of this configuration can be determined as in I, § 5, and we find that

$$\frac{R_0(\beta_c\,;\,\delta)}{l} = \left(\frac{n+1}{4}\right)^{1/2}\left(\frac{960}{\pi^4}\frac{1-\beta_c}{\beta_c^4}\right)^{1/6}\frac{1}{x_0}\frac{\xi_1(\theta_{n_{\beta_c}})}{\xi_1(\theta_3)}. \tag{53}$$

Comparing this with I, equation (34), we deduce that

$$R_0(\beta_c\,;\,\delta) = R_0(\beta_c\,;\,1/2)\mathfrak{J}_R(n_{\beta_c}), \tag{54}$$

where

$$\mathfrak{J}_R(n) = \left(\frac{n+1}{4}\right)^{1/2}\frac{\xi_1(\theta_n)}{\xi_1(\theta_3)}. \tag{55}$$

(53) now defines a curve in the $(R, 1-\beta_c)$-plane. The region enclosed by the two axes and the $(R_0, 1-\beta_c)$-curve now defines our domain of degeneracy in this plane.

13. In our numerical work we shall confine ourselves exclusively to the case $\delta = 0$. Then we have

$$n_{\beta_c} = 3 + \frac{4-3\beta_c}{1+3\beta_c}. \tag{56}$$

The quantities $\mathfrak{J}_R(n)$ and $\mathfrak{J}_M(n)$ are known for certain values of n, and for the intermediate values recourse was made to methods of interpolation.

* It may be noticed here that $\mathfrak{M}_{(1/2)}$ is our original "$\mathfrak{M}$."

14. From (56) we now deduce that

$$n_{\beta_\omega}=3\cdot343\,;\quad \mathcal{J}_M(3\cdot343)=1\cdot080. \tag{57}$$

Hence

$$\mathfrak{M}_{(0)}=1\cdot080\mathfrak{M}_{(1/2)}=1\cdot249M_3, \tag{58}$$

or putting in numerical values

$$\mathfrak{M}_{(0)}=7\cdot153\mu^{-2}\odot. \tag{59}$$

15. In Table III we have tabulated for this model ($\delta=0$) the corresponding sets of values for x, $1-\beta_c$, R_0 and M. The (R_0, $1-\beta_c$)-curve is shown in fig. 2.

TABLE III

x	$1-\beta_c$	$n_{\beta c}$	R_0/l	M/M_3
0	0	3·250	∞	0
0·2	·00039	3·250	2·381	·066
0·4	·00282	3·253	1·655	·178
0·6	·00793	3·257	1·320	·301
0·8	·01505	3·264	1·113	·421
1·0	·02305	3·272	·967	·531
1·2	·03101	3·280	·856	·627
1·4	·03839	3·287	·769	·710
1·6	·04496	3·294	·697	·780
1·8	·05068	3·299	·638	·839
2·0	·05561	3·304	·587	·889
2·2	·05983	3·309	·544	·931
2·4	·06344	3·312	·506	·967
2·6	·06653	3·316	·473	·998
2·8	·06919	3·318	·444	1·024
3·0	·07149	3·321	·418	1·046
3·5	·07598	3·326	·365	1·090
4·0	·07920	3·329	·323	1·122
4·5	·08158	3·331	·289	1·145
5·0	·08337	3·333	·262	1·163
6·0	·08583	3·336	·220	1·187
7·0	·08739	3·338	·190	1·202
8·0	·08844	3·339	·167	1·212
9·0	·08918	3·340	·149	1·220
10·0	·08972	3·340	·134	1·225
20·0	·09150	3·342	·067	1·242
30·0	·09185	3·342	·045	1·246
∞	·09212	3·343	0	1·249

16. *The Nature of the Curves of Constant Mass for $M \leqslant M_3$ in the Domain of Degeneracy.*—Let $\beta_c^{\dagger}$ be the value of β_c for a wholly gaseous configuration (of mass less than M_3) which in its completely collapsed state has a central density corresponding to $y=y_0$. Then by equation (46)

$$\frac{\Omega(y_0)}{\omega_3^{\,0}}=\left(\frac{960}{\pi^4}\,\frac{1-\beta_c^{\dagger}}{\beta_c^{\dagger 4}}\right)^{1/2}\mathcal{J}_M(n_{\beta_c}{}^{\dagger}). \tag{60}$$

Now the line through $(1-\beta_c^\dagger)$ parallel to the R-axis will intersect the $(R_0, 1-\beta_c)$-curve at $(R_0(M(y_0)), 1-\beta_c^\dagger)$. In the domain of degeneracy the continuation of the curve must in some way connect the point $(R_0(M(y_0)), 1-\beta_c^\dagger)$ and the point R_1 on the R-axis where

$$\frac{R_1}{l}=\frac{1}{y_0(M)}\frac{\eta_1(\phi(y_0(M)))}{\xi_1(\theta_3)}. \tag{61}$$

In I (Table II) we have already tabulated the values of R_1 for ten different values for y_0 and for these configurations $(1-\beta_c^\dagger)$ was obtained by interpolating among the figures given in Table III. The corresponding pairs of points on the $(R_0, 1-\beta_c)$-curve and the R-axis are shown in fig. 2. (The points marked 5 to 15 on the R-axis and also on the $(R_0, 1-\beta_c)$-curve.) It is of course clear that for $M=M_3$ the associated curve of constant mass must pass through the origin of our system of axes. If we denote by $\beta_c(0)$ the value of β_c which M_3 has in the wholly gaseous state then we should have by (60) that

$$\left(\frac{960}{\pi^4}\frac{1-\beta_c(0)}{\beta_c^4(0)}\right)^{1/2}\mathfrak{J}_M(n_{\beta_c(0)})=1. \tag{62}$$

Numerically $(1-\beta_c(0))$ is found to be 0·0668.

17. *The Nature of the Curves of Constant Mass for $M_3 \leqslant M \leqslant \mathfrak{M}_{(\delta)}$ in the Domain of Degeneracy.*—The discussion of this case will naturally depend on the assumption one makes regarding the opacity in the degenerate core. We shall assume that " $(\kappa\eta)_2$ " is constant in the core. Then we have (*cf.* I, equation (63))

$$P=\frac{1}{\beta_2}A_2 f(x)-D_2\frac{1-\beta_2}{\beta_2}, \tag{63}$$

where D_2 is a constant and

$$\beta_2=\left(1-\frac{(\kappa\eta)_2 L}{4\pi cGM}\right). \tag{63'}$$

The reduction to our differential equation (1) for the degenerate core follows at once.

In our present scheme " β " is of course allowed to vary in the gaseous envelope, and we shall in the first instance consider the case where β_2 is just equal to the value of β_1 at the interface between the degenerate core and the gaseous envelope. For this case the discussion can be carried out as in I, §9.

A completely relativistically degenerate configuration has a mass given by (H.C. I, page 463)

$$M=M_3\beta^{-3/2}, \tag{64}$$

and is of zero radius. Since we have the further relation

$$\mathfrak{M}_{(1/2)}=M_3\beta_\omega^{-3/2}, \tag{65}$$

it follows that the curve of constant mass for $\mathfrak{M}_{(1/2)}$ must connect the point $(R_0, 1-\beta_c^\dagger(\mathfrak{M}_{(1/2)}))$ on the $(R_0, 1-\beta_c)$-curve and the point $(0, 1-\beta_\omega)$ on the $(1-\beta_c)$-axis. It is therefore clear that for $M_3 \leqslant M \leqslant \mathfrak{M}_{(1/2)}$ the

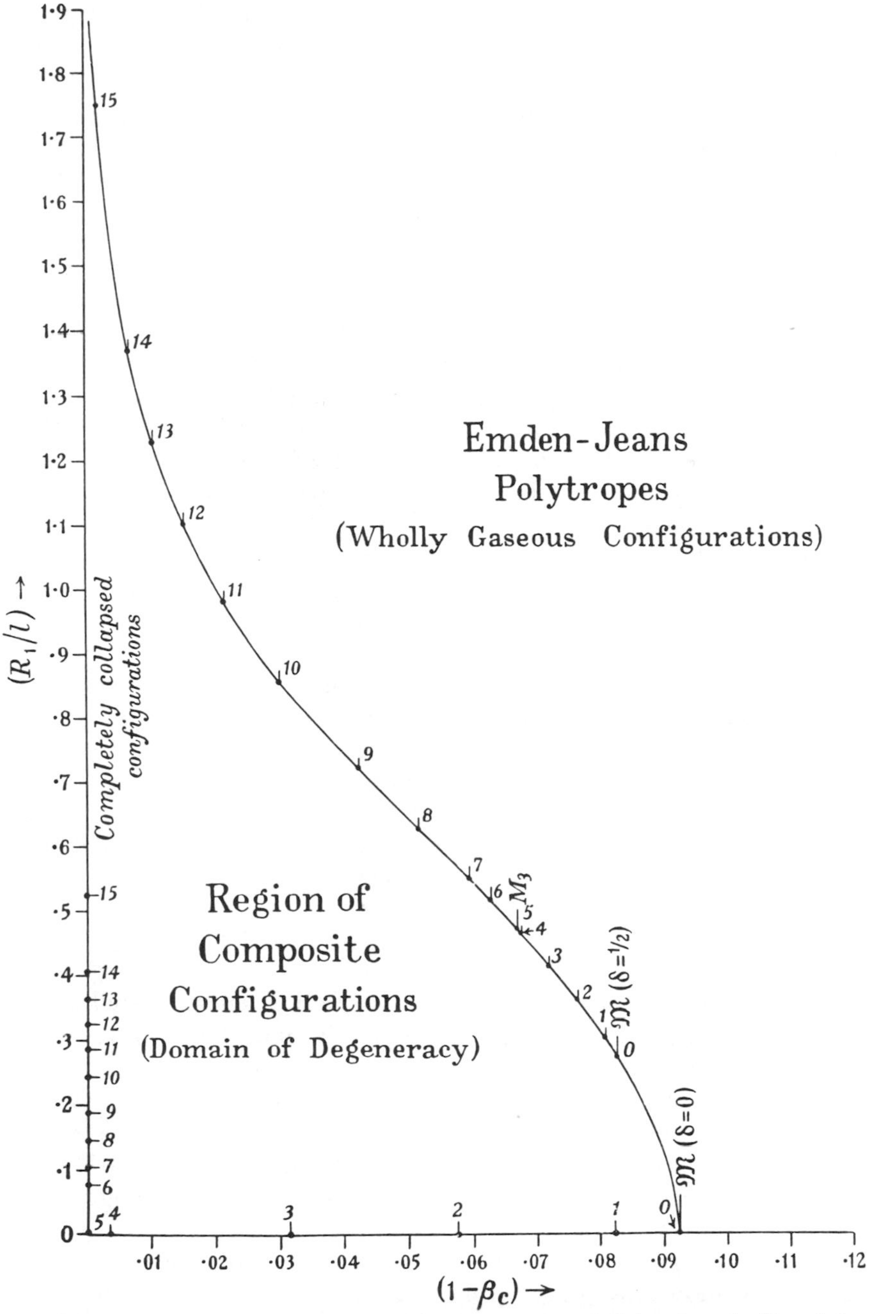

FIG. 2.—*The curve running from* $1-\beta_c = 0{\cdot}92$ . . . *to infinity along the* R_1*-axis is the* $(R_0(\beta_c\,;\ \delta=0),\ 1-\beta_c)$*-curve (see equation* (53)). *The points marked* (5 . . . 15) *on the* $(R_0,\ 1-\beta_c)$*-curve and on the* R_1*-axis are the end points in the domain of degeneracy for the curves of constant mass for the values of* M *for which the* ϕ*-integrals are known (H.C. II). The points marked* (0, 1 . . . 4) *on the* $(R_0,\ 1-\beta_c)$*-curve and on the* $(1-\beta_1)$*-axis are the corresponding end points for some curves of constant mass in the domain of degeneracy on the model discussed in* 17 *(see equation* (67)). *The point* 0 *corresponds to* $\mathfrak{M}_{(\delta=\frac{1}{2})}$, i.e. *our earlier* "$\mathfrak{M}$."

curves of constant mass must cross the $(1-\beta_c)$-axis at a point $(1-\beta^*)$ such that

$$M = M_3 \beta^{*-3/2}. \tag{66}$$

If $\beta_c^\dagger$ denotes the value of β_c in the wholly gaseous state then by equating (66) and (46) we have

$$\beta^* = \left(\frac{\pi^4}{960}\frac{\beta_c^{\dagger 4}}{1-\beta_c^\dagger}\right)^{1/3} (\mathcal{J}_M(n_{\beta_c}{}^\dagger))^{-2/3}. \tag{67}$$

Some corresponding pairs of points on the $(R_0, 1-\beta_c)$-curve and the $(1-\beta_c)$-axis are shown in fig. 2.

Finally, if $\mathfrak{M}_{(1/2)} \leqslant M \leqslant \mathfrak{M}_{(\delta)}$ it is immediately clear that the curves of constant mass consist simply of segments connecting the point $(R_0, 1-\beta_c^\dagger(M))$ on the $(R_0, 1-\beta_c)$-curve to the point $(0, 1-\beta_\omega)$ on the $(1-\beta_c)$-axis.

We thus see that on this model the curves of constant mass in the $(R_0, 1-\beta_c)$-plane combine in the same diagram some of the features of both figs. 3 and 4 of I, obtained on the basis of the two extreme cases of the generalized standard model.

18. It may finally be pointed out that if one assumed that the opacity of the degenerate core is zero then the general qualitative features of the system of the curves of constant mass in the $(R, 1-\beta_c)$-plane must be *exactly* the same as in I, fig. 4.

19. A complete discussion on the basis of Jeans's model will require a study of the composite configurations. The formal theory (which would run similar to I, §§ 11 to 15) can easily be sketched, but as such discussions are not of much interest without the necessary numerical work (which would be considerable) we shall not go into these details here. However, it is clear that the general results derived on the basis of the standard model are fully retained even in this more general analysis.

Section III

20. *The Wolf-Rayet Phenomenon.*—It has already been suggested in I that the Wolf-Rayet phenomenon of the radial ejection of matter may be indirectly due to the fact that the stars of mass greater than $\mathfrak{M}$ (or its equivalent $\mathfrak{M}_{(\delta)}$ on the more general stellar models discussed in Section II) cannot pass *directly* into the white-dwarf stage.* This suggestion is confirmed by observation in so far as general estimates do indicate that Wolf-Rayet stars are massive and dense. On the theoretical side the suggestion gains further support from the following argument:—

Consider a mass greater than $\mathfrak{M}_{(1/2)}$. On the standard model the star must necessarily be wholly gaseous, and we have in a certain system of "natural units" (*cf.* equation (14))

$$L^* = (M^*\beta_1)^{7/2}(1-\beta_1)R^{*-1/2}. \tag{68}$$

* This suggestion (in a rather different form) was independently made to the writer by Dr. W. H. McCrea, to whom the author had earlier communicated a preliminary statement of the main results in the form later published in *Observatory*, **57**, 373, 1934.

Further, the value of the surface gravity g is given by

$$g = \frac{GM}{R^2}. \tag{69}$$

From (68) and (69) we have for the ratio X between the integrated flux of radiation πF at the surface of the star to the value of gravity, the expression

$$X = \frac{\pi F}{g} = \frac{L}{4\pi cGM} = \frac{L_1}{4\pi cGM_3} M^{*5/2} \beta_1^{7/2} (1 - \beta_1) R^{*-1/2}. \tag{70}$$

From our definition of L_1 in (13) we have

$$X = \left(\frac{1}{\alpha \kappa_1}\right) M^{*5/2} \beta_1^{7/2} (1 - \beta_1) R^{*-1/2}. \tag{71}$$

From (71) we see that for a given mass ($> \mathfrak{M}$) the ratio X steadily increases with decreasing R and in fact tends to infinity. This suggests that at some stage in the process of contraction the radiation pressure (a measure of which is given by πF) must overbalance gravity. Ejection of matter must necessarily ensue. In drawing this inference it is of course realized that the deduction from (71) cannot be regarded as a rigorous proof in so far as in our analysis the equations of equilibrium have been integrated up to the boundary. But if one takes this last boundary condition seriously then one cannot also strictly speak of a "mass-luminosity-effective temperature" relation as the temperature has been made zero at the boundary. That this involves no real contradiction was shown in the early writings of Eddington, Jeans and Russell and more recently by Cowling. Bearing this in mind it is now clear that the fact that $X \to \infty$, as $R \to 0$ merely means that the approximations underlying the deduction of the mass-luminosity-radius relation (68) should cease to be valid at some stage. Our conclusion that the ejection of matter must ensue since $X \to \infty$, as $R \to 0$ is now seen to be equivalent to the suggestion that the Wolf-Rayet phenomenon should set in precisely in the region of the Russell diagram where the mass-luminosity-radius relation for the massive wholly gaseous configurations ceases to be valid on the perfect-gas hypothesis itself. It is of course necessary that the star should be massive ($M > \mathfrak{M}$ or its equivalent on more general stellar models) for otherwise we could not extrapolate (68) to high mean densities—degeneracy would have set in earlier for the less massive stars.

21. *The Hydrogen Content of the Massive Stars.*—In § 20 we have used the term "massive stars" to denote those with $M > \mathfrak{M}_{(\delta)}$. It was found on Jeans's model ($\delta = 0$) that we have

$$\mathfrak{M}_{(0)} = 7 \cdot 15 \mu^{-2} \odot. \tag{72}$$

To define $\mathfrak{M}_{(\delta)}$ more precisely we need to know the hydrogen content. Depending on the hydrogen content $\mathfrak{M}_{(\delta)}$ can be varied numerically by a factor 16 ($\mu = \frac{1}{2}$ to $\mu = 2$), and it becomes necessary therefore to know at least the minimum hydrogen contents of stars as a function of their mass.

Fortunately we have for our guidance here Strömgren's systematic investigations of this problem.* From Strömgren's work it appears that the molecular weight of the massive B stars already tends towards the lower limit 0·5. Thus we can conclude that for our purposes a star of mass greater than about 25☉ can be regarded as "massive."

It is of interest to recall in this connection that in his "Interpretation of the Hertzsprung-Russell diagram" Strömgren says: "With an appreciable overcompressible nucleus the predicted luminosities would be appreciably larger than the observed, and increasing the hydrogen content—as is usually possible to remove the difference—is not possible in these cases, as the limit has already been reached. We conclude then that for the B stars in question there cannot be any appreciable nucleus." We now see that Strömgren's conclusions receive further indirect confirmation from our analysis.

22. *The Hydrogen Content of the White Dwarfs.*—The hydrogen content of the white dwarfs had been investigated earlier by various writers on the Emden polytrope $n=3/2$ approximation for them. In our H.C. II we have made an exact study of these completely degenerate configurations, and it is now possible to make a more reliable estimate for the appropriate molecular weights for the white dwarfs.

The necessary data required for this calculation are given in Table III of H.C. II. The following table is due to Strömgren :—

TABLE IV

μ *for Sirius B*

μ	1·91	1·74	1·58	1·44	1·29	0·95
T_{eff} (calculated)	18,300°	13,800°	11,300°	9700°	8300°	6200°

From the above table Strömgren concludes that the value of μ for Sirius *B* should be about 1·6, which means relatively low hydrogen content. The low hydrogen content of the white dwarfs has already been discussed by Strömgren (*loc. cit.*).

23. *Some Remarks on Figure* 1.—Figure 1 is of course the domain of the Hertzsprung-Russell diagram. From an examination of this diagram it is immediately clear that the white dwarfs are placed in their right positions in the Russell diagram. The two essential observational results concerning the white dwarfs, namely, their small mass and low luminosity, receive their natural explanations. The region of the diagram in which we should expect the Wolf-Rayet stars is also indicated. The region of the ordinary stars is indicated by "perfect gas stars." Presumably stars like Krueger 60 are representatives of the "incipiently degenerate" region of our diagram.

The above general conclusions, so far as they go, should clarify the present position regarding stellar structure.

* B. Strömgren, *Z. f. Astrophysik*, **7**, 222, 1933.

24. *Deviations from Perfect Gas Laws arising from Causes other than Degeneracy.*—In our discussion we have so far considered only deviations from perfect gas laws which are due to degeneracy. However, Dirac's theory of the electron predicts a further different type of deviation from the perfect gas laws due to the production of electron pairs at very high temperatures. The bearing of this phenomenon on the theory of stellar structure has been examined in a preliminary communication by L. Rosenfeld and the writer.* As we have indicated in that letter, the deviations from the perfect gas laws arising from this cause are of quite negligible importance for stars with $M < \mathfrak{M}_{(\delta)}$; however, they become increasingly important for the very massive stars. The detailed results of this study will be published separately by Rosenfeld and the writer, but it may be mentioned here that it follows from that study that the production of electron pairs will be of importance in considerations of the structure of stars of masses about 80⊙ and more. The existence of such very massive stars is indicated by the work of J. S. Plaskett, O. Struve, Bottlinger, Trumpler and others, and it seems very probable that the discussion of their structure will lead to some essentially new considerations in the studies on stellar structure.

Finally, it is necessary to point out in this connection that J. von Neumann has recently shown that the *very* ultimate equation of state for matter should *always* be

$$P = \frac{1}{3}c^2\rho. \tag{73}$$

The considerations of this new equation of state does not, however, introduce any essential modifications in our present scheme.

Concluding Remarks.—In two earlier papers (*M.N.*, **95**, 207–260, 1935) a first systematic attack was made on the problem of how the conclusions regarding stellar constitution and stellar evolution that have been drawn on the perfect gas hypothesis for the stars have to be modified by the physical possibility of degeneracy in stellar interiors. In this paper the discussion is carried one stage further. Firstly, the physical results have been made more explicit by considering the curves of constant mass in the $(\log L, \log R)$-diagram, which is essentially the domain of the Hertzsprung-Russell diagram. Secondly, the analysis has been extended to include other stellar models more general than the standard model. Thirdly, the bearing of the Wolf-Rayet phenomenon on the evolution of massive stars is examined a little more closely. Certain other miscellaneous questions have also been considered.

In conclusion, I wish to record here my best thanks to Dr. W. H. McCrea, Professor J. von Neumann, Dr. L. Rosenfeld and Dr. B. Strömgren for the encouraging interest they have taken in these studies and for many stimulating discussions.

Trinity College, Cambridge :
1935 *June* 7.

* *Nature*, **135**, 999, 1935.

RELATIVISTIC DEGENERACY

CHR. MØLLER, D.PHIL., AND S. CHANDRASEKHAR, PH.D.

1. It has generally been accepted that the pressure P exerted by a gas is simply the mean rate of transfer of momentum across an ideal surface of unit area in the gas. From this definition it follows that

$$PV = \frac{1}{3}\int_0^\infty N(p) p v_p dp, \tag{1}$$

where $N(p)$ is the number of particles having momenta between p and $p+dp$, and v_p is the velocity associated with the momentum p. For an electron assembly which is completely degenerate ($T=0$) (1) reduces to

$$P = \frac{8\pi}{3h^3}\int_0^{p_0} p^3 \frac{dE}{dp} dp, \tag{2}$$

where E is the kinetic energy of the electron which has a momentum p. If we substitute in (2) the relation between E and p given by relativistic mechanics, we immediately get the relation between PV and N (the total number of electrons in the assembly) for completely degenerate matter.

2. However, Eddington has recently suggested † that (2) is based on a misunderstanding and that we cannot simply insert in (2) the relativistic relation between E and p. We are quite unable to follow his arguments in his

† *M.N.*, **95**, 197, 1935.

first paper, but in his second paper * he has made a more direct derivation of his result, using the energy-stress tensor $T_{\mu\nu}$. Since the use of the energy-stress tensor $T_{\mu\nu}$ in the quantum theory is not new it is now possible to see why Eddington obtains a result different from the usual treatment of a degenerate gas. Eddington defines the energy-stress tensor $T_{\mu\nu}$ as "the expectation value of the differential operator

$$T_{\mu\nu} = -\frac{1}{m}\frac{\partial^2}{\partial x_\mu \partial x_\nu}, \tag{3}$$

where the right-hand side is to be summed in an appropriate (invariant) way." Eddington does not justify his choice of (3) except by the statement "to satisfy tensor conditions." However, in relativistic quantum mechanics one defines the energy-stress tensor in a different way. In fact, the energy-stress tensor for one particle in free space is defined by (*cf.* W. Pauli, *Handbuch der Physik*, **24** (1), 255)

$$T_{\mu\nu} = \frac{c\hbar}{2i}\left(\psi^* \alpha^\nu \frac{\partial \psi}{\partial x_\mu} - \frac{\partial \psi^*}{\partial x_\mu}\alpha^\nu \psi\right), \tag{4}†$$

where α^1, α^2, α^3 are the Dirac matrices and α^4 is here defined as i times the unit matrix; also $x_4 = ict$. In (4) ψ is a solution of the Dirac equation and ψ^* is its conjugate complex.

To go over to the case of N electrons in a finite volume V which satisfy the exclusion principle *one has to consider* ψ *and* ψ^* *as non-commuting quantities satisfying the commutation rules established by Jordan and Wigner.*‡

Let $\psi_{\mathbf{p},s}$ be a suitably normalised eigen-solution satisfying the Dirac equation § and representing a plane wave with a definite value for the momentum $\mathbf{p}$ and spin s. Since the electrons are confined in a finite volume, $\mathbf{p}$ takes on only discrete values and, further, s can take two different values corresponding to the two different directions for the spin.

[The discreteness is obtained as usual by imposing the following periodicity condition:—

$$\psi_{\mathbf{p},s}(x + l_x, y + l_y, z + l_z) = \psi_{\mathbf{p},s}(x, y, z), \tag{i}$$

where l_x, l_y, l_z define the sides of a rectangular "box" and the volume V is clearly given by

$$V = l_x l_y l_z.$$

From the above condition (i) we find that the eigen values of the components of momentum are given by

$$p_x = \frac{2\pi n_x \hbar}{l_x}; \qquad p_y = \frac{2\pi n_y \hbar}{l_y}; \qquad p_z = \frac{2\pi n_z \hbar}{l_z}, \tag{ii}$$

where n_x, n_y, n_z are arbitrary integers.]

* We are indebted to Sir A. S. Eddington for allowing us to see a manuscript copy of his paper.

† The tensor $T_{\mu\nu}$ was introduced by Tetrode, *Z. f. Physik*, **49**, 858, 1928.

‡ P. Jordan and E. Wigner, *Z. f. Physik*, **47**, 631, 1928.

§ For the explicit expressions see C. G. Darwin, *P.R.S.*, **118**, 654, 1928.

We now expand ψ and its conjugate complex ψ^* in terms of $\psi_{\mathbf{p},s}$ and $\psi^*_{\mathbf{p},s}$ respectively:

$$\psi=\sum_{\mathbf{p},s}a_{\mathbf{p},s}\,\psi_{\mathbf{p},s},\tag{5}$$

$$\psi^*=\sum_{\mathbf{p},s}a^*_{\mathbf{p},s}\,\psi^*_{\mathbf{p},s}.\tag{6}$$

In (5) and (6) $a_{\mathbf{p},s}$ and $a^*_{\mathbf{p},s}$ are q-numbers satisfying the commutability relations (*cf.* Jordan and Wigner, *loc. cit.*).

$$\left.\begin{aligned}a^*_{\mathbf{p},s}\,a_{\mathbf{p}',s'}+a_{\mathbf{p}',s'}\,a^*_{\mathbf{p},s}&=\delta_{\mathbf{p},\mathbf{p}'}\,\delta_{s,s'}\\ a_{\mathbf{p},s}\,a_{\mathbf{p}',s'}+a_{\mathbf{p}',s'}\,a_{\mathbf{p},s}&=0\\ a^*_{\mathbf{p},s}\,a^*_{\mathbf{p}',s'}+a^*_{\mathbf{p}',s}\,a_{\mathbf{p},s}&=0\end{aligned}\right\}.\tag{7}$$

From (4) we now have

$$S=\tfrac{1}{3}(T_{11}+T_{22}+T_{33})$$

$$=\frac{c\hbar}{6i}\sum_{\mathbf{p},s}\sum_{\mathbf{p}',s}\psi^*_{\mathbf{p},s}\left\{a^*_{\mathbf{p},s}\left(\sum_{\nu=1}^{3}\alpha^{\nu}\frac{i}{\hbar}p_{\nu}'\right)a_{\mathbf{p}',s'}+a^*_{\mathbf{p},s}\left(\sum_{\nu=1}^{3}\alpha^{\nu}\frac{i}{\hbar}p_{\nu}\right)a_{\mathbf{p}',s'}\right\}\psi_{\mathbf{p}',s}\tag{8}$$

or

$$S=\frac{c}{6}\sum_{\mathbf{p},s}\sum_{\mathbf{p}',s}a^*_{\mathbf{p},s}\,a_{\mathbf{p}',s'}\sum_{\nu=1}^{3}(p_{\nu}+p_{\nu}')\psi^*_{\mathbf{p},s}\,\alpha^{\nu}\psi_{\mathbf{p}',s}.\tag{9}$$

We shall denote the expectation value of an operator by putting a bar over it. Thus the pressure P is given by the expectation value $\bar{S}$ of S for a state of the whole assembly in which the N lowest states of the particles are occupied. It now follows from the commutation rules that

$$\overline{a^*_{\mathbf{p},s}\,a_{\mathbf{p}',s'}}=N_{\mathbf{p},s}\delta_{\mathbf{p},\mathbf{p}'}\,\delta_{s,s'},\tag{10}^*$$

where $N_{\mathbf{p},s}=1$ if the state $(\mathbf{p},s)$ is occupied and zero otherwise. From (9) we now have

$$P=\frac{c}{3}\sum_{\mathbf{p},s}N_{\mathbf{p},s}\sum_{\nu=1}^{3}\psi^*_{\mathbf{p},s}\,\alpha^{\nu}\psi_{\mathbf{p},s}p_{\nu}.\tag{11}$$

Since $c\alpha^{\nu}$ has the meaning of velocity in Dirac's theory we clearly have

$$c\psi^*_{\mathbf{p},s}\,\alpha^{\nu}\psi_{\mathbf{p},s}=\frac{p_{\nu}c^2}{EV}=\frac{v_{\nu}}{V},\tag{12}$$

where

$$E=c\sqrt{m^2c^2+\sum_{\nu=1}^{3}p_{\nu}^2}.\tag{13}$$

Relations (12) and (13) are easily verified using the explicit expressions for $\psi_{\mathbf{p},s}$ as given by Darwin (*loc. cit.*).

From (11) and (12) we now have

$$PV=\frac{1}{3}\sum_{\mathbf{p},s}N_{\mathbf{p},s}\frac{|\mathbf{p}|^2c^2}{E}\tag{14}$$

or

$$PV=\frac{1}{3}\sum_{\mathbf{p},s}N_{\mathbf{p},s}(\mathbf{p},\mathbf{v}),\tag{15}$$

* This is clear from the matrix representations for the $a_{\mathbf{p},s}$ (*cf.*, for instance, H. Weyl, *Gruppen Theorie und Quanten Mechanik*, pp. 223, 224).

which is precisely our earlier expression (2) and therefore leads to the same final result.

3. Since Eddington prefers to work with standing waves, it may be mentioned here that we could just as well have taken for the set of orthogonal functions the following set of standing waves instead of our $\psi_{\mathbf{p},s}$:—

$$\left.\begin{aligned}\phi_{\mathbf{p},s}&=\frac{1}{\sqrt{2}}(\psi_{\mathbf{p},s}+\psi_{-\mathbf{p},s})\\ \phi_{-\mathbf{p},s}&=\frac{1}{\sqrt{2}}(\psi_{\mathbf{p},s}-\psi_{-\mathbf{p},s})\end{aligned}\right\}. \qquad (16)$$

It is easily verified that (12) continues to be true if the $\psi_{\mathbf{p},s}$'s are replaced by the $\phi_{\mathbf{p},s}$'s. Thus (15) continues to be true even if the fundamental set of orthogonal functions were taken to correspond to standing waves.

4. In conclusion we wish to state that we do not intend this note as a reply in any sense to Eddington's papers. We thought it of interest, however, to point out that, starting with the energy-stress tensor as is defined in *relativistic* quantum mechanics and following Eddington's own procedure for calculating the pressure, we are simply led back to the relation between P and N one had earlier derived from (2) by directly inserting in it the relation between E and p given by relativistic mechanics.

Cambridge:
1935 *June* 7.

Production of Electron Pairs and the Theory of Stellar Structure

In the discussion of problems of stellar structure, only the deviations from the perfect gas laws arising from degeneracy due to the exclusion principle for the electrons have so far been considered. In fact, as has recently been shown by one of us[1], these deviations involve far-reaching limitations on the possible stellar configurations under given conditions. Thus, it can be deduced from the form of the equation of state of a degenerate gas, taking due account of relativity, that in order that degeneracy should develop in any part of a star, the ratio β of gas pressure to total pressure at that point must satisfy the condition

$$\frac{960}{\pi^4} \cdot \frac{1-\beta}{\beta} < 1 ; \qquad (1)$$

and on the standard model, in which β is assumed to be constant throughout the star, this implies the existence of a critical mass

$$\mathfrak{M} = 6{\cdot}6 \odot \mu^{-2}, \qquad (2)$$

($\odot$ denoting the mass of the sun and μ the molecular weight) above which degeneracy cannot set in at all. A study of the equilibrium of completely degenerate gas spheres leads further to the result that there is an upper limit

$$M_3 = 5{\cdot}7 \odot \mu^{-2} \qquad (3)$$

to the masses of such configurations ; this affords the possibility, for stars of mass $\leqslant M_3$, of a course of evolution leading to complete degeneracy through intermediate stages comparable to the observed white dwarf configurations.

Quite another type of deviations from the perfect gas laws, however, arises from the existence of a definite distribution of positrons as well as electrons in equilibrium with temperature radiation, and in this note we desire to point out the bearing of this fact on the theory of stellar structure, and especially to indicate to what extent the conclusions summarised above have to be modified.

In the first place, no effect of the latter type can take place if the electron assembly is completely degenerate ; for in that case all the states of negative energy will necessarily be occupied, which on Dirac's well-known picture implies the total absence of positrons. For the theory of stellar structure this obvious remark has the consequence that, under white dwarf conditions, the influence of pair production on the configuration will be entirely negligible, and the possibility of evolution mentioned above, for stars of mass $\leqslant M_3$, can be upheld without modification.

More generally, the presence of an equilibrium distribution of pairs in addition to the 'excess' of electrons, which is proportional to the material density, will give rise to a correction term in the equation of state, and the effect of this term on the stellar structures may conveniently be surveyed on the standard model. It is found that for a fixed value of the ratio β, the correction increases with temperature, tending to a finite limit as the temperature tends to infinity. When the condition (1) is fulfilled, the maximum deviation from the perfect gas law is less than 2 per cent, which means that the effect is altogether negligible for stars of mass $< \mathfrak{M}$, in which degeneracy of the electron assembly is able to occur. For more massive stars, however, the correction term becomes increasingly important. Thus already when $1 - \beta = 0{\cdot}2$, corresponding on the standard model to a mass of $12{\cdot}6 \odot \mu^{-2}$, the maximum effect amounts to 7 per cent. For very massive stars, say, of mass greater than $30 \odot \mu^{-2}$, equilibrium configurations analogous to the white dwarf configurations for masses $< M_3$—but differing from the white dwarfs in that the deviations from the perfect gas laws now arise from the production of pairs and not from degeneracy—are therefore formally possible, and the question suggests itself: Do such configurations exist in Nature ?

A detailed derivation of the results here summarised is to be published elsewhere.

S. CHANDRASEKHAR.

Trinity College, Cambridge.

L. ROSENFELD.

Institut for teoretisk Fysik,
Copenhagen.

April 25.

[1] S. Chandrasekhar, *Mon. Not. R.A.S.*, **95**, 207–260; Jan., 1935.

The White Dwarfs and Their Importance for Theories of Stellar Evolution

by S. CHANDRASEKHAR.

1. According to current ideas the structure of the White Dwarfs if to be understood in terms of the deviations from the perfect gas law ($p \propto \rho T$) which quantum statistics predicts. The application of the quantum theory to the statistical mechanics of an electron assembly indicates that the classical treatment of the same problem ignores two factors. The first is the consequences of the Pauli principle which requires the wave functions describing the system to be antisymmetrical in the coordinates (x, y, z, and spin) of the different electrons; the second is the effect of the variation of mass with velocity predicted by the special theory of relativity. Both of these effects can be taken into account and the application of standard methods leads to the following parametric form for the equation of state of an electron gas:

$$N = \frac{8\pi V m^3 c^3}{h^3} \int_0^\infty \frac{\sinh^2\theta \cosh\theta \, d\theta}{\frac{1}{\Lambda} e^{\theta m c^2 \cosh\theta} + 1} \qquad (1)$$

$$P = \frac{8\pi m^4 c^5}{3 h^3} \int_0^\infty \frac{\sinh^4\theta \, d\theta}{\frac{1}{\Lambda} e^{\theta m c^2 \cosh\theta} + 1} \qquad (2)$$

where

$$\theta = 1/kT \qquad (3)$$

In equations (1), (2), and (3), m, c, and h denote the mass of the electron, the velocity of light, and the Planck constant, respectively. Further these equations refer to the pressure P exerted by the electrons in an enclosure of volume V containing N electrons.

Equations (1) and (2) can be written in the forms :

$$N = V f_1 (\Lambda, \theta) ; \qquad P = f_2 (\Lambda, \theta). \qquad (4)$$

and the elimination of Λ will give the required equation of state. This elimination cannot be effected satisfactorily except in two limiting cases, namely when (i) $\Lambda << 1$ and (ii) $\Lambda >> 1$. For these two cases we have respectively,

$$N = PV \qquad (5)$$

and

$$N = \frac{8\pi m^3 c^3}{3h^3} x^3 = A x^3$$

$$P = \frac{\pi m^4 c^5}{3h^3} f(x) = B f(x) \qquad (6)$$

where

$$f(x) = x(2x^2 - 3)(x^2 + 1)^{1/2} + 3 \sinh^{-1} x \qquad (6')$$

and

$$A = \frac{8\pi m^3 c^3}{3h^3} ; \qquad B = \frac{\pi m^4 c^5}{3h^3} \qquad (6'')$$

These two limiting forms for the equation of state defined by (1) and (2) are said to correspond to the non-degenerate and degenerate cases respectively [1].

2. It is now generally agreed that the equation of state appropriate for the discussion of the structure of White Dwarfs is the degenerate form of the equation of state. The main argument for this conclusion comes from the application of the theory of stellar envelopes which shows that the gaseous outer regions of the White Dwarfs will constitute only a thin outer fringe. Consequently almost the entire mass of the White Dwarfs must be degenerate in the sense already indicated.

The equations governing the equilibrium of the White Dwarfs are therefore (in the standard notation) :

$$\frac{1}{r^2} \frac{d}{dr} \left(\frac{r^2}{\rho} \frac{dP}{dr} \right) = -4 \pi G \rho \qquad (7)$$

where P and ρ are related according to equation (6).
By the transformations

$$r = \left(\frac{2A}{\pi G} \right)^{1/2} \frac{1}{By_0} \eta ; \qquad y = y_0 \Phi \qquad (8)$$

where

$$y_0^2 = x_0^2 + 1 \qquad (9)$$

equation (7) reduces to

$$\frac{1}{\eta^2}\frac{d}{d\eta}\left(\eta^2\frac{d\Phi}{d\eta}\right) = -\left(\Phi^2 - \frac{1}{y_0^2}\right)^{3/2} \qquad (10)$$

Equation (10) has to be solved with the boundary conditions

$$\Phi = 1\,; \quad \frac{d\Phi}{d\eta} = 0 \quad \text{at} \quad \eta = 0 \qquad (11)$$

For each specified value of y_0 we have one such solution. The boundary is defined at the point where the density vanishes, and this by (6) and (9) means that if η_1 specified the boundary,

$$\Phi(\eta_1) = 1/y_0 \qquad (12)$$

The integrations for the function Φ have been carried out for ten different values of y_0 and the physical characteristics of the resulting configuration are shown in Table 1.

The most important characteristic of these configurations is that they possess a natural limit, i. e. as

$$y^0 \rightarrow \infty \qquad (13)$$

TABLE I.

THE PHYSICAL CHARACTERISTICS
OF COMPLETELY DEGENERATE CONFIGURATIONS

$1/y_0^2$	$M/\odot$	ρ_0 in grams per cubic centimeter	ρ mean in grams per cubic centimeter	Radius in centimeters
—	—	—	—	—
0	5,75	∞	∞	0
0,01	5,51	$9{,}85 \times 10^8$	$3{,}70 \times 10^7$	$4{,}13 \times 10^8$
0,02	5,32	$3{,}37 \times 10^8$	$1{,}57 \times 10^7$	$5{,}44 \times 10^8$
0,05	4,87	$8{,}13 \times 10^7$	$5{,}08 \times 10^6$	$7{,}69 \times 10^8$
0,1	4,33	$2{,}65 \times 10^7$	$2{,}10 \times 10^6$	$9{,}92 \times 10^8$
0,2	3,54	$7{,}85 \times 10^6$	$7{,}9 \times 10^5$	$1{,}29 \times 10^9$
0,3	2,95	$3{,}50 \times 10^6$	$4{,}04 \times 10^5$	$1{,}51 \times 10^9$
0,4	2,45	$1{,}80 \times 10^6$	$2{,}29 \times 10^5$	$1{,}72 \times 10^9$
0,5	2,02	$9{,}82 \times 10^5$	$1{,}34 \times 10^5$	$1{,}93 \times 10^9$
0,6	1,62	$5{,}34 \times 10^5$	$7{,}7 \times 10^4$	$2{,}15 \times 10^9$
0,8	0,88	$1{,}23 \times 10^5$	$1{,}92 \times 10^4$	$2{,}79 \times 10^9$
1,0	0	0	0	∞

(The values given in this table differ slightly from the published values (S. Chandrasekhar, *M. N. R. A. S.*, 95, 208, 1935, Table III). The difference is due to the change in the accepted value of the fundamental physical constants. The calculations are for $\mu_e = 1$. For other values of μ_e, M should be multiplied by μ_e^{-2}, R by μ_e^{-1}, and ρ_0 by μ.)

and

$$\Phi \rightarrow \theta_3 \qquad (14)$$

(where θ_3 is the Lane-Emden function of index 3), and the mass tends to a finite limit M_3. Numerically it is found that

$$M_3 = 5.75\ \mu_e^{-2} \odot \qquad (15)$$

A glance at Table I shows that the mean density, the mass, and the radius of these degenerate configurations are all of the right order of magnitude to provide the basis for the theoretical discussion of the White Dwarfs. However, a really satisfactory test of the theory will be capable of providing an observational basis for the existence of a mass such that as we approach it the mean density increases several times, even for a slight increase in mass. The observational evidence which supports this theoretical prediction is discussed in Dr. Kuiper's report to this Colloquium.

3. The Mass $\mathfrak{M} = M_3\, \beta_\omega^{-3/2}$, ($\beta_\omega = 0.908\ldots$).

Closely connected with the existence of M_3 are the circumstances which enable us to find an upper limit to the mass of a stellar configuration which, consistent with the physics of degenerate matter, can be regarded as wholly degenerate. This limit arises in the following way :

Consider an electron assembly of N electrons in a volume V at temperature T. Then, on the basis of the perfect gas law, the electron pressure p_e would be given by

$$p_e = \left(\frac{N}{V}\right) kT \qquad (16)$$

At temperature T we also have the radiation pressure of amount given by the Stefan-Boltzmann law :

$$p_r = 1/3 \quad a \quad T^4 \qquad (17)$$

where the radiation constant a is given by

$$a = \frac{8\,\pi^5\, k^4}{15\ h^3 c^3} \qquad (18)$$

Let us denote by P the total pressure ($= p_e + p_r$) and introduce a parameter β_e, defined as follows :

$$P = p_r + p_e = \frac{1}{\beta_e}\, p_e = \frac{1}{1 - \beta_e}\, p_r \tag{19}$$

Eliminating T between the relations (16) and (17), we find

$$p_e = \left[k^4 \frac{3}{a} \frac{1 - \beta_e}{\beta_e}\right]^{1/3} n^{4/3} \tag{20}$$

where we have used n for (N/V). Let

$$n = \frac{8\pi\, m^3 c^3}{3h^3}\, x^3 \tag{21}$$

Then, equation (20) can be transformed into

$$p_e = \frac{\pi m^4 c^5}{3h^3} \left(\frac{512\,\pi\, k^4}{h^3 c^3 a}\, \frac{1 - \beta_e}{\beta_e}\right)^{1/3} 2x^4 \tag{22}$$

or, using (18), we have

$$p_e = A \left(\frac{960}{\pi^4}\, \frac{1 - \beta_e}{\beta_e}\right)^{1/3} 2x^4 \tag{23}$$

where A is defined as in equation (6″).

Now for an assembly having the same number N of electrons in the volume V we can formally calculate the electron pressure that would be given by the degenerate formula, namely

$$p_{deg} = A\, f(x) \tag{24}$$

for f (x) as defined in (6) it is readily shown that

$$f(x) < 2x^4 \qquad (x < \infty) \tag{25}$$

Hence, comparing (23) and (24), we have the result that if for a prescribed N and T, the value of β_e calculated on the basis of the perfect gas equation (16) be such that

$$\frac{960}{\pi 4}\, \frac{1 - \beta_e}{\beta_e} \geqslant 1 \tag{26}$$

then under these circumstances matter can *never* become degenerate. Inequality (26) is readily seen to be equivalent to

$$\beta_e < 0.90788\ldots = \beta_\omega \text{ (say)} \tag{27}$$

Let us now "introduce" radiation into the completely degenerate configurations. If we consider a degenerate configuration built on the standard model, then

$$P = \beta_e^{-1} p_e \tag{28}$$

where p_e is given by (6). For such configuration we easily find that

$$M(\beta_e ; y_0) = M(1 ; y_0)\, \beta_e^{-3/2} \tag{29}$$

in an obvious notation In particular

$$M(\beta_e ; \infty) = M_3 \beta_e^{-3/2} \tag{30}$$

From (30) it would appear at first sight that by allowing $\beta_e \rightarrow 0$ we can obtain degenerate configurations of any mass. This is, however, incorrect. For if $\beta_e < \beta_\omega$ then matter cannot be regarded as degenerate. Hence the maximum mass of the configuration which can be regarded as degenerate is given by

$$\mathfrak{M} = M_3\, \beta_\omega^{-3/2} \tag{31}$$

The result just stated is extremely general and can be proved as follows :

Consider a completely degenerate configuration of mass M slightly less than M_3. The density will everywhere be so great that we can increase the radiation pressure from zero to a value only slightly less than $(1-\beta_\omega)$ at each point of the configuration and still regard the matter as degenerate. The mass of the new configuration so obtained will be approximately $M\beta_\omega^{-3/2}$. When $M \rightarrow M_3$ the result becomes exact.

We have thus proved that the maximum mass of a stellar configuration which, consistent with the physics of degenerate matter, can be regarded as wholly degenerate, is $M_3\, \beta_\omega^{-3/2}$. Numerically,

$$\mathfrak{M} = 6.65\, \mu_e^{-2}\, \odot \tag{32}$$

3. Evolutionary Significance of M_3 and of $\mathfrak{M}$.

It should be clear now that a discussion of the role which the White Dwarfs are likely to play in any theory of stellar evolution must be necessarily linked with the evolutionary significance which we attach to the two critical masses M_3 and $\mathfrak{M}$.

For stars of mass less than M_3 we can tentatively assume that the completely degenerate state represents the last stage in the evolution of the stars — the stage of complete darkness and extinction. These completely degenerate configurations with $M < M_3$ are of course characterized by the finite radii.

For $M > M_3$ no such simple interpretation is possible. The problem that we are faced with can be stated as follows :

Consider a star of mass greater than $\mathfrak{M}$ and suppose that it has exhausted all its sources of subatomic energy — hydrogen in this connection. The star must then contract according to the Helmholtz-Kelvin time scale. Since degeneracy cannot set in, in the interior of such stars, continued and unrestricted contraction is possible, in theory.

However, we may expect instability of one kind or another (e. g. rotational) to set in long before, resulting in the " explosion " of the star into smaller fragments. It is also conceivable that the star may decrease its mass below M_3 by a process of continual ejection of matter. The Wolf Rayet phenomenon is suggestive in this connection.

For stars with masses in the range $M_3 < M < \mathfrak{M}$ there exist other possibilities. During the contractive stage, such stars are likely to develop degenerate cores. If the degenerate cores attain sufficiently high densities (as is possible for these stars) the protons and electrons will combine to form neutrons. This would cause a sudden diminution of pressure resulting in the collapse of the star onto a neutron core giving rise to an enormous liberation of gravitational energy. This may be the origin of the Supernova phenomenon.

The above remarks on the evolutionary significance of M_3 and $\mathfrak{M}$ are made with due reserve and no definiteness is claimed for them.

4. The Rotation of the White Dwarfs.

The effects of rotation on the structure of White Dwarfs are likely to be of considerable importance in connection with the remarks made in the preceding section. The theory of such rotationally distorted White Dwarfs can be developed on lines analogous to the authors's earlier investigations on the theory of distorted polytropes.

It can be shown that the structure of such rotating White Dwarfs will be governed by the differential equation

$$\frac{1}{\eta^2}\frac{\partial}{\partial\eta}\left(\eta^2\frac{\partial\Phi}{\partial\eta}\right)+\frac{1}{\eta^2}\frac{\partial}{\partial\mu}\left((1-\mu^2)\frac{\partial\Phi}{\partial\mu}\right)=\left(\Phi^2-\frac{1}{y_o^2}\right)^{3/2}+V \quad (33)$$

where

$$y=y_o\,\Phi\,;\quad y^2=x^2+1 \quad (34)$$

(y_o referring to the central value) and $\mu=\cos\theta$ (θ being the polar angle). Further

$$v=\frac{\omega^2}{2\pi\,GBy_o^3} \quad (35)$$

For small v (i. e. for small rotational velocities) the solution of (33) is found to be

$$\Phi=\Phi(y_o)+v\left\{\psi_o-\frac{5}{6}\frac{\eta_1^2\psi_2}{3\psi_2(\eta_1)+\eta_1\psi_2^1(\eta_1)}P_2(\mu)\right\} \quad (36)$$

where η_1 refers to the boundary of $\Phi(y_o)$ and ψ_o and ψ_2 satisfy the differential equations

$$\frac{1}{\eta_2}\frac{d}{d\eta}\left(\eta^2\frac{d\psi_o}{d\eta}\right)=-3\Phi\left(\Phi^2-\frac{1}{y_o^2}\right)^{1/2}\psi_o+1 \quad (37)$$

$$\frac{1}{\eta_2}\frac{d}{d\eta}\left(\eta^2\frac{d\psi_2}{d\eta}\right)=\left\{-3\Phi\left(\Phi^2-\frac{1}{y_o^2}\right)^{1/2}+\frac{6}{\eta_2}\right\}\psi_2 \quad (38)$$

and P_2 is the second Legendre plynomial in μ. The equation of the boundary of the rotating configuration is given by

$$\eta=\eta_1+\frac{v}{|\Phi'|}\left[\psi_o(\eta_1)-\frac{5}{6}\frac{\eta_1^2\psi_2(\eta_1)P_2(\mu)}{3\psi_2(\eta_1)+\eta_1\psi'(\eta_1)}\right] \quad (39)$$

giving for the oblateness of the configuration the expression

$$\sigma=\frac{5}{4}\frac{v}{|\Phi'^1|}\frac{\eta_1\psi_2(\eta_1)}{3\psi_2(\eta_1)+\eta_1\psi_2'(\eta_1)} \quad (40)$$

The discussion of these results must await the integration of the equations (37) and (38).

1. Detailed derivations of the formulae quoted will be found in the author's monograph « An Introduction to the Study of Stellar Structure », Chapter X. More recently more rigorous derivations of the same results have been given by D. van Dantzig.

Discussion of Dr. Chandrasekhar's Communication.

Professor Russell and Dr. Chandrasekhar consider a star whose mass is smaller than that of a White Dwarf, of the order of 0,20 or less. Such a star would become less luminous as it approached degeneracy, but the distribution of its density and temperature would be the same as for a White Dwarf. Its radius would be larger than for stars of larger mass, and, for the same surface temperature it would be brighter than a normal White Dwarf. We could then explain the region mentioned by Dr. Kuiper where no objects are observed.

Sir Arthur Eddington states that White Dwarfs do not have any particular luminosity anyway until their radii reached rather large values; the larger the luminosity, the greater the expenditure of energy, and the shorter the life.

Professor Russell agrees that the star would pass rapidly through such a stage, and would show a much larger surface temperature and bolometric magnitude, although its visual magnitude would not be very much larger.

Dr. Strömgren brings up the question of a star on the verge of becoming degenerate. Dr. Chandrasekhar says that one of his pupils and Dr. Stoner are working on the problem of what happens in a region in which the equation of state of degenerate matter approaches the equation of state of a perfect gas. No conclusions can be drawn yet, since the computations are incomplete. If degeneracy is just beginning to appear in the center, the radius is three times that of a completely degenerate star. Since the matter is almost wholly gaseous, the luminosity is high, and with such a small radius the star goes to a remote portion of the Hertzsprung-Russell diagram.

Sir Arthur wishes to know if Dr. Chandrasekhar's statement that the total pressure is the sum of the gaseous pressure and the degenerate pressure is true or if it is merely a simplification. Dr. Chandrasekhar shows that it is true.

Answering a question of Dr. Atkinson's, Dr. Chandrasekhar states that in an incompletely degenerate star the degenerate nucleus must have a mass less than $\mathfrak{M}$, but that the mass of the gaseous portion may be considerable. In practice, however, a degenerate nucleus cannot exist unless the total mass of the star is less than $\mathfrak{M}$. The matter has been studied in detail for various models.

Professor Russell requests a value of the density at which the equation of state for degeneracy differs appreciably from that of a perfect gas. Dr. Chandrasekhar says that the perfect gas laws would hold up to densities of 10^3 or 10^4; in Sirius B it is 1700, with a temperature of $10^{7\circ}$. Then, suggests Professor Russell, the perfect gas laws would hold for an ordinary red dwarf.

Sir Arthur Eddington and Dr. Chandrasekhar answer that their temperature is only $10^{6\circ}$, so that ionization would not be complete, and there would be corrections. In some cases other types of degeneracy exist where the ionization is high. Dr. Kuiper adds that for red dwarfs the density would be nearer 150 than 1000, as Dr. Chandrasekhar quoted, and that furthermore the temperatures are probably too low. Little is known for M dwarfs later than M2; the evidence of their temperatures indicates that as we go from M0 to M6 the density stays nearly constant, while the central temperatures decrease. Sir Arthur Eddington questions the adopted effective temperatures. In the case of Wolf 359, says Dr. Kuiper, the large color index of 5, measured as far as 8500 A, is due to the great intensity of the titanium bands. Professor Russell emphasizes the importance of observations on the spectral type and the color index of eclipsing variables in this connection, particularly those with high density, but Dr. Kuiper has some doubts on the feasibility of such observations, since there are only a few such stars with great enough luminosity to permit spectroscopic observation.

PARTIALLY DEGENERATE STELLAR CONFIGURATIONS

As the author has shown,[1] the structure of completely degenerate configurations can be fully described by the differential equation

$$\frac{1}{\eta^2}\frac{d}{d\eta}\left(\eta^2\frac{d\varphi}{d\eta}\right) = -\left(\varphi^2 - \frac{1}{y_0^2}\right)^{3/2} . \tag{I}$$

This equation describes exactly the structure of stellar configurations based on the exact equation of state of a degenerate electron gas, and in particular takes into account the gradual change of the equation of state from the law $p = K_1\rho^{5/3}$ to the law $p = K_2\rho^{4/3}$, with increasing density. On the theory of white dwarfs based on equation (I), the gaseous fringe is neglected as being insignificant, as may indeed be readily verified.[2] However, for such stars in the interior of which we have only partial degeneracy at not too high densities, the relativistic effects can be properly neglected; and the problem now is to take into account the gradual change of the equa-

[1] *M.N.*, **95**, 207, 1934.

[2] B. Strömgren, *Erg. d. Exact. Natur.*, **16**, 508, 1937.

tion of state from the law $p = K_1\rho^{5/3}$ to the law $p = (k/\mu H)\rho T$. The exact equation of state, neglecting relativity effects, can be written as

$$\rho = 2\,\frac{(2\pi mkT)^{3/2}}{h^3}\,\mu H\,U_{1/2}(A)\,, \tag{1}$$

$$p = 2\,\frac{(2\pi mkT)^{3/2}}{h^3}\,kT\,U_{3/2}(A)\,, \tag{2}$$

where $U_\rho(A)$ is a function of A defined by

$$U_\rho = \frac{1}{\Gamma(\rho+1)}\int_0^\infty \frac{u^\rho du}{\frac{1}{A}e^u + 1}\,. \tag{3}$$

To be able to study the structure of stellar configurations based on the equation of state parametrically expressed by (1) and (2), we should have to make an assumption concerning the circumstances determining the temperature gradient and the energy-source distribution. We shall consider two cases: (*a*) an isothermal gas sphere, and (*b*) the standard model.

a) We can show that the structure of an isothermal gas sphere based on (1) and (2) is governed by the differential equation

$$\frac{1}{\xi^2}\frac{d}{d\xi}\left(\xi^2\frac{d\psi}{d\xi}\right) = -\,U_{1/2}(e^\psi)\,, \tag{II}$$

where

$$\psi = \log A \tag{4}$$

and ξ is the radius vector in a suitably chosen scale. We easily verify that, as $A \to 0$, equation (II) reduces to the usual Emden isothermal equation. On the other hand, if $\log A \gg 1$, equation (II) reduces to the Emden equation of index $n = 3/2$. A study of the configurations based on (II) should be of importance in following up certain cosmological speculations by Eddington.[3]

[3] *Proc. Roy. Soc., London*, A, **111**, 424, 1925.

b) For the "standard model" ($p_{\rm gas} : p_{\rm rad} = \beta : 1 - \beta =$ Constant), we find that the structure will be governed by solutions of the differential equation

$$\frac{1}{\xi^2}\frac{d}{d\xi}\left(\xi^2 U_{3/2}^{2/3}\frac{d\psi}{d\xi}\right) = -U_{3/2}U_{1/2}, \tag{III}$$

where ψ is defined as in (4) and ξ in a suitably chosen scale is the radius vector. Again one can show that, as $A \to 0$, equation (III) reduces to the Emden equation of index 3; while for $\log A \gg 1$, equation (III) reduces to the Emden equation of index $n = 3/2$. Equation (III), then, generalizes the usual standard model for stars of small mass and with incipient degeneracy in the central regions.

Finally, we may consider an isothermal gas sphere at such high temperatures that, if and when degeneracy sets in, it is already relativistic. For such configurations we can show that the differential equation governing the structure is

$$\frac{1}{\xi^2}\frac{d}{d\xi}\left(\xi^2\frac{d\psi}{d\xi}\right) = -U_2. \tag{IV}$$

Configurations described by (IV) may be of importance in considerations relating to the ultimate fate of massive stars.

The problems suggested in this preliminary note will be examined in detail in a future communication.

S. CHANDRASEKHAR

YERKES OBSERVATORY
November 1937

19

RALPH HOWARD FOWLER

1889–1944

In the death of Sir Ralph Fowler in Cambridge on July 28, 1944, astronomy has lost one of its great pioneers in theoretical astrophysics. We owe to Fowler some of the very fundamental ideas in modern astrophysics; and his work, characterized by a rare combination of physical insight and mathematical precision, shows what theoretical astrophysics at its best can be.

Ralph Howard Fowler was born on January 17, 1889, the eldest son of Howard Fowler, of Burnham, Somerset. He was educated at Winchester (of which he later became a fellow) and at Trinity College, Cambridge, where he was elected to a prize fellowship for research in pure mathematics.

During the war of 1914–1918 he served as a lieutenant in the Royal Marine Artillery and was seriously wounded at Gallipoli. Later during the war he, together with A. V. Hill, E. A. Milne, and W. Hartree, organized the A. A. Experimental Section of the Munitions Inventions Department. After the war, in 1919, he returned to Trinity as a fellow anu a member of its mathematical staff.

In Cambridge the years following 1919 were ones of intense activity in physics consequent to the return of the late Lord Rutherford as Cavendish professor. It was natural that Fowler should have been drawn into this activity, and physics in the broadest sense was to be the central theme of his activity for the rest of his life. By his remarkable versatility and scholarship Fowler naturally became the leader of a growing school, and among his students and collaborators can be counted practically all the well-known names in theoretical or mathematical physics which have come out of Cambridge during the last twenty or more years. And so it was no surprise when, in 1932, Fowler was appointed to the newly created Plummer Chair of Mathematical Physics at Cambridge.

In 1938 Fowler was appointed to the directorship of the National Physical Laboratory, but an unexpected illness prevented his taking up this new appointment. But Cambridge was happy to reappoint him to the chair he had resigned. During the present war he undertook important liaison work between the wartime scientific establishments in Great Britain, Canada, and the United States. For this work he was created a knight in 1942. Unfortunately, his illness returned in 1943. He could not, however, be persuaded to reduce his efforts, and he threw himself into further work for the Admiralty. He was attending important conferences up to within a few weeks of his death.

Fowler's contributions to science range over a very wide domain and include many diverse fields. Here we shall be concerned with only those of his contributions which have a bearing on astronomy.

In many ways it was fortunate that, when Saha's investigations accounting for the principal features of stellar spectra in terms of the thermodynamic theory of ionization equilibrium began to appear in the pages of the *Philosophical Magazine* for the years 1920–1921, Fowler should have been occupied, in collaboration with C. G. Darwin, with the development of new and powerful methods for studying the properties of matter in equilibrium. For, the success of Saha's work had revealed the urgency of a theory which will be adequate for dealing properly with the statistical mechanics of assemblies of atoms, atomic ions, and electrons. Fowler and Darwin had already laid the foundations for dealing with precisely such situations, and the complete theory of such assemblies had been published during the years 1922–1923.[1] Once this theory was perfected, it was natural that Fowler's interest should have turned to the most spectacular field of its application at the time, namely, to stellar atmospheres. And so it was that the classical papers of Fowler and Milne on "The Intensities of Absorption Lines in Stellar Spectra and the Temperature and Pressures in the Reversing Layers"[2] and "On the Maxima of Absorption lines in Stellar Spectra"[3] came into being. The methods which were originated in these papers and the results to which they led have now become so much a part of our common knowledge that we might pause for a while to restate the basic problems as formulated in these papers.

In Saha's investigations attention was focused on the relative intensities of absorption lines arising from the successive stages of ionization of an atom, and this was correlated with the relative numbers of atoms in the various stages of ionization in the reversing layers. However, in attempting to deduce the temperature of the reversing layers, the "first" and the "last" (or the "marginal") appearances of a line along the spectral sequence were used as the principal criteria. But, as Fowler and Milne pointed out, the use of this notion of marginal appearance makes the precision of the early calculations doubtful, for we do not know, a priori, how small the "very small" fraction of the atoms must be at marginal appearance. Among other things, this will depend on the relative abundances of the various *elements* giving rise to the lines. In view of these difficulties, Fowler and Milne start with the following assumption: "Other things being equal, the intensity of a given absorption line in a stellar atmosphere varies always in the same sense as the concentration of the atoms in the reversing layer capable of absorbing the line." With this formulation and with attention focused on the place along the spectral sequence at which a given line attains its *maximum intensity*, the vagueness associated with the concept of marginal appearance is avoided, for, on the assumption which has been made, a line will attain its maximum intensity when the concentration of the atoms in the particular stage of ionization and excitation capable of absorbing the line reaches its maximum. Consequently, the position of the maximum will depend only on the temperature and the electron pressure and will be independent of the absolute abundance of the particular element under consideration. It is, moreover, apparent that the temperature at which, for a given electron pressure, a given line arising from a known stage of ionization and excitation will reach its maximum intensity can be readily derived from the properties of the equilibrium state. Thus, on this method each observed maximum of a line in the spectral sequence relates the temperature and the electron pressure of the reversing layer at the point of the sequence. This is the well-known "method of maxima." A major result of this investigation was to reveal, for the first time, the correct order of magnitude of the electron pressures which prevail in stellar atmospheres.

The method of Fowler and Milne rapidly became the principal basis for analyzing stellar spectra in the twenties, and we can relive the enthusiasm which it inspired and the impetus which it gave to astrophysics during these years by reading again C. H. Payne's brilliant tract on "Stellar Atmospheres."[4] It is remarkable how, in spite of the enormous

[1] Fowler, *Phil. Mag.*, **45**, 1, 1923, and subsequent papers.

[2] *M.N.*, **83**, 403, 1923.

[3] *Ibid.*, **84**, 499, 1924.

[4] *Harvard Mono.*, No. 1, 1925.

progress which has been achieved in recent years on the mechanism of the formation of the absorption lines themselves, the original method of Fowler and Milne still retains its pre-eminence for a first analysis of complex stellar spectra; for by their simple assumption (though crude from a more advanced point of view) they sweep the problem of its inessentials and bring into relief the basic physical factors which are in operation.

In a later paper of considerable interest Fowler shows[5] that the conspicuous contrast between the persistence of the hydrogen lines with increasing temperature beyond the maximum and the very rapid decline of the metallic arc lines beyond their maxima are simple consequences of the basic difference between the "last spectrum" of an atom and the earlier ones.

Having thus explored the applications of statistical mechanics to the problems of stellar atmospheres, Fowler next turns his attention to the more complicated problem of the physical state of matter in the interior of the stars. By 1924 Eddington's investigations on the interior of the stars had shown that the perfect gas laws could be expected to be approximately valid for densities up to at least the order of 400 gm/cm^3 and for temperatures of the order of twenty or more millions of degrees. But the precise evaluation of the degree of ionization of any kind of atom at the temperatures and densities involved and of the distribution among the various stationary states of the atoms in the different stages of ionization; the specification of the pressure, density, and temperature relationship under these conditions, with a view to studying the departures from the perfect gas laws; and the determination of the (P, V) adiabatic curves of the stellar material—all these are matters which require the fullest resources of modern statistical mechanics. The problems are not easy, particularly when allowances have to be made for "excluded volumes," electrostatic corrections, and so forth. However, in two papers,[6] written in collaboration with E. A. Guggenheim, Fowler set down the principles of such calculation and showed the feasibility of the theory by making a number of model calculations.

The investigations summarized above and a great deal more are assembled in his *Statistical Mechanics*,[7] for which he was awarded the Adams Prize in 1924.

We now come to what is perhaps the most spectacular of Fowler's discoveries, namely, his recognition of the state of matter in the white-dwarf stars. This discovery, which must certainly be counted among the more important of the astronomical discoveries of our time, could not have been made except by one whose grasp of the theories both of physics and of astrophysics was of the highest. But there is even more to the discovery than this. Dirac's paper, which contains the derivation of what has since come to be called the "Fermi-Dirac distribution," was communicated by Fowler to the Royal Society on August 26, 1926.[8] On November 3, Fowler communicated a paper of his own[9] in which the application of the laws of the "new quantum theory" to the statistical mechanics of assemblies consisting of similar particles is systematically developed and incorporated into the general scheme of the Darwin-Fowler method. And by December 10 (i.e., before a month had elapsed) his paper entitled "Dense Matter" was read before the Royal Astronomical Society.[10] In this paper Fowler drew attention to the fact that the electron gas in matter as dense as in the white-dwarf stars, as in the case of the companion of Sirius, must be degenerate in the sense of the Fermi-Dirac statistics. Thus, to Fowler belongs also the credit for first recognizing a field of application for the then "very new" statistics of Fermi and Dirac.

5 "Notes on the Theory of Absorption Lines in Stellar Spectra," *M.N.*, **85**, 970, 1925.

6 "Applications of Statistical Mechanics To Determine the Properties of Matter in Stellar Interiors, Part I: The Mean Molecular Weight"; Part II: "The Adiabatics," *M.N.*, **85**, 939, 961, 1925.

7 A second edition of this book was published in 1936. It has been further translated into German.

8 *Proc. R. Soc.*, A, **112**, 661, 1926.

9 *Proc. R. Soc.*, A, **113**, 432, 1926.

10 *M.N.*, **87**, 114, 1926.

In view of the importance of Fowler's work for all discussions relating to stellar structure, it is perhaps of interest to recall the chain of ideas which led Fowler to this discovery.

It was known by 1924 that matter in the white-dwarf stars must exist at densities of at least 100,000 gm/cm^3, or greater. As these stars go on radiating, they can, if anything, only contract still further. We would consequently expect them eventually to lose all power of radiating energy and fall to zero temperature. Fowler aptly describes such a state as belonging to the "black-dwarf stages." In this state they must consist of matter at low temperature and high density. This can clearly be achieved only if the constituents were free nuclei and free electrons forming a neutral mixture. Fowler immediately recognized that under these conditions matter must be degenerate, and he went on to show how this resolves a paradox that had been reached on the basis of classical statistics. The paradox was this: On the basis of classical statistics, matter in the black-dwarf stage would contain far less energy than the same matter expanded in the form of free complete atoms at rest at infinite separation. If such matter were removed from the interior of the black dwarf, it could not then resume its ordinary state. But Fowler showed that in the completely degenerate state the internal energy of the electron gas will all be in the form of zero-point energy when the pressure and the internal energy become (in the limit) functions of density only. He further showed that this zero-point energy is so great that, after allowing for the negative potential energy, the matter (if removed from the black dwarf and so free to expand) could reconstitute itself as an expanded gas of complete atoms at an extremely high temperature. Fowler expresses this conclusion in the following terms:

> The black dwarf material is best likened to a single gigantic molecule in its lowest quantum state. On the Fermi-Dirac statistics, its high density can be achieved in one and only one way, in virtue of a correspondingly great energy content. But this energy can no more be expended in radiation than the energy of a normal atom or molecule. The only difference between black dwarf matter and a normal molecule is that the molecule can exist in a free state while the black dwarf matter can only so exist under very high external pressure.

And how well expressed!

To pass from Fowler's discovery to the modern theory of white dwarfs, only two further steps are needed. The first is to find the equation of state of a completely degenerate gas, taking into account the relativistic mass variation with velocity. The second is to use this equation of state in conjunction with the equation of hydrostatic equilibrium and derive the physical characteristics of stellar configurations in equilibrium under their own gravitation. There are no difficulties in carrying out these steps, and the theory of white-dwarf stars along these lines has been perfected and refined. But the clue to all this was provided by Fowler.

Fowler's last incursion into astrophysical problems was in 1930, when the discussion of composite stellar models came to the forefront. Recalling his own early studies on the asymptotic behavior of continuous solutions of certain differential equations of the second order, Fowler discussed, with a detail and with a thoroughness characteristic of him, the complete arrangement of the solutions of the Lane-Emden equation. While a large part of the discussion relating to this problem had already been made by Emden in his *Gaskugeln*, Fowler put the finishing touches and enunciated the final comprehensive theorems of the subject.[11] Attention may particularly be drawn to his last paper on the subject in the *Oxford Quarterly*, where a complete tabulation is made of the arrangement of the solutions for the various possible cases. These theorems have proved to be of the utmost value in all discussions relating to composite stellar models.

We have now surveyed Fowler's principal contributions to theoretical astrophysics.

[11] "The Solutions of Emden's and Similar Equations," *M.N.*, **91**, 63, 1930, and "Further Studies on Emden's and Similar Differential Equations," *Quart. J. Math.* (Oxford series), **2**, 259, 1931.

But, as was stated at the outset, these form only a small fraction of his entire contribution to mathematical physics. This is not the place to review these larger contributions of Fowler, but it may be said that in all of them he showed the same quickness of apprehension and the same power to get abreast of the details.

Fowler was elected to the Royal Society in 1925 and was awarded its Royal Medal in 1936. He married Eileen, the only daughter of the late Lord Rutherford. He leaves two sons and two daughters.

Fowler was big and powerful of frame and applied his strength with success to a variety of ball games. He had claims to distinction as a cricketer, both in batting and in bowling. He played excellent games of both lawn and real tennis. He represented Cambridge at golf, and he was also a rock-climber.

We may conclude by quoting Professor E. A. Milne, who was also one of his early collaborators: "Fowler was the whole man of many parts. His life was one of unsparing devotion to high scientific ideals. We cannot overestimate the loss his untimely death means to Great Britain and to science generally."

S. CHANDRASEKHAR

Ramsey & Muspratt, Post-office Terrace, Cambridge

RALPH HOWARD FOWLER

1889–1944

PART TWO

The Equilibrium of Distorted Polytropes

20

THE EQUILIBRIUM OF DISTORTED POLYTROPES

(I). The Rotational Problem.

S. Chandrasekhar.

(Communicated by E. A. Milne)

§ 1. Emden's well-known researches on the equilibrium of polytropic gas spheres has been of fundamental importance in its repercussions on the modern theories of stellar structure. But it is a matter of some surprise that scarcely any serious attempt has been made to extend Emden's researches to the case of rotating gas spheres which in their non-rotating states have polytropic distributions described by the so-called Emden functions.

The problem is exceedingly simple in its classical severity and can be formulated as follows :—

We have a gas sphere in gravitational equilibrium. It is given that the total pressure P is related to the density ρ by means of the relation

$$P = K\rho^{1+\frac{1}{n}}. \tag{1}$$

This defines for the non-rotating gas sphere a distribution of density and pressure governed by the differential equation

$$\frac{1}{\xi^2}\frac{d}{d\xi}\left(\xi^2\frac{d\theta}{d\xi}\right) = -\theta^n \qquad (2)^*$$

—*Emden's differential equation of index n.* Now the gas sphere is set rotating at a constant small angular velocity ω. *The problem is to determine the shape and density distribution in such a gas sphere and explicitly to relate the structure of such rotating masses with the Emden functions which describe the polytropic non-rotating state.*

§ 2. In treating this problem we shall assume that the rotation is so slow that the configurations are only slightly oblate. In other words, the purpose of this paper is to specify *completely* those configurations for the polytropic model which correspond to the Maclaurin spheroids † in the case of the "incompressible rotating stellar masses."

It is thus clear that the point of view and the problem here is different from that considered by Jeans under the heading "Adiabatic-model" in his book on *The Problems of Cosmogony and Stellar-dynamics* (p. 165). The problem treated by Jeans is to enumerate the complete sequence of the geometry of the configurations for the whole range of ω. Further, the analysis of Jeans is to establish a general result that a gas sphere rotating as a rigid body can break up in two distinct ways—either by fission into two detached masses or by a process of equatorial break-up after assuming

* The significance of ξ and θ are explained later.

† For small ω.

a lenticular shape as in the Roche model—according as the central condensation is "weak" or "pronounced." We shall not enter into such questions—which of course are of fundamental significance in any problem of cosmogony—but confine ourselves to the comparatively more simple and elementary problem of specifying completely the "polytropic Maclaurin spheroids." Also the fundamental mathematical point of view will be different from that of Jeans. We will explicitly relate the geometry and the physical properties of these configurations with the Emden functions describing the polytropic non-rotating gas spheres.

Substantially the case $n=3$ has been treated already by Milne * and von Zeipel.† Indeed it was in this connection that von Zeipel discovered his fundamental theorem.‡ The method of solution adopted in the sequel will in main be that of Milne, to whose memoir the present paper owes a great deal.

§ 3. *Equations of the Problem.*—The equations of mechanical equilibrium are, taking the Z-axis as the axis of rotation,

$$\left.\begin{aligned} \frac{\partial P}{\partial x} &= \rho\frac{\partial V}{\partial x} + \rho\omega^2 x, \\ \frac{\partial P}{\partial y} &= \rho\frac{\partial V}{\partial y} + \rho\omega^2 y, \\ \frac{\partial P}{\partial z} &= \rho\frac{\partial V}{\partial z}, \end{aligned}\right\} \tag{3}$$

where V is the gravitational potential, which of course satisfies Poisson's equation

$$\sum_{x,y,z} \frac{\partial}{\partial x}\left(\frac{\partial V}{\partial x}\right) = -4\pi G\rho. \tag{4}$$

Introducing polar co-ordinates r, θ, ϕ and neglecting ϕ on account of symmetry we find that

$$\left.\begin{aligned} \frac{\partial P}{\partial r} &= \rho\frac{\partial V}{\partial r} + \rho\omega^2 r(1-\mu^2), \\ \frac{\partial P}{\partial \mu} &= \rho\frac{\partial V}{\partial \mu} - \rho\omega^2 r^2\mu, \end{aligned}\right\} \tag{3'}$$

where $\mu = \cos\theta$. Also (4) in these variables reduces to

$$\frac{1}{r^2}\frac{\partial}{\partial r}\left(r^2\frac{\partial V}{\partial r}\right) + \frac{1}{r^2}\frac{\partial}{\partial \mu}\left((1-\mu^2)\,.\,\frac{\partial V}{\partial \mu}\right) = -4\pi G\rho. \tag{4'}$$

From (3′) and (4′) we deduce the fundamental equation of the problem:

$$\frac{1}{r^2}\frac{\partial}{\partial r}\left(\frac{r^2}{\rho}\frac{\partial P}{\partial r}\right) + \frac{1}{r^2}\frac{\partial}{\partial \mu}\left(\frac{1-\mu^2}{\rho}\frac{\partial P}{\partial \mu}\right) = -4\pi G\rho + 2\omega^2, \tag{5}$$

* E. A. Milne, *M.N.*, **83**, 118, 1923. † H. von Zeipel, *M.N.*, **84**, 665, 684, 1924.

‡ For a general discussion of the whole problem we refer the reader to Milne's article in the *Handbuch der Astrophysik*, Band III/1, p. 235.

where

$$P = K\rho^{1+\frac{1}{n}}. \tag{6}$$

§ 4. If we introduce the new variables

$$\rho = \lambda\Theta^n\,; \qquad P = \lambda^{1+\frac{1}{n}} \,.\, K \,.\, \Theta^{n+1}, \tag{7}$$

(5) reduces to

$$\frac{\lambda^{\frac{1}{n}-1} K(n+1)}{4\pi G}\left\{\frac{1}{r^2}\frac{\partial}{\partial r}\left(r^2\frac{\partial\Theta}{\partial r}\right) + \frac{1}{r^2}\frac{\partial}{\partial\mu}\left((1-\mu^2)\frac{\partial\Theta}{\partial\mu}\right)\right\} = -\Theta^n + \frac{\omega^2}{2\pi G\lambda}. \tag{8}$$

Put

$$r = \left[\frac{(n+1)K}{4\pi G}\lambda^{\frac{1}{n}-1}\right]^{\frac{1}{2}}\xi \tag{9}$$

and

$$v = \frac{\omega^2}{2\pi G\lambda}. \tag{10}$$

We get

$$\frac{1}{\xi^2}\frac{\partial}{\partial\xi}\left(\xi^2\frac{\partial\Theta}{\partial\xi}\right) + \frac{1}{\xi^2}\frac{\partial}{\partial\mu}\left((1-\mu^2)\frac{\partial\Theta}{\partial\mu}\right) = -\Theta^n + v. \tag{11}$$

§ 5. *Non-rotating Configuration.*—If the gas sphere is non-rotating then ω and therefore v is zero. If, further, now

$$\rho = \lambda\theta^n, \tag{7'}$$

we see that θ satisfies the differential equation

$$\frac{1}{\xi^2}\frac{d}{d\xi}\left(\xi^2\frac{d\theta}{d\xi}\right) = -\theta^n, \tag{12}$$

—Emden's differential equation of index n.

§ 6. *Solution for the Rotating Gas Sphere.*—We will now seek a solution of (11) in terms of those of (12), and indeed we will assume the following form for our solution:—

$$\Theta = \theta + v\Psi + v^2\Phi + \ldots \tag{13}$$

We shall work consistently only up to the first order in v, i.e. we consider only such slow rotations that the effects arising from ω^4 can be neglected. Ψ then should satisfy the differential equation, remembering that θ is a spherically symmetrical function and therefore independent of μ,

$$\frac{1}{\xi^2}\frac{\partial}{\partial\xi}\left(\xi^2\frac{\partial\Psi}{\partial\xi}\right) + \frac{1}{\xi^2}\frac{\partial}{\partial\mu}\left((1-\mu^2)\frac{\partial\Psi}{\partial\mu}\right) = -n\theta^{n-1}\Psi + 1. \tag{14}$$

Now we shall assume for Ψ the following form: *

$$\Psi = \psi_0(\xi) + \sum_{j=1}^{\infty} A_j\psi_j(\xi)P_j(\mu), \tag{15}$$

* For a formal justification of this assumption see von Zeipel (*loc. cit.*), p. 691.

where the $P_j(\mu)$'s are the Legendre functions of the various indices, the function with index j satisfying the differential equation

$$\frac{\partial}{\partial\mu}\left((1-\mu^2)\frac{\partial P_j}{\partial\mu}\right)+j(j+1)P_j=0. \tag{16}$$

Further, the ψ_j's are functions of ξ only.

Substituting (15) in (14), and using (16) and equating coefficients of P_j, we get

$$\frac{1}{\xi^2}\frac{d}{d\xi}\left(\xi^2\frac{d\psi_0}{d\xi}\right)=-n\theta^{n-1}\psi_0+1, \tag{17}$$

$$\frac{1}{\xi^2}\frac{d}{d\xi}\left(\xi^2\frac{d\psi_j}{d\xi}\right)=\left(\frac{j(j+1)}{\xi^2}-n\theta^{n-1}\right)\psi_j, \tag{18}$$

$$(j=1, 2, \ldots).$$

So far the A_j's are arbitrary. To determine them we must evaluate the potential V. For equation (5) (from which (14) is deduced) contains no explicit reference to the potential and is the same whatever the external gravitational field. To remove this indeterminateness we must use the solution found (with the arbitrary A_j's) to calculate the potential arising from the matter and then determine the A_j's such that the equations of equilibrium (3′) are satisfied.

Poisson's equation in the ξ, μ variables takes the form

$$\frac{1}{\xi^2}\frac{\partial}{\partial\xi}\left(\xi^2\frac{\partial V}{\partial\xi}\right)+\frac{1}{\xi^2}\frac{\partial}{\partial\mu}\left((1-\mu^2)\frac{\partial V}{\partial\mu}\right)$$
$$=-(n+1)K\lambda^{\frac{1}{n}}\left[\theta^n+n\theta^{n-1}v\left\{\psi_0+\sum_j A_j\psi_jP_j\right\}\right]. \tag{19}$$

We develop V in the form (to the first order in v)

$$V=U+v\left\{V_0(\xi)+\sum_j V_j(\xi)P_j(\mu)\right\}, \tag{20}$$

where U is the potential of the non-rotating configuration. Substitution of (20) in (19) yields on equating the coefficients of $P_j(\mu)$,

$$\frac{1}{\xi^2}\frac{d}{d\xi}\left(\xi^2\frac{dU}{d\xi}\right)=-R\theta^n, \tag{21}$$

$$\frac{1}{\xi^2}\frac{d}{d\xi}\left(\xi^2\frac{dV_0}{d\xi}\right)=-Rn\theta^{n-1}\psi_0, \tag{22}$$

$$\frac{1}{\xi^2}\frac{d}{d\xi}\left(\xi^2\frac{dV_j}{d\xi}\right)=\frac{j(j+1)}{\xi^2}V_j-Rn\theta^{n-1}A_j\psi_j, \tag{23}$$

where we have used the abbreviation

$$R=(n+1)K\lambda^{\frac{1}{n}}.$$

Remembering that θ satisfies the Emden equation with index n we deduce from (21) that

$$U=R\theta+\text{constant}. \tag{24}$$

Using the differential equation (17) defining ψ_0, we derive from (22) that

$$\frac{1}{\xi^2}\frac{d}{d\xi}\left(\xi^2\frac{dV_0}{d\xi}\right)=R\left[\frac{1}{\xi^2}\frac{d}{d\xi}\left(\xi^2\frac{d\psi_0}{d\xi}\right)-1\right]$$
$$=R\left[\frac{1}{\xi^2}\frac{d}{d\xi}\left(\xi^2\frac{d}{d\xi}(\psi_0-\tfrac{1}{6}\xi^2)\right)\right].$$

Hence

$$V_0=R(\psi_0-\tfrac{1}{6}\xi^2)+\text{constant}. \tag{25}$$

Similarly from (23) and (18) we get

$$\frac{1}{\xi^2}\frac{d}{d\xi}\left(\xi^2\frac{dV_j}{d\xi}\right)-\frac{j(j+1)}{\xi^2}V_j=A_jR\left[\frac{1}{\xi^2}\frac{d}{d\xi}\left(\xi^2\frac{d\psi_j}{d\xi}\right)-\frac{j(j+1)}{\xi^2}\psi_j\right]. \tag{26}$$

A particular solution of (26) is

$$V_j=RA_j\psi_j+\text{constant}. \tag{27}$$

The general solution is obtained by adding any *regular* solution of the equation

$$\frac{1}{\xi^2}\frac{d}{d\xi}\left(\xi^2\frac{dV_j}{d\xi}\right)-\frac{j(j+1)}{\xi^2}V_j=0, \tag{28}$$

which is $RB_j\xi^j$ where B_j is arbitrary. Hence

$$V_j=R(A_j\psi_j+B_j\xi^j)+\text{constant}. \tag{29}$$

Combining the results (24), (25) and (29) we get after some minor rearrangement of the terms (to the first order in v)

$$V=R\left[\Theta+v\left\{\sum_{j=1}^{\infty}B_j\xi^jP_j(\mu)-\tfrac{1}{6}\xi^2\right\}\right]. \tag{30}$$

We have to substitute now (30) in (3′), which in the ξ, μ variables takes the form

$$\frac{dP}{d\xi}=\rho\frac{\partial V}{\partial\xi}+\frac{2}{3}\rho\omega^2\xi\left[\frac{(n+1)K}{4\pi G}\lambda^{\frac{1}{n}-1}\right](1-P_2(\mu)).$$

Substituting (30) in the above equation and remembering that

$$P=\lambda^{1+\frac{1}{n}}\,.\,K\,.\,\Theta^{n+1},$$

and equating coefficients of $P_j(\mu)$ we find that

$$B_j=0,\qquad j\neq 2$$

and

$$B_2=\tfrac{1}{6}. \tag{31}$$

Finally, we obtain

$$V=R(\Theta-\tfrac{1}{6}v(\xi^2-P_2(\mu)\xi^2))+\text{constant}. \tag{32}$$

We have still to ensure that V is the *actual* potential arising from the mass. This will determine the A_j's.

A little consideration shows that (compare Milne (*loc. cit.*), p. 134) to the order of accuracy we are working, the potential (and its derivative) given by (32) should be continuous with an expression of the type

$$V_{\text{external}} = R\left[\frac{C_0}{\xi} + v\sum_{j=1}^{\infty} \frac{C_j}{\xi^{j+1}} P_j(\mu)\right] + \text{constant} \tag{33}$$

on a sphere of radius ξ_1 the first zero of the Emden's function with index n. Comparing the "inner" and the "external" potentials at $\xi = \xi_1$ and also their derivatives we obtain

$$A_j = C_j = 0, \qquad j \neq 2.$$

But, if $j = 2$, we get

$$\left.\begin{aligned} \frac{C_2}{\xi_1^3} &= A_2\psi_2(\xi_1) + \tfrac{1}{6}\xi_1^2, \\ -\frac{3C_2}{\xi_1^4} &= A_2\psi_2'(\xi_1) + \tfrac{1}{3}\xi_1. \end{aligned}\right\} \tag{34}$$

This gives

$$A_2 = -\frac{5}{6}\,\frac{\xi_1^2}{3\psi_2(\xi_1) + \xi_1\psi_2'(\xi_1)}. \tag{35}$$

Hence the solution to the problem is given by

$$\Theta = \theta + v\left[\psi_0(\xi) - \frac{5}{6}\,\frac{\xi_1^2}{3\psi_2(\xi_1) + \xi_1\psi_2'(\xi_1)}\psi_2(\xi)P_2(\mu)\right], \tag{36}$$

where ψ_0 and ψ_2 satisfy the differential equations

$$\frac{1}{\xi^2}\frac{d}{d\xi}\left(\xi^2\frac{d\psi_0}{d\xi}\right) = -n\theta^{n-1}\psi_0 + 1, \tag{37$_1$}$$

$$\frac{1}{\xi^2}\frac{d}{d\xi}\left(\xi^2\frac{d\psi_2}{d\xi}\right) = \left(-n\theta^{n-1} + \frac{6}{\xi^2}\right)\psi_2, \tag{37$_2$}$$

which are for purposes of numerical integration more conveniently written as

$$\frac{d^2\eta_0}{d\xi^2} = -n\theta^{n-1}\eta_0 + \xi; \qquad \frac{d^2\eta_2}{d\xi^2} = \left(-n\theta^{n-1} + \frac{6}{\xi^2}\right)\eta_2, \tag{37$'$}$$

where

$$\eta_{0,2} = \xi\psi_{0,2}. \tag{37$''$}$$

§ 7. *Expansion, Ellipticity and Oblateness.*—The boundary ξ_0 is given by $\Theta = 0$, and hence by (36)

$$\xi_0 = \xi_1 + \frac{v}{|\theta_1'|}\left[\psi_0(\xi_1) - \frac{5}{6}\,\frac{\xi_1^2\psi_2(\xi_1)P_2(\mu)}{3\psi_2(\xi_1) + \xi_1\psi_2'(\xi_1)}\right]. \tag{38}$$

Thus there is an expansion of the star as a whole of amount $v\psi_0(\xi_1)/|\theta_1'|$ and superposed on this an ellipticity. At the equator $P_2(\mu) = -\frac{1}{2}$ and at the poles $P_2(\mu) = +1$. Hence we get for the *oblateness* of the boundary the general expression

$$\sigma = \frac{5}{4}\,\frac{v}{|\theta_1'|}\cdot\frac{\xi_1\psi_2(\xi_1)}{3\psi_2(\xi_1) + \xi_1\psi_2'(\xi_1)}. \tag{39}$$

§ 8. *Mass Relation.*—The mass is given by

$$M = 2\pi \iint \rho r^2 dr d\mu.$$

The ellipticity term does not clearly contribute to the mass on the average. Introducing the ξ, θ variables in the above integral we have

$$M = 4\pi \left[\frac{(n+1)K}{4\pi G}\lambda^{\frac{1}{n}-1}\right]^{3/2} \lambda \int_0^{\xi_1 + d\xi_1} (\theta^n + n\theta^{n-1} v\psi_0)\xi^2 d\xi,$$

which to the first order in v can be replaced by

$$M = 4\pi \left[\frac{(n+1)K}{4\pi G}\lambda^{\frac{1}{n}-\frac{1}{3}}\right]^{3/2} \left\{\int_0^{\xi_1} \theta^n \xi^2 d\xi + v\int_0^{\xi_1} n\theta^{n-1}\psi_0 \xi^2 d\xi\right\}.$$

Now

$$\int_0^{\xi_1} \theta^n \xi^2 d\xi = -\int_0^{\xi_1} \frac{d}{d\xi}\left(\xi^2 \frac{d\theta}{d\xi}\right) = -\xi_1^2\left(\frac{d\theta}{d\xi}\right)_1,$$

$$\int_0^{\xi_1} n\theta^{n-1}\psi_0 \xi^2 d\xi = -\int_0^{\xi_1}\left\{\frac{d}{d\xi}\left(\xi^2 \frac{d\psi_0}{d\xi}\right) - \xi^2\right\}d\xi$$

$$= -\xi_1^2 \psi_0'(\xi_1) + \tfrac{1}{3}\xi_1^3.$$

Hence

$$M = -4\pi \left[\frac{(n+1)K}{4\pi G}\lambda^{\frac{3-n}{3n}}\right]^{3/2} \xi_1^2 \left(\frac{d\theta}{d\xi}\right)_1 \cdot \left[1 + v\frac{\frac{1}{3}\xi_1 - \psi_0'(\xi_1)}{|\theta_1'|}\right]. \tag{40}$$

If $v = 0$, we have for the case of non-rotating stars

$$M_0 = -4\pi \left[\frac{(n+1)K}{4\pi G}\lambda^{\frac{3-n}{3n}}\right]^{3/2} \xi_1^2 \left(\frac{d\theta}{d\xi}\right)_1. \tag{41}$$

Hence the "mass relation" for two gas spheres with *equal central densities*—one rotating with an angular velocity ω and the other non-rotating—is

$$M_\omega = M_0 \cdot \left[1 + v\frac{\frac{1}{3}\xi_1 - \psi_0'(\xi_1)}{|\theta_1'|}\right]. \tag{42}$$

Hence the rotating configuration has a *greater mass*, as indeed we should expect on general grounds.*

§ 9. *Volume and Relation between Mean and Central Density.*—The volume N of the configuration is clearly given by

$$N = \frac{4\pi}{3}\left[\frac{(n+1)K}{4\pi G}\lambda^{\frac{1}{n}-1}\right]^{3/2} \xi_1^3 + \left[\frac{(n+1)K}{4\pi G}\lambda^{\frac{1}{n}-1}\right]^{3/2} \cdot \int_0^{2\pi}\int_{-1}^{+1} (\xi_0 - \xi_1)\xi_1^2 d\mu d\phi,$$

which by (38) yields

$$N = \frac{4}{3}\pi \left[\frac{(n+1)K}{4\pi G}\lambda^{\frac{1}{n}-1}\right]^{3/2} \xi_1^3 \cdot \left[1 + v \cdot \frac{3\psi_0(\xi_1)}{\xi_1 |\theta_1'|}\right]. \tag{43}$$

* Compare Milne's article in the *Handbuch* ("General Effects of Rotation," p. 236).

Using the mass relation (40) we get for the mean density ρ_m the formula

$$\rho_m = -3\lambda\frac{1}{\xi_1}\left(\frac{d\theta}{d\xi}\right)_1 \cdot \frac{1+v\,.\,\dfrac{\frac{1}{3}\xi_1 - \psi_0'(\xi_1)}{|\,\theta_1'\,|}}{1+v\,.\,\dfrac{3\psi_0(\xi_1)}{\xi_1|\,\theta_1'\,|}},$$

or to the order of accuracy we are working we have, remembering that $\lambda=\rho_c$ (θ and Θ are both unity at the centre),

$$\rho_m = -\frac{3}{\xi_1}\left(\frac{d\theta}{d\xi}\right)_1 \rho_c \cdot \left[1+v\frac{\frac{1}{3}\xi_1^2 - \psi_0'(\xi_1)\xi_1 - 3\psi_0(\xi_1)}{\xi_1|\,\theta_1'\,|}\right]. \tag{44}$$

If $v=0$, we get the well-known formula for the non-rotating polytropes

$$\rho_m = -\frac{3}{\xi_1}\left(\frac{d\theta}{d\xi}\right)_1 \rho_c. \tag{45}$$

From (44) we deduce (to the first order) that

$$v = \frac{\omega^2}{2\pi G\rho_c} = -\frac{3}{\xi_1}\left(\frac{d\theta}{d\xi}\right)_1 \cdot \frac{\omega^2}{2\pi G\rho_m}. \tag{46}$$

Formulæ (38, (39), (40), (42), (43) can all now be expressed in terms of ρ_m (instead of ρ_c) by means of (46).

§ 10. *Numerical Integration of the Differential Equations.*—Thus we see that the structure of a slowly rotating polytropic gas configuration is completely specified when the pair of differential equations ($37_{1,\,2}$) are solved. There is just one case where integration can be effected at once. The equation for ψ_0 for the index $n=1$ can be written as

$$\frac{1}{\xi^2}\frac{d}{d\xi}\left(\xi^2\frac{d(\psi_0-1)}{d\xi}\right) = -(\psi_0-1), \tag{47}$$

which is just Emden's equation with index 1. Remembering the boundary conditions that ψ_0 and ψ_0' are to be zero at the origin, we see that

$$\psi_0 = 1 - \frac{\sin\xi}{\xi}, \tag{48}$$

is the required solution. [By (48) we see that

$$\psi_0(\xi_1) = 1\,; \qquad \psi_0'(\xi_1) = \frac{1}{\pi}\,; \qquad \xi_1 = \pi. \tag{48'}$$

The mass relation (42) and the volume relation (43) therefore take for this case the neat forms

$$M_\omega = M_0[1 + v(\tfrac{1}{3}\pi^2 - 1)], \tag{48''}$$

$$N_\omega = N_0[1 + 3v]. \tag{48'''}$$

In all other cases numerical integration must be adopted. To do this we must have a power series in ξ for ψ_0 and ψ_2 at the origin, and with a start thus made the integration has to be continued by any of the known standard methods.

Now it is seen that near the origin Emden's equation with index n has the expansion *

$$\theta = 1 - \frac{1}{3!}\xi^2 + \frac{n}{5!}\xi^4 - \frac{8n^2 - 5n}{3 \cdot 7!}\xi^6 + \dots \qquad (49)$$

which yields for θ^{n-1} the expansion

$$\theta^{n-1} = 1 - \frac{(n-1)}{6}\xi^2 + \frac{(n-1)(4n-5)}{180}\xi^4 - \dots \qquad (50)$$

Substituting this in equations ($37_{1,\,2}$), and assuming for ψ_0 and ψ_2 power series of the type

$$\psi_{0,\,2} = a\xi^2 + b\xi^3 + c\xi^4 + \dots, \qquad (51)$$

and solving, we obtain the following expansions for ψ_0 and ψ_2 near the origin :—

$$\psi_0 = \frac{1}{6}\xi^2 - \frac{n}{120}\xi^4 + \frac{n(13n-10)}{42 \cdot 360}\xi^6 - \frac{n(90n^2 - 157n + 70)}{72 \cdot 42 \cdot 360}\xi^8 + \dots \qquad (52)$$

$$\psi_2 = \xi^2 - \frac{n}{14}\xi^4 + \frac{n(10n-7)}{42 \cdot 36}\xi^6 - \frac{n(308n^2 - 503n + 210)}{42 \cdot 36 \cdot 330}\xi^8 + \dots \qquad (53)$$

The numerical integration was carried out for the cases $n = 1, 1{\cdot}5, 2, 3, 4$, and the functions ψ_0 and ψ_2 are tabulated in Tables I–V appended to the end of this paper. At the bottom of each table the value of ψ_0 and ψ_2 and their derivatives at ξ_1—the first zero of Emden's equation—are given. The method of integration adopted is the one attributed to Adams and sketched at the end of the second of von Zeipel's papers referred to at the outset. In the following Table VI the values of $\psi_0(\xi_1)$, $\psi_0'(\xi_1)$, $\psi_2(\xi_1)$, $\psi_2'(\xi_1)$, ξ_1 and θ_1' are given.

Table VI

n	1	1·5	2	3	4
ξ_1	3·14159	3·6538	4·3529	6·8968	14·9715
$-\theta_1'$	0·31831	0·20330	0·12725	0·04243	0·0080181
$\psi_0(\xi_1)$	1·00000	1·2942	1·9153	5·8380	33·5327
$\psi_0'(\xi_1)$	0·31831	0·6364	0·9961	2·0391	4·8812
$\psi_2(\xi_1)$	4·55940	4·7820	5·6431	11·2780	46·5444
$\psi_2'(\xi_1)$	0·42060	1·1495	1·7559	3·0409	6·1766

§ 11. With the use of the values given in the above table we can now numerically evaluate many of the "coefficients" occurring in the formulæ

* See the introduction (by D. H. Sadler) to the *Mathematical Tables*, vol. ii, on "Emden Functions," issued by the British Association for the Advancement of Science.

of §§ 7, 8 and 9. We shall now give the precise forms for the cases where the numerical integration has been effected.

Firstly, equation (36), expressing Θ, takes the following forms :—

$$\left.\begin{aligned}\Theta &= \theta + v[\psi_0(\xi) - 0{\cdot}5483\psi_2(\xi)P_2(\mu)], & n&=1\\ \Theta &= \theta + v[\psi_0(\xi) - 0{\cdot}5999\psi_2(\xi)P_2(\mu)], & n&=1{\cdot}5\\ \Theta &= \theta + v[\psi_0(\xi) - 0{\cdot}6426\psi_2(\xi)P_2(\mu)], & n&=2\\ \Theta &= \theta + v[\psi_0(\xi) - 0{\cdot}72325\psi_2(\xi)P_2(\mu)], & n&=3\\ \Theta &= \theta + v[\psi_0(\xi) - 0{\cdot}8048\psi_2(\xi)P_2(\mu)], & n&=4\end{aligned}\right\}. \tag{54}$$

The equations of the boundary ξ_0 are :

$$\left.\begin{aligned}\xi_0 &= 3{\cdot}1416 + 3{\cdot}1416v[1 - 2{\cdot}5000P_2(\mu)], & n&=1\\ \xi_0 &= 3{\cdot}6538 + 4{\cdot}9188v[1{\cdot}2942 - 2{\cdot}8687P_2(\mu)], & n&=1{\cdot}5\\ \xi_0 &= 4{\cdot}3529 + 7{\cdot}8585v[1{\cdot}9153 - 3{\cdot}6260P_2(\mu)], & n&=2\\ \xi_0 &= 6{\cdot}8968 + 23{\cdot}568v[5{\cdot}8380 - 8{\cdot}1568P_2(\mu)], & n&=3\\ \xi_0 &= 14{\cdot}9715 + 124{\cdot}718v[33{\cdot}533 - 37{\cdot}460P_2(\mu)], & n&=4\end{aligned}\right\}. \tag{55}$$

Remembering that $P_2(\mu) = -\frac{1}{2}$ at the equator and $+1$ at the poles we obtain from (55) the following table giving *the fractional elongation at the equator, the fractional contraction at the poles and the oblateness* as expressed by equation (39):—

TABLE VII

Geometry of the Boundary

n	Fractional Elongation at the Equator	Fractional Contraction at the Poles	Oblateness σ
1	$2{\cdot}2500v$	$1{\cdot}5000v$	$3{\cdot}7500v$
1·5	$3{\cdot}6730v$	$2{\cdot}1195v$	$5{\cdot}7926v$
2	$6{\cdot}7309v$	$3{\cdot}0884v$	$9{\cdot}8194v$
3	$33{\cdot}888v$	$7{\cdot}9240v$	$41{\cdot}811v$
4	$435{\cdot}35v$	$32{\cdot}714v$	$468{\cdot}07v$

The mass relations are :

$$\left.\begin{aligned}M_\omega &= M_0\,.\,(1 + 2{\cdot}290v), & n&=1\\ M_\omega &= M_0\,.\,(1 + 2{\cdot}860v), & n&=1{\cdot}5\\ M_\omega &= M_0\,.\,(1 + 3{\cdot}575v), & n&=2\\ M_\omega &= M_0\,.\,(1 + 6{\cdot}123v), & n&=3\\ M_\omega &= M_0\,.\,(1 + 13{\cdot}632v), & n&=4\end{aligned}\right\}. \tag{56}$$

The volume relations are :

$$\left.\begin{aligned}N_\omega &= N_0\,.\,(1 + 3v), & n&=1\\ N_\omega &= N_0\,.\,(1 + 5{\cdot}227v), & n&=1{\cdot}5\\ N_\omega &= N_0\,.\,(1 + 10{\cdot}373v), & n&=2\\ N_\omega &= N_0\,.\,(1 + 59{\cdot}849v), & n&=3\\ N_\omega &= N_0\,.\,(1 + 838{\cdot}00v), & n&=4\end{aligned}\right\}. \tag{57}$$

The relations between the mean and the central densities are :

$$\left.\begin{aligned} \rho_c &= \rho_m \times 3\cdot 2899[1+0\cdot 710v], & n&=1 \\ \rho_c &= \rho_m \times 5\cdot 9907[1+2\cdot 367v], & n&=1\cdot 5 \\ \rho_c &= \rho_m \times 11\cdot 4025[1+6\cdot 797v], & n&=2 \\ \rho_c &= \rho_m \times 54\cdot 1825[1+53\cdot 724v], & n&=3 \\ \rho_c &= \rho_m \times 622\cdot 408[1+814\cdot 37v], & n&=4 \end{aligned}\right\}. \tag{58}$$

§ 12. *Comparison of Configurations with equal Mass.*—The above sections (in particular § 11) give a complete answer to the problem formulated at the outset in § 1. But it is of interest to compare two configurations, one "stationary" and the other rotating with a slow angular velocity ω, both having the *same mass*. Now for this problem to have a meaning, the radius of the non-rotating polytrope must be determined when the mass is given. As is well known (and as is clear from equation (41)) this is not the case when $n=3$. Hence we should expect (and indeed as it turns out subsequently) that $n=3$ must be of the nature of a singularity with respect to this problem.

We have already seen that if a rotating and a non-rotating configuration are to have the same central density, then the non-rotating configuration has a smaller mass. To secure the same mass we must alter the central density of one of them—say the rotating one. The problem then is to find the fractional increase (or decrease) in λ (*i.e.* the central density) such that the masses of the two configurations are the same, *i.e.* we have to find $\delta\lambda$ such that

$$M(\lambda+\delta\lambda,\ \omega)=M(\lambda,\ 0). \tag{59}$$

From (41) we easily find by differentiation with respect to λ that

$$\frac{\delta\lambda}{\lambda}=-\frac{2n}{\dfrac{(3-n)}{v}\dfrac{|\theta_1'|}{\frac{1}{3}\xi_1-\psi_0'(\xi_1)}-3(n-1)}. \tag{60}$$

If $n\neq 3$ we can rewrite the above relation approximately to the first order as

$$\frac{\delta\lambda}{\lambda}=-v\cdot\frac{2n}{3-n}\frac{\frac{1}{3}\xi_1-\psi_0'(\xi_1)}{|\theta_1'|}. \tag{61}$$

It is at once clear from the above formula that *if $n>3$ the central density of the rotating configuration is greater than that of the non-rotating configuration, while if $n<3$ the converse is true.*

Of course the above result is precisely what we should expect. For if the central densities of the two configurations are equal the non-rotating configuration has a *smaller* mass, and as we have

$$M_0\propto\lambda^{\frac{3-n}{2n}}, \tag{62}$$

to *increase* the mass of the non-rotating sphere we should increase or decrease λ according as n is less than or greater than 3. This is just what (61) shows. When $n=3$ the radius of the non-rotating configuration is indeterminate and the question "the fractional change in the central density" has no meaning, for the radius of the polytrope $n = 3$ is first determined only when the mean density (and therefore also the central density) is given (in addition to M).

If $n \neq 3$ the change in the central density is proportional to v and hence *formulæ* (54), (55), (57), (58) *and the results of* Table VI *continue to be true (to the first order in v) when we are comparing two configurations with equal mass.*

Finally, the following short table gives precise values for $\frac{\delta\lambda}{\lambda}$:—

n	1	1·5	2	3	4
$-\delta\lambda/\lambda$	$2{\cdot}290v$	$5{\cdot}720v$	$14{\cdot}299v$	...	$-109{\cdot}06v$

§ 13. *General Discussion.*—There are a number of points which can now be discussed in connection with the above calculations and their bearing on actual physical facts, but I shall make only some brief comments and reserve more detailed references to possible applications. for a separate communication after treating the closely related problem of the double stars.

(1) Though the problem has been formulated in an abstract form and the whole calculation proceeds on the assumption that there exists a relation $P = K\rho^{1+\frac{1}{n}}$ between the total pressure and density, it can actually be shown that the whole investigation applies to the more general "standard model."

(2) A glance at Table VII shows that as the central condensation of the configuration increases the fractional elongation at the equator and the fractional contraction at the poles increase rapidly for a given angular velocity ω. But what is noteworthy is that the fractional elongation at the equator increases much more rapidly than the fractional contraction at the poles. Thus the ratio of the fractional elongation at the equator to the fractional contraction at the poles increases from 1·5 for $n = 1$ to 13·3 for $n = 4$. This is precisely what one would expect if for *large values* of ω the configurations with large central condensations should tend to assume lenticular forms for equatorial break-up.

(3) In § 12 we considered the fractional change in the central density ($\delta\lambda/\lambda$) which takes place when a gas sphere is set rotating with a slow angular velocity ω. It was found that if $n < 3$, $\delta\lambda/\lambda$ is *negative*, while if $n > 3$, $\delta\lambda/\lambda$ is *positive*. Thus from this standpoint configurations with "large" central condensations behave quite differently from configurations with comparatively weak central condensations. One is tempted to conjecture that these two distinct different *initial* behaviours of rotating masses correspond to the two distinct types of break-up—namely, the fissional break-up and the equatorial break-up—which occur when ω gets large. But this conjecture, though fascinating, is by no means well founded.

In conclusion, I wish to record my thanks to Professor N. Bohr for allowing me the very valuable privileges of his Institute, where the above work was carried out. My thanks are also due to Professor E. A. Milne for his interest and advice.

APPENDIX*

Solutions of the differential equations

$$\frac{1}{\xi^2}\frac{d}{d\xi}\left(\xi^2\frac{d\psi_0}{d\xi}\right) = -n\theta^{n-1}\psi_0 + 1,$$

and

$$\frac{1}{\xi^2}\frac{d}{d\xi}\left(\xi^2\frac{d\psi_2}{d\xi}\right) = \left(-n\theta^{n-1} + \frac{6}{\xi^2}\right)\psi_2,$$

for n = 1, 1·5, 2, 3 *and* 4.

In the following tables, in addition to ψ_0 and ψ_2, $n\theta^{n-1}$ is also tabulated. The values of the Emden functions were taken from *Mathematical Tables*, vol. ii, on "Emden Functions," issued by L. J. Comrie on behalf of the British Association for the Advancement of Science. Only as many decimal figures are retained as are regarded to be reliable.

TABLE I

Index n = 1

ξ	$\frac{6}{\xi^2} - 1$	ψ_0	ψ_2
0·	∞	0·	0·
0·2	149·0000	0·00665	0·039886
0·4	36·5000	0·02645	0·158180
0·6	15·6667	0·05893	0·350835
0·8	8·3750	0·10330	0·611258
1·0	5·0000	0·15853	0·93053
1·2	3·16666	0·22330	1·2977
1·4	2·06122	0·28611	1·7001
1·6	1·34375	0·37527	2·1239
1·8	0·85185	0·45897	2·5544
2·0	0·50000	0·54535	2·9767
2·2	0·23967	0·63250	3·3759
2·4	0·04167	0·71856	3·7380
2·6	−0·11243	0·80173	4·0499
2·8	−0·23469	0·88036	4·3002
3·0	−0·33333	0·95296	4·4795
3·2	−0·41406	...	4·5804
3·4	−0·48097	...	4·5982

$\xi_1 = 3·14159$,

$\psi_0(\xi_1) = 1·00000$, $\psi_2(\xi_1) = 4·55940$,

$\psi_0'(\xi_1) = 0·31831$, $\psi_2'(\xi_1) = 0·42060$.

* More accurate tabulations of these functions appear in S. Chandrasekhar and Norman R. Lebovitz, "On the Oscillations and the Stability of Rotating Gaseous Masses: 3, The Distorted Polytropes" (*The Astrophysical Journal* 136, no. 3 [1962]: 1082–1104), reprinted as paper 21 in volume 4 of these selected papers.

TABLE II

Index $n = 1{\cdot}5$

ξ	$1{\cdot}5\theta^{\frac{1}{2}}$	ψ_0	ψ_2
0·0	1·5000	0·	0·
0·2	1·4950	0·0066467	0·0398291
0·4	1·4801	0·026351	0·157289
0·6	1·4556	0·058423	0·346477
0·8	1·4217	0·101785	0·59813
1·0	1·3790	0·155061	0·90040
1·2	1·3281	0·21669	1·23978
1·4	1·2698	0·28506	1·6022
1·6	1·2046	0·35862	1·9741
1·8	1·1332	0·43601	2·3430
2·0	1·0564	0·51612	2·6985
2·2	0·97449	0·59822	3·0324
2·4	0·88802	0·68194	3·3401
2·6	0·79699	0·76739	3·6197
2·8	0·70067	0·85511	3·8714
3·0	0·59787	0·94618	4·0992
3·2	0·48403	1·04226	4·3095
3·4	0·35286	1·14581	4·5121
3·6	0·15797	1·26074	4·7218

$$\xi_1 = 3{\cdot}6538,$$
$$\psi_0(\xi_1) = 1{\cdot}2942, \qquad \psi_2(\xi_1) = 4{\cdot}7820,$$
$$\psi_0'(\xi_1) = 0{\cdot}6364, \qquad \psi_2'(\xi_1) = 1{\cdot}1495.$$

TABLE III

Index $n = 2$

ξ	2θ	ψ_0	ψ_2
0·	2·00000	0·	0·
0·2	1·98672	0·0066401	0·039773
0·4	1·94751	0·026249	0·15641
0·6	1·88419	0·057935	0·34226
0·8	1·79959	0·100359	0·58572
1·0	1·69731	0·152110	0·87277
1·2	1·58134	0·21126	1·1886
1·4	1·45582	0·27629	1·5190
1·6	1·32472	0·34605	1·8519
1·8	1·19165	0·41957	2·1778
2·0	1·05967	0·49641	2·4904
2·2	0·93130	0·57652	2·7862
2·4	0·80842	0·66030	3·0643
2·6	0·69237	0·74845	3·3263
2·8	0·58398	0·84195	3·5751
3·0	0·48365	0·94200	3·8151
3·2	0·39145	1·0500	4·0515
3·4	0·30720	1·1675	4·2898
3·6	0·23051	1·2963	4·5361
3·8	0·16086	1·4381	4·7966
4·0	0·09768	1·5949	5·0776
4·2	0·04032	1·7691	5·3855
4·4	...	1·9629	5·7274

$$\xi_1 = 4{\cdot}3529,$$
$$\psi_0(\xi_1) = 1{\cdot}9153, \qquad \psi_2(\xi_1) = 5{\cdot}6431,$$
$$\psi_0'(\xi_1) = 0{\cdot}9961, \qquad \psi_2'(\xi_1) = 1{\cdot}7559.$$

TABLE IV

Index $n=3$

ξ	$3\theta^2$	ψ_0	ψ_2
0·	3·000	0·	0·
0·2	2·960	0·0066271	0·039657
0·4	2·846	0·026050	0·15470
0·6	2·668	0·057010	0·33425
0·8	2·444	0·09768	0·56286
1·0	2·194	0·14637	0·8237
1·2	1·932	0·20142	1·1015
1·4	1·676	0·26168	1·3841
1·6	1·434	0·32655	1·6632
1·8	1·215	0·39596	1·9341
2·0	1·019	0·47021	2·195
2·2	0·8486	0·5499	2·448
2·4	0·7023	0·6360	2·693
2·6	0·5785	0·7292	2·935
2·8	0·4743	0·8306	3·175
3·0	0·3870	0·9412	3·419
3·2	0·3147	1·0618	3·668
3·4	0·2549	1·1932	3·924
3·6	0·2055	1·3364	4·192
3·8	0·1648	1·4920	4·472
4·0	0·1314	1·6607	4·767
4·2	0·1040	1·8431	5·078
4·4	0·08168	2·0398	5·407
4·6	0·06352	2·2513	5·753
4·8	0·04876	2·4780	6·121
5·0	0·03682	2·7203	6·506
5·2	0·02730	2·9787	6·916
5·4	0·01971	3·2532	7·346
5·6	0·01376	3·5443	7·798
5·8	0·00853	3·8520	8·272
6·0	0·00574	4·1765	8·768
6·2	0·00325	4·5180	9·288
6·4	0·00155	4·8760	9·829
6·6	0·00052	5·2510	10·39
6·8	0·00005	5·6429	10·98
7·0	...	6·0510	11·59

$$\xi_1 = 6{\cdot}8969,$$

$$\psi_0(\xi_1) = 5{\cdot}8380, \qquad \psi_2(\xi_1) = 11{\cdot}2780,$$

$$\psi_0'(\xi_1) = 2{\cdot}0391, \qquad \psi_2'(\xi_1) = 3{\cdot}0409.$$

TABLE V

Index $n=4$

ξ	$4\theta^3$	ψ_0	ψ_2
0·	4·0000	0·	0·
0·2	3·9209	0·0066140	0·0395434
0·4	3·6976	0·025857	0·153027
0·6	3·3651	0·056144	0·32671
0·8	2·9682	0·095460	0·54239
1·0	2·5514	0·141909	0·78215
1·2	2·1490	0·19413	1·03149
1·4	1·7824	0·25144	1·2809
1·6	1·4625	0·31378	1·5255
1·8	1·1915	0·37044	1·7636
2·0	0·96690	0·45543	1·9962
2·2	0·78328	0·53624	2·2254
2·4	0·63448	0·62487	2·4536
2·6	0·51458	0·72217	2·6837
2·8	0·41824	0·82892	2·9181
3·0	0·34076	0·94583	3·1591
3·2	0·27853	1·07355	3·4087
3·4	0·22845	1·2126	3·6683
3·6	0·18807	1·3636	3·9394
3·8	0·15533	1·5268	4·2230
4·0	0·12869	1·7027	4·5200
4·2	0·10695	1·8916	4·8311
4·4	0·08916	2·0938	5·1569
4·6	0·07454	2·3096	5·4979
4·8	0·06251	2·5390	5·8546
5·0	0·05253	2·7825	6·2270
5·2	0·04424	3·0400	6·6161
5·6	0·03159	3·5980	7·4434
6·0	0·02271	4·2140	8·3380
6·4	0·01639	4·8886	9·3008
6·8	0·01190	5·6223	10·3326
7·2	0·008648	6·4156	11·4337
7·6	0·006285	7·2685	12·6045
8·0	0·004566	8·1811	13·8451
8·4	0·003306	9·1535	15·1558
8·8	0·002384	10·1856	16·5364
9·2	0·001707	11·2773	17·9870
9·6	0·001211	12·4301	19·5076
10·0	0·000850	13·6407	21·0980
10·4	0·000588	14·9118	22·7509
10·8	0·000399	16·2414	24·4756
11·2	0·000264	17·6294	26·2663
11·6	0·000170	19·0754	28·1281
12·0	0·000105	20·5792	30·0592
12·4	0·000062	22·1405	32·0591
12·8	0·000034	23·7589	34·1276
13·2	0·000017	25·4344	36·2645
13·6	0·000007	27·1664	38·4695
14·0	0·000002	28·9549	40·7424
14·4	0·0000004	30·7996	43·0830
14·8	0·0000000	32·7003	45·4911
15·2	...	34·6567	47·9667

$$\xi_1 = 14{\cdot}9715,$$
$$\psi_0(\xi_1) = 33{\cdot}5327, \qquad \psi_2(\xi_1) = 46{\cdot}5444,$$
$$\psi_0'(\xi_1) = 4{\cdot}8812, \qquad \psi_2'(\xi_1) = 6{\cdot}1766.$$

Institut For Teoretisk Fysik,
Copenhagen :
1933 *January* 4.

THE EQUILIBRIUM OF DISTORTED POLYTROPES.

(II) The Tidal Problem.

S. Chandrasekhar.

§ 1. In a paper * recently communicated to the Society the equilibrium of slowly rotating polytropes was studied. It was shown there that the structure of such configurations can be specified completely when two functions ψ_0 and ψ_2 related in a special way to the Emden function θ are numerically integrated. In this communication another class of distorted polytropes will be studied, namely, the class of *tidally distorted polytropes.*

§ 2. The problem which will be studied here can be formulated as follows:—

We have a gas configuration of mass M—which we shall call the *primary* —acted on by tidal forces originating from a second gas configuration of mass M'—which we shall call the *secondary.* Let the centre of gravity of the primary be taken as the origin, and let the centre of gravity of the

* "The Equilibrium of Distorted Polytropes: (I) The Rotational Problem," *M.N.*, **93**, 390, 1933. Referred to here as I.

secondary be at a distance R from the origin on the Z-axis. The potential of the secondary at great distances can be expressed in the form

$$V' = G\left(\frac{M'}{r'} + \frac{A'+B'+C'-3I'}{2r'^3} + \ldots\right). \tag{1}$$

The term in r'^{-2} is lacking if the radius vector r' is reckoned from the centre of gravity of the secondary. A', B', C' denote the principal moments of inertia of the secondary at its centre of gravity, and I' the moment of inertia about the radius vector. These moments of inertia are of the order of magnitude $M'a'^2$ where a' is the mean radius of the secondary. Now for a spherically symmetrical distribution of matter $(A'-I')$, $(B'-I')$, $(C'-I')$ are all identically zero, and since, as is well known, the distortions arising from tidal influences is of the third order in the ratio of the radius of configuration to the distance between the components, we have the result that $(A'-I')$, $(B'-I')$, $(C'-I')$ for the secondary distorted by the tidal field due to the primary * is of the order $M'a'^2(a'R^{-1})^3$. Hence the second term in (1) is of the order of magnitude

$$G' \cdot \left(\frac{M'}{a'}\right) \cdot \left(\frac{a'}{R}\right)^6. \tag{2}$$

We shall neglect quantities of this order, *i.e. we shall consistently carry out our approximation to that order of accuracy in which the effects of the secondary on the primary are the same as that of a mass point.*†

The potential V' at a point (x, y, z) in the neighbourhood of the primary is therefore

$$V' = \frac{GM'}{r'} = \frac{GM'}{R}\left(1 - 2\frac{z}{r}\left(\frac{r}{R}\right) + \frac{r^2}{R^2}\right)^{-\frac{1}{2}}, \tag{3}$$

or writing $z/r = \cos\theta = \mu$ and expanding (3) in terms of the Legendre polynomials, $P_j(\mu)$ we have

$$V' = \frac{GM'}{R}\sum_{j=1}^{\infty}\left(\frac{r}{R}\right)^j P_j(\mu) + \text{constant}. \tag{4}$$

Consistently with our scheme of approximation where we neglect quantities of the order (2) we have to retain only the first four terms in the above summation, *i.e.*

$$V' = \frac{GM'}{R^2} r\cos\theta + \frac{GM'}{R}\sum_{j=2}^{4}\left(\frac{r}{R}\right)^j P_j(\mu) + \text{constant}. \tag{5}$$

The first term on the right gives a uniform field of force of intensity M'/R^2, which produces a Newtonian acceleration M'/R^2 in the primary. We can

* Which means that we are *not* considering an *arbitrary* secondary but one which (like the primary) is also in equilibrium under the combined effect of its own gravitation and the tidal forces due to the *primary*.

† Another point in this connection may be explicitly stated. The next term in the expansion (1) is of order involving r'^{-5} (*i.e.*) R^{-5}. This term is zero for a sphere. Since the distortions are of the order $(a'/R)^3$ we see that we can neglect these terms, consistently with our scheme of approximation. (I am indebted to the referee of my paper for this remark.)

neutralise this term by supposing the axes of reference to move with this acceleration. The centre of gravity of the primary will then always be at the origin.*

We are thus left with a *tide-generating potential*

$$V_T = \frac{GM'}{R} \sum_{j=2}^{4} \left(\frac{r}{R}\right)^j P_j(\mu) + \text{constant}, \tag{6}$$

and we have to consider the effect of such a tidal potential on a gas configuration in which the density distribution is originally polytropic. In other words, we are given that in the primary we have a relation of the type

$$P = K\rho^{1+\frac{1}{n}}, \tag{7}$$

where P is the total pressure and ρ is the density. In the absence of the tidal field V_T the density distribution is governed by the Emden function θ of index $-n$. *The tidal problem is to determine the shape and the density distribution in this configuration when distorted by a tidal field of the form* V_T *and explicitly relate its structure with the Emden function describing the undistorted state.*

Formulated in this way the problem and the standpoint taken here are different from the earlier investigations on the effect of tidal forces on " compressible stellar masses."

The problem considered here does not seem to have been attempted before, even for the special case $n = 3$.

§ 3. *The Equations of the Problem.*—The equations of mechanical equilibrium are

$$\text{grad } P = \rho \text{ grad } (V + V_T), \tag{8}$$

$$\text{div grad } V = -4\pi G\rho, \tag{9}$$

$$\text{div grad } V_T = 0. \tag{10}$$

From (8), (9) and (10) we deduce that

$$\text{div}\left(\frac{1}{\rho} \text{ grad } P\right) = -4\pi G\rho. \tag{11}$$

Changing to polar co-ordinates r, μ, ϕ and omitting the term in ϕ from symmetry and introducing the Θ and ξ variables defined by

$$r = \alpha\xi = \left[\frac{(n+1)K}{4\pi G}\lambda^{\frac{1}{n}-1}\right]^{\frac{1}{2}} \xi, \tag{12}$$

$$\rho = \lambda\Theta^n; \qquad P = K\lambda^{1+\frac{1}{n}}\Theta^{n+1}, \tag{13}$$

we find that (11) reduces to

$$\frac{1}{\xi^2}\frac{\partial}{\partial\xi}\left(\xi^2\frac{\partial\Theta}{\partial\xi}\right) + \frac{1}{\xi^2}\frac{\partial}{\partial\mu}\left((1-\mu^2)\frac{\partial\Theta}{\partial\mu}\right) = -\Theta^n. \tag{14}$$

* Compare J. H. Jeans, *Problems of Stellar Dynamics and Cosmogony* (Cambridge, 1919), p. 43, § 47.

§ 4. *Solution.*—We assume a solution of the form

$$\Theta = \theta + \Psi(\xi, \mu), \tag{15}$$

where θ is Emden's function of index n and Ψ is a function numerically small compared with θ, so that we can neglect quantities in Ψ^2 and higher orders. Substituting (15) in (14) and neglecting terms in Ψ^2, we have for the differential equation defining Ψ:

$$\frac{1}{\xi^2}\frac{\partial}{\partial\xi}\left(\xi^2\frac{\partial\Psi}{\partial\xi}\right) + \frac{1}{\xi^2}\frac{\partial}{\partial\mu}\left((1-\mu^2)\frac{\partial\Psi}{\partial\mu}\right) = -n\theta^{n-1}\Psi. \tag{16}$$

Now, we take for Ψ the following form :

$$\Psi = \sum_{j=1}^{\infty}\psi_j(\xi)A_jP_j(\mu), \tag{16'}$$

where A_j's are at present arbitrary, to be determined later. In the case of the tidal problem there is no purely radial function analogous to $\psi_0(\xi)$ which appeared in the rotational problem, for it is clear that "$\psi_0(\xi)$" refers to an expansion of the configuration as a whole (*cf.* I, § 7), and such an expansion is not possible in the tidal problem. (The formal justification is given later. See footnote on page 453.)

Substitution of (16′) in (16) yields for ψ_j the differential equation

$$\frac{1}{\xi^2}\frac{d}{d\xi}\left(\xi^2\frac{d\psi_j}{d\xi}\right) = \left(\frac{j(j+1)}{\xi^2} - n\theta^{n-1}\right)\psi_j, \qquad j = 1, 2, 3, \ldots \tag{17}$$

For the determination of the coefficients A_j we have to calculate the potential for reasons stated in I. The calculation proceeds exactly as in I (§ 6), and we shall not go into the details. The result is that

$$V = (n+1)K\lambda^{\frac{1}{n}}\left[\Theta + \sum_{j=1}^{\infty}B_j\xi^jP_j(\mu)\right], \tag{18}$$

where the B_j's are at present arbitrary. To determine first the B_j's, we have to substitute the above expression for V in the equation of mechanical equilibrium

$$\frac{\partial P}{\partial\xi} = \rho\left(\frac{\partial V}{\partial\xi} + \frac{\partial V_T}{\partial\xi}\right), \tag{19}$$

where the tidal potential V_T (given by (6)) expressed in the ξ-variable takes the form

$$V_T = \frac{GM'}{R}\sum_{2}^{4}\left(\frac{\alpha}{R}\right)^jP_j(\mu)\,.\,\xi^j. \tag{20}$$

On going through the algebra and equating the coefficients of the various Legendre-polynomials we find that

$$B_j = 0 \quad \text{if} \quad j \neq 2, 3, \text{ or } 4,$$

and

$$B_j = -\frac{GM'}{R}\frac{1}{(n+1)K\lambda^{\frac{1}{n}}}\left(\frac{\alpha}{R}\right)^j, \qquad j = 2, 3, 4. \tag{21}$$

We can simplify the above expression for B_j by using the equation expressing the mass (see equation (31) below), namely,

$$M = 4\pi a^3 \xi_1^2 \lambda \mid \theta_1' \mid. \tag{22}$$

(21) now simplifies to

$$B_j = -\frac{M'}{M}\xi_1^2 \mid \theta_1' \mid \left(\frac{a}{R}\right)^{j+1}, \qquad j = 2, 3, 4. \tag{23}$$

Hence our expression for the potential is by (18)

$$V = (n+1)K\lambda^{\frac{1}{n}}\left[\Theta - \sum_{j=2}^{4}\frac{M'}{M}\xi_1^2 \mid \theta_1' \mid \left(\frac{a}{R}\right)^{j+1}\xi^j P_j(\mu)\right]. \tag{18'}$$

Now the above expression for V and its derivative will have to be continuous on a sphere of radius ξ_1 (the first zero of Emden's function θ) with an expression of the type,

$$V_{\text{external}} = (n+1)K\lambda^{\frac{1}{n}}\left[\frac{C_0}{\xi} + \sum_{j=1}^{\infty}\frac{C_j}{\xi^{j+1}}P_j(\mu)\right]. \tag{24}$$

This will determine the A_j's. We find that

$$A_j = 0 \quad \text{if} \quad j \neq 2, 3 \text{ or } 4, \tag{25}$$

and for $j = 2$, 3 and 4 we have

$$A_j = (2j+1)\xi_1^{j+2} \cdot \frac{M'}{M} \cdot \mid \theta_1' \mid \cdot \left(\frac{a}{R}\right)^{j+1} \cdot \frac{1}{(j+1)\psi_j(\xi_1) + \xi_1 \psi_j'(\xi_1)}. \tag{26}$$

Hence the solution to the problem * is expressed by

$$\Theta = \theta + \xi_1 \mid \theta_1' \mid \cdot \nu^3 \cdot \frac{M'}{M}\sum_{j=2}^{4}(2j+1)\nu^{j-2} \cdot \frac{\psi_j(\xi)P_j(\mu)}{(j+1)\psi_j(\xi_1) + \xi_1\psi_j'(\xi_1)}, \tag{27}$$

where

$$\nu = \frac{a\xi_1}{R}. \tag{28}$$

ν has the simple meaning, that it is the *ratio of the radius of the undistorted configuration to the distance between the centre of gravities of the two components.* On introducing the quantities Δ_j defined by

$$\Delta_j = \frac{(2j+1)\psi_j(\xi_1)}{(j+1)\psi_j(\xi_1) + \xi_1\psi_j'(\xi_1)}, \qquad j = 2, 3, 4, \tag{29}$$

we can rewrite (27) in the form

$$\Theta = \theta + \xi_1 \mid \theta_1' \mid \cdot \frac{M'}{M}\sum_{j=2}^{4}\nu^{j+1} \cdot \Delta_j \cdot \frac{\psi_j(\xi)}{\psi_j(\xi_1)} \cdot P_j(\mu). \tag{30}$$

* The reason why we cannot have any purely radial function can be seen as follows: The radial function can be expressed as $A_0\psi_0(\xi)$ where $\psi_0(\xi)$ will satisfy the differential equation

$$\frac{1}{\xi^2}\frac{d}{d\xi}\left(\xi^2\frac{d\psi_0}{d\xi}\right) = -n\theta^{n-1}\psi_0,$$

but A_0 will be left undetermined even when all the conditions of the problem are satisfied. Going to the limit $R \to \infty$, we see that $A_0 = 0$.

(30) *shows that consistently with our approximation in treating the secondary as a mass point, we can solve the tidal problem correctly to the fifth order in* ν.

§ 5. *Mass and Volume Relations.*—The mass is given by

$$M = 2\pi \iint \rho r^2 dr d\mu.$$

Remembering that

$$\int_{-1}^{+1} P_j(\mu) d\mu = 0,$$

we see that the terms in $P_j(\mu)$ in (30) do not contribute to the mass to the order of accuracy with which we are concerned. Hence (*cf.* I, equation (41)).

$$M = -4\pi a^3 \lambda \xi_1^2 \left(\frac{d\theta}{d\xi}\right)_1. \tag{31}$$

Similarly the volume N of the configuration is given by

$$N = \frac{4\pi}{3} a^3 \xi_1^3, \tag{32}$$

the "ellipticity-terms" again not contributing to the volume. The relation between the mean and the central density is given by

$$\rho_{\text{mean}} = 3\xi_1 \cdot | \theta_1' |^{-1} \cdot \rho_{\text{central}}, \tag{33}$$

the same relation as for undistorted polytropes. Relations (31), (32) and (33) express the following fact:—

*If we consider a gas sphere and slowly bring towards it a tidally disturbing secondary, the configuration then gets distorted in shape, but in such a way that the volume and the central density remain constant.**

§ 6. *External Shape.*—The boundary ξ_0 is given by $\Theta = 0$ or by (30) we have for the equation of the boundary

$$\xi_0 = \xi_1 \left\{ 1 + \frac{M'}{M} \sum_{j=2}^{4} \Delta_j \cdot \nu^{j+1} \cdot P_j(\mu) \right\}, \tag{34}$$

or differently

$$\frac{\xi_0 - \xi_1}{\xi_1} = \frac{M'}{M} \sum_{j=2}^{4} \Delta_j \cdot \nu^{j+1} \cdot P_j(\mu). \tag{35}$$

It may be noted here that

$$\left.\begin{aligned} P_2(\mu) &= \tfrac{1}{2}(3\mu^2 - 1) \\ P_3(\mu) &= \tfrac{1}{2}(5\mu^3 - 3\mu) \\ P_4(\mu) &= \tfrac{1}{8}(35\mu^4 - 30\mu^2 + 3) \end{aligned}\right\}. \tag{36}$$

* This theorem shows that the conjecture made in I (§ 13, remark 3) is perhaps not correct. For it is known that in the tidal problem we also meet with two different types of break-up analogous to the two types of break-up met with in the rotational problem. Actually the fact that the effect of rotation is to decrease the central density when $n < 3$ and increase it when $n > 3$ appears more to have a bearing on questions of "*stability*" than on questions of "*break-up*." A more specific discussion on this point will be attempted in a separate paper.

The surface of the configuration as expressed by (34) is symmetrical with respect to the Z-axis but unsymmetrical with respect to the other two axes. There is an elongation at either of the poles ($\theta = 0$ or π) but of different amounts. The elongation at the pole nearer the secondary is greater than that at the opposite end. At the equator, on the other hand, there is a *uniform contraction*. We shall denote by σ_{+1}, σ_{-1} and σ_0 the fractional elongation at the pole $\theta = 0 (\mu = +1)$, the fractional elongation at the opposite pole $\theta = \pi (\mu = -1)$ and the fractional *contraction* at the equator $\theta = \pi/2 (\mu = 0)$ respectively. Remembering that $P_j(\mu)$ $(j = 2, 3, 4)$ takes for $\mu = 0, +1$ and -1 the values

μ	0	+1	−1
$P_2(\mu)$	$-\frac{1}{2}$	+1	+1
$P_3(\mu)$	0	+1	−1
$P_4(\mu)$	$\frac{3}{8}$	+1	+1

given in the table, we derive from (35) the following explicit relations for σ_{+1}, σ_{-1} and σ_0 :—

$$\left.\begin{aligned} \sigma_{+1} &= \frac{M'}{M}(\Delta_2 \nu^3 + \Delta_3 \nu^4 + \Delta_4 \nu^5) \\ \sigma_{-1} &= \frac{M'}{M}(\Delta_2 \nu^3 - \Delta_3 \nu^4 + \Delta_4 \nu^5) \\ \sigma_0 &= \frac{M'}{M}(\tfrac{1}{2}\Delta_2 \nu^3 - \tfrac{3}{8}\Delta_4 \nu^5) \end{aligned}\right\}. \tag{37}$$

We note that

$$\sigma_{+1} - \sigma_{-1} = 2\frac{M'}{M} \cdot \Delta_3 \nu^4. \tag{38}$$

Further, we see from (37) that

$$\frac{\sigma_{+1}}{\sigma_0} = \frac{\sigma_{-1}}{\sigma_0} + O(\nu) = 2 + O(\nu), \tag{39}$$

or in words (39) expresses the following fact :—

The fractional elongations at the two poles are to a first approximation equal and are to the same approximation just twice the uniform fractional contraction at the equator. Further, this fact is true for all polytropes.

The above is, of course, nothing new in principle and is primarily a numerical consequence of the values of $P_2(\mu)$ for $\mu = 0$ and 1. Indeed (39) must be true for all density distributions in the configuration, because in the tidal problem there is no expansion of the configuration as a whole, and to a first order the equation of the boundary must be of the following form when slightly distorted by a tidal field:—

$$\xi_0 = \xi_1 + \text{constant} \cdot P_2(\mu). \tag{40}$$

Attention may here be drawn to the fact that in the rotational problem (as against the tidal problem) the ratio of the elongation at the equator to the contractions at the poles is rather sensitive to the polytropic index, increasing from 1·5 for $n=1$ to 13·3 for $n=4$.* This difference between the two problems arises because in the rotational problem there is a general expansion of the configuration as a whole and this expansion is rather sensitive to the undisturbed density distribution with which we start.

Again, from (35) we see that *if we neglect terms in* ν^4, *the configurations are spheroids with a small ellipticity* (*or oblateness*) ϵ *given by*

$$\epsilon = 1{\cdot}5 \,.\, \frac{M'}{M} \,.\, \Delta_2 \nu^3. \tag{41}$$

§ 7. *Numerical Results.*—The structure of the tidally distorted polytrope will thus be specified completely when the functions ψ_2, ψ_3 and ψ_4 are known. The functions ψ_2 for indices of n equal to 1, 1·5, 2, 3, 4 have been already tabulated in I (Appendix). The functions ψ_3 and ψ_4 have also been integrated for the same values of n and are tabulated in an appendix to this paper. At the bottom of each table the values of ψ_3 and ψ_4 and their derivatives at ξ_1, the first zero of θ, are given.

The series expansions with which the numerical integrations were started may be noted here.

$$\psi_3 = \xi^3\left[1 - \frac{n}{18}\xi^2 + \frac{n(4n-3)}{18{\cdot}44}\xi^4 - \frac{n(434n^2-749n+330)}{18{\cdot}44{\cdot}1170}\xi^6 + \,.\,.\,.\right], \tag{42}$$

$$\psi_4 = \xi^4\left[1 - \frac{n}{22}\xi^2 + \frac{n(14n-11)}{66{\cdot}52}\xi^4 - \frac{n(1744n^2-3129n+1430)}{1350{\cdot}66{\cdot}52}\xi^6 + \,.\,.\,.\right]. \tag{43}$$

In Table VI the values of ψ_2, ψ_3, ψ_4 and also their derivatives at ξ_1 are collected.

TABLE VI

n	1	1·5	2	3	4
ξ_1	3·14159	3·6538	4·3529	6·8968	14·9715
$\psi_2(\xi_1)$	4·55940	4·7820	5·6431	11·2780	46·5444
$\psi_2'(\xi_1)$	0·42060	1·1495	1·7559	3·0409	6·1766
Δ_2	1·51985	1·2892	1·1482	1·0289	1·00267
$\psi_3(\xi_1)$	17·3743	23·369	35·572	124·814	1177·24
$\psi_3'(\xi_1)$	9·7959	14·828	21·852	53·366	235·65
Δ_3	1·2129	1·1079	1·0488	1·00736	1·00047
$\psi_4(\xi_1)$	61·1506	100·088	188·075	1086·30	22684·42
$\psi_4'(\xi_1)$	59·023	96·44	164·06	626·08	6058·63
Δ_4	1·1205	1·0562	1·0231	1·00281	1·00014
$\xi_1 \,.\, \lvert\theta_1'\rvert$	1·0000	0·74283	0·55391	0·29263	0·12004

* Compare I, § 13, remark (2).

The values of Δ_2, Δ_3, Δ_4 (which occur in the equations (30) and (34) expressing Θ and the boundary respectively) are also included in heavy type. In the last row the values of $\xi_1 \mid \theta_1' \mid$ are also inserted so that the exact numerical forms of (30) and (34) for different values of n can be read out at once.

It is of interest to note the inequality

$$\Delta_2 > \Delta_3 > \Delta_4, \tag{44}$$

which is seen to exist for all indices.

It is a rather curious fact that Δ_2, Δ_3, Δ_4 (the coefficients of the third, fourth and the fifth powers of ν in the equation for the boundary) tend rapidly to unity as n approaches 5. As a consequence of this we can predict at once that the *expression*

$$\xi_0 \equiv \xi_1 \left\{ 1 + \frac{M'}{M} \sum_{j=2}^{4} \nu^{j+1} \, . \, P_j(\mu) \right\}, \tag{45}$$

will represent to a high degree of approximation the equation of the boundary of tidally distorted polytropes with $5 > n > 4$.

§ 8. *The Formation of a "Furrow" on the Boundary.*—There is still another feature of the boundary that remains to be pointed out.

If we take the equation of the boundary (34) then the equation

$$\frac{\partial \xi_0}{\partial \mu} = 0 \tag{46}$$

has a solution apart from $\mu = \pm 1$ or 0. Equation (46), when written out fully, takes the form

$$3\Delta_2 \cos\theta + \Delta_3 \, . \, (\tfrac{15}{2}\cos^2\theta - \tfrac{3}{2}) \, . \, \nu + \Delta_4(\tfrac{35}{2}\cos^3\theta - \tfrac{15}{2}\cos^2\theta)\nu^2 = 0. \tag{47}$$

An approximate solution can easily be obtained by setting $\theta = \frac{\pi}{2} - \chi$ and neglecting terms in $\sin^2\chi \, . \, \nu$, $\sin\chi \, . \, \nu^2$ and higher orders. We find that

$$\chi = \tfrac{1}{2} \, . \, \frac{\Delta_3}{\Delta_2} \, . \, \nu + O(\nu^3). \tag{48}$$

This angle χ corresponds to a "*furrow*" having developed on the surface. The following table gives this angle (in radians) :—

n	1	1·5	2	3	4	$n > 4$
χ	$0{\cdot}3995\nu$	$0{\cdot}4297\nu$	$0{\cdot}4567\nu$	$0{\cdot}4895\nu$	$0{\cdot}4989\nu$	$0{\cdot}5\nu$

This concludes the formal analysis of the simple tidal problem. But as both the rotational and the simple tidal problems are very intimately related to the "double-star" problem it is better to study this problem before attempting any detailed discussion with references to actual physical applications.

APPENDIX

Solutions of the Differential Equations

$$\frac{1}{\xi^2}\frac{d}{d\xi}\left(\xi^2\frac{d\psi_3}{d\xi}\right)=\left(-n\theta^{n-1}+\frac{12}{\xi^2}\right)\psi_3$$

and

$$\frac{1}{\xi^2}\frac{d}{d\xi}\left(\xi^2\frac{d\psi_4}{d\xi}\right)=\left(-n\theta^{n-1}+\frac{20}{\xi^2}\right)\psi_4$$

for $n=1$, 1·5, 2, 3 *and* 4.

The values of the Emden functions were taken from Comrie and Sadler's tables. Only as many decimal figures are retained as are regarded to be reliable.

TABLE I

Index $n=1$

ξ	ψ_3	ψ_4
0·	0·	0·
0·2	0·0079822	0·0015971
0·4	0·063362	0·025415
0·6	0·21171	0·12750
0·8	0·49406	0·39783
1·0	0·94569	0·95541
1·2	1·5942	1·94152
1·4	2·4582	3·5118
1·6	3·5463	5·8272
1·8	4·8566	9·0440
2·0	6·3760	13·3043
2·2	8·0813	18·7252
2·4	9·9390	25·3918
2·6	11·9066	33·3475
2·8	13·9343	42·5878
3·0	15·9660	53·0560
3·2	17·9419	64·6402
3·4	19·8003	77·1728

$\xi_1=3·14159$

$\psi_3(\xi_1)=17·3743$, $\psi_4(\xi_1)=61·1506$,

$\psi_3'(\xi_1)=9·7959$, $\psi_4'(\xi_1)=59·023$.

TABLE II—*Index* $n=1.5$

ξ	ψ_3	ψ_4
0·	0·	0·
0·2	0·0079732	0·0015957
0·4	0·063156	0·025323
0·6	0·20968	0·12649
0·8	0·48584	0·39244
1·0	0·92205	0·93596
1·2	1·5395	1·8874
1·4	2·3501	3·3867
1·6	3·3565	5·5758
1·8	4·5543	8·5928
2·0	5·9334	12·5691
2·2	7·4806	17·6281
2·4	9·1825	23·8883
2·6	11·0281	31·4685
2·8	13·0112	40·497
3·0	15·1341	51·127
3·2	17·4109	63·552
3·4	19·8730	78·045
3·6	22·5836	95·025

$$\xi_1 = 3{\cdot}6538$$

$$\psi_3(\xi_1) = 23{\cdot}3689, \qquad \psi_4(\xi_1) = 100{\cdot}088,$$

$$\psi_3'(\xi_1) = 14{\cdot}828, \qquad \psi_4'(\xi_1) = 96{\cdot}44.$$

TABLE III—*Index* $n=2$

ξ	ψ_3	ψ_4
0·	0·	0·
0·2	0·0079764	0·0015942
0·4	0·062883	0·025234
0·6	0·20770	0·12552
0·8	0·47808	0·38734
1·0	0·90039	0·91815
1·2	1·49116	1·83962
1·4	2·2580	3·2801
1·6	3·2015	5·3701
1·8	4·3178	8·2392
2·0	5·6012	12·0163
2·2	7·0465	16·8328
2·4	8·6512	22·8270
2·6	10·4167	30·1502
2·8	12·3497	39·0073
3·0	14·4621	49·5550
3·2	16·7722	62·023
3·4	19·3044	76·681
3·6	22·0894	93·842
3·8	25·1646	113·877
4·0	28·5745	137·219
4·2	32·3714	164·378

$$\xi_1 = 4{\cdot}3529$$

$$\psi_3(\xi_1) = 35{\cdot}572, \qquad \psi_4(\xi_1) = 188{\cdot}075,$$

$$\psi_3'(\xi_1) = 21{\cdot}852, \qquad \psi_4'(\xi_1) = 164{\cdot}06.$$

TABLE IV

Index $n=3$

ξ	ψ_3	ψ_4
0·	0·	0·
0·2	0·0079471	0·0015913
0·4	0·06235	0·025059
0·6	0·20395	0·12367
0·8	0·46376	0·37791
1·0	0·86199	0·88643
1·2	1·4092	1·75815
1·4	2·1091	3·1073
1·6	2·9623	5·0521
1·8	3·9686	7·7159
2·0	5·1296	11·2290
2·2	6·4491	15·7440
2·4	7·9350	21·4078
2·6	9·5973	28·3959
2·8	11·4492	36·9024
3·0	13·5063	47·1426
3·2	15·7859	59·3540
3·4	18·3073	73·797
3·6	21·0906	90·755
3·8	24·1571	110·537
4·0	27·5291	133·473
4·2	31·2290	159·921
4·4	35·280	190·259
4·6	39·706	224·894
4·8	44·529	264·251
5·0	49·774	308·781
5·2	55·465	358·957
5·4	61·624	415·272
5·6	68·274	478·244
5·8	75·439	548·41
6·0	83·140	626·31
6·2	91·399	712·53
6·4	100·236	807·64
6·6	109·670	912·24
6·8	119·721	1026·94
7·0	130·402	1152·35

$\xi_1 = 6{\cdot}8968$

$\psi_3(\xi_1) = 124{\cdot}814, \qquad \psi_4(\xi_1) = 1086{\cdot}30,$

$\psi_3'(\xi_1) = 53{\cdot}366, \qquad \psi_4'(\xi_1) = 626{\cdot}08.$

TABLE V—*Index* $n=4$

ξ	ψ_3	ψ_4
0·	0·	0·
0·2	0·00797297	0·00158850
0·4	0·061827	0·024889
0·6	0·20037	0·12192
0·8	0·45091	0·36945
1·0	0·82912	0·85931
1·2	1·34245	1·69182
1·4	1·9938	2·9731
1·6	2·7854	4·8160
1·8	3·7209	7·3423
2·0	4·8062	10·6857
2·2	6·0499	14·9943
2·4	7·4632	20·4322
2·6	9·0586	27·1805
2·8	10·8508	35·4385
3·0	12·8550	45·4237
3·2	15·0876	57·3723
3·4	17·5653	71·5386
3·6	20·3053	88·1959
3·8	23·3249	107·636
4·0	26·6418	130·168
4·2	30·2736	156·122
4·4	34·2383	185·844
4·6	38·5536	219·700
4·8	43·2376	258·074
5·0	48·3080	301·369
5·2	53·7829	350·003
5·6	66·0178	465·068
6·0	80·0856	607·00
6·4	96·1293	779·81
6·8	114·292	987·82
7·2	134·716	1235·61
7·6	157·542	1528·06
8·0	182·914	1870·35
8·4	210·978	2267·90
8·8	241·849	2726·46
9·2	275·693	3252·02
9·6	312·630	3850·9
10·0	352·829	4529·6
10·4	396·395	5295·0
10·8	443·475	6154·2
11·2	494·207	7114·5
11·6	548·724	8183·7
12·0	607·163	9369·6
12·4	669·657	10680·5
12·8	736·341	12124·8
13·2	807·349	13711·2
13·6	882·813	15448·8
14·0	962·867	17346·8
14·4	1047·646	19414·9
14·8	1137·281	21662·8

$$\xi_1 = 14{\cdot}9715$$

$$\psi_3(\xi_1) = 1177{\cdot}244, \qquad \psi_4(\xi_1) = 22684{\cdot}42,$$

$$\psi_3'(\xi_1) = 235{\cdot}648, \qquad \psi_4'(\xi_1) = 6058{\cdot}63.$$

Institut for Teoretisk Fysik, Copenhagen,
1933 *March* 29.

22

THE EQUILIBRIUM OF DISTORTED POLYTROPES.

(III) The Double-star Problem.

S. Chandrasekhar.

§ 1. In two communications * the equilibrium of two types of distorted polytropes was studied. In this paper a third class of distorted polytropes will be considered, namely, the components of "double-star" systems which are both tidally and rotationally distorted. As will be shown in the following, no further quadratures are necessary to solve this problem.

The double-star problem can be formulated as follows :—

We have a system of two gas configurations revolving about their common centre of gravity. We assume that the ratio of the mean radii of the components to their distance apart is so small that we can neglect quantities of the sixth order in this ratio, since this enables one to consider (as shown in II, § 4) one of the configurations as a mass point when studying the equilibrium of the other. We fix our attention on one of the components of our "double-star" system. It is given that in this the total pressure P is related to the density ρ by means of the relation

$$P = K\rho^{1+\frac{1}{n}}. \tag{1}$$

If we disregard the tidal influence arising from the secondary and also the effects of rotation, the structure of the configuration is completely specified when the associated Emden function of index n is known. We now take into account the effects of distortion due to the tidal and the rotational forces. *The problem is to determine the shape and the density distribution in such tidally and rotationally distorted polytropes and explicitly to relate the structure of such configurations with the Emden functions which describe the polytropic state.*

Substantially the case $n = 3$ has already been treated by von Zeipel,† but, as will be seen, the solution given here is a direct consequence of the methods developed in I and II.

§ 2. *Equations of the Problem.*—We take as the origin of our system of co-ordinates the centre of gravity of the primary. The Z-axis is parallel to the axis of rotation, and the X-axis is directed towards the centre of gravity of the secondary. Let the masses of the primary and the secondary be respectively M and M'. Let R be the distance between the centres of gravity. By Kepler's third law we have the relation

$$\omega^2 = \frac{G(M+M')}{R^3}(1+\epsilon), \tag{2}$$

* "The Equilibrium of Distorted Polytropes: (I) The Rotational Problem," *M.N.*, **93**, 390, 1933 ; "The Equilibrium of Distorted Polytropes : (II) The Tidal Problem," the previous paper. These two are referred to as I and II in the sequel.

† H. von Zeipel, *M.N.*, **84**, 702, 1924.

where ϵ is a quantity of the order of magnitude $(\bar{a}/R)^8$,* where $\bar{a}$ represents the mean radii of the components. As we are going to carry our approximation only to the fifth order in quantities of this magnitude, we shall accordingly use the relation

$$\omega^2 = \frac{G(M+M')}{R^3}. \tag{2'}$$

From (2′) it is clear that neglecting quantities which are of the sixth order in $(\bar{a}/R)$ is equivalent to neglecting quantities of the second order in ω^2.

Let V be the inner attraction potential in the primary and V' the outer attraction potential due to the secondary. The equations of mechanical equilibrium are

$$\left.\begin{aligned} \frac{\partial P}{\partial x} &= \rho\left\{\frac{\partial}{\partial x}(V+V') + \omega^2 x - \frac{M'}{M+M'}R\omega^2\right\} \\ \frac{\partial P}{\partial y} &= \rho\left\{\frac{\partial}{\partial y}(V+V') + \omega^2 y\right\} \\ \frac{\partial P}{\partial z} &= \rho\left\{\frac{\partial}{\partial z}(V+V')\right\} \end{aligned}\right\}, \tag{3}$$

where P is related to ρ by (1). Also V and V' satisfy respectively Poisson's and Laplace's equations:

$$\nabla^2 V = -4\pi G\rho\,; \qquad \nabla^2 V' = 0. \tag{4}$$

From (3) and (4) we deduce that

$$\operatorname{div}\left(\frac{1}{\rho}\operatorname{grad} P\right) = -4\pi G\rho + 2\omega^2. \tag{5}$$

Changing over to polar co-ordinates and introducing the ξ and Θ variables as defined in I and II we find that (5) reduces to

$$\nabla^2\Theta = -\Theta^n + v, \tag{6}$$

where

$$v = \frac{\omega^2}{2\pi G\lambda}. \tag{7}$$

§ 3. We seek a solution of (6) in the form

$$\Theta = \theta + v\Psi + v^2\Phi + \ldots, \tag{8}$$

where θ is Emden's function of index n. The differential equation defining Ψ is easily found to be

$$\nabla^2\Psi = -n\theta^{n-1}\Psi + 1. \tag{9}$$

We take for Ψ the following form:

$$\Psi = \psi_0(\xi) + \sum_{j=1}^{\infty}\psi_j(\xi)S_j(\mu, \phi), \tag{10}$$

* Compare G. H. Darwin, *Scientific Papers*, 3, § 15, p. 479.

where $S_j(\mu, \phi)$, ($\mu = \cos\theta$) is at present an arbitrary surface-harmonic of degree j satisfying the differential equation

$$\frac{\partial}{\partial\mu}\left((1-\mu^2)\frac{\partial S_j}{\partial\mu}\right)+\frac{1}{1-\mu^2}\frac{\partial^2 S_j}{\partial\phi^2}+j(j+1)S_j=0. \tag{11}$$

Substituting (10) in (9) and using (11) and equating the coefficients of the successive orders of the surface harmonics we find that the differential equations defining the radial functions $\psi_0(\xi)$ and $\psi_j(\xi)$ are

$$\frac{1}{\xi^2}\frac{d}{d\xi}\left(\xi^2\frac{d\psi_0}{d\xi}\right)=-n\theta^{n-1}\psi_0+1, \tag{12}$$

$$\frac{1}{\xi^2}\frac{d}{d\xi}\left(\xi^2\frac{d\psi_j}{d\xi}\right)=\left(-n\theta^{n-1}+\frac{j(j+1)}{\xi^2}\right)\psi_j, \qquad j=1, 2, 3, \ldots \tag{13}$$

To determine the precise forms of the surface-harmonics S_j we have, as usual, to calculate the potential. The result is

$$V=(n+1)\,.\,K\,.\,\lambda^{\frac{1}{n}}\left[\Theta+v\left\{\sum_{j=1}^{\infty}\xi^j S_j'(\mu, \phi)-\tfrac{1}{6}\xi^2\right\}\right], \tag{14}$$

where S_j''s are again surface harmonics of the successive orders, arbitrary at present and different from the S_j's introduced in (10). To determine the precise forms of these surface-harmonics S_j' we have to substitute the above expression for V in the equation of equilibrium

$$\frac{\partial P}{\partial\xi}=\rho\left(\frac{\partial V}{\partial\xi}+\frac{\partial V'}{\partial\xi}+\frac{\partial\Omega}{\partial\xi}\right), \tag{15}$$

where Ω is used to denote the centrifugal potential.

§ 4. *The Centrifugal Potential.*—The equations of mechanical equilibrium (3) can be written in the form

$$\text{grad}\, P=\rho\,\text{grad}\left\{V+V'+\tfrac{1}{2}\omega^2(x^2+y^2)-\frac{M'}{M+M'}R\omega^2 x\right\}. \tag{16}$$

We see at once that the centrifugal potential can be written as

$$\Omega=\tfrac{1}{3}\omega^2 r^2(1-P_2(\mu))-\frac{M'}{M+M'}R\omega^2 r P_1(\sin\theta\cos\phi), \tag{17}$$

where $P_j(x)$ is used to denote the Legendre polynomial of order j in x. Using equation (2) defining ω^2 we can rewrite the above as

$$\Omega=\tfrac{1}{3}\omega^2 a^2\xi^2(1-P_2(\mu))-\frac{GM'}{R}\left(\frac{a}{R}\right)\xi P_1(\sin\theta\cos\phi)+\text{constant}. \tag{18}$$

§ 5. *The Outer Potential* V'.—It is shown in II (§ 2) that consistent with our present scheme of approximation we can treat the secondary as a mass point and that, further, it is sufficient to retain the first four terms in the expansion of r'^{-1} * in terms of the Legendre polynomials. We have (*cf.* II, equation (5))

$$V' = \frac{GM'}{R}\sum_{j=1}^{4}\left(\frac{a}{R}\right)^j \xi^j P_j(\sin\theta\cos\phi) + \text{constant}. \tag{19}$$

By (18) and (19) we have

$$V' + \Omega = \frac{GM'}{R}\sum_{j=2}^{4}\left(\frac{a}{R}\right)^j . \xi^j . P_j(\sin\theta\cos\phi) + \tfrac{1}{3}\omega^2 a^2 \xi^2 (1 - P_2(\mu)). \tag{20}$$

§ 6. Introducing (20) in (15) with the expression for V given by (14) we find that in the summation

$$\Sigma \xi^j S_j'(\mu, \phi)$$

only the surface harmonics

$$P_2(\cos\theta)\,; \qquad P_j(\sin\theta\cos\phi), \qquad j = 2, 3, 4, \tag{21}$$

occur and with the numerical coefficients

$$\tfrac{1}{6}\,; \qquad -\tfrac{1}{2}\frac{M'}{M+M'}\left(\frac{a}{R}\right)^{j-2}, \qquad j = 2, 3, 4, \tag{22}$$

respectively.

Hence our expression for V takes finally the form

$$V = (n+1) . K . \lambda^{\frac{1}{n}}\left[\Theta + v\left\{\tfrac{1}{6}\xi^2(P_2(\mu) - 1) - \sum_{j=2}^{4}\frac{1}{2}\frac{M'}{M+M'}\cdot\left(\frac{a}{R}\right)^{j-2} . \xi^j . P_j(\sin\theta\cos\phi)\right\}\right]. \quad ($$

To determine the precise forms of the surface harmonics $S_j(\mu, \phi)$ occurring in the expression for Θ, we have to make the above expression for V and also its derivative continuous with an expression of the type

$$V_{\text{outer}} = (n+1) . K . \lambda^{\frac{1}{n}}\left[\frac{C_0}{\xi} + v\sum_{j=1}^{\infty}\frac{S_j''(\mu, \phi)}{\xi^{j+1}}\right], \tag{24}$$

on a sphere of radius ξ_1, the first zero of θ. On going through the procedure we find that among the S_j's only the surface harmonics

$$P_2(\mu)\,; \qquad P_j(\sin\theta\cos\phi), \qquad j = 2, 3, 4, \tag{25}$$

occur and with the numerical coefficients

$$-\tfrac{5}{6}\frac{\xi_1^2}{3\psi_2(\xi_1) + \xi_1\psi_2'(\xi_1)} \tag{26}$$

and

$$\tfrac{1}{2}\frac{M'}{M+M'}\cdot\left(\frac{a}{R}\right)^{j-2}\cdot\frac{(2j+1)\xi_1^j}{(j+1)\psi_j(\xi_1) + \xi_1\psi_j'(\xi_1)}, \qquad j = 2, 3, 4, \tag{26'}$$

respectively.

We finally obtain therefore as the solution of the problem

$$\Theta = \theta + v\left\{\psi_0(\xi) - \tfrac{5}{6} . \frac{\xi_1^2 P_2(\cos\theta)\psi_2(\xi)}{3\psi_2(\xi_1) + \xi_1\psi_2'(\xi_1)} + \sum_{j=2}^{4}(j+\tfrac{1}{2}) . \frac{M'}{M+M'} . \nu^{j-2} . \frac{\xi_1^2 P_j(\sin\theta\cos\phi)\psi_j(\xi)}{(j+1)\psi_j(\xi_1) + \xi_1\psi_j'(\xi_1)}\right\}, \tag{27}$$

* r' is the distance of the centre of gravity of the secondary to the point (x, y, z) in the primary.

where (as in II) we have introduced the quantity ν defined as the ratio of the radius of the undistorted configuration to the distance R between the components, *i.e.*,

$$\nu = \frac{\alpha \xi_1}{R}. \tag{28}$$

We therefore see that the structure of the component will be specified completely to the order of accuracy with which we are concerned (i.e. *to the fifth order in* ν) when the functions ψ_0, ψ_2, ψ_3, and ψ_4 are known. ψ_0 and ψ_2 are tabulated in I for certain values of n, and ψ_3, ψ_4 are tabulated in II for the same values of n. Further quadratures are unnecessary.

§ 6. *Mass and Volume Relations.*—The mass is given by

$$M = \iiint \rho r^2 dr d\mu d\phi.$$

All the terms in the Legendre polynomials occurring in our solution for Θ do not contribute to the mass. The actual calculation proceeds exactly as in I, § 8, and yields

$$M = 4\pi \alpha^3 \xi_1^2 \,.\, |\theta_1'| \,.\, \lambda \,.\left[1 + v \frac{\frac{1}{3}\xi_1 - \psi_0'(\xi_1)}{|\theta_1'|}\right]$$

or a "mass-relation"

$$M_\omega = M_0 \,.\left[1 + v \,.\, \frac{\frac{1}{3}\xi_1 - \psi_0'(\xi_1)}{|\theta_1'|}\right], \tag{29}$$

an expression identical in form with the corresponding one obtained for the simple rotational problem—the Kepler angular velocity ω here replacing the uniform angular velocity occurring in that problem. Similarly we have a volume-relation and a relation between the mean and the central density of the same form as in the rotational problem (*cf.* I, § 9):

$$N_\omega = N_0 \,.\left[1 + v \frac{3\psi_0(\xi_1)}{\xi_1 |\theta_1'|}\right], \tag{30}$$

$$\rho_{\text{mean}} = -\lambda \,.\, \frac{3}{\xi_1}\left(\frac{d\theta}{d\xi}\right)_1 .\left[1 + v \frac{\frac{1}{3}\xi_1^2 - \psi_0'(\xi_1)\xi_1 - 3\psi_0(\xi_1)}{\xi_1 \,.\, |\theta_1'|}\right]. \tag{31}$$

Again, by definition

$$v = \frac{\omega^2}{2\pi G \lambda}, \tag{32}$$

and since we are working only to the first order in v, we can rewrite the above as

$$v = \frac{(M+M')}{MR^3} \,.\, \frac{M}{2\pi\lambda} = 2\left(1 + \frac{M'}{M}\right) .\, \nu^3 \,.\, \frac{|\theta_1'|}{\xi_1}. \tag{33}$$

§ 7. *Comparison with the Rotation and Tidal Problems.*—Using the above expression for v (equation (33)), we can rewrite our solution (27) for Θ as

$$\Theta = \theta + v\left\{\psi_0(\xi) - \tfrac{5}{6}\frac{\xi_1^2 P_2(\cos\theta) \,.\, \psi_2(\xi)}{3\psi_2(\xi_1) + \xi_1 \psi_2'(\xi_1)}\right\}$$

$$+ \xi_1 \,.\, |\theta_1'| \,.\, \frac{M'}{M}\sum_{j=2}^{4} \Delta_j \,.\, \nu^{j+1} \,.\, P_j(\sin\theta\cos\phi) \,.\, \frac{\psi_j(\xi)}{\psi_j(\xi_1)}, \tag{34}$$

where (as in II) we have introduced the quantities Δ_j defined by

$$\Delta_j = \frac{(2j+1)\psi_j(\xi_1)}{(j+1)\psi_2(\xi_1)+\xi_1\psi_j'(\xi_1)}, \qquad j=2, 3, 4. \tag{35}$$

On comparing * the above solution with the corresponding solutions for the simple rotational and the tidal problems (I, equation (36) and II, equation (30)), we notice that the "distortion terms" (*i.e.* $\Theta-\theta$) occurring in (34) are exactly the sum of the corresponding terms in those two problems. In other words, *the distortion of a "double-star" component is the same as if it rotated like a rigid body about its own axis with the Kepler angular velocity* ω *and then tidally influenced by a secondary at a distance* R *from its centre of gravity, the two effects being simply added. This superposition theorem is true to the fifth order in the ratio of the radius of the component to the distance between the two components.*†

§ 8. *External Shape.*—The boundary ξ_0 is given by $\Theta=0$, or by (34)

$$\xi_0 = \xi_1 + \frac{v}{|\theta_1'|}\left\{\psi_0(\xi_1) - \frac{5}{6}\frac{\xi_1^2\psi_2(\xi_1)P_2(\mu)}{3\psi_2(\xi_1)+\xi_1\psi_2'(\xi_1)}\right\} + \xi_1\frac{M'}{M}\sum_{j=2}^{4}\Delta_j\,.\,\nu^{j+1}\,.\,P_j(\sin\theta\cos\phi). \tag{36}$$

On comparing this equation of the boundary with those obtained for the simple rotational and tidal problems, we again meet here another example of the superposition theorem stated in § 7. (36) can be written more conveniently as (on using (33))

$$\frac{\xi_0-\xi_1}{\xi_1} = 2\left(1+\frac{M'}{M}\right).\,\nu^3\,.\,\frac{\psi_0(\xi_1)}{\xi_1^2} - \frac{1}{3}\left(1+\frac{M'}{M}\right).\,\Delta_2\,.\,\nu^3\,.\,P_2(\cos\theta) + \frac{M'}{M}\sum_{j=2}^{4}\Delta_j\,.\,\nu^{j+1}\,.\,P_j(\sin\theta\cos\phi). \tag{37}$$

We thus see that there is an expansion of the configuration as a whole of amount

$$2\left(1+\frac{M'}{M}\right)\nu^3\,.\,\frac{\psi_0(\xi_1)}{\xi_1^2}, \tag{38}$$

and superposed on this general expansion are the "ellipticity" and "harmonic" terms. The *deviations* from a spherical shape are given therefore only by the terms in $P_2(\cos\theta)$ and $P_j(\sin\theta\cos\phi)$. We shall therefore introduce the function $\sigma(\theta, \phi)$ ‡ defined as

* In comparing the solution with that of the tidal problem we have to remember that the secondary in the tidal problem was placed on the Z-axis, while now the other component of our double-star system is on the X-axis. Naturally the "$\cos\theta$" in the tidal problem is now replaced by "$\sin\theta\cos\psi$."

† This theorem is *not* likely to be true for higher orders in ν, for the sixth and higher order terms ought to take explicitly into account the density distribution in the secondary. I am indebted to Dr. B. Strömgren for this remark.

‡ $\sigma(\theta)$ has the simple meaning that it represents the fractional deviations from a *spherical* volume having the same volume as our distorted configuration.

$$\sigma(\theta, \phi) = \frac{\xi_0 - \xi_1}{\xi_1} - 2\left(1 + \frac{M'}{M}\right) . \nu^3 . \frac{\psi_0(\xi_1)}{\xi_1^2}$$
$$= -\tfrac{1}{3}\left(1 + \frac{M'}{M}\right) . \Delta_2 . \nu^3 . P_2(\cos\theta) + \frac{M'}{M}\sum_{j=2}^{4} \Delta_j . \nu^{j+1} . P_j(\sin\theta\cos\phi). \quad (39)$$

A general idea of the nature of the distortion in shape suffered by the configuration can be obtained from a study of the function $\sigma(\theta, \phi)$. (A reference

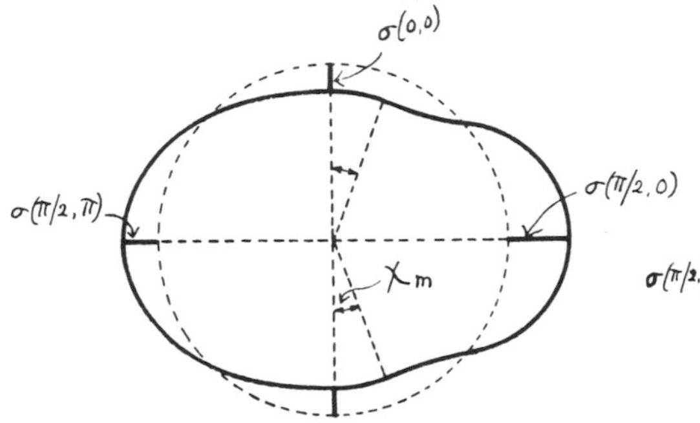

Fig. 1
The Meridian Plane, $\sigma(\theta,0)$

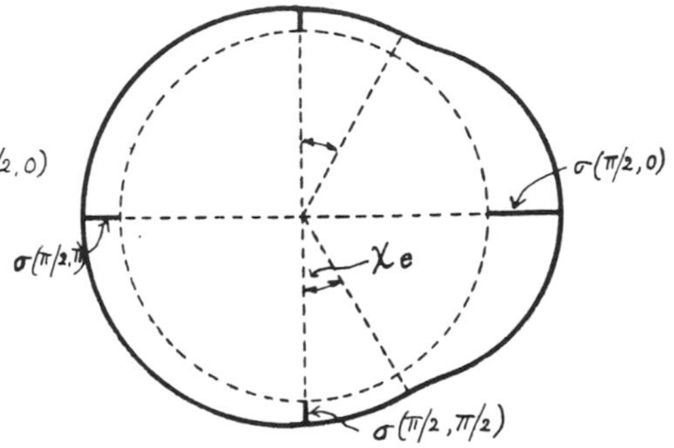

Fig. 2
The Equatorial Plane, $\sigma(\pi/2,\phi)$

to figs. 1 and 2 may help to visualise the nature of the surface we are considering.) In the *equatorial plane* $\sigma(\theta, \phi)$ has the form

$$\sigma\left(\frac{\pi}{2}, \phi\right) = \tfrac{1}{6}\left(1 + \frac{M'}{M}\right)\nu^3 + \frac{M'}{M}\sum_{j=2} \Delta_j . \nu^{j+1} . P_j(\cos\phi). \quad (40)$$

In the *principal meridian* through the centre of gravity of the secondary we have

$$\sigma(\theta, 0) = -\tfrac{1}{3}\left(1 + \frac{M'}{M}\right) . \Delta_2 . \nu^3 . P_2(\cos\theta) + \frac{M'}{M} . \sum_{j=2}^{4} \Delta_j . \nu^{j+1} . P_j(\sin\theta), \quad (41)$$

and finally in the *diametrical plane* along the axis of y we have

$$\sigma\left(\theta, \frac{\pi}{2}\right) = -\tfrac{1}{3}\left(1 + \frac{M'}{M}\right) . \Delta_2 . \nu^3 . P_2(\cos\theta) - \tfrac{1}{2} . \frac{M'}{M} . \Delta_2 . \nu^3 + \tfrac{3}{8} . \frac{M'}{M} . \Delta_4 . \nu^5. \quad (42)$$

From the above equations we easily deduce that

$$\left.\begin{aligned}
\sigma\left(\frac{\pi}{2}, 0\right) &= \tfrac{1}{6}\left(1 + 7\frac{M'}{M}\right) . \Delta_2 . \nu^3 + \frac{M'}{M} . \Delta_3 . \nu^4 + \frac{M'}{M} . \Delta_4 . \nu^5, \\
\sigma\left(\frac{\pi}{2}, \pi\right) &= \tfrac{1}{6}\left(1 + 7\frac{M'}{M}\right) . \Delta_2 . \nu^3 - \frac{M'}{M} . \Delta_3 . \nu^4 + \frac{M'}{M} . \Delta_4 . \nu^5, \\
\sigma\left(\frac{\pi}{2}, \frac{\pi}{2}\right) &= \tfrac{1}{6}\left(1 - 2\frac{M'}{M}\right) . \Delta_2 . \nu^3 + \tfrac{3}{8} . \frac{M'}{M} . \Delta_4 . \nu^5, \\
\sigma(0, 0) &= -\tfrac{1}{3}\left(1 + \tfrac{5}{2}\frac{M'}{M}\right) . \Delta_2 . \nu^3 + \tfrac{3}{8} . \frac{M'}{M} . \Delta_4 . \nu^5.
\end{aligned}\right\} \quad (43)^*$$

* We may notice the inequality $\sigma\left(\frac{\pi}{2}, 0\right) > \sigma\left(\frac{\pi}{2}, \pi\right) > \sigma\left(\frac{\pi}{2}, \frac{\pi}{2}\right) > \sigma(0, 0)$. At $(\pi/2, 0)$ we have the *absolute maximum* and at $(0, 0)$ or $(\pi, 0)$ the *absolute minimum*.

The exact numerical forms of equations (39) to (43) for the various polytropes ($n = 1$, 1·5, 2, 3 and 4) can be read out at once on referring to Table VI in II where the values of Δ_2, Δ_3, Δ_4 for some values of n are given. To obtain the actual fractional elongations or contractions at the various points $\left(\theta = \frac{\pi}{2}, \phi = 0, \text{etc.}\right)$ we have to add to the various values of σ obtained from (43) the general expansion term (38). To obtain this (as also the precise numerical forms of the equations of the boundary for different values of n) we need the value of $\psi_0(\xi_1)/\xi_1^2$. Values of $\psi_0(\xi_1)$ are given in Table VI, I. For convenience the following short table is added:—

n	1	1·5	2	3	4
$\dfrac{\psi_0(\xi_1)}{\xi_1^2}$	0·10132	0·09694	0·10108	0·12273	0·14961

(44)

Again from (43) we see that *if we neglect quantities of the order ν^4 and ν^5 then the equilibrium configurations are ellipsoids. The ellipticities ϵ_m, ϵ_e and ϵ_d in the principal meridian, the equatorial plane and in the diametrical plane along the axis of* y *respectively are given by*

$$\left.\begin{aligned} \epsilon_m &= \sigma\left(\frac{\pi}{2}, 0\right) - \sigma(0, 0) = \tfrac{1}{2}\left(1 + 4\frac{M'}{M}\right)\Delta_2 \,.\, \nu^3 \\ \epsilon_e &= \sigma\left(\frac{\pi}{2}, 0\right) - \sigma\left(\frac{\pi}{2}, \frac{\pi}{2}\right) = \tfrac{3}{2}\frac{M'}{M}\Delta_2 \,.\, \nu^3 \\ \epsilon_d &= \sigma\left(\frac{\pi}{2}, \frac{\pi}{2}\right) - \sigma(0, 0) = \tfrac{1}{2}\left(1 + \frac{M'}{M}\right)\Delta_2 \,.\, \nu^3 \end{aligned}\right\}. \tag{43'}$$

Also we notice that

$$\left.\begin{aligned} \epsilon_m > \epsilon_d > \epsilon_e \quad &\text{if} \quad 2M' < M \\ \epsilon_m > \epsilon_e \geqslant \epsilon_d \quad &\text{if} \quad 2M' \geqslant M \end{aligned}\right\}. \tag{43''}$$

There is still another feature of the external shape that has to be discussed. Now if we take the equation of the boundary in the meridian plane specified by (41) (apart from the constant additive term (38)) then we find that the equation

$$\frac{d\sigma(\theta, 0)}{d\theta} = 0, \tag{45}$$

has a solution apart from $\theta = 0$, $\dfrac{\pi}{2}$ or π. Equation (45) explicitly takes the form

$$\left(1 + \frac{M'}{M}\right) . \Delta_2 . \sin\theta + \frac{M'}{M}\{3\Delta_2 \sin\theta + \tfrac{1}{2}\Delta_3 . \nu . (15 \sin^2\theta - 3) + \tfrac{1}{2}\Delta_4 . \nu^2(35 \sin^3\theta - 15 \sin\theta)\} = 0. \tag{45'}$$

An approximate solution of (45′) when θ is small can be obtained by omitting in the above all terms in $\sin^2\theta \,.\, \nu$ or ν^3 and writing θ for $\sin\theta$. If χ_m denotes the solution, then

$$\chi_m = \frac{3}{2}\frac{\Delta_3}{\Delta_2}\frac{M'}{M+4M'} \,.\, \nu + O(\nu^3). \tag{46}$$

Similarly, if we take the equatorial plane we again have a solution for

$$\frac{d\sigma\left(\frac{\pi}{2}, \phi\right)}{d\phi} = 0, \tag{47}$$

apart from $\phi = 0, \pi/2, \pi, 3\pi/2$. The equation determining this angle is found to be

$$3\Delta_2 \cos\phi + \tfrac{1}{2} \,.\, \Delta_3 \,.\, \nu \,.\, (15\cos^2\phi - 3) + \tfrac{1}{2}\Delta_4 \,.\, \nu^2(35\cos^3\phi - 15\cos\phi) = 0. \tag{47′}$$

Denoting the solution by $(\pi/2 - \chi_e)$ we obtain as an approximate solution:

$$\chi_e = \tfrac{1}{2} \,.\, \frac{\Delta_3}{\Delta_2} \,.\, \nu + O(\nu^3). \tag{48}$$

This angle is the same as the angle at which the furrow is formed in the tidal problem (*cf.* II, § 8). In the following table the values of $\frac{\Delta_3}{\Delta_2}$ for different values of n are given so that χ_m and χ_e can be obtained from (46) and (48) in their numerical forms:—

n	1	1·5	2	3	4
Δ_3/Δ_2	0·79803	0·85937	0·91344	0·97904	0·99780

(49)

§ 9. Finally, one rather curious fact may be noted. In the double-star problem there can actually be a genuine fractional contraction in the equatorial plane, if the ratio of the mass of the secondary to that of the primary is greater than a certain quantity which we will now determine. Now, the fractional change at $(\theta = \pi/2,\ \phi = \pi/2)$ in the equatorial plane is given by (*cf.* equation (43))

$$\sigma\left(\frac{\pi}{2}, \frac{\pi}{2}\right) + 2\left(1 + \frac{M'}{M}\right) \,.\, \nu^3 \,.\, \frac{\psi_0(\xi_1)}{\xi_1^{\,2}}$$
$$= \left\{2\left(1 + \frac{M'}{M}\right) \,.\, \frac{\psi_0(\xi_1)}{\xi_1^{\,2}} + \tfrac{1}{6}\left(1 - \frac{2M'}{M}\right) \,.\, \Delta_2 + O(\nu)\right\}\nu^3. \tag{50}$$

We will neglect terms of $O(\nu)$ compared to unity. Then the above expression is negative if

$$\frac{M'}{M} > \frac{1}{2}\,\frac{\Delta_2 + 12\dfrac{\psi_0(\xi_1)}{\xi_1^{\,2}}}{\Delta_2 - 6\dfrac{\psi_0(\xi_1)}{\xi_1^{\,2}}}. \tag{51}$$

It is easily verified that the quantity on the right is just the ratio of the fractional elongation at the equator to the fractional contraction at the poles that would occur in the simple rotational problem where the configuration is rotating about the Z-azis with a constant, uniform angular velocity. Or more precisely stated, *if the ratio of the mass of the secondary to that of the primary in our double-star system is equal to the ratio of the fractional elongation at the equator to the fractional contraction at the poles, which would be found if the primary rotated about itself as a rigid body with a uniform angular velocity, then the radius of the configuration in the equatorial plane and in a direction at right angles to the line joining the centres of gravity of the two components suffers no change in length.* If the ratio M'/M is greater than this, then there is contraction in this direction in the equatorial plane.

§ 10. *Change in the Central Density.*—In I, § 13, we considered the change in the central density as the configuration is set rotating. A similar question can be asked in connection with the double-star problem as well. Remembering that the mass-relation in the double-star problem is identical in form with that in the simple rotational problem, we find that the fractional change in the central density is given by (I, equation (61)),

$$\frac{\delta\lambda}{\lambda} = -v\frac{2n}{3-n}\frac{\frac{1}{3}\xi_1 - \psi_0'(\xi_1)}{|\,\theta_1'\,|}. \tag{52}$$

Numerical forms of (52) have already been obtained in I. Again, by (33) we can rewrite (52) as

$$\frac{\delta\lambda}{\lambda} = -\left(1+\frac{M'}{M}\right)\cdot\frac{4n}{3-n}\cdot\frac{\frac{1}{3}\xi_1 - \psi_0'(\xi_1)}{\xi_1}\cdot\nu^3. \tag{53}$$

Thus in the double-star problem we meet with the same difference in the behaviour of polytropes with $n > 3$ and those with $n < 3$, as was met with in the simple rotational problem. There is an increase in the central density if $n > 3$ and a decrease if $n < 3$.

In a separate paper the general properties of the three classes of distorted polytropes we have studied will be more closely examined to obtain information on the qualitative behaviour of the tidal and rotational distortions regarded as functions of the complete march of the density distribution in the configuration.

In conclusion, I have great pleasure in recording my thanks to Professor N. Bohr for allowing the very valuable privileges of his Institute, where the above work was carried out.

Institut for Teoretisk Fysik,
Copenhagen :
1933 *April* 2.

THE EQUILIBRIUM OF DISTORTED POLYTROPES

(IV) The Rotational and the Tidal Distortions as Functions of the Density Distribution

S. CHANDRASEKHAR

§ 1. The formal mathematical theory of the equilibrium of (i) the rotationally distorted polytropes, (ii) the tidally distorted polytropes, and (iii) the rotationally and tidally distorted polytropes, which are components of double-star systems, have already been worked out in three recent communications.* In this paper we shall more closely examine the general properties of these three classes of distorted polytropes and see what information can be obtained regarding the general nature of the functional dependence of the rotational and tidal distortions on the complete march of the density distribution in the undistorted configuration.

In this connection it is of interest to consider the two extreme limits of density distribution, namely, (i) the uniform distribution of density ($\rho = \lambda =$ constant), and (ii) the complete concentration of the mass towards the centre. The former will be referred to here as the *Maclaurin model* and the latter is of course the usual *Roche model*. We shall consider the equilibrium of these models as well, and relate their theory to the theory of polytropes. There is a consequent simplification in the theory of these configurations, and in some cases (as in the tidal problem for instance) solutions of decidedly greater accuracy than have been given so far are obtained by very elementary methods.

The following two sections are devoted to obtain the solutions for the Maclaurin and the Roche models for the three problems. The three subsequent sections consider the rotational, the tidal and the double-star problems in turn.

I. *Solutions for the Maclaurin Model*

§ 2. The Maclaurin model corresponds to a polytrope with index $n = 0$. This polytrope differs from the other polytropes in this that the density does not vanish on the boundary. If the density vanishes on the boundary, then when the configuration is distorted by a slow uniform rotation, for instance, the density in the "distorted regions" is of order v ($= \omega^2/2\pi G\lambda$), and since the volume distortion is also of order v, the mass enclosed in these regions is of order v^2, and consequently the effects arising from this mass of order v^2 can be neglected when we are working only to the first order in v. In the case of the Maclaurin model, however, since the density does not vanish on the boundary, the mass enclosed in the "distorted regions" is of order v, and we cannot therefore straightaway use the solutions obtained in I for this special case. We shall treat the problem *de novo*.

§ 3. *The General Procedure.*—The general procedure is always to determine the internal and the external potentials V and U, and making the solution for the density distribution determinate by making these potentials and their derivatives continuous on the boundary of the configuration. If $\mathbf{V}_0$ and $\mathbf{U}_0$ are the corresponding potentials for the undistorted configuration we then have

$$\left.\begin{aligned} V &= \mathbf{V}_0 + \epsilon \mathbf{V}_1, \\ U &= \mathbf{U}_0 + \epsilon \mathbf{U}_1, \end{aligned}\right\} \tag{1}$$

where $\epsilon = v$ in the rotational and the double-star problems and $\epsilon = (a/R)^3$

* "The Equilibrium of Distorted Polytropes, I," *M.N.*, **93**, 390, 1933; "The Equilibrium of Distorted Polytropes, II," *M.N.*, **93**, 449, 1933; "The Equilibrium of Distorted Polytropes, III," *M.N.*, **93**, 462, 1933. These three papers will be referred to as I, II and III in the sequel.

for the simple tidal problem. Now the boundary ξ_0 is given by an equation of the form

$$\xi_0 = \xi_1 + \epsilon\xi_2 \quad \text{(say)}. \tag{2}$$

The boundary conditions can now be expressed as

$$\left.\begin{aligned} (\mathbf{V}_0 - \mathbf{U}_0)_{\xi_1} + \epsilon\{\xi_2(\mathbf{V}_0' - \mathbf{U}_0') + (\mathbf{V}_1 - \mathbf{U}_1)\}_{\xi_1} = 0, \\ (\mathbf{V}_0' - \mathbf{U}_0')_{\xi_1} + \epsilon\{\xi_2(\mathbf{V}_0'' - \mathbf{U}_0'') + (\mathbf{V}_1' - \mathbf{U}_1')\}_{\xi_1} = 0. \end{aligned}\right\} \tag{3}$$

The terms in the different powers of ϵ must vanish separately. Hence we should have

$$(\mathbf{V}_0 - \mathbf{U}_0)_{\xi_1} = 0 ; \qquad (\mathbf{V}_0' - \mathbf{U}_0')_{\xi_1} = 0, \tag{4}$$

and

$$\left.\begin{aligned} (\mathbf{V}_1 - \mathbf{U}_1)_{\xi_1} = 0, \\ \xi_2(\mathbf{V}_0'' - \mathbf{U}_0'')_{\xi_1} + (\mathbf{V}_1' - \mathbf{U}_1')_{\xi_1} = 0. \end{aligned}\right\} \tag{5}$$

Now, in *general* for polytropes the Poisson and the Laplace equations defining $\mathbf{V}_0$ and $\mathbf{U}_0$ reduce to (as one easily verifies)

$$\left.\begin{aligned} \frac{1}{\xi^2}\frac{d}{d\xi}\left(\xi^2\frac{d\mathbf{V}_0}{d\xi}\right) &= -(n+1)K\lambda^{\frac{1}{n}}\theta^n, \\ \frac{1}{\xi^2}\frac{d}{d\xi}\left(\xi^2\frac{d\mathbf{U}_0}{d\xi}\right) &= 0. \end{aligned}\right\} \tag{6}$$

From (6) we see that

$$\frac{d^2}{d\xi^2}(\mathbf{V}_0 - \mathbf{U}_0) + \frac{2}{\xi}\frac{d}{d\xi}(\mathbf{V}_0 - \mathbf{U}_0) + (n+1)K\lambda^{\frac{1}{n}}\theta^n = 0. \tag{7}$$

Since $\theta = 0$ for $\xi = \xi_1$ we see that $(\mathbf{V}_0'' - \mathbf{U}_0'')_{\xi_1}$ is also zero on the sphere $\xi = \xi_1$. Hence in general the conditions (5) simplify to

$$(\mathbf{V}_1 - \mathbf{U}_1)_{\xi_1} = 0 ; \qquad (\mathbf{V}_1' - \mathbf{U}_1')_{\xi_1} = 0. \tag{8}$$

It was the condition (8) which was used in the analysis in I, II and III. But the arguments which simplify (5) to (8) break down when $n = 0$, and the boundary conditions to be used for the Maclaurin model are those given in (5).

(A) The Rotational Problem :—

§ 4. Equation (5), I, continues to hold good. Let the constant value of the density ρ be λ, and put

$$P = P_c\Theta, \tag{9}$$

where P_c is the central pressure. Introduce further the variable ξ defined by

$$r = \left[\frac{P_c}{4\pi G\lambda^2}\right]^{\frac{1}{2}}\xi. \tag{10}$$

In place of (8), I, we have now

$$\frac{1}{\xi^2}\frac{d}{d\xi}\left(\xi^2\frac{d\Theta}{d\xi}\right)+\frac{1}{\xi^2}\frac{\partial}{\partial\mu}\left((1-\mu^2)\frac{\partial\Theta}{\partial\mu}\right)=-1+v, \tag{11}$$

where v has the usual meaning.

§ 5. *Non-rotating Configuration.*—When there is no rotation let $P=P_c\theta$. Then the differential equation for θ is

$$\frac{1}{\xi^2}\frac{d}{d\xi}\left(\xi^2\frac{d\theta}{d\xi}\right)=-1, \tag{12}$$

the required solution of which is

$$\theta=1-\tfrac{1}{6}\xi^2, \tag{13}$$

giving

$$\xi_1=\sqrt{6}\,; \qquad \theta_1'=-\tfrac{1}{3}\sqrt{6}. \tag{14}$$

If $\mathbf{V}_0$ and $\mathbf{U}_0$ are the potentials, then the Poisson and the Laplace equations defining them are readily found to be

$$\left.\begin{aligned}\frac{1}{\xi^2}\frac{d}{d\xi}\left(\xi^2\frac{d\mathbf{V}_0}{d\xi}\right)&=-\frac{P_c}{\lambda},\\ \frac{1}{\xi^2}\frac{d}{d\xi}\left(\xi^2\frac{d\mathbf{U}_0}{d\xi}\right)&=0.\end{aligned}\right\} \tag{15}$$

For definiteness we shall assume that the zero of the potential is on the boundary of the undistorted configuration, *i.e.* on a sphere of radius ξ_1. From (15) and the boundary conditions (4) we obtain

$$\mathbf{V}_0=\frac{P_c}{\lambda}(1-\tfrac{1}{6}\xi^2)\,; \qquad \mathbf{U}_0=\frac{P_c}{\lambda}\left(-2+\frac{2\sqrt{6}}{\xi}\right). \tag{16}$$

§ 6. *Solution for the Rotating Configuration.*—To solve (11) we take as usual the following form for Θ:—

$$\Theta=\theta+v\{\psi_0(\xi)+\sum_{j=1}^{\infty}A_j\psi_j(\xi)P_j(\mu)\}+O(v^2). \tag{17}$$

The differential equations defining ψ_0 and ψ_j are

$$\left.\begin{aligned}\frac{1}{\xi^2}\frac{d}{d\xi}\left(\xi^2\frac{d\psi_0}{d\xi}\right)&=1,\\ \frac{1}{\xi^2}\frac{d}{d\xi}\left(\xi^2\frac{d\psi_j}{d\xi}\right)&=\frac{j(j+1)}{\xi^2}\psi_j.\end{aligned}\right\} \tag{18}$$

Equations (18) can be integrated as they stand, and yield

$$\psi_0=\tfrac{1}{6}\xi^2\,; \qquad \psi_j=\xi^j. \tag{19}$$

The calculation of the potential proceeds exactly as in I, § 6, and yields

$$V=\frac{P_c}{\lambda}[\Theta-\tfrac{1}{6}v\xi^2(1-P_2(\mu))]. \tag{20}$$

We rewrite (20) in the form

$$V=\frac{P_c}{\lambda}\left[(1-\tfrac{1}{6}\xi^2)+v\left\{\tfrac{1}{6}\xi^2P_2(\mu)+\sum_{j=1}^{\infty}A_j\xi^jP_j(\mu)\right\}\right]. \tag{21}$$

Similarly the outer potential U is found to be

$$U=\frac{P_c}{\lambda}\left[\left(-2+\frac{2\sqrt{6}}{\xi}\right)+v\left\{C_0'+\frac{C_0}{\xi}+\sum_{j=1}^{\infty}\frac{C_j}{\xi^{j+1}}P_j(\mu)\right\}\right], \tag{22}$$

where C_0, C_0' and the C_j's are arbitrary. The boundary conditions (5) will determine these as well as the A_j's.

To proceed further we must also know the equation of the boundary in terms of the A_j's. The boundary is given by $\Theta=0$, and since Θ is given by

$$\Theta=\theta+v\left\{\tfrac{1}{6}\xi^2+\sum_{j=1}^{\infty}A_j\xi^jP_j(\mu)\right\}, \tag{23}$$

we have

$$\xi_0=\xi_1+\frac{3v}{\sqrt{6}}\left\{1+\sum_{j=1}^{\infty}A_j\xi_1^jP_j(\mu)\right\}. \tag{24}$$

Hence ξ_2 occurring in (2) is given by

$$\xi_2=\tfrac{1}{2}\xi_1\left\{1+\sum_{j=1}^{\infty}A_j\xi_1^jP_j(\mu)\right\}. \tag{25}$$

The boundary conditions (5) reduce to the following set of equations on equating the coefficients of $P_j(\mu)$:—

$$\left.\begin{aligned}&C_0'+\frac{C_0}{\xi_1}=0,\\ &A_j\xi_1^j=\frac{C_j}{\xi_1^{j+1}},\qquad (j\neq 2);\qquad 6A_2+1=\frac{C_2\,.\,\sqrt{6}}{36},\qquad (j=2),\end{aligned}\right\} \tag{26}$$

and

$$\left.\begin{aligned}&\frac{C_0}{\xi_1^2}=\tfrac{1}{2}\xi_1;\qquad (j-3)A_j\xi_1^{j-1}=-\frac{(j+1)C_j}{\xi_1^{j+2}},\qquad (j\neq 2)\\ &-A_2\xi_1+\tfrac{1}{3}\xi_1=-\frac{3C_2}{36},\qquad (j=2).\end{aligned}\right\} \tag{27}$$

Solving these sets of equations we obtain

$$C_0=\tfrac{1}{2}\xi_1^3=3\sqrt{6};\qquad C_0'=-\tfrac{1}{2}\xi_1^2=-3. \tag{28}$$

If $j\neq 2$,

$$(j-1)A_j=0, \tag{29}$$

which shows that $A_j=C_j=0$ if $j\neq 2,\ \neq 1$. A_1 is left undetermined. Going to the limit $v=0$ we see that $A_1=C_1=0$. Finally if $j=2$,

$$A_2=-\tfrac{5}{12};\qquad C_2=-9\sqrt{6}. \tag{30}$$

Hence our solution is

$$\Theta=1-\tfrac{1}{6}\xi^2+v\tfrac{1}{6}\xi^2(1-\tfrac{5}{2}P_2(\mu)). \tag{31}$$

Also we note that

$$\left.\begin{aligned} V &= \frac{P_c}{\lambda}\{1 - \tfrac{1}{6}\xi^2 - \tfrac{1}{4}v\xi^2 P_2(\mu)\}, \\ U &= \frac{P_c}{\lambda}\left\{\left(-2+\frac{2\sqrt{6}}{\xi}\right)+v\left(-3+\frac{3\sqrt{6}}{\xi}-\frac{9\sqrt{6}}{\xi^3}P_2(\mu)\right)\right\}. \end{aligned}\right\} \tag{32}$$

By (24) the equation of the boundary is given by

$$\xi_0 = \sqrt{6}\{1 + \tfrac{1}{2}v(1 - \tfrac{5}{2}P_2(\mu))\}. \tag{33}$$

From (33) we deduce that

$$\left.\begin{aligned} &\text{Fractional Elongation at Equator} = 9/8.v, \\ &\text{Fractional Contraction at Poles} = 3/4.v, \\ &\text{Oblateness } \sigma = 15/8.v. \end{aligned}\right\} \tag{34}$$

The eccentricity e^2 of an ellipse is defined by $(a^2-b^2)/a^2$, where a denotes the major axis and b the minor axis. When the eccentricity is small it is equal to twice the oblateness σ (to the first order). Hence from (34) we obtain

$$e^2 = \frac{15}{4}\frac{\omega^2}{2\pi G\lambda}. \tag{35}$$

§ 7. *Comparison with the Known Solution.*—The configurations of equilibrium of rotating masses of uniform density can be worked out accurately and the "evolutionary sequence"—Maclaurin spheroids, Jacobian ellipsoids—exactly followed. For the Maclaurin spheroids the exact solution is *

$$\frac{\omega^2}{2\pi G\lambda} = \frac{3-2e^2}{e^3}(1-e^2)^{\frac{1}{2}}\sin^{-1}e - 3\left(\frac{1}{e^2}-1\right), \tag{36}$$

or, when e^2 is small,

$$\frac{\omega^2}{2\pi G\lambda} = \frac{4}{15}e^2 + O(e^4), \tag{36'}$$

in agreement with (35).

§ 8. *Some Relations for the Polytrope* "n = 0."—The volume N_ω of the configuration is easily found. We have

$$N_\omega = N_0(1 + 1\cdot 5v). \tag{37}$$

Since the density is supposed to remain constant we have clearly

$$M_\omega = M_0(1 + 1\cdot 5v). \tag{38}$$

Again when we are comparing configurations with equal mass, we require the fractional change in the central pressure when the configuration is set rotating; *i.e.* we have to find δP_c such that

$$M(P_c, \omega) = M(P_c - \delta P_c, 0). \tag{39}$$

* *Cf.*, for instance, J. H. Jeans, *Stellar Dynamics and Problems of Cosmogony* (1919), § 38.

By (10) M_0 is proportional to $P_c^{3/2}$. We clearly have therefore

$$\frac{\delta P_c}{P_c} = -v. \tag{40}$$

(B) The Tidal Problem :—

§ 9. Equation (11), II, continues to hold good. Introducing as in § 4 the variables ξ and Θ, we find that the equation we have to solve is

$$\frac{1}{\xi^2}\frac{\partial}{\partial\xi}\left(\xi^2\frac{\partial\Theta}{\partial\xi}\right)+\frac{1}{\xi^2}\frac{\partial}{\partial\mu}\left((1-\mu^2)\frac{\partial\Theta}{\partial\mu}\right) = -1. \tag{41}$$

To solve (41) we assume for Θ the form *

$$\Theta = \theta + \left(\frac{a}{R}\right)^3 \sum_{j=1}^{\infty} A_j\psi_j(\xi)P_j(\mu) + \left(\frac{a}{R}\right)^6 \Phi + \cdots \tag{42}$$

The differential equation for ψ_j is

$$\frac{1}{\xi^2}\frac{d}{d\xi}\left(\xi^2\frac{d\psi_j}{d\xi}\right) = \frac{j(j+1)}{\xi^2}\psi_j. \tag{43}$$

The required solution is therefore

$$\psi_j = \xi^j. \tag{44}$$

The calculation of the potential proceeds exactly as in II, § 4, and yields

$$V = \frac{P_c}{\lambda}\left[\theta + \left(\frac{a}{R}\right)^3\left\{\sum_{j=1}^{\infty} A_j\xi^jP_j(\mu) - \frac{GM'}{R}\frac{\lambda}{P_c}\sum_{j=2}^{4}\left(\frac{a}{R}\right)^{j-3}\xi^jP_j(\mu)\right\}\right],$$

which we rewrite in the form

$$V = \frac{P_c}{\lambda}\left[(1-\tfrac{1}{6}\xi^2) + \left(\frac{a}{R}\right)^3\left\{\sum_{j=1}^{\infty} A_j\xi^jP_j(\mu) - 2\sqrt{6}\,.\,\frac{M'}{M}\sum_{j=2}^{4}\left(\frac{a}{R}\right)^{j-2}\xi^jP_j(\mu)\right\}\right]. \tag{45}$$

Similarly

$$U = \frac{P_c}{\lambda}\left[\left(-2+\frac{2\sqrt{6}}{\xi}\right) + \left(\frac{a}{R}\right)^3\sum_{j=1}^{\infty}\frac{C_j}{\xi^{j+1}}P_j(\mu)\right]. \tag{46}$$

By (42) the equation of the boundary takes the form

$$\xi_0 = \xi_1 + \left(\frac{a}{R}\right)^3 . \frac{3}{\sqrt{6}}\sum_{j=1}^{\infty} A_j\,.\,\xi_1{}^jP_j(\mu). \tag{47}$$

---

* The differential equation (41) contains no reference to any small parameter, and it is not necessary in fact to anticipate our solution to be a series in powers of $(a/R)^3$ as we have done in (42). Actually the $(a/R)^3$ in front of the second term could have been absorbed in the A_j's, and we could assume a solution (as in II) of the form $\theta + \Psi$ and neglect squares of Ψ. The result will be the same, and as such (42) is the really more convenient form to work with.

The boundary conditions (5) resolve into the following sets of simultaneous equations:—

$$\left.\begin{aligned} A_j\xi_1^{\,j} &= \frac{C_j}{\xi_1^{\,j+1}},\\ (j-3)A_j\xi_1^{\,j-1} &= -(j+1)\cdot\frac{C_j}{\xi_1^{\,j+2}},\end{aligned}\right\} \tag{48}$$

for $j \neq 2, \neq 3, \neq 4$, and

$$\left.\begin{aligned} A_j\xi_1^{\,j} - 2\sqrt{6}\cdot\frac{M'}{M}\left(\frac{a}{R}\right)^{j-2}\cdot\xi_1^{\,j} &= \frac{C_j}{\xi_1^{\,j+1}},\\ (j-3)A_j\xi_1^{\,j-1} - 2\sqrt{6}\cdot\frac{M'}{M}\left(\frac{a}{R}\right)^{j-2}\cdot j\cdot\xi_1^{\,j-1} &= -(j+1)\frac{C_j}{\xi_1^{\,j+2}},\end{aligned}\right\} \tag{49}$$

for $j=2$, 3 or 4.

Solving we find

$$A_j = C_j = 0, \qquad j \neq 2, \neq 3, \neq 4,$$

$$\left.\begin{aligned} A_j &= \frac{(2j+1)}{2(j-1)}\cdot\xi_1^2\,|\theta_1'|\cdot\frac{M'}{M}\left(\frac{a}{R}\right)^{j-2},\\ C_j &= \frac{3}{2(j-1)}\cdot\xi_1^{2j+3}\,|\theta_1'|\cdot\frac{M'}{M}\left(\frac{a}{R}\right)^{j-2},\end{aligned}\right\} \tag{50}$$

for $j=2, 3, 4$.

Hence our solution is given by

$$\Theta = 1 - \tfrac{1}{6}\xi^2 + 2\frac{M'}{M}\cdot\sum_{j=2}^{4}\nu^{j+1}\cdot\frac{2j+1}{2(j-1)}\cdot\frac{\xi^j}{6^{j/2}}P_j(\mu), \tag{51}$$

where $\nu(=a\xi_1/R)$ has the same meaning as in II and III.

The equation of the boundary ξ_0 can now be expressed in the "standard form" (*cf.* II, equation (35)),

$$\frac{\xi_0-\xi_1}{\xi_1} = \frac{M'}{M}\sum_{j=2}^{4}\Delta_j\cdot\nu^{j+1}\cdot P_j(\mu), \tag{52}$$

if we *define* for the polytrope "$n=0$" the Δ_j's in the following manner:—

$$\Delta_j = \frac{2j+1}{2(j-1)}, \qquad j=2, 3, 4. \tag{53}$$

This *definition* for the Δ_j's has *always* to be used whenever we are dealing with the Maclaurin model. (For the other polytropes the general definition of the Δ_j's (equation (29), II) has to be used.)

Explicitly written out, (52) takes the form

$$\frac{\xi_0-\xi_1}{\xi_1} = \frac{M'}{M}\cdot\nu^3[2{\cdot}5P_2(\mu) + 1{\cdot}75\nu P_3(\mu) + 1{\cdot}5\nu^2 P_4(\mu)]. \tag{54}$$

When the secondary is so far away that we could consider the configuration as a spheroid, then we have to retain only the first term in (54). We then have

$$\left.\begin{aligned}\text{Fractional Elongation at Poles} &= 2{\cdot}5\nu^3 \,.\, \frac{M'}{M},\\ \text{Fractional Contraction at Equator} &= 1{\cdot}25\nu^3 \,.\, \frac{M'}{M}.\end{aligned}\right\} \tag{55}$$

Also,

$$\text{Eccentricity, } e^2 = 7{\cdot}5\nu^3\frac{M'}{M} + E(\nu^4), \tag{56}$$

where by $E(\nu^4)$ we mean that the "error term" is of $O(\nu^4)$.

§ 10. *Comparison with the Known Solution.*—If the tidal field is *strictly* of the form (*cf.* equation (6), II)

$$\frac{M'r^2}{R^3}P_2(\cos\theta), \tag{57}$$

then the tidal problem admits of exact solution and is given by (*cf.* J. H. Jeans, *loc. cit.*, § 49)

$$\frac{M'}{\pi R^3\lambda} = \frac{1-e^2}{e^3}\log\left(\frac{1+e}{1-e}\right) - \frac{6(1-e^2)}{e^2(3-e^2)}. \tag{58}$$

For small e^2 we find

$$\frac{M'}{\pi R^3\lambda} = \frac{8}{45}e^2 + O(e^4), \tag{59}$$

or

$$e^2 = 7{\cdot}5\frac{M'}{M} \,.\, \nu^3 + O(\nu^6). \tag{60}$$

(60) agrees with (56) in the principal term, but the order of the omitted terms in (56) and (60) do not agree. This difference arises because in obtaining the solution (54) we have not merely used the first term in the expansion (6), II, as is done in obtaining the "exact solution" (58).

Actually consistent with the approximation of treating the secondary as a mass point, we should take into account not only the term in $P_2(\mu)$ but also the two subsequent terms involving $P_3(\mu)$ and $P_4(\mu)$ which occur in the expansion for the outer potential of the secondary. These terms arising from the nearness of the secondary distort the configuration from the true spheroidal form, and these effects are of $O(\nu^4)$.—On the whole the solution as expressed by (54) is preferable to the "exact solution" (58).

§ 11. In II, § 8, we considered the angle at which the furrow in the external shape develops. Since the equation of the boundary (52) is of the standard form, the analysis contained there applies to the Maclaurin model as well. Thus

$$\chi = \tfrac{1}{2}\frac{\Delta_3}{\Delta_2}\nu + O(\nu^3),$$

or by (53)

$$\chi = 0{\cdot}35\nu + O(\nu^3). \tag{61}$$

(C) THE DOUBLE-STAR PROBLEM :—

§ 12. There is no need to go into details to obtain the solution. We can write this at once by an application of the superposition theorem (III, § 7), which clearly holds good for the Maclaurin model as well. Thus by (31) and (51) we have

$$\Theta = \theta + v \cdot \tfrac{1}{6}\xi^2(1 - \tfrac{5}{2}P_2(\mu)) + 2\frac{M'}{M}\sum_{j=2}^{4} \nu^{j+1} \cdot \frac{2j+1}{2(j-1)} \cdot \frac{\xi^j}{6^{j/2}} \cdot P_j\,(\sin\theta\cos\phi), \quad (62)$$

or differently as

$$\Theta = 1 - \tfrac{1}{6}\xi^2 + \tfrac{2}{3}\left(1 + \frac{M'}{M}\right) \cdot \nu^3 \cdot \tfrac{1}{6}\xi^2(1 - \tfrac{5}{2}P_2(\mu))$$
$$+ 2\frac{M'}{M}\sum_{j=2}^{4} \nu^{j+1} \cdot \frac{2j+1}{2(j-1)} \cdot \frac{\xi^j}{6^{j/2}} P_j(\sin\theta\cos\phi). \quad (63)$$

The equation of the boundary is

$$\frac{\xi_0 - \xi_1}{\xi_1} = \tfrac{1}{3}\left(1 + \frac{M'}{M}\right)\nu^3 - \tfrac{5}{6}\left(1 + \frac{M'}{M}\right)\nu^3 P_2(\cos\theta)$$
$$+ \frac{M'}{M}\{2{\cdot}5\nu^3 P_2(\sin\theta\cos\phi) + 1{\cdot}75\nu^4 P_3(\sin\theta\cos\phi) + 1{\cdot}5\nu^5 P_4(\sin\theta\cos\phi)\}. \quad (64)$$

Thus there is an expansion of amount

$$\tfrac{1}{3}\left(1 + \frac{M'}{M}\right) \cdot \nu^3, \quad (65)$$

and superposed on this are the "ellipticity" and "harmonic" terms. To study the *deviations* from a spherical volume we can define as in III, equation (39),

$$\sigma(\theta, \phi) = \frac{\xi_0 - \xi_1}{\xi_1} - \tfrac{1}{3}\left(1 + \frac{M'}{M}\right) \cdot \nu^3. \quad (66)$$

We readily verify that we can express $\sigma(\theta, \phi)$ in the form:

$$\sigma(\theta, \phi) = -\tfrac{1}{3}\left(1 + \frac{M'}{M}\right) \cdot \Delta_2 \cdot \nu^3 P_2(\mu) + \frac{M'}{M} \cdot \sum_{j=2}^{4} \Delta_j \cdot \nu^{j+1} \cdot P_j(\sin\theta\cos\phi), \quad (67)$$

with the Δ_j's defined by (53). (67) is of the same form as (39), III, and the subsequent discussions in III apply to the Maclaurin model as well.

We may note explicitly that to the order of accuracy where the equilibrium configurations are ellipsoids the eccentricities in the three principal planes are (*cf.* equation (43′), III)

$$\left.\begin{aligned} e_e^2 &= 7{\cdot}5\frac{M'}{M} \cdot \nu^3, \\ e_m^2 &= 2{\cdot}5\left(1 + 4\frac{M'}{M}\right) \cdot \nu^3, \\ e_d^2 &= 2{\cdot}5\left(1 + \frac{M'}{M}\right) \cdot \nu^3. \end{aligned}\right\} \quad (68)$$

Further

$$\left.\begin{aligned}\chi_m &= \frac{M'}{M+4M'} \times 1{\cdot}05\nu,\\ \chi_e &= 0{\cdot}35\nu.\end{aligned}\right\} \tag{69}$$

II. *Solutions for the Roche Model*

§ 13. The Roche model corresponds to the polytrope $n=5$. Since, however, the actual Emden functions employed are those which take the value unity at the origin, the configuration extends to infinity, and one has therefore to be rather careful in performing the limiting process.

§ 14. *The Rotational Problem.*—Now the equation of the boundary for a polytrope of index $n \neq 5$ is given by equation (38), I:

$$\xi_0 = \xi_1 + \frac{v}{|\theta_1'|}\left[\psi_0(\xi_1) - \frac{5}{6}\frac{\xi_1^2\psi_2(\xi_1)}{3\psi_2(\xi_1)+\xi_1\psi_2'(\xi_1)}P_2(\mu)\right]. \tag{70}$$

When we are comparing the distortions as a function of the density distribution we keep the mean density constant. Instead of v we introduce therefore the quantity ζ defined by (*cf.* equation (46), I)

$$\zeta = \frac{\omega^2}{2\pi G\rho_m} = \frac{\omega^2}{2\pi G\lambda} \cdot \frac{\xi_1}{3|\theta_1'|}, \tag{71}$$

or

$$v = \frac{3|\theta_1'|}{\xi_1} \,.\, \zeta. \tag{71'}$$

Introducing this in (70) we can rewrite the equation of the boundary in the following way :—

$$\frac{\xi_0 - \xi_1}{\xi_1} = \zeta\left[\frac{3\psi_0(\xi_1)}{\xi_1^2} - \tfrac{1}{2}\Delta_2 \,.\, P_2(\mu)\right]. \tag{72}$$

Now ψ_0 and ψ_2 satisfy the differential equations :

$$\frac{1}{\xi^2}\frac{d}{d\xi}\left(\xi^2\frac{d\psi_0}{d\xi}\right) = -n\theta^{n-1}\psi_0 + 1, \tag{73}$$

$$\frac{1}{\xi^2}\frac{d}{d\xi}\left(\xi^2\frac{d\psi_2}{d\xi}\right) = \left(\frac{6}{\xi^2} - n\theta^{n-1}\right)\psi_2. \tag{74}$$

Now, as is known, when the polytropic index n approaches 5 the radius ξ_1 tends to infinity and at the same time θ is very nearly zero long before it actually crosses the ξ-axis at ξ_1. Hence as $n \to 5$ the differential equation satisfied by ψ_0 will approximate to

$$\frac{1}{\xi^2}\frac{d}{d\xi}\left(\xi^2\frac{d\psi_0}{d\xi}\right) = 1, \tag{73'}$$

the solution of which is

$$\psi_0 = \tfrac{1}{6}\xi^2. \tag{73''}$$

Hence in (72) we have to put

$$\frac{3\psi_0(\xi_1)}{\xi_1^2}=\tfrac{1}{2}, \qquad \text{for} \qquad n=5. \tag{75}$$

Similarly from (74) we see that as $n \to 5$ the differential equation for ψ_2 for a long way from the boundary is

$$\frac{1}{\xi^2}\frac{d}{d\xi}\left(\xi^2\frac{d\psi_2}{d\xi}\right)=\frac{6\psi_2}{\xi^2}, \tag{74$'$}$$

the solution of which is of the form:

$$\psi_2=B\xi^2, \tag{74$''$}$$

the constant B depending on the whole march of θ. But Δ_2 is found to be unity independent of B. Hence we have to put

$$\Delta_2=1 \tag{76}$$

for $n=5$. Hence our solution is

$$\frac{\xi_0-\xi_1}{\xi_1}=\tfrac{1}{2}\zeta(1-P_2(\mu))=\tfrac{3}{4}\zeta\sin^2\theta. \tag{77}$$

The solution (77) is obtained by decidedly non-rigorous mathematical arguments, but it would be very difficult to make the arguments more exact, as it would require, for instance, the way, $\xi_1 \to \infty$ as $n \to 5$, and similar theorems where we regard the Emden functions as a function of the two variables ξ and n. It is, however, an easy matter to verify that the solution (77) is consistent with the usual treatment of the Roche model which is the following:—

If we regard the whole mass M as being concentrated at the origin, then the sum of the gravitational and the centrifugal potentials is given by

$$W=\frac{GM}{r}+\tfrac{1}{2}\omega^2(x^2+y^2). \tag{78}$$

The surfaces $W=$ constant are equipotentials, and if there is a configuration of equilibrium we should have a closed equipotential, which we will always have when ω is smaller than a certain critical value. The boundary of the configuration must also be an equipotential, and we select from the infinity of the equipotentials, $W=$ constant, the one which encloses a volume just adequate to contain the whole amount of matter.

Now if R_1 is the radius of the original configuration, then the equation of the boundary given by (77) is of the form:

$$R_0=R_1(1+\tfrac{3}{4}\zeta\sin^2\theta). \tag{79}$$

Hence

$$\frac{GM}{R_0}=\frac{GM}{R_1}(1-\tfrac{3}{4}\zeta\sin^2\theta),$$

or substituting the zero order approximation $M = 4/3 \, . \, \pi R_1^3 \rho_{\text{mean}}$ for the term in ζ we find that

$$\frac{GM}{R_0} = \frac{GM}{R_1} - \tfrac{1}{2} R_1^2 \omega^2 \sin^2 \theta. \tag{80}$$

Again,

$$\tfrac{1}{2}\omega^2(x^2+y^2) = \tfrac{1}{2} R_1^2 \omega^2 \sin^2 \theta. \tag{80'}$$

Hence

$$W(R_0) = \frac{GM}{R_1} = \text{constant}. \tag{81}$$

Thus the surface (79) is an equipotential and encloses a volume just sufficient to contain the whole amount of matter, thus verifying our solution (77). This consistency of (77) with the correct solution of the Roche model makes it very likely—if not certain—that the type of arguments used to derive it will in any case give the correct results when we want to consider the polytrope $n = 5$.

Finally we note that for the Roche model

$$\left.\begin{array}{ll} \text{Fractional Elongation at Equator} & = 0{\cdot}75\zeta, \\ \text{Fractional Contraction at Poles} & = 0, \\ \text{Oblateness } \sigma & = 0{\cdot}75\zeta. \end{array}\right\} \tag{82}$$

§ 15. *The Tidal and the Double-star Problems.*—To study the external shape for the tidally distorted configuration we need only the Δ_j's. By means of arguments similar to those used to derive (76) we infer that for the Roche model we have to set all the Δ_j's to be equal to unity. The numerical results in II (Table VI) convinces us of the correctness of this procedure (*cf.* the remarks in italics at the end of § 7 in II). Thus for the Roche model the equation of the boundary is given by (*cf.* equation (45), II)

$$\xi_0 = \xi_1 \left\{ 1 + \frac{M'}{M} \sum_{j=2}^{4} \nu^{j+1} \, . \, P_j(\mu) \right\}. \tag{83}$$

By the methods used in the second half of § 14 we can easily verify that the solution (83) is consistent with the usual treatment of the Roche model.

The angle χ at which the furrow develops in the external shape for the Roche model is given by

$$\chi = 0{\cdot}5\nu + O(\nu^3). \tag{84}$$

The double-star problem requires no special remarks, but we may note that to the order of accuracy, where the equilibrium configurations are ellipsoids, the eccentricities in the three principal planes are given by

$$\left.\begin{aligned} e_e^2 &= 3\frac{M'}{M} \, . \, \nu^3, \\ e_d^2 &= \left(1 + \frac{M'}{M}\right) . \, \nu^3, \\ e_m^2 &= \left(1 + \frac{4M'}{M}\right) . \, \nu^3. \end{aligned}\right\} \tag{85}$$

Further

$$\left.\begin{aligned} \chi_m &= 1{\cdot}5\frac{M'}{M+4M'}\nu + O(\nu^3), \\ \chi_e &= 0{\cdot}5\nu \qquad\qquad + O(\nu^3). \end{aligned}\right\} \tag{86}$$

III. *The Rotational Problem*

§ 16. *Results for the Polytrope* "n = 3·5."—The necessary numerical integrations for the polytropes $n=1$, 1·5, 2, 3 and 4 have been made in I. However, to draw unambiguous conclusions regarding the dependence of the rotational distortions on the density distribution it was found necessary to have the results for one more polytrope with $3 < n < 4$. The functions ψ_0 and ψ_2 for $n=3{\cdot}5$ are tabulated in the Appendix II.

The values of ψ_0 and ψ_2 and their derivatives at ξ_1 are

$$\left.\begin{aligned} &\xi_1 = 9{\cdot}5358, \\ \psi_0(\xi_1) &= 12{\cdot}4466, \qquad \psi_2(\xi_1) = 20{\cdot}1390, \\ \psi_0'(\xi_1) &= \ \ 2{\cdot}9611, \qquad \psi_2'(\xi_1) = \ \ 4{\cdot}0706. \end{aligned}\right\} \tag{87}$$

The numerical forms for the various formulæ may be noted :

$$\Theta = \theta + v[\psi_0(\xi) - 0{\cdot}7636\psi_2(\xi)P_2(\mu)], \tag{88}$$

$$\xi_0 = \xi_1 + 48{\cdot}098v[12{\cdot}4466 - 15{\cdot}379P_2(\mu)]. \tag{89}$$

$$\left.\begin{aligned} &\text{Fractional Elongation at Equator} = 101{\cdot}57\ v, \\ &\text{Fractional Contraction at Poles} \ \ = \ \ 14{\cdot}789v, \\ &\text{Oblateness, } \sigma \qquad\qquad\qquad\qquad = 116{\cdot}36\ v. \end{aligned}\right\} \tag{90}$$

$$\left.\begin{aligned} M_\omega &= M_0[1 + 10{\cdot}463v], \\ N_\omega &= N_0[1 + 188{\cdot}34v], \\ \rho_c &= \rho_m \times 152{\cdot}884[1 + 177{\cdot}88v]. \end{aligned}\right\} \tag{91}$$

§ 17. *Distortions of the External Shape as a Function of the Polytropic Index.*—In I all the results were expressed in terms of v as it is this small parameter which appears explicitly in the differential equation for Θ. But when we want to compare the distortions in the different polytropes we should not compare configurations with the same central density (λ), since polytropes with differing n but equal λ are characterised by different masses and different radii. To compare polytropes having the same mass and the same radius we must compare configurations with the same mean density. Hence we must rewrite all our equations in terms of ζ (equation (71)).

Now the equation of the boundary can be expressed in the form (equation (72)) :

$$\frac{\xi_0 - \xi_1}{\xi_1} = \zeta\left[\frac{3\psi_0(\xi_1)}{\xi_1^2} - \tfrac{1}{2}\Delta_2 P_2(\mu)\right]. \tag{92}$$

From (92) we obtain the results summarised in Table I.

TABLE I

Geometry of the Boundary

n	General Expansion Term $\frac{3\psi_0(\xi_1)}{\xi_1^2} \cdot \zeta^{-1}$	Fractional Elongation at Equator $\times \zeta^{-1}$	Fractional Contraction at Poles $\times \zeta^{-1}$	Ratio of Elongation to Contraction	Oblateness, $\sigma \times \zeta^{-1}$
0	0·5	1·125	0·75	1·5	1·875
1	0·3040	0·6839	0·4560	1·5000	1·1399
1·5	0·2908	0·6131	0·3538	1·7329	0·9669
2	0·3032	0·5903	0·2709	2·1792	0·8612
3	0·3682	0·6254	0·1462	4·2762	0·7716
3·5	0·4106	0·6643	0·0967	6·868	0·7611
4	0·4488	0·6995	0·0525	13·31	0·7520
5	0·5	0·75	0	∞	0·75

The results of Table I are graphically illustrated in figs. 1 and 2. We notice the following features :—

(i) The fractional elongation at the equator as a function of n has a *minimum* somewhere in the region $n = 2$.

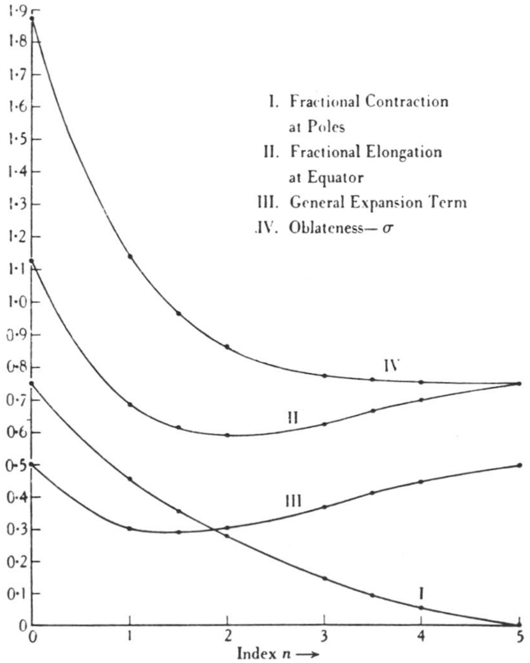

Figure 1.

(ii) The general expansion term also passes through a minimum, but for a smaller value of n.

(iii) The fractional contraction at the poles monotonically tends to zero as n tends to 5.

(iv) The ratio of the elongation at the equator to the contraction at the poles tends to infinity as $n \to 5$.* Also, this ratio has *exactly* (to the first order in ζ) the value 1·5 for $n=0$. By numerical calculations this ratio is found to be 1·5000 for $n=1$. This exact coincidence may be an accident, but it is clear that this ratio has a minimum for $0 < n < 1$.

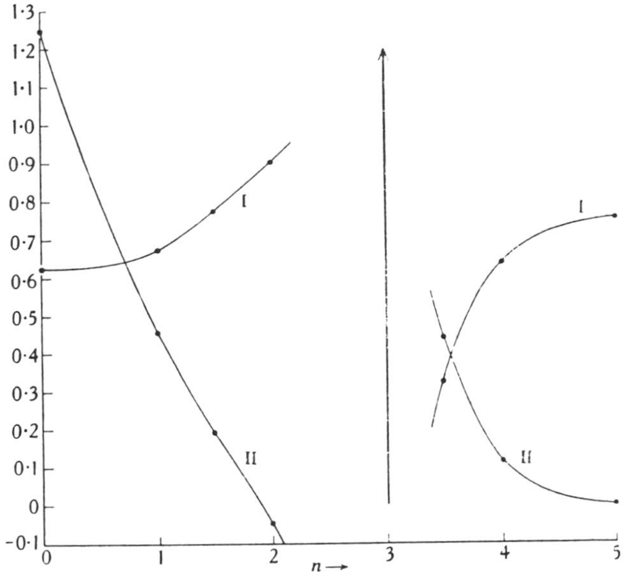

Figure 1 *a*.

The Fractional Changes in Length as a Configuration of a Given Mass is set Rotating

I. At Equator

II. At Poles

Further from (92) we see that the oblateness of the boundary of the configuration is given by

$$\sigma(\xi_0) = 0{\cdot}75\Delta_2\zeta. \tag{92'}$$

In the inner parts of the configuration strata of equal density are also characterised by ellipticities different from but *less* than that given by (92′). The oblateness for a specified value of ξ is easily found to be (*cf.* equation (39), I)

$$\sigma(\xi) = -\frac{5}{4}\frac{v}{\theta'}\frac{\xi\psi_2(\xi)}{3\psi_2(\xi)+\xi\psi_2'(\xi)}.$$

When $\xi \to 0$ we have the limiting forms

$$\theta' = -\tfrac{1}{3}\xi + O(\xi^3),$$
$$\psi_2(\xi) = \xi^2 + O(\xi^4),$$
$$\psi_2'(\xi) = 2\xi + O(\xi^3).$$

* This fact, combined with the theorem stated in III, § 9, shows that if a component of a double-star system is a polytrope with $n=5$, then there can *never* be a contraction in its equatorial plane however large the ratio M'/M may be.

Hence

$$\sigma(\xi) \to 0{\cdot}75v \quad \text{as} \quad \xi \to 0. \tag{92''}$$

By (92′) and (92″) we see that

$$\frac{\sigma(\xi_0)}{\sigma(0)} = \Delta_2 \cdot \frac{\zeta}{v} = \Delta_2 \cdot \frac{\rho_{\text{central}}}{\rho_{\text{mean}}}. \tag{92'''}$$

We easily verify that (92‴) as it stands is also valid for the Maclaurin model when we consider the oblateness of strata of equal pressure (*i.e.*) of equal Θ.

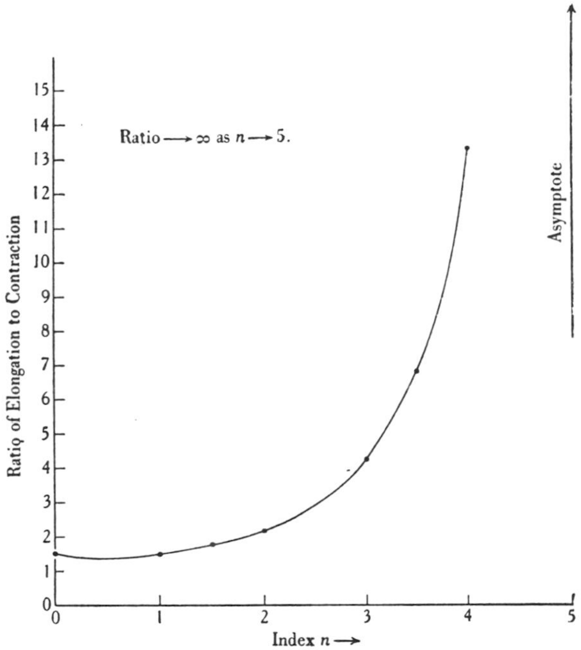

Figure 2.

§ 18. *The Oblateness as a Function of the Central Density.*—The quantity of interest in the rotational problem is the oblateness. In fig. 1 this was plotted against the polytropic index. But to obtain the qualitative nature of the dependence of the distortions on the march of the density distribution we shall plot the oblateness against the logarithm of the central density, regarding the mean density as unity (fig. 3).

Now, for the polytropic model the oblateness must be a function only of the central density, since for the polytropes the specification of the central density uniquely determines the density distribution, as the distribution function "θ" would have to be a solution of Emden's equation for some index n (and hence characterised by a unique ratio between the central and the mean densities).

Actually, however, the oblateness must be a function of the complete march of the density distribution. However that may be, we expect that the oblateness curve in fig. 3 gives us the general nature of its dependence on the march of the density distribution. For the larger the central density, the larger is the concentration of the mass towards the centre, and conversely. Though the density distribution is not unambiguously determined by the specification of the central density, it is clear, however, that since the density would have to decrease monotonically from the centre outwards, the deviations from two different density distributions, which enclose in a prescribed spherical volume a prescribed mass and at the same time are characterised by equal densities and density gradients $\partial\rho_c/\partial r(=0)$ at the centre, cannot be

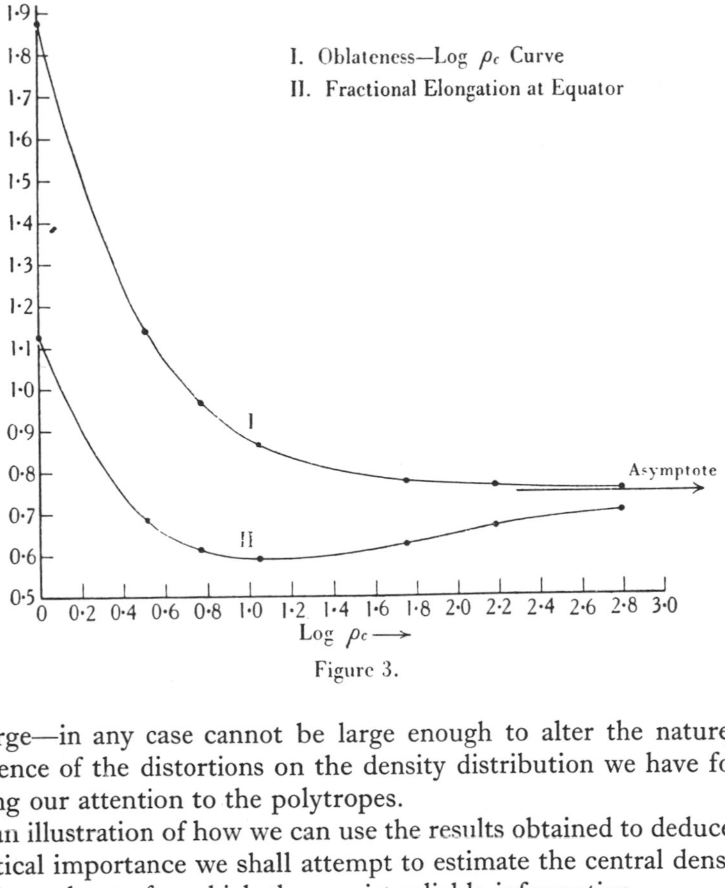

Figure 3.

very large—in any case cannot be large enough to alter the nature of the dependence of the distortions on the density distribution we have found by confining our attention to the polytropes.

As an illustration of how we can use the results obtained to deduce results of practical importance we shall attempt to estimate the central densities for the various planets for which there exist reliable information.

§ 19. *An Application to the Figure of the Earth.*—The *ellipticity* ϵ of a rotating configuration is defined as $(a-b)/a$, where a and b are the major and the minor axes of the spheroid of equilibrium. To the order of accuracy to which we are working, the ellipticity ϵ equals the quantity we have called oblateness and denoted by σ. Now for all polytropes we can express ϵ as

$$\epsilon = C \times \frac{\omega^2}{2\pi G\rho_{\text{mean}}}, \tag{93}$$

where C is a constant tabulated in the last column of Table I. Now the value of the surface gravity g is, to the zero order approximation, given by

$$g = \tfrac{4}{3}\pi a G \rho_{\text{mean}}. \qquad (94)$$

From (93) and (94) we deduce that

$$\epsilon = \tfrac{2}{3} C \, . \, \frac{\omega^2 a}{g}. \qquad (95)$$

The usual notation of the subject is to denote by m the quantity $\omega^2 a/g$, which is simply the ratio of the centrifugal force of the "planet's" rotation at the equator to the mean pure gravity.

Now for the Earth G. H. Darwin * gave $m = (289{\cdot}66)^{-1}$ or $m = 0{\cdot}003452$. Hence for a configuration having an "m" equal to that of the Earth we have

$$\epsilon = 0{\cdot}002301 \times C. \qquad (96)$$

From the above equation the following table was constructed:—

TABLE II

ϵ for Configurations having $m = 0{\cdot}003452$

n	0	1	1·5	2	3	5
ρ_c	1	3·290	5·991	11·403	54·182	∞
$10^3 \times \epsilon$	4·314	2·623	2·225	1·982	1·776	1·726

Actually the observed ellipticity is

$$\epsilon = \tfrac{1}{297} = 3{\cdot}367 \times 10^{-3}. \qquad (97)$$

By interpolation among the values given in Table II a value of ρ_c for $\epsilon = 0{\cdot}003367$ was found to be 1·8. It is a matter of doubt as to the amount of extrapolation that is being made if we actually did identify $\rho_c = 1{\cdot}8$ as indicating the kind of central densities that are to be found inside the Earth. In any case a crude estimate can be made this way.

The mean density of the Earth is 5·5 grm.-cm.$^{-3}$. Hence the estimated central density is

$$\rho_c = 1{\cdot}8 \times 5{\cdot}5 = 9{\cdot}9 \text{ grm.-cm.}^{-3}. \qquad (98)$$

This agrees very well with the usual estimates for the internal densities in the Earth.†

§ 20. *Applications to other Planets.*—A quantity which is useful in this connection is $5m/2\epsilon$. By (95)

$$\frac{5m}{2\epsilon} = \frac{15}{4} \times C^{-1}. \qquad (99)$$

* *Scientific Papers*, **3**, 57–68.

† Compare H. Jeffreys, *The Earth* (Cambridge, 1929), pp. 218–220, where estimates ranging from 8 to 12 grm.-cm.$^{-3}$ are given.

In Table III the values for $5m/2\epsilon$ for the different polytropes are collected.

TABLE III

"$5m/2\epsilon$"

n	0	1	1·5	2	3	4	5
$5m/2\epsilon$	2	3·290	3·878	4·355	4·860	4·986	5

It is of interest to note that $5m/2\epsilon$ *must always lie between the extreme limits* 2 *and* 5 :

$$2 \leqslant \frac{5m}{2\epsilon} < 5. \tag{100}$$

The existence of an *absolute* upper limit (namely 5) to this ratio does not seem to have been noticed before. The existence of the lower limit was, however, first recognised by Laplace.

In fig. 4*a* the results of Table III are plotted. The curve increases rather sharply in the range $0 < n < 2$. In fig. 4*b* $5m/2\epsilon$ is plotted against

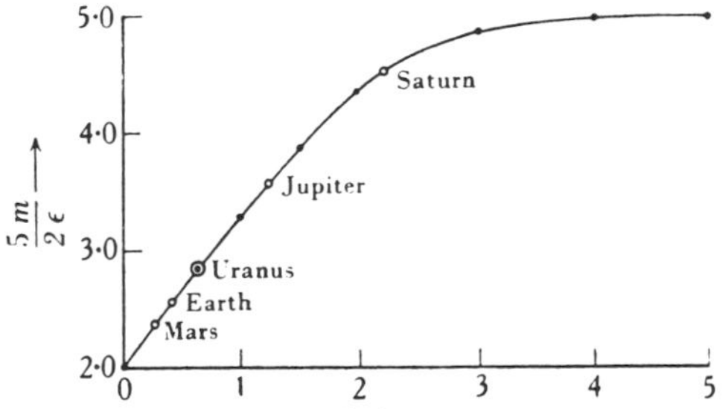

Figure 4 *a*.

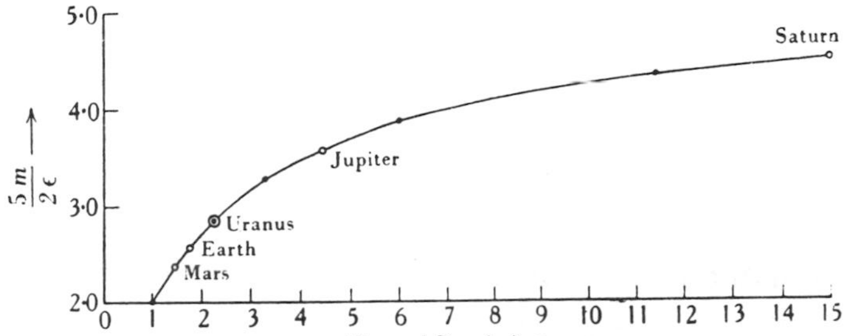

Figure 4 *b*.

ρ_c for the interesting part of the curve. If the value of $5m/2\epsilon$ for a rotating configuration is known we can estimate the central densities from this curve.

This was done for those planets for which we have reliable data.* The results are summarised in Table IV.

TABLE IV

Estimates of the Central Densities for some Planets

Planet	Period of Rotation	ϵ^{-1}	Mean Density (Density of Earth as Unity)	$5m/2\epsilon$	ρ_c/ρ_m	ρ_c in grm.-cm.$^{-3}$
	h					
Mars	24·6	190	0·65	2·38	1·5	5·3
Jupiter	9·8	16	0·23	3·57	4·45	5·6
Saturn	10·2	11	0·115	4·53	~15	~9·5
Uranus	10·8	12 − 15	0·2	2·5 − 3·2	1·6 − 3·05	3·3 − 1·8

The values of the "ρ_c/ρ_m" ratio were read off from the curve in fig. 4*b*. The actual estimates for the central densities are given in the last column. Though much significance is not to be attached to the precise values given in the last column, yet the estimates are in no way unreasonable. Jupiter is of special interest inasmuch as the ratio ρ_c/ρ_m for this planet is rather large compared with the Earth. This, however, confirms an earlier suggestion of Darwin. He says: "The surface density of Jupiter is far less than the mean, and is it not possible that Jupiter may still be in a semi-nebulous condition?" Darwin further conjectured that "the like may be true of Saturn." But at that time there was no reliable information regarding $5m/2\epsilon$, and Darwin gives values for this ratio ranging from 2·7 to 4·25. The more reliable data available now indicates that ρ_c/ρ_m is the largest for this planet.† In agreement as this is with the still gaseous nature of Saturn, it seems possible that it may have a density distribution similar to that of an Emden polytrope with index $n = 2{\cdot}2$.

§ 21. *Comparison of Configurations with Equal Mass.*—The "mass-relation" for two gas configurations with equal central densities—one rotating with an angular velocity ω and the other non-rotating—is (*cf.* equation (42), I)

$$M_\omega = M_0\left[1 + \zeta\left(1 - \frac{3\psi_0'(\xi_1)}{\xi_1}\right)\right]. \tag{101}$$

The coefficients of ζ in the above expression for different values of n are tabulated in the second column of Table V. From (101) we easily derive

* For the observational data, see Russell, Dugan and Stewart, *Astronomy*, 1, Appendix, Table IV, 1926; also *Handbuch der Astrophysik*, 4, 359, 1929.

† The conclusions drawn here from an application of the theory of polytropes appear to be in general agreement with some deductions of Jeffreys regarding Saturn from quite different considerations (*cf.* H. Jeffreys, *M.N.*, 84, 534, 1924).

that the fractional change in the central density which takes place when a gas sphere of a given mass is set rotating is (*cf.* equation (61), I)

$$\frac{\delta\lambda}{\lambda} = -\frac{2n}{3-n}\left(1 - \frac{3\psi_0'(\xi_1)}{\xi_1}\right) . \zeta. \tag{102}$$

Another quantity which is more fundamental than $\delta\lambda/\lambda$ is the fractional change in the central pressure ($\delta P_c/P_c$) when a gas sphere is set rotating. From (102) we easily derive that

$$\frac{\delta P_c}{P_c} = -\frac{2(n+1)}{3-n}\left(1 - \frac{3\psi_0'(\xi_1)}{\xi_1}\right) . \zeta. \tag{103}$$

For $n=0$ the result is given in (40). Precise values of (102) and (103) can be obtained from the third and the fourth columns of Table V.

When we are comparing configurations of the same mass, then ξ_0 and ξ_1 do *not* represent the boundary of the two configurations on the same scale. If $r(\omega, \mu)$ and r_0 represent the actual boundaries for the rotating and the non-rotating configurations, then

$$r(\omega, \mu) = \left(\alpha + \frac{\partial\alpha}{\partial\lambda}\delta\lambda\right)\xi_0, \qquad \left(\alpha = \left[\frac{(n+1)K}{4\pi G}\lambda^{\frac{1}{n}-1}\right]^{\frac{1}{2}}\right), \tag{104}$$

α being the scale in which ξ represents the radius and $\delta\lambda$ is equal to that given by (102). From (104) we easily deduce that

$$\frac{r(\omega, \mu) - r_0}{r_0} = \zeta\left[\frac{3\psi_0(\xi_1)}{\xi_1^2} + \frac{n-1}{3-n}\left(1 - \frac{3\psi_0'(\xi_1)}{\xi_1}\right) - \tfrac{1}{2}\Delta_2 P_2(\mu)\right].$$

The case $n=0$ has to be treated separately. We find that

$$\frac{r(\omega, \mu) - r_0}{r_0} = -\tfrac{5}{4}\zeta P_2(\mu). \tag{105}$$

From (105) we can calculate the fractional changes in lengths that take place

TABLE V

n	$\frac{M_\omega - M_0}{M_0} . \zeta^{-1}$ (Equation (101))	$\frac{\delta\lambda}{\lambda} . \zeta^{-1}$	$\frac{\delta P_c}{P_c} . \zeta^{-1}$	$\frac{r(\omega, -\frac{1}{2}) - r_0}{r_0} . \zeta^{-1}$, Elongation at Equator	$\frac{-r(\omega, 1) + r_0}{r_0} . \zeta^{-1}$, Contraction at Poles
0	1.5	...	−1	0.625	1.25
1	$0.6960\left(=1-\frac{3}{\pi^2}\right)$	−0.6960	−1.3921	0.6839	0.4560
1.5	0.4775	−0.9550	−1.5916	0.7723	0.1947
2	0.3135	−1.2539	−1.8809	0.9038	−0.0426
3	0.1130	...	...	...	...
3.5	0.06842	+0.9579	+1.2316	0.3222	0.4389
4	0.02188	0.1750	0.2188	0.6339	0.1182
5	0	0	0	0.75	0

at the equator and at the poles when a gas sphere of a given mass is set rotating.* The results are given in the fifth and the sixth columns in Table V and are further graphically illustrated in fig. 1*a* (p. 554).

Considering first the fractional change in length that occurs at the equator we see that there can actually be a *contraction* provided n is sufficiently near 3. This would happen, for instance, for a polytrope with $n=3\cdot 1$. As for the fractional change in length that occurs at the poles, we see that we can have an *elongation* for polytropes with n less than 3 but greater than a certain critical value. Already for $n=2$ we have an *elongation* at the poles.

Now the volume relation for configurations with equal central densities is given by (*cf.* equation (43), I)

$$N_\omega = N_0\left(1+\frac{9\psi_0(\xi_1)}{\xi_1^2}\zeta\right). \tag{106}$$

The volume relations for two configurations with equal mass is found to be

$$N_\omega(M) = N_0\left[1+\zeta\left\{\frac{9\psi_0(\xi_1)}{\xi_1^2}+\frac{3(n-1)}{3-n}\left(1-\frac{3\psi_0'(\xi_1)}{\xi_1}\right)\right\}\right]. \tag{107}$$

The coefficients of ζ occurring in (106) and (107) are tabulated in Table VI.

TABLE VI

Volume Relations

n	$\dfrac{N_\omega - N_0}{N_0}\cdot\zeta^{-1}$	$\dfrac{N_\omega(M) - N_0}{N_0}\cdot\zeta^{-1}$
0	1·5	0
1	0·9119 $\left(=\frac{9}{\pi^2}\right)$	0·9119
1·5	0·8725	1·3499
2	0·9097	1·8501
3	1·1046	...
3·5	1·2319	0·2056
4	1·3465	1·1496
5	1·5	1·5

We notice (i) *The fractional change in volume at constant λ for the Maclaurin and the Roche models are equal.* (ii) *For the polytropes* n = 1 *and* n = 5 *the fractional change in volume at constant λ equals the fractional change in volume at constant* M.

Further, by interpolation among the values in the third column of Table VI it was found that for a polytrope with $3 < n < 3\cdot 325$ there occurs a *diminution* in volume when the configuration is set rotating. This happens

* It may be stated here explicitly that, since the change in λ is of order ζ, the expression for the oblateness ($\sigma = 0\cdot 75\Delta_2\zeta$) is the same whether we are comparing two polytropes (with the *same n*) of equal mass or with equal central densities.

because for these polytropes the rotation is followed by such large modifications in the density distribution, in the sense of increasing the central density, that there is a consequent "shrinkage" of the whole mass.

§ 22. *The Variation of the Apparent Gravity on the Boundary of the Configuration.*—It was shown by H. von Zeipel * that *the variation of brightness* (i.e. *of the emergent flux of radiation*) *over the surface of a rotating star exactly corresponds to the variation of apparent gravity.* We shall denote by $g_{\omega,\,\mu}$ the apparent gravity at latitude θ when the configuration is rotating with an angular velocity ω. Then, to the first order,

$$g_{\omega,\,\mu} = -\left[\frac{d(V+\Omega)}{dr}\right]_{\text{Boundary}}, \tag{108}$$

where V is the gravitational potential and Ω the centrifugal potential. By equations (32), (36), I,

$$g_{\omega,\,\mu} = -\frac{(n+1)K\lambda^{\frac{1}{n}}}{a}\left(\frac{\partial\Theta}{\partial\xi}\right)_{\xi_0}, \tag{109}$$

where ξ_0 is the boundary given by (38), I. We find that

$$\left(\frac{\partial\Theta}{\partial\xi}\right)_{\xi_0} = \theta_1' - \frac{v}{\theta_1'}\left\{\psi_0(\xi_1) - \frac{5}{6}\frac{\xi_1^2\psi_2(\xi_1)P_2(\mu)}{3\psi_2(\xi_1)+\xi_1\psi_2'(\xi_1)}\right\}\left(\frac{d^2\theta}{d\xi^2}\right)_1$$
$$+v\left\{\psi_0'(\xi_1) - \frac{5}{6}\frac{\xi_1^2\psi_2'(\xi_1)P_2(\mu)}{3\psi_2(\xi_1)+\xi_1\psi_2'(\xi_1)}\right\}. \tag{110}$$

Remembering that $\theta_1'' = 2\xi_1^{-1}\,.\,|\,\theta_1'\,|$ we obtain from (109) and (110) after some elementary transformations

$$\gamma(\mu) = \frac{g_{\omega,\,\mu}-g_0}{g_0} = -\zeta\left\{\frac{6\psi_0(\xi_1)}{\xi_1^2} + \frac{3\psi_0'(\xi_1)}{\xi_1} - \tfrac{1}{2}(5-\Delta_2)P_2(\mu)\right\}, \tag{111}$$

where

$$g_0 = \frac{(n+1)K\lambda^{\frac{1}{n}}}{a}\,|\,\theta_1'\,|. \tag{112}\dagger$$

For $n=5$, (111) takes the form:

$$\gamma(\mu) = -2\zeta(1-P_2(\mu)) = -3\zeta\sin^2\theta. \tag{113}$$

* *M.N.*, **84**, 665, 1924.

† Using relations (101) and (106) we can rewrite our expression for $g_{\omega,\,\mu}$ in the form

$$g_{\omega,\,\mu} = \frac{GM}{a^2}[1-\zeta\{1-\tfrac{1}{2}(5-\Delta_2)P_2(\mu)\}], \tag{112'}$$

where a is the radius of a sphere which will have the same volume as the distorted configuration. By (92′) and (99) we can express (112′) differently as

$$g_{\omega,\,\mu} = \frac{GM}{a^2}[1-\tfrac{2}{3}m-(\tfrac{5}{2}m-\sigma)(\tfrac{1}{3}-\cos^2\theta)], \tag{112''}$$

where the quantity m is the same as that we have already introduced in § 20. (112″) expresses merely the well-known *Clairaut's relation* for rotating fluid masses. (*Cf.* Thomson and Tait, *Natural Philosophy*, **2**, § 794, p. 362, 1903.)

The case $n=0$ has to be treated separately. From (31) and (33) we easily deduce that

$$\frac{g_{\omega,\mu}-g_0}{g_0}=-\zeta(0.5-1.25P_2(\mu)). \tag{111'}$$

From (111) and (111′) we deduce the general formula:

$$\Delta g=\frac{g_{\omega,1}-g_{\omega,-\frac{1}{2}}}{g_0}=\tfrac{3}{4}(5-\Delta_2)\,.\,\zeta. \tag{114}$$

For $n=0$ and $n=5$ we have

$$\left.\begin{aligned}\Delta g&=1.875\zeta, & n&=0\\ \Delta g&=3\zeta, & n&=5\end{aligned}\right\}. \tag{114'}$$

Values of Δg and the related quantities are tabulated in Table VII. In fig. 5 the variation of Δg with n is illustrated. We notice that the fractional increase in "g" at the poles passes through a maximum for some value of n between $n=0$ and $n=1$ (see fig. 6).

In connection with the variation of surface gravity over the surface, it is of interest to consider the fractional changes in surface gravity that occur when a configuration of a given mass is set rotating. If we denote by $g'_{\omega,\mu}$ the surface gravity on the boundary of the rotating configuration, then we easily verify that $g'_{\omega,\mu}$ is related to $g_{\omega,\mu}$ by means of the relation

$$g'_{\omega,\mu}=g_{\omega,\mu}\left(1+\frac{1}{2}\frac{\delta P_c}{P_c}\right). \tag{115}$$

Numerical values for $\delta P_c/P_c$ are given in Table V. From (115) we can calculate the fractional *decrease* in g at the equator and the fractional *increase* at the poles. These quantities are tabulated in last two columns of Table VII and are illustrated in figs. 7 and 8.

TABLE VII

Variation of Apparent Gravity on the Surface

n	$\Delta g\,.\,\zeta^{-1}$	Comparison of Configurations with Equal Central Density		Comparison of Configurations with Equal Mass	
		Fractional Decrease in g at Equator $\times\zeta^{-1}$	Fractional Increase in g at Poles $\times\zeta^{-1}$	Fractional Decrease in g at Equator $\times\zeta^{-1}$	Fractional Increase in g at Poles $\times\zeta^{-1}$
0	1.875	1.125	0.75	1.625	0.25
1	2.610	1.7819	0.8281	2.4779	0.1322
1.5	2.783	2.0318	0.7512	2.8277	−0.0446
2	2.889	2.2559	0.6327	3.1964	−0.3078
3	2.978	2.6161	0.3623	...	...
3.5	2.989	2.7492	0.2398	2.1334	+0.8556
4	2.998	2.8751	0.1229	2.7657	0.2323
5	3	3	0	3	0

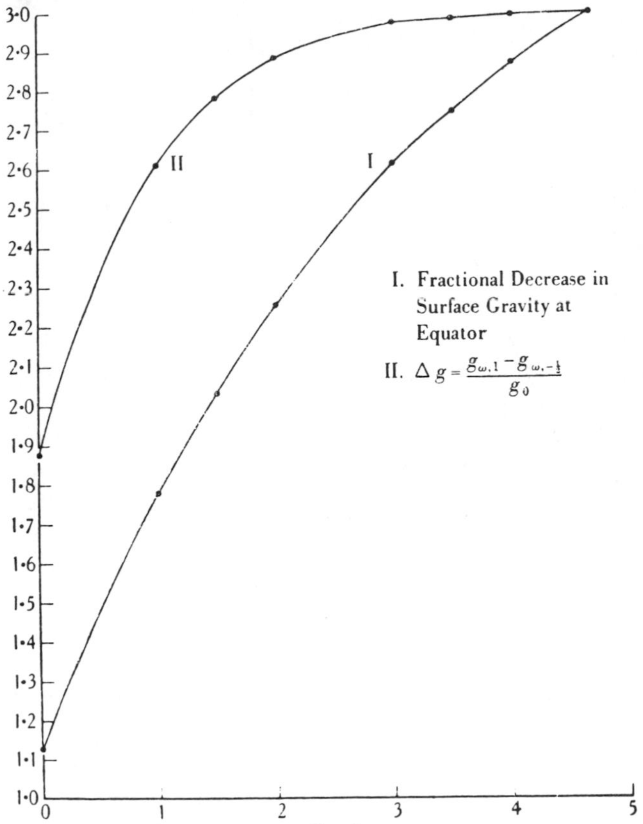

Figure 5.

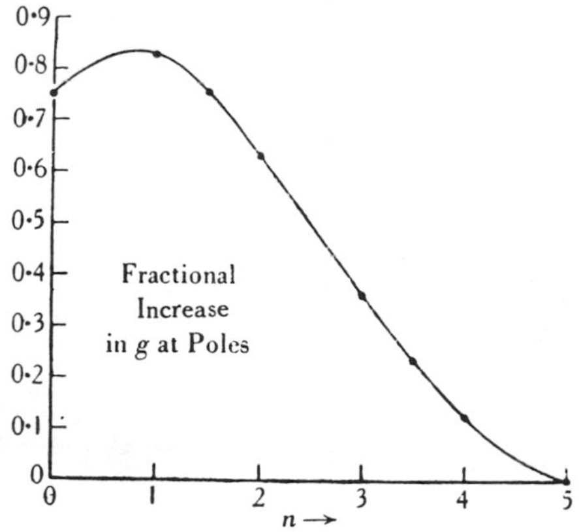

Figure 6.

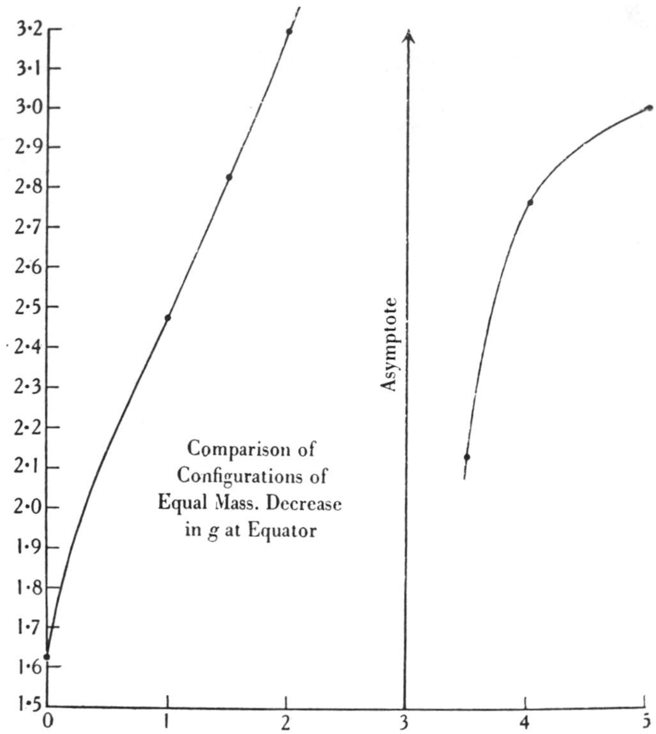
3·2
3·1
3·0
2·9
2·8
2·7
2·6
2·5
2·4
2·3
2·2
2·1
2·0
1·9
1·8
1·7
1·6
1·5
0
1
2
3
4
5
Asymptote
Comparison of Configurations of Equal Mass. Decrease in g at Equator

Figure 7.

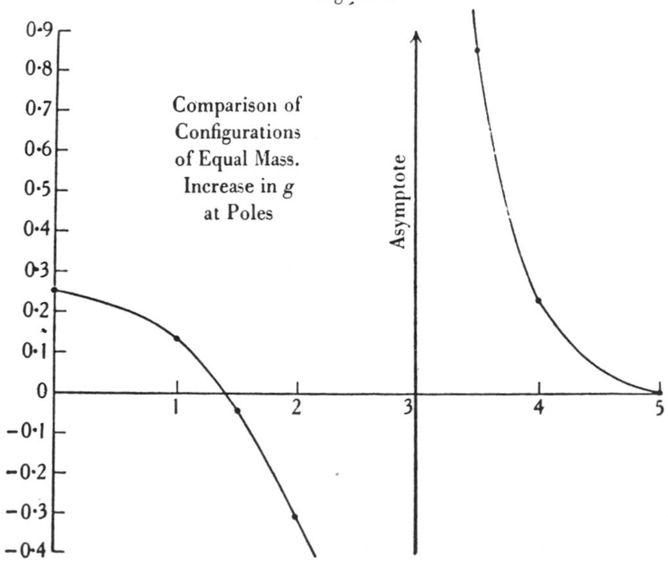
0·9
0·8
0·7
0·6
0·5
0·4
0·3
0·2
0·1
0
−0·1
−0·2
−0·3
−0·4
1
2
3
4
5
Comparison of Configurations of Equal Mass. Increase in g at Poles
Asymptote

Figure 8.

Considering first the change in the surface gravity at the equator given by (115) we see that we can under certain circumstances have actually an *increase* in g at the equator. This will happen for polytropes with their index very near but greater than 3. As for the change in the apparent gravity at the poles, we see that there exist configurations which are characterised by a *diminution* in the value of the apparent gravity at the poles when it is set rotating. From fig. 8 we see that $n \sim 1{\cdot}41$ corresponds to a polytrope for which, associated with rotation, there is no change in gravity at the poles. *Polytropes with* $1{\cdot}41 < n < 3$ *are characterised by a decrease in* g *at the poles when they are set rotating.* These interesting properties of the variation of g on the boundary lends itself to much speculation, but this is not the place for them.

IV. *The Tidal Problem*

§ 23. *Distortions in External Shape as a Function of the Density Distribution.*—As a consequence of the theorem stated at the end of § 5, II, the only

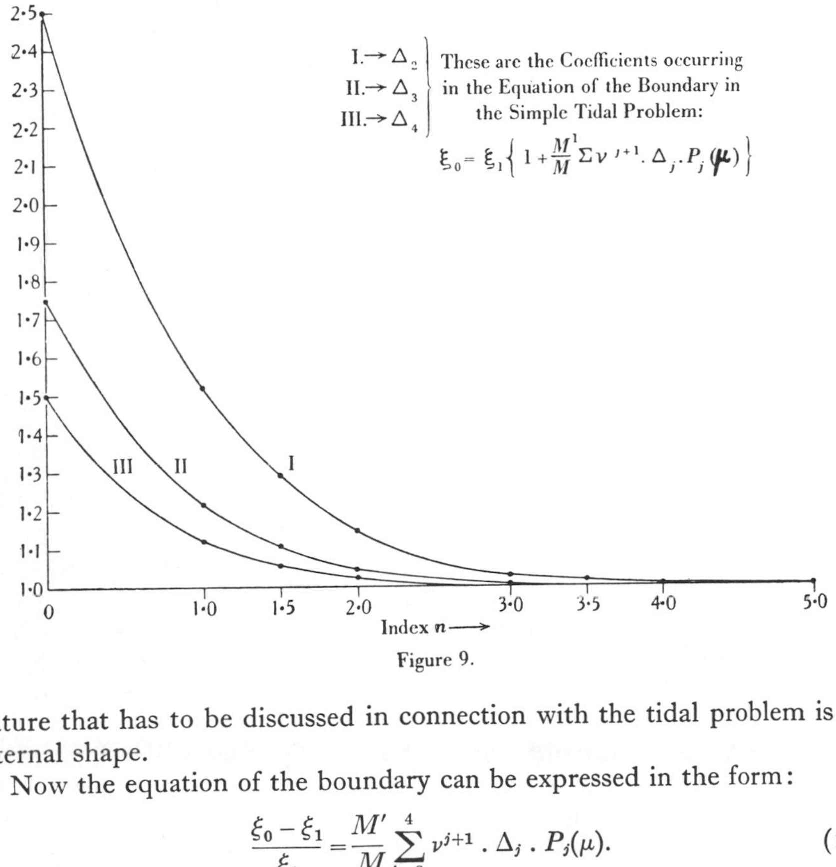

Figure 9.

feature that has to be discussed in connection with the tidal problem is the external shape.

Now the equation of the boundary can be expressed in the form:

$$\frac{\xi_0 - \xi_1}{\xi_1} = \frac{M'}{M}\sum_{j=2}^{4} \nu^{j+1} \,.\, \Delta_j \,.\, P_j(\mu). \tag{116}$$

The coefficients Δ_j occurring in (116) for different values of n are collected in Table VIII, and their variation with n illustrated in fig. 9.

TABLE VIII

n	Δ_2	Δ_3	Δ_4
0	2·5	1·75	1·5
1	1·51985	1·2129	1·1205
1·5	1·2892	1·1079	1·0562
2	1·1482	1·0488	1·0231
3	1·0289	1·00736	1·00281
3·5	1·0147	...	...
4	1·0027	1·00047	1·00014
5	1	1	1

For the case $M' = M$ and $\nu = 0{\cdot}1$ the exact values of the distortions σ_{+1}, σ_{-1}, σ_0 at the poles and at the equator were calculated and are summarised in Table IX.

TABLE IX

External Shape for a Tidally Distorted Polytrope, $M' = M$, $\nu = 0{\cdot}1$

n	$\sigma_{+1} \times 10^3$	$\sigma_{-1} \times 10^3$	$\sigma_0 \times 10^3$
0	2·690	2·340	1·244
1	1·652	1·410	0·756
1·5	1·411	1·189	0·641
2	1·263	1·054	0·570
3	1·140	0·938	0·511
4	1·113	0·913	0·498
5	1·110	0·910	0·497

The results of Table IX are illustrated in fig. 10. It is rather remarkable to see how the least deviation from the uniform density distribution affects enormously the tidal distortions. This is a consequence of the Δ_j's (the coefficients of the third, fourth and fifth powers of ν occurring in the equation of the boundary) so rapidly falling down to the value unity as n deviates from zero. It would be of interest in this connection to study polytropes with $0 < n < 1$. But Emden functions for proper fractional indices are not available.

Further, when $\nu = 0{\cdot}1$ the angle χ at which the furrow develops on the external shape is about $2°$ (in circular measure) for $n = 0$ and about $3°$ for $n = 5$. The variation is not large.

§ 24. *The Variation of Apparent Gravity on the Boundary.*—In § 22 we stated a theorem of von Zeipel regarding the variation of brightness on the surface of a rotating star. We shall prove in Appendix I that an exactly similar theorem can be formulated for a tidally distorted star. We shall therefore examine here the variation of apparent gravity on the boundary.

If we denote by $g(R, \mu)$ the apparent gravity on the tidally distorted

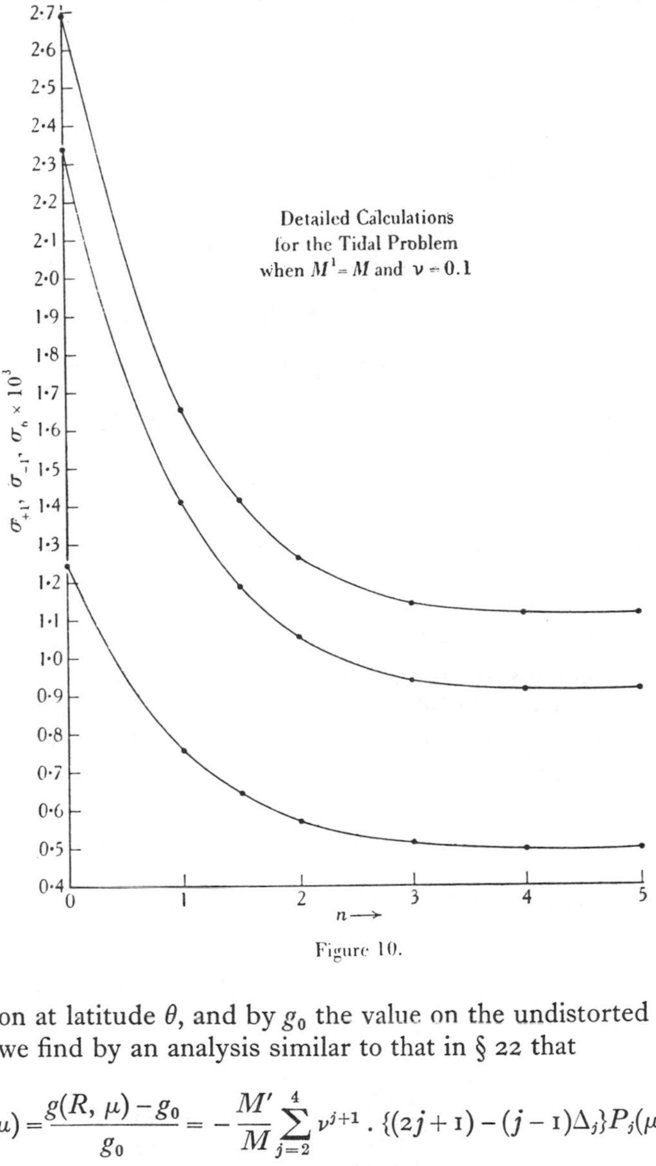

Figure 10.

configuration at latitude θ, and by g_0 the value on the undistorted configuration, then we find by an analysis similar to that in § 22 that

$$\gamma(\mu) = \frac{g(R, \mu) - g_0}{g_0} = -\frac{M'}{M}\sum_{j=2}^{4} \nu^{j+1} \,.\, \{(2j+1) - (j-1)\Delta_j\} P_j(\mu). \tag{117}$$

Two special cases of (117) may be noted

$$\left.\begin{aligned} \gamma(\mu) &= -\frac{M'}{M}\sum_{j=2}^{4} \nu^{j+1} \,.\, (j+\tfrac{1}{2}) \,.\, P_j(\mu), \qquad n = 0, \\ \gamma(\mu) &= -\frac{M'}{M}\sum_{j=2}^{4} \nu^{j+1} \,.\, (j+2) \,.\, P_j(\mu), \qquad n = 5. \end{aligned}\right\} \tag{118}$$

Noting that the expression for $\gamma(\mu)$ is exactly of the same form as the equation of the boundary (116), we see that, similar to the formation of a furrow in the external shape (II, § 8), there is an analogous phenomenon occurring with the variation of gravity on the surface. The equation

$$\frac{\partial \gamma(\mu)}{\partial \mu} = 0 \tag{119}$$

has a "non-trivial" solution. If we denote the solution of (119) by χ_g, then we clearly have (equation (48), II)

$$\chi_g = \tfrac{1}{2}\frac{7-2\Delta_3}{5-\Delta_2}\nu + O(\nu^3). \tag{120}$$

For $n=0$

$$\chi_g = 0{\cdot}7\nu + O(\nu^3), \tag{121}$$

and for $n=5$

$$\chi_g = 0{\cdot}625\nu + O(\nu^3). \tag{122}$$

Comparing (121) with (61) we notice the following peculiarity of the Maclaurin model :—

For the Maclaurin model the furrow in the variation of gravity is formed at angle (measured from the equator) which is exactly twice the angle at which the actual furrow in the external shape develops.

V. *The Double-star Problem*

§ 25. The double-star problem has already been discussed fully in III, but it is of interest to have all the principal distortions (III, equations (38) and (43)) numerically evaluated for one special case. As a representative example we choose $M' = 2M$ and $\nu = 0{\cdot}1$. The results are tabulated in Table X.

TABLE X

Distortions in a Double-star Component, $M' = 2M$, $\nu = 0{\cdot}1$

n	Elongation at $(\pi/2, 0) \times 10^3$	Elongation at $(\pi/2, \pi) \times 10^3$	Elongation at $(\pi/2, \pi/2) \times 10^3$	Contraction at $(0, 0) \times 10^3$
0	7·630	6·930	−0·239	3·989
1	4·673	4·187	−0·144	2·423
1·5	4·047	3·604	−0·055	1·989
2	3·707	3·288	+0·040	1·682
3	3·530	3·127	0·229	1·315
4	3·624	3·224	0·404	1·100
5	3·720	3·320	0·505	0·995

The variations of the quantities in the second, third and fifth columns of Table X with n are graphically shown in fig. 11.

Further, for this case ($M' = 2M$), $\chi_m = 2/3\chi_e$, and χ_e has already been tabulated in II (§ 8).

§ 26. *Variation of Surface Gravity.*—Without calculations we can obtain the results for the double-star problem by an application of the superposition

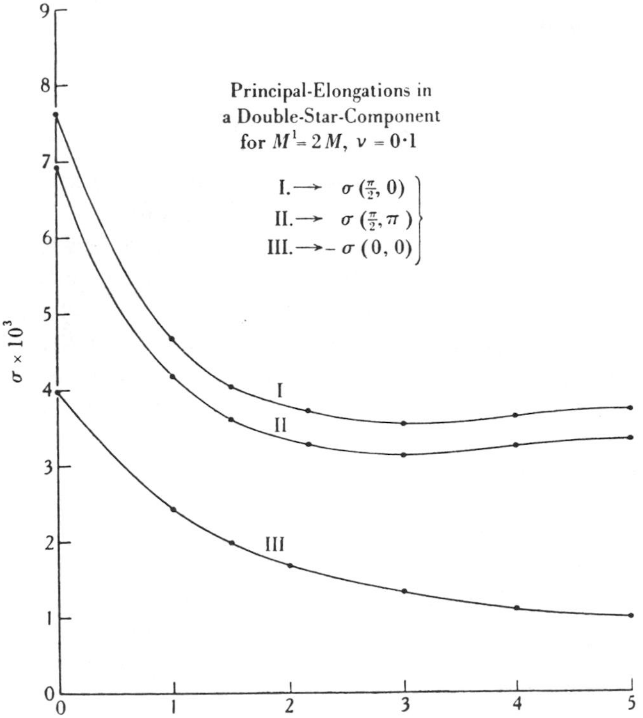

Figure 11.
(Double-Star Problem)

theorem (III, § 7). If we denote by $g(\mu, \phi)$ the apparent gravity on the boundary, then we deduce from the results of §§ 22, 24 that

$$\gamma(\mu, \phi) = \frac{g(\mu, \phi) - g_0}{g_0} = -2\left(1 + \frac{M'}{M}\right)\nu^3\left(\frac{2\psi_0(\xi_1)}{\xi_1^2} + \frac{\psi_0'(\xi_1)}{\xi_1}\right)$$
$$+ \tfrac{1}{3}\left(1 + \frac{M'}{M}\right)(5 - \Delta_2) \,.\, \nu^3 P_2(\cos\theta)$$
$$- \frac{M'}{M}\sum_{j=2}^{4} \nu^{j+1}\{(2j+1) - (j-1)\Delta_j\} P_j(\sin\theta\cos\phi). \qquad (123)$$

The terms in the Legendre polynomials in (123) are of exactly the same form as the function $-\sigma(\theta, \phi)$ introduced in III (§ 8). Analogous to the angles χ_m and χ_e we have now two other angles $\chi_{g,m}$ and $\chi_{g,e}$. The expressions for these latter angles can be obtained from those of the former (equations (46) and (48) in III) by writing $((2j+1) - (j-1)\Delta_j)$ wherever Δ_j occurs. Thus we have

$$\left.\begin{aligned}\chi_{g,m}&=\frac{3}{2}\frac{M'}{M+4M'}\frac{7-2\Delta_3}{5-\Delta_2}\nu+O(\nu^3),\\ \chi_{g,e}&=\frac{1}{2}\frac{7-2\Delta_3}{5-\Delta_2}\nu+O(\nu^3).\end{aligned}\right\} \tag{124}$$

Finally we notice the relation

$$\frac{\chi_{g,m}}{\chi_{g,e}}=\frac{\chi_m}{\chi_e}=3\frac{M'}{M+4M'}. \tag{125}$$

It would be interesting to settle whether the relation (125) is a peculiarity of the polytropic model or whether it is generally valid.

Conclusion.—The underlying physical problem in these studies on "Distorted Polytropes" was to obtain some general idea regarding the nature of the dependence of the distortions in a spherically symmetrical distribution of matter consequent to a slow uniform rotation or to the presence of a tidally disturbing secondary on the original density distribution. Actually, we have solved the problem for the polytropic model mainly for this reason, that the comparatively short range for the polytropic index n {(0, 5)} includes a very large variety of density distributions including the two limiting cases. In answering the physical problem we have argued that the general nature of the dependence of the distortions on the density distribution is in effect given by their dependence on the polytropic index n (§ 18).

It is, however, quite possible that the density distributions governed by the Emden function θ have such special features that it may not be quite safe to draw general conclusions from the study of polytropes alone. In any case it was the hope that the study of the distorted polytropes have an interest of their own that made the writer undertake the rather laborious numerical work that underlies these studies.

One final remark may be permitted.

It has been found that the structure of the distorted polytropes are completely specified when four functions which we have denoted by ψ_0, ψ_2, ψ_3 and ψ_4 are known. Because of the important rôle these functions play in the theory it would perhaps be worth while to give them some special name. Personally I should suggest the name *Associated Emden Functions of the Different Orders*, with their index specified. Thus $(1-\sin\xi/\xi)$ (*cf.* equation (48), I) would on this nomenclature be called "the associated Emden function of order 0 and index 1."

APPENDIX I

A General Theorem in the Theory of Tidally Distorted Stars.—I shall prove the following theorem :—

If a mass of material in radiative equilibrium is in static equilibrium under the combined influence of its own gravitation and the tidal forces due to neighbouring masses, and if further the rate of liberation of energy $4\pi\epsilon$, *the absorption coefficient* κ *and the gas-pressure* p *are functions of the density* ρ *and the temperature* T *only, then* ϵ *is a constant.*

For the theorem to be valid, the presence of tidally disturbing neighbouring masses is necessary. In the absence of a tidal field the theorem is not true.

The above theorem is analogous to von Zeipel's theorem in the theory of rotating stars, which states that under the same physical hypotheses as the above theorem

$$\epsilon = \text{constant} \times \left(1 - \frac{\omega^2}{2\pi G\rho}\right). \tag{1}$$

Proof.—The equations of mechanical equilibrium (II, equation (8)) can be written as

$$dP = \rho dW, \tag{2}$$

where

$$W = V + V_T. \tag{3}$$

From (2) and (3) we easily deduce (as in the proof of von Zeipel's theorem) that surfaces of constant W (*level surfaces*) are also surfaces of constant p, ρ and T. If p' is the radiation-pressure, the equation of radiative equilibrium is

$$\operatorname{div}\left(\frac{1}{\kappa\rho}\operatorname{grad} p'\right) = -\frac{4\pi\epsilon\rho}{c}. \tag{4}$$

Since P and p' are constant over a level surface we may consider p' as a function P and write (4) as

$$\sum_{x,y,z} \frac{\partial}{\partial x}\left(\frac{1}{\kappa}\frac{dp'}{dP} \cdot \frac{1}{\rho}\frac{dP}{dx}\right) = -\frac{4\pi\epsilon\rho}{c}, \tag{5}$$

which on differentiating yields

$$\frac{d}{dP}\left(\frac{1}{\kappa}\frac{dp'}{dP}\right) \cdot \frac{1}{\rho}\left(\frac{dP}{dn}\right)^2 + \frac{1}{\kappa}\frac{dp'}{dP}\operatorname{div}\left(\frac{1}{\rho}\operatorname{grad} P\right) = -\frac{4\pi\epsilon\rho}{c}, \tag{6}$$

where we have used the identity

$$\left(\frac{dP}{dn}\right)^2 = \left(\frac{\partial P}{\partial x}\right)^2 + \left(\frac{\partial P}{\partial y}\right)^2 + \left(\frac{\partial P}{\partial z}\right)^2. \tag{7}$$

The differential dn in (6) corresponds to the normal distance between neighbouring level surfaces. But as we have (II, equation (11))

$$\operatorname{div}\left(\frac{1}{\rho}\operatorname{grad} P\right) = -4\pi G\rho, \tag{8}$$

we can rewrite (6) in the form:

$$\frac{d}{dP}\left(\frac{1}{\kappa}\frac{dp'}{dP}\right)\frac{1}{\rho}\left(\frac{dP}{dn}\right)^2 = -\frac{4\pi\epsilon\rho}{c} + \frac{1}{\kappa}\frac{dp'}{dP} \cdot 4\pi G\rho. \tag{9}$$

The right-hand side of (9) is a constant over a level surface. Hence the left-hand side is either constant over a level surface or is identically zero.

If it were a constant over a level surface it would follow that the differential element dn is also constant over a level surface, or, in other words, that the equipotentials are parallel surfaces. This is impossible for a tidally distorted configuration. Hence

$$\frac{d}{dP}\left(\frac{1}{\kappa}\frac{dp'}{dP}\right)=0, \tag{10}$$

or

$$\frac{1}{\kappa}\frac{dp'}{dP}=\text{constant}=\beta \text{ (say)}. \tag{11}$$

(9) now reduces to

$$\epsilon=cG\beta, \tag{12}$$

which proves the theorem.*

Q.E.D.

As a corollary to the theorem we prove :—

The variation of brightness (i.e. *of the emergent flux of radiation*) *over the surface of a tidally distorted star corresponds exactly to the variation of apparent gravity.*

We referred to this fact in § 24 before studying the variation of "g" on the surface of a tidally distorted configuration.

Proof.—Let F_n be the flux of radiant energy across a level surface at any point per unit area. Then

$$F_n=-\frac{c}{\kappa\rho}\frac{dp'}{dn}, \tag{13}$$

or by (11)

$$F_n=-\frac{c\beta}{\rho}\frac{dP}{dn}, \tag{14}$$

or again by (2)

$$F_n=-c\beta\frac{dW}{dn}.$$

Q.E.D.

APPENDIX II

Solutions of the Differential Equations :—

$$\frac{1}{\xi^2}\frac{d}{d\xi}\left(\xi^2\frac{d\psi_0}{d\xi}\right)=-n\theta^{n-1}\psi_0+1$$

and

$$\frac{1}{\xi^2}\frac{d}{d\xi}\left(\xi^2\frac{d\psi_2}{d\xi}\right)=\left(\frac{6}{\xi^2}-n\theta^{n-1}\right)\psi_2$$

for $n=3\cdot5$.

* Even as we do not believe that von Zeipel's condition (1) would be fulfilled in an actual star, so also we should not expect that our new condition $\epsilon=$constant will be fulfilled in a tidally distorted star. A consequence of this would be the setting up of a *permanent* slow system of circulating currents (principally in the meridian planes), since no kind of static equilibrium is possible which is inconsistent with the condition $\epsilon=$constant. We shall not go into these delicate matters here, as it will lead us too far from our main object.

TABLE XI *

Index $n=3\cdot5$

ξ	ψ_0	ψ_2	ξ	ψ_0	ψ_2
0	0	0	4·8	2·5107	5·9809
0·2	0·0066205	0·039604	5·0	2·7532	6·3595
0·4	0·025952	0·15386	5·2	3·0106	6·7562
0·6	0·05657	0·33038	5·4	3·2829	7·1713
0·8	0·09658	0·55231	5·6	3·5704	7·6050
1·0	0·14409	0·80214	5·8	3·8732	8·0572
1·2	0·19761	1·0649	6·0	4·1915	8·5293
1·4	0·25623	1·3296	6·2	4·5253	9·0202
1·6	0·31962	1·5898	6·4	4·8747	9·5304
1·8	0·38800	1·8425	6·6	5·2398	10·0601
2·0	0·46189	2·0877	6·8	5·6206	10·6093
2·2	0·54206	2·3269	7·0	6·0171	11·1779
2·4	0·62939	2·5627	7·2	6·4294	11·7661
2·6	0·77477	2·7979	7·4	6·8572	12·3739
2·8	0·82907	3·0354	7·6	7·3010	13·0011
3·0	0·94311	3·2778	7·8	7·7603	13·6478
3·2	1·06765	3·5275	8·0	8·2352	14·3139
3·4	1·2034	3·7867	8·2	8·7256	14·9993
3·6	1·3509	4·0570	8·4	9·2315	15·7039
3·8	1·5107	4·3399	8·6	9·7528	16·4277
4·0	1·6835	4·6366	8·8	10·2894	17·1705
4·2	1·8694	4·9483	9·0	10·8412	17·9323
4·4	2·0691	5·2758	9·2	11·4082	18·7126
4·6	2·2827	5·6198	9·4	11·9901	19·5121

$$\xi_1=9\cdot5358,$$
$$\psi_0(\xi_1)=12\cdot4466, \qquad \psi_2(\xi_1)=20\cdot1390,$$
$$\psi_0'(\xi_1)=\ 2\cdot9611, \qquad \psi_2'(\xi_1)=\ 4\cdot0706.$$

Institut for Teoretisk Fysik,
Copenhagen, and
Trinity College, Cambridge:
1933 *May* 31.

* More accurate tabulations of these functions appear in S. Chandrasekhar and Norman R. Lebovitz, "On the Oscillations and the Stability of Rotating Gaseous Masses: 3, The Distorted Polytropes" (*The Astrophysical Journal* 136, no. 3 [1962]: 1082–1104), reprinted as paper 21 in volume 4 of these selected papers.

24

Discussion of Papers 20–23 by E. A. Milne, H. N. Russell, and A. S. Eddington

Mr. S. Chandrasekhar. In my communications on "Distorted Polytropes" I have attempted to obtain some information on the following physical problem: we start with a spherically symmetrical distribution of matter with the density decreasing monotonically outwards from the centre. We perturb the configuration either by a slow uniform rotation about its axis, or by the tidal forces arising from a distant secondary, or by making it a component of a double-star system. The configuration gets distorted and is no longer spherically symmetrical. Now the distortion induced in the configuration by the rotational and tidal forces is a function of the complete march of the density-distribution in the undistorted configuration. The problem is: what is the nature of this functional dependence?

It is clear that to answer this problem comprehensively we must consider all the possible density-distributions which can be associated with a prescribed mass of material M and a prescribed spherical volume V. There are two extreme limiting forms of density-distribution. We can either distribute the mass uniformly with a constant density, or concentrate it completely at the centre. Intermediate between these two extremes are a variety of other possible density-distributions with different degrees of central condensation. A quantity which gives us a general idea of the nature of the density-distribution is the ratio of the central to the mean density

(I shall refer to this ratio simply as the central density). The larger the central density the more centrally condensed is the configuration and *vice versa*. It is, however, clear that the specification of the central density does not uniquely determine the density-distribution. Hence if we really want to determine *completely* the functional dependence of the distortions on the density-distribution we must consider the different possible density-distributions which can be associated with a given mass, a given radius and a given central density. But, if, instead, we choose a special form for the density-distribution for each specified value of the central density, then the distortions become a function of the one variable—the central density. Restricting ourselves to a special class of density-distribution, we can for instance in the case of the rotational problem plot a curve between the central density and ellipticity of the configuration. If we restricted ourselves to a different class of density-distributions we shall not get an identical curve, but it appears fair to argue that the general form of the curve would be preserved for all *classes* of density-distribution.

I have restricted myself to the study of the polytropic class of density-distributions, *i. e.*, if we write the density in the form $\rho=\lambda\theta^n$, then we make θ satisfy Emden's differential equation with index n. Now the great merit of the polytropes is that a comparatively small range $(0\leq n\leq 5)$ for the polytropic index includes a very large variety of density-distributions, including the two limiting forms, the uniform distribution ($n=0$) and the complete concentration of the mass towards the centre ($n=5$).

In solving the hydrodynamical problem associated with the rotational, the tidal, and the double star problems we determine a function Θ which plays a role identical with the Emden function θ for the undistorted polytropes. In fact, we express Θ as the sum of two functions—the Emden function θ and another small function Ψ. In this way it is found that we can solve the rotational problem correct to the first order in ω^2 and the tidal and the double star problems correct to the fifth order in the ratio of the radius of the undistorted configuration to the distance between the centre of gravities of the components.

In the rotational problem a quantity of importance

is $5m/2\epsilon$ (m is the ratio of the centrifugal force at the equator to the mean purely gravitational force and ϵ is the ellipticity). This is a pure number and is 2 for $n=0$ and 5 for $n=5$. For the intermediate values of n we have the values for this ratio ranging from 2 to 5. Now if $5m/2\epsilon$ were known for a rotating configuration then we can find the "effective polytropic index" for this configuration and in this way estimate the central densities. I have done this for the different planets. For the Earth for instance the central density appears to be about twice the mean density.

In the case of the tidal problem one finds that the least departure from the uniform distribution has rather an enormous effect on the distortions, and the polytrope $n=3$ behaves not very differently from the Roche-model ($n=5$).

The double-star problem is just a little more complicated, but once we have fully analysed the rotational and the tidal problems the discussion of the double-star problem becomes quite straightforward.

Prof. Milne. I have had the opportunity of looking at Mr. Chandrasekhar's work in detail, and I think the beauty of it is in the systematic march of the distortions. A particularly good point is that he has improved the accuracy of the approximations by taking the disturbing body to be itself a polytrope.

Prof. H. N. Russell. I ought to apologize for talking so much, but I don't get a chance often! First I must congratulate Mr. Chandrasekhar on the lucid presentation he has given of this intricate problem. His work is going to be of great practical value, for we all know how critical in present theories of stellar structure is the question of the degree of central condensation.

Now in certain eclipsing binaries the ellipticity which Mr. Chandrasekhar has determined can lead to a rotation of the line of apsides. In one star of this type, Y Cygni, my colleague, Professor Dugan, and Dr. Redman have definitely shown that an advance of periastron occurs. The period of the star is 3 days and of the apsidal rotation 54 years. RU Monocerotis is another such star with period of about 4 days and an apsidal rotation with period of the order of 1000 years.

The actual rate of motion of periastron in both cases is what you would get with homogeneous stars half the size of the actual stars. When you substitute in the equations the condensation to the centre comes out less than that in Eddington's $n=3$ polytrope. This is only approximate, and I want to see Mr. Chandrasekhar's work before I say more. There are four or five other eclipsing variables of this kind, and we are arranging for them to be studied at Harvard and at Princeton, so that we may hope eventually to have some really definite information on the question of the actual degree of central condensation in these stars.

Sir Arthur Eddington. It is difficult to follow this highly complicated subject, but the question is raised as to how closely actual stars will follow models of this kind. I think I am right in saying that the polytropic approximation is more difficult to justify in the case of a rotating star than in a stationary star, because it represents as it were a smoothing out in two dimensions instead of one. It should be possible, consistent with the mechanical equations, to have any density distribution along the polar axis and any other along an equatorial axis, with any rotation. The polytropic approximation introduces a correlation between these which exists only in the mathematical equations. One speaks of taking a star with a given density distribution and then setting it to rotate with a certain angular velocity, but that is not the way in which a star is actually set rotating. In addition to the polytropic classification one could introduce another classification in another dimension, and the question then arises as to how far the variation in the cross direction would be important in dealing with actual stars.

Mr. Chandrasekhar. The reduction to polytropes is of course made by the relation $P=K\rho^{1+\frac{1}{n}}$ which brings in the correlation along the equator and at the poles referred to by Sir Arthur Eddington. But there exists an "equation of state" for every configuration of equilibrium. We go radially from the boundary to the centre and measure the pressure P and the density ρ at each point and plot a curve. We thus get an "equation of state" which is independent of any theory of the properties of stellar material. If we can approximate to this curve fairly closely by a parabola of the type P=constant ρ^{x}, then I think the analysis of the distorted polytrope would give us the required information. If the approximation is not possible, then the " two-dimensional " effects referred to by Sir Arthur Eddington will have to be considered *.

* An examination of the curves in my " Distorted Polytropes.—IV." will show that such " two-dimensional " effects are not likely to be important for the more centrally condensed polytropes ($n > 2{\cdot}5$).—S. C.

A NOTE ON THE PERTURBATION THEORY FOR DISTORTED STELLAR CONFIGURATIONS

S. CHANDRASEKHAR AND WASLEY KROGDAHL

ABSTRACT

In this paper we relate the general perturbation theory developed in the preceding paper with the earlier general discussion of distorted equilibrum configurations.

1. In the preceding paper[1] one of us has developed a general theory of perturbations for describing stellar configurations distorted by tidal and (or) centrifugal forces. The general method consists of simply expressing the changes in the physical parameters caused by the perturbing forces in terms of the density and pressure distributions in an undistorted configuration with the same central density. Thus, considering the purely tidal problem, for example, it is found that the pressure distribution can be expressed in the form

$$\eta = \eta_0 + \chi_0'' \sum_{j=z}^{4} a_{1,j} \eta_{1,j}^* P_j(\mu) , \tag{1}$$

where η denotes the pressure (expressed in units of the central pressure) and the $\eta_{1,j}^*$'s are solutions of the differential equations (cf. *op. cit.*, eq. [140])

$$\frac{1}{\xi^2}\left\{\frac{d}{d\xi}\left(\xi^2 \frac{d}{d\xi}\left[\frac{\eta_{1,j}^*}{\zeta_0}\right]\right) - j(j+1)\frac{\eta_{1,j}^*}{\zeta_0}\right\} = -\eta_{1,j}^* \frac{\dfrac{d\zeta_0}{d\xi}}{\dfrac{d\eta_0}{d\xi}} \qquad (j = 2, 3, 4) \tag{2}$$

together with the boundary conditions

$$\eta_{1,j}^* = \xi^j + O(\xi^{j+2}) \qquad (\xi \to 0). \tag{3}$$

Further, in equations (1) and (2) η_0 and ζ_0 correspond to solutions for the pressure and density distributions in the corresponding undisturbed configuration, i.e., they satisfy the differential equation

$$\frac{1}{\xi^2}\frac{d}{d\xi}\left(\frac{\xi^2}{\zeta_0}\frac{d\eta_0}{d\zeta}\right) = -\zeta_0 . \tag{4}$$

The $a_{1,j}$'s in equation (1) are certain numbers defined in terms of the values which $\eta_{1,j}^*$ and $d/d\xi(\eta_{1,j}^*/\zeta_0)$ take at the boundary of the configuration.[2] Finally,

$$\chi_0'' = \frac{M''}{4\pi\rho_c a^2 R}, \tag{5}$$

where M'' denotes the mass of the secondary, R the distance between the centers of gravity of the two stars, ρ_c the central density of the primary, and a the chosen unit of distance. The solutions for the rotational and the combined rotational and tidal problems take similar forms.

[1] W. Krogdahl, *Ap. J.*, **96**, 124, 1942.

[2] Cf. *op. cit.*, eq. (135).

2. Now there exists a general theory of distorted equilibrium configurations in which the emphasis is on the variation in the forms of the surfaces of constant pressure (or density) through the star.[3] On this theory the isobaric surfaces are written in the form

$$\xi = \bar{\xi}\left\{1 + \sum_{j=2}^{4} Y_j(\xi) P_j(\mu)\right\}, \tag{6}$$

where $\bar{\xi}$ is the mean value of ξ for a given value of the pressure; and it is shown that the functions S_j ($j = 2, 3, 4$), defined as

$$S_j = \xi \frac{d \log Y_j}{d\xi}, \tag{7}$$

are solutions of the first-order equation

$$\xi \frac{dS_j}{d\xi} + S_j^2 - S_j - j(j+1) + \frac{6\rho}{\bar{\rho}}(S_j + 1) = 0, \tag{8}$$

with the boundary conditions

$$S_j = j - 2 \qquad (\xi = 0). \tag{9}$$

In equation (8) $\bar{\rho}$ is the mean density interior to ξ.

While the objectives of this theory are more limited than those of the general perturbation theory of the preceding paper, it is clear that equation (8) must be a simple mathematical consequence of equation (2). In this note we shall show that this is actually the case and thus establish the formal equivalence of the two theories.

3. From equation (1) it readily follows that the equation of the isobaric surfaces can be written in the form

$$\xi(\zeta_0) = \bar{\xi}(\zeta_0)\left[1 - \chi_0'' \sum_{j=2}^{4} a_{1,j} \frac{\overset{*}{\eta}_{1,j}}{\xi \dfrac{d\eta_0}{d\xi}} P_j(\mu)\right]. \tag{10}$$

Comparing equations (6) and (10), we conclude that

$$Y_j = -\chi_0'' a_{1,j} \frac{\overset{*}{\eta}_{1,j}}{\xi \dfrac{d\eta_0}{d\xi}}. \tag{11}$$

Hence,

$$S_j = \xi \frac{d \log Y_j}{d\xi} = -1 + \xi \frac{\overset{*}{\eta}{}'_{1,j}}{\overset{*}{\eta}_{1,j}} - \xi \frac{\eta_0''}{\eta_0'}, \tag{12}$$

where we have used primes to denote differentiation with respect to ξ.

We have to show that S_j satisfies the differential equation (8) in virtue of its definition and the differential equation (2) which $\overset{*}{\eta}_{1,j}$ satisfies. To show this, we shall first transform equation (2) by introducing the variable

$$\psi_j = \xi \frac{\overset{*}{\eta}_{1,j}}{\zeta_0}, \tag{13}$$

[3] H. Jeffreys, *The Earth*, chap. xiii, Cambridge, England, 1929; T. E. Sterne, *M.N.*, **99**, 451, 1939.

whence equation (2) becomes

$$\frac{\psi_j''}{\psi_j} - \frac{j(j+1)}{\xi^2} + \zeta_0 \frac{\zeta_0'}{\eta_0'} = 0 . \tag{14}$$

Now, let

$$\varphi_j = \xi \frac{\psi_j'}{\psi_j} . \tag{15}$$

Then

$$\frac{1}{\psi_j} \frac{d}{d\xi}\left(\frac{\varphi_j \psi_j}{\xi}\right) - \frac{j(j+1)}{\xi^2} + \zeta_0 \frac{\zeta_0'}{\eta_0'} = 0 ; \tag{16}$$

or, since

$$\frac{d}{d\xi}\left(\frac{\varphi_j \psi_j}{\xi}\right) = -\frac{\varphi_j \psi_j}{\xi^2} + \frac{\psi_j}{\xi}\varphi_j' + \frac{\varphi_j}{\xi}\psi_j' , \tag{17}$$

equation (16) becomes, after some further reductions,

$$\xi \frac{d\varphi_j}{d\xi} + \varphi_j^2 - \varphi_j - j(j+1) + \zeta_0 \xi^2 \frac{\zeta_0'}{\eta_0'} = 0 . \tag{18}$$

4. We shall now express S_j in terms of φ_j. According to equations (13) and (15),

$$\varphi_j = \xi \frac{d \log \psi_j}{d\xi} = 1 + \xi \frac{\eta_{1,j}^{*\prime}}{\eta_{1,j}^{*}} - \xi \frac{\zeta_0'}{\zeta_0} . \tag{19}$$

Hence, combining equations (12) and (19), we have

$$\varphi_j = 2 + \xi \left(\frac{\eta_0''}{\eta_0'} - \frac{\zeta_0'}{\zeta_0}\right) + S_j . \tag{20}$$

We can eliminate η_0'' from the foregoing equation by using equation (4); according to this equation,

$$2\xi \frac{\eta_0'}{\zeta_0} + \xi^2 \frac{\eta_0''}{\zeta_0} - \xi^2 \frac{\eta_0' \zeta_0'}{\zeta_0^2} = -\xi^2 \zeta_0 , \tag{21}$$

or

$$\xi \left(\frac{\eta_0''}{\eta_0'} - \frac{\zeta_0'}{\zeta_0}\right) = -2 - \xi \frac{\zeta_0^2}{\eta_0'} . \tag{22}$$

We can therefore re-write equation (20) as

$$\varphi_j = S_j - \xi \frac{\zeta_0^2}{\eta_0'} . \tag{23}$$

We can express the foregoing relation somewhat differently, using the following relation between the actual and the mean densities:

$$\frac{\rho(\xi)}{\bar{\rho}(\xi)} = -\frac{1}{3} \xi \frac{\zeta_0^2}{\eta_0'} . \tag{24}$$

Thus,

$$\varphi_j = S_j + 3\,\frac{\rho(\xi)}{\bar{\rho}(\xi)}\,. \tag{25}$$

We now substitute the foregoing relation in equation (18). We find

$$\xi\,\frac{d}{d\xi}\left(S_j - \xi\,\frac{\zeta_0^2}{\eta_0'}\right) + \left(S_j + 3\,\frac{\rho}{\bar{\rho}}\right)^2 - S_j - 3\,\frac{\rho}{\bar{\rho}} - j(j+1) + \zeta_0 \xi^2\,\frac{\zeta_0'}{\eta_0'} = 0 \tag{26}$$

or, after some reductions,

$$\left.\begin{aligned} &\xi\,\frac{dS_j}{d\xi} + S_j^2 - S_j - j(j+1) + \frac{6\rho}{\bar{\rho}}\,(S_j+1) \\ &\qquad = \xi\,\frac{d}{d\xi}\left(\xi\,\frac{\zeta_0^2}{\eta_0'}\right) + 9\frac{\rho}{\bar{\rho}} - 9\left(\frac{\rho}{\bar{\rho}}\right)^2 - \zeta_0 \xi^2\,\frac{\zeta_0'}{\eta_0'}\,. \end{aligned}\right\} \tag{27}$$

Expanding the right-hand side of equation (27) and using the relation (24) we readily obtain

$$\left.\begin{aligned} &\xi\,\frac{\zeta_0^2}{\eta_0'} + 2\xi^2\,\frac{\zeta_0\zeta_0'}{\eta_0'} - \xi^2\,\frac{\zeta_0^2}{\eta_0'^2}\,\eta_0'' - 3\xi\,\frac{\zeta_0^2}{\eta_0'} - \xi^2\,\frac{\zeta_0^4}{\eta_0'^2} - \xi^2\,\frac{\zeta_0'\zeta_0}{\eta_0'} \\ &\qquad = -2\xi\,\frac{\zeta_0^2}{\eta_0'} - \xi^2\,\frac{\zeta_0^4}{\eta_0'^2} - \xi^2\,\frac{\zeta_0^2}{\eta_0'}\left(\frac{\eta_0''}{\eta_0'} - \frac{\zeta_0'}{\zeta_0}\right), \end{aligned}\right\} \tag{28}$$

which vanishes identically in virtue of equation (22). Hence,

$$\xi\,\frac{dS_j}{d\xi} + S_j^2 - S_j - j(j+1) + 6\,\frac{\rho}{\bar{\rho}}(S_j+1) = 0\,. \tag{29}$$

Finally, we verify that, according to equations (3), (13), (15), and (23),

$$S_j = j - 2 \qquad \text{at} \qquad \xi = 0\,. \tag{30}$$

This proves the formal equivalence of the two perturbation theories now available.

YERKES OBSERVATORY
May 18, 1942

PART THREE

Stellar Evolution

STELLAR MODELS WITH ISOTHERMAL CORES

LOUIS R. HENRICH AND S. CHANDRASEKHAR

ABSTRACT

This paper is devoted to the study of stellar models with isothermal cores. Two types of such configurations have been studied: (1) models with isothermal cores and polytropic envelopes ($n = 3$) and (2) models with isothermal cores and radiative point-source envelopes with a law of opacity $\kappa = \kappa_0 \rho T^{-3.5}$. The most important characteristic of these models is the existence of an upper limit to the fraction ν of the total mass which can be contained in the core. For models of type 2, $\nu_{max} \sim 35\%$. Also it appears that, as ν increases, the radius of the star first decreases to a minimum value and then increases. Further, the luminosity of the star is found to increase by about a factor 3 from the stage when it has no isothermal core to the stage when the core contains the maximum possible mass.

1. *Introduction.*—The possible physical importance of stellar models with isothermal cores was first indicated by Gamow[1] who suggested that these models may have their counterparts in nature if resonance penetration of charged particles into nuclei should become the main source of energy. For under such circumstances the energy may be thought of as being generated in a spherical shell, in which case the regions interior to the shell would be isothermal. Again, if, following Gamow and Teller,[2] we suppose that the proton disintegration of the light-nuclei (D, Li, Be, and B) provides the energy source for the giants, it is conceivable that the available element at a particular time, lithium say, becomes exhausted in the central regions; also the physical conditions may be such that, before the temperature rises sufficiently for the disintegration of the next element, beryllium, to become effective, a situation may arise when the disintegration of lithium in the outer parts becomes the primary source of energy. Under these circumstances, also, the stellar configurations will have isothermal cores. A situation similar to what we have described may prevail quite generally with the exhaustion of hydrogen in the central regions of stars, during the course of their normal evolution. A study of the physical characteristics of stellar models with isothermal cores becomes, therefore, a matter of some interest. A first attempt in this direction has already been made by Critchfield and Gamow.[3] But the essential peculiarities of the model arising from the isothermal nature of the core has been overlooked by these authors.[4] In this paper we therefore propose to study these models under varying conditions to elucidate their physical characteristics.

2. *The equations of the isothermal core.*—We shall consider first the equilibrium of the isothermal core. In the core we can write

$$P = K_2 \rho + D \, , \tag{1}$$

where

$$K_2 = \frac{k}{\mu H} T_c \, ; \qquad D = \tfrac{1}{3} a T_c^4 \, , \tag{2}$$

[1] *Ap. J.*, **87**, 206, 1938; *Phys. Rev.*, **53**, 595, 1938.

[2] *Phys. Rev.*, **53**, 608, 1938.

[3] *Ap. J.*, **89**, 244, 1939.

[4] Thus Critchfield and Gamow assume series expansions for $M(r)$, P, etc., which are valid in the immediate neighborhood of the center, no matter what the equation of state is (cf. eqs. [7] and [8] in the paper referred to in n. 3). Further, it appears that the parts of the isothermal function which are necessary to describe the core cannot be satisfactorily expressed by any kind of series expansion (see §§ 3, 4, and 5 in the present paper).

where K_2 and D are constants. The reduction to the isothermal equation is made by the substitutions[5]

$$\rho = \lambda_2 e^{-\psi} ; \qquad P = K_2 \lambda_2 e^{-\psi} + D , \tag{3}$$

and

$$r = \left(\frac{K_2}{4\pi G \lambda_2} \right)^{1/2} \xi . \tag{4}$$

Further, we have the mass relation

$$M(\xi) = 4\pi \left(\frac{K_2}{4\pi G} \right)^{3/2} \lambda_2^{-1/2} \xi^2 \frac{d\psi}{d\xi} . \tag{5}$$

3. *Stellar models with isothermal cores and polytropic* ($n = 3$) *envelopes.*—As a first example of stellar models with isothermal cores we shall consider the case where the structure of the envelope is governed by the isothermal equation of index $n = 3$. Physically, this assumption implies that in the envelope we have the standard model approximation "$\kappa\eta$ = constant." Under these circumstances we can write

$$P = \left[\left(\frac{k}{\mu H} \right)^4 \frac{3}{a} \frac{1 - \beta}{\beta^4} \right]^{1/3} \rho^{4/3} = K_1 \rho^{4/3} , \tag{6}$$

where K_1 is a constant. The reduction to the polytropic equation is made by the substitutions

$$\left. \begin{aligned} \rho = \lambda_1 \theta^3 ; \qquad P = K_1 \lambda_1^{4/3} \theta^4 ; \\ r = \left(\frac{K_1}{\pi G} \right)^{1/2} \lambda_1^{-1/3} \eta . \end{aligned} \right\} \tag{7}$$

Also, we have the relation

$$M(\eta) = -4\pi \left(\frac{K_1}{\pi G} \right)^{3/2} \eta^2 \frac{d\theta}{d\eta} . \tag{8}$$

The mass M of the whole configuration is given by

$$M = 4\pi \left(\frac{K_1}{\pi G} \right)^{3/2} \omega_3 , \tag{9}$$

where

$$\omega_3 = - \left(\eta^2 \frac{d\theta}{d\eta} \right)_1 , \tag{10}$$

the subscript 1 indicating that the quantity in parenthesis is evaluated at the point where θ has its zero. It may be noted that ω_3 is a homology-invariant constant.[6]

Now, at the interface where the isothermal core joins the polytropic envelope the values of P, ρ, r, and $M(r)$, given by the two sets of formulae (3), (4) and (5), and (7)

[5] See S. Chandrasekhar, *An Introduction to the Study of Stellar Structure*, p. 155, Chicago, 1939.

[6] See *ibid.*, p. 149.

and (8) should be identical. The resulting four equations of fit can be reduced to two equations involving only the homology-invariant combinations

$$u_\infty = \frac{\xi e^{-\psi}}{\psi'}; \qquad v_\infty = \xi \frac{d\psi}{d\xi}; \tag{11}$$

and

$$u_3 = -\frac{\eta\theta^3}{\theta'}; \qquad v_3 = -\frac{\eta\theta'}{\theta}. \tag{12}$$

We find[7]

$$\left.\begin{aligned} u_\infty(\xi_i) &= u_3(\eta_i), \\ \tfrac{1}{4}\beta v_\infty(\xi_i) &= v_3(\eta_i), \end{aligned}\right\} \tag{13}$$

where the subscript i denotes that the respective quantities are evaluated at the interface. According to equations (13) every intersection of a (u_3, v_3)-curve with the $(u_\infty, \frac{1}{4}\beta v_\infty)$-curve derived from the complete isothermal function gives a solution of the equations of fit and corresponds to a definite configuration of the type we are looking for. An examination of the general arrangement of the (u, v)-curves for $n = 3$ and $n = \infty$ readily shows that solutions for equations (13) exist only for (u_3, v_3)-curves derived from M-solutions;[8] however, it may be noted that not all M-solutions provide solutions to equations (13).

According to the views expressed in § 1, in considering stellar models with isothermal cores we are primarily interested in the changes which occur in the parameters describing a star, as the isothermal core at some fixed temperature "grows" at the expense of the envelope. We shall now obtain the relations necessary for this purpose.

Suppose that a (u_3, v_3)-curve labeled by a certain value for the homology-invariant constant ω_3 intersects the $(u_\infty, \frac{1}{4}\beta v_\infty)$-curve derived from an E-solution of the isothermal equation at a point where $\xi = \xi_i$ and $\eta = \eta_i$. At this point the equations of fit (13) are therefore satisfied. The fraction q of the radius R occupied by the core is clearly given by

$$q = \frac{\eta_i}{\eta_1}, \tag{14}$$

where $\eta = \eta_1$ defines the boundary of the particular solution $\theta(\eta, \omega_3)$. The fraction ν of the mass M contained in the core is also readily found. We have

$$\nu = \frac{M(\eta_i)}{M(\eta_1)} = -\frac{(\eta^2\theta')_i}{\omega_3}, \tag{15}$$

where the subscript i indicates that the quantity in parenthesis is evaluated at the interface. Using the definitions of u_3 and v_3 we can re-write equation (15) more conveniently as

$$\nu = \frac{(u_3 v_3^3)_i^{1/2}}{\omega_3}. \tag{16}$$

[7] See *ibid.*, pp. 170–76.

[8] For the classification of the solutions of the Lane-Emden equation see *ibid.*, chap. iv.

The ratio of the central to the mean density is given by[9]

$$\frac{\rho_c}{\bar{\rho}} = \frac{\lambda_2}{\left(\frac{M}{\frac{4}{3}\pi R^3}\right)} = -\frac{\lambda_2}{3\lambda_1\left(\frac{1}{\eta}\frac{d\theta}{d\eta}\right)_1}; \tag{17}$$

or, using the formula

$$\lambda_2 e^{-\psi_i} = \lambda_1\theta_i^3, \tag{18}$$

which expresses the equality of the density at the interface, we have

$$\frac{\rho_c}{\bar{\rho}} = -\frac{\theta_i^3 e^{\psi_i}}{3\left(\frac{1}{\eta}\frac{d\theta}{d\eta}\right)_1} = -\frac{\theta_i^3\eta_i^3 e^{\psi_i}}{3\left(\eta^2\frac{d\theta}{d\eta}\right)_1\left(\frac{\eta_i}{\eta_1}\right)^3}. \tag{19}$$

Hence,

$$\frac{\rho_c}{\bar{\rho}} = \frac{(u_3 v_3)_i^{3/2} e^{\psi_i}}{3\omega_3 q^3}. \tag{20}$$

Finally, to determine the $R(q)$ relation for a given mass and T_c, we start from the relation

$$K_2 = \beta K_1 \lambda_1^{1/3}\theta_i, \tag{21}$$

expressing the equality of p_{gas}/ρ on the two sides of the interface, and eliminate λ_1 from the equation (cf. eq. [7])

$$R = \left(\frac{K_1}{\pi G}\right)^{1/2}\lambda_1^{-1/3}\eta_1. \tag{22}$$

We obtain

$$R = \pi G\beta\left(\frac{K_1}{\pi G}\right)^{3/2}\frac{1}{K_2}\left(\frac{\eta_1}{\eta_i}\right)\eta_i\theta_i, \tag{23}$$

or, using equations (2), (9), and (12), we have

$$R = Q(q)\beta\frac{\mu H}{k}\frac{GM}{T_c}, \tag{24}$$

where we have written

$$Q(q) = \frac{(u_3 v_3)_i^{1/2}}{4\omega_3 q}. \tag{25}$$

Equation (24) is an important relation which determines the dependence of R on q for a configuration of a given mass and fixed central temperature.

Restricting ourselves to the most important case of negligible radiation pressure and putting $\beta = 1$, four solutions of the equations of fit (13) were obtained, using the two

[9] In writing these equations we have assumed that the particular solution of the isothermal equation used is the one for which $\psi = 0$ at $\xi = 0$, i.e., the solution commonly denoted by $\Psi(\xi)$ (see *ibid.*, p. 156).

M-solutions ($\omega_3 = 1.90$ and 1.50) integrated by Fairclough.[10] The results of the fitting are summarized in Table 1. Further, in Figure 1 we have illustrated the $(M(\text{core})/M, q)$ and the (R, q) relations. We shall return to the physical meanings to be attached to these relationships in § 5.

4. *Stellar models with isothermal cores and point-source envelopes with the law of opacity* $\kappa = \kappa_0 \rho T^{-3.5}$.—The standard model approximation for the envelopes which we have considered in § 3, while giving an insight into the general behavior of these models, is not in strict conformity with the physical circumstances under which we might expect isothermal cores. For, consistent with the views expressed in § 1, we should rather suppose that the energy is generated in a thin spherical shell (of thickness Δr_i, say) at the

TABLE 1

STELLAR MODELS WITH ISOTHERMAL CORES AND $n=3$ ENVELOPES

ω_3	q	ν	$\rho_c/\bar{\rho}$	$Q(q)$
2.018	0	0	54.2	0.854
1.90	0.151	0.180	101	0.767
1.50	.148	.349	708	0.954
1.50	.094	.209	6.7×10^5	1.207
1.50	.104	.235	2.9×10^6	1.158
1.42(?)*	0.100	0.250	∞	1.250

* The figures in this row are not reliable. They give very rough estimates of the points about which the respective curves spiral.

interface between the isothermal core and the outer envelope. Under these circumstances the luminosity of the star will be given by

$$L = 4\pi r_i^2 \rho_i \Delta r_i \epsilon_0 , \tag{26}$$

where ϵ_0 denotes the rate of generation of energy per gram of the material. Accordingly, the regions of the star outside $r = r_i$ will be governed by the same equations as those for the point-source model. The equations of equilibrium for these regions are, therefore,

$$\frac{d}{dr}\left(\frac{k}{\mu H}\rho T + \tfrac{1}{3}aT^4\right) = -\frac{GM(r)}{r^2}\rho \tag{27}$$

and

$$\frac{d}{dr}(\tfrac{1}{3}aT^4) = -\frac{\kappa_0 L}{4\pi c r^2}\frac{\rho^2}{T^{3.5}} , \qquad (28)^{11}$$

where we have assumed for the coefficient of opacity the law

$$\kappa = \kappa_0 \frac{\rho}{T^{3.5}} . \tag{29}$$

[10] *M.N.*, **93**, 40, 1932.

[11] It is conceivable that circumstances may arise which require the replacement of equation (28), valid under conditions of radiative equilibrium, by another equation, valid under conditions of convective equilibrium. However, in the models we shall be primarily concerned with, this is not of much significance. Actually, apart from one possible exception, in the models considered the conditions for the validity of radiative equilibrium are not violated.

In equation (29), κ_0 (which is a constant throughout the configuration) may depend on the chemical composition (in particular on the hydrogen and helium abundances).

We shall now consider the method of fitting an isothermal core to a solution of equations (27) and (28): At the interface the quantities ρ, P, $M(r)$, and r as known along a solution of equations (27) and (28) must join continuously with the respective quantities

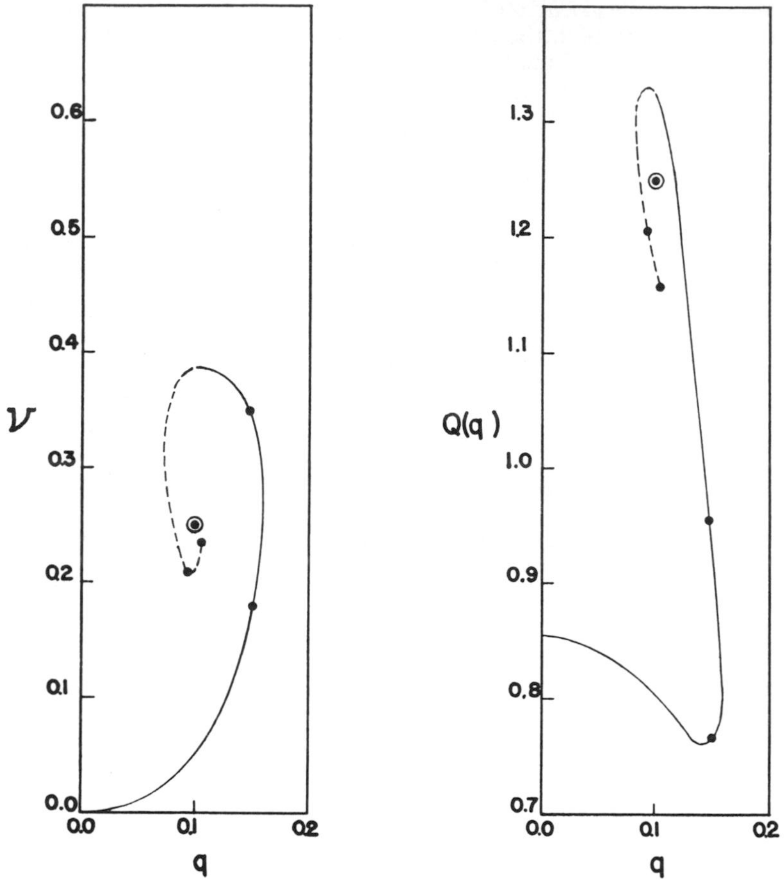

FIG. 1

determined by equations (3), (4), and (5) in terms of an appropriate E-solution of the isothermal equation.

According to equations (3), (4), and (5),

$$u_\infty = \frac{\xi e^{-\psi}}{\psi'} = 4\pi \frac{r^3\rho(r)}{M(r)}, \tag{30}$$

and

$$v_\infty = \xi\psi' = \frac{G}{\beta(r)} \frac{\rho M(r)}{rP(r)} = \frac{\mu H}{k} \frac{GM(r)}{rT(r)}. \tag{31}$$

Hence, an isothermal core can be fitted to a solution of equations (27) and (28) whenever the curve

$$\left[4\pi\,\frac{r^3\rho(r)}{M(r)}\,; \qquad \frac{\mu H}{k}\,\frac{GM(r)}{rT(r)}\right] \tag{32}$$

derived from such a solution intersects the (u_∞, v_∞)-curve associated with an E-solution of the isothermal equation. At such an intersection

$$\left.\begin{aligned} u_\infty(\xi_i) &= 4\pi\left(\frac{r^3\rho}{M(r)}\right)_{r=r_i}, \\ v_\infty(\xi_i) &= \frac{\mu H}{k}\,G\left(\frac{M(r)}{rT}\right)_{r=r_i}, \end{aligned}\right\} \tag{33}$$

and the values of r/R and $M(r)/M$ at this point will determine, at once, the ratios q and ν of the radius and the mass of the core to the radius and the mass of the star, respectively. Further, according to the second of equations (33),

$$T_c = Q(q)\,\frac{\mu H}{k}\,\frac{GM}{R}\,; \qquad Q = \frac{\nu}{v_\infty(\xi_i)q}\,. \tag{34}$$

The ratio of the mean to the central density is also readily found. We have (cf. eq. [17])

$$\frac{\rho_c}{\bar{\rho}} = \frac{\lambda_2}{\dfrac{M}{\frac{4}{3}\pi R^3}}\,. \tag{35}$$

On the other hand, according to equations (4) and (5), we have identically

$$\lambda_2 = \frac{M(r_i)}{4\pi r_i^3}\,\frac{\xi_i}{\psi_i'}\,. \tag{36}$$

Hence, combining equations (35) and (36),

$$\frac{\rho_c}{\bar{\rho}} = \frac{\nu\xi_i}{3q^3\psi_i'} = \frac{\nu u_\infty(\xi_i)}{3q^3}\,e^{\psi_i}\,. \tag{37}$$

Now, there exist five integrations of equations (27) and (28) which can be used for our present purposes. Three of these integrations (due to Miss I. Nielsen[12]) are for the solar values of L, M, and R, with $\mu = 1$ and for values of $\log \kappa_0 = 24.792$, 24.892, and 24.992. Further, in these integrations of Miss Nielsen the radiation pressure as a factor in the equation of hydrostatic equilibrium has been ignored. The two other integrations (which were found to give solutions for the equations of fit) are due to Strömgren.[13] These integrations also refer to the solar values of L, M, and R, but with $\mu = 2.2$ and $\log \kappa_0 = 27.4$ and 27.8. Further, in these integrations the effect of the term $aT^4/3$ in equation (27) has also been taken into account.

[12] Under the supervision of B. Strömgren.

[13] *Zs. f. Ap.*, **2**, 345, 1931.

The results of fitting isothermal cores to the five integrations of equations (27) and (28) referred to, are summarized in Table 2.

It is known that the point-source model with negligible radiation pressure and with a law of opacity of the form $\kappa = \kappa_0 \rho^n T^m$ is a homology-invariant configuration.[14] It follows, therefore, that among the models with negligible radiation pressure, which consist of isothermal cores and point-source envelopes, those with a constant q form a homologous family. Hence, the physical relations derived from the three integrations I of Table 2 are invariant to homologous transformations. In particular the relations

$$R(q) = Q(q) \frac{\mu H}{k} \frac{GM}{T_c} ; \qquad \nu(q) = \frac{M(\text{core})}{M} , \tag{38}$$

will be valid for all stars. In practice, however, the foregoing relations will give sufficient accuracy only for stars of mass less than, say, $5\odot$. This is confirmed, for example,

TABLE 2

STELLAR MODELS WITH ISOTHERMAL CORES AND POINT-SOURCE ENVELOPES

	Integration of Equations (27) and (28)	q	ν	$Q(q)$	ρ_c/ρ	$\frac{L_0(q)}{[Q(q)]^{1/2}} \times 10^{-24}$	$\Delta r_i/R$ in an Arbitrary Scale	$1-\beta_i$	Remarks
	$\log \kappa_0 = 24.735$; $\mu = 1$	0	0	0.900	37.0	5.73	....	....	Cowling model
I...	$\log \kappa_0 = 24.792$; $\mu = 1$	0.139	0.103	0.791	54	6.97	1.00	0.0043	Inger Nielsen's integrations for the point-source envelope. Radiation pressure neglected in equation (27)
	$\log \kappa_0 = 24.892$; $\mu = 1$	.158	.176	0.778	80	8.85	0.98	.0040	
	$\log \kappa_0 = 24.992$; $\mu = 1$	.166	.231	0.788	115	11.1	1.11	.0042	
II...	$\log \kappa_0 = 27.4$; $\mu = 2.2$	.158	.239	0.776	160	7.71	0.77	.078	Stromgren's integrations for the point-source envelope. Radiation pressure accurately taken into account. The results of fitting valid for configurations having a mass $4.84\,\mu^{-2}\,\odot$
	$\log \kappa_0 = 27.8$; $\mu = 2.2$	.119	.319	1.07	175	16.5	1.51	.103	
	$\log \kappa_0 = 27.8$; $\mu = 2.2$	0.081	0.224	1.33	635	14.8	1.70	0.115	

by the results of the last three rows of Table 2: These have been derived for the solar mass with $\mu = 2.2$, taking full account of the radiation pressure in the equation of hydrostatic equilibrium. It is, however, clear that we shall obtain the same results for a star of mass $M = (2.2)^2\odot = 4.84\odot$ and $\mu = 1$. From the column "$1 - \beta_i$" in Table 2 we notice that the radiation pressure, while it is appreciable in these models II, is still not of primary importance. This is reflected, for instance, in the fact that $Q(q)$ and $\nu(q)$ for these models fall roughly on the same curve as those for the three other cases in which the radiation pressure in equation (27) has been treated as negligible (see Fig. 2 where the models I are indicated by dots and models II by crosses).

5. *The physical characteristics of stellar models with isothermal cores.*—An examination of the results of §§ 3 and 4 (particularly Tables 1 and 2 and Figs. 1 and 2) brings out the following essential features of these models:

a) At a fixed central temperature, the radius R of the star first decreases as q increases from $q = 0$. For a value of $q \sim 0.15$–0.16 the radius passes through a minimum. Further, there exists also a maximum possible value for q ($q_{\max} \sim 0.16$–0.17). As q decreases after passing through $q_{\max}$, R increases very rapidly, reaches a maximum, and begins spiraling about a determinate point.

b) Again, at a fixed central temperature, the fraction of the total mass, ν, contained in the core increases slowly at first and soon very rapidly as q approaches $q_{\max}$. How-

[14] See Chandrasekhar, *op. cit.*, pp. 234–39.

ever, this increase of ν does not continue indefinitely; ν soon attains a maximum value $\nu_{\max}$. There exists, therefore, an upper limit to the mass which can be contained in the isothermal core. For the models with point-source envelopes and inappreciable radiation pressure, $\nu_{\max} \sim 0.32$ and occurs for $q \sim 0.12$. The curve $\nu(q)$ also shows the spiraling characteristic.[15]

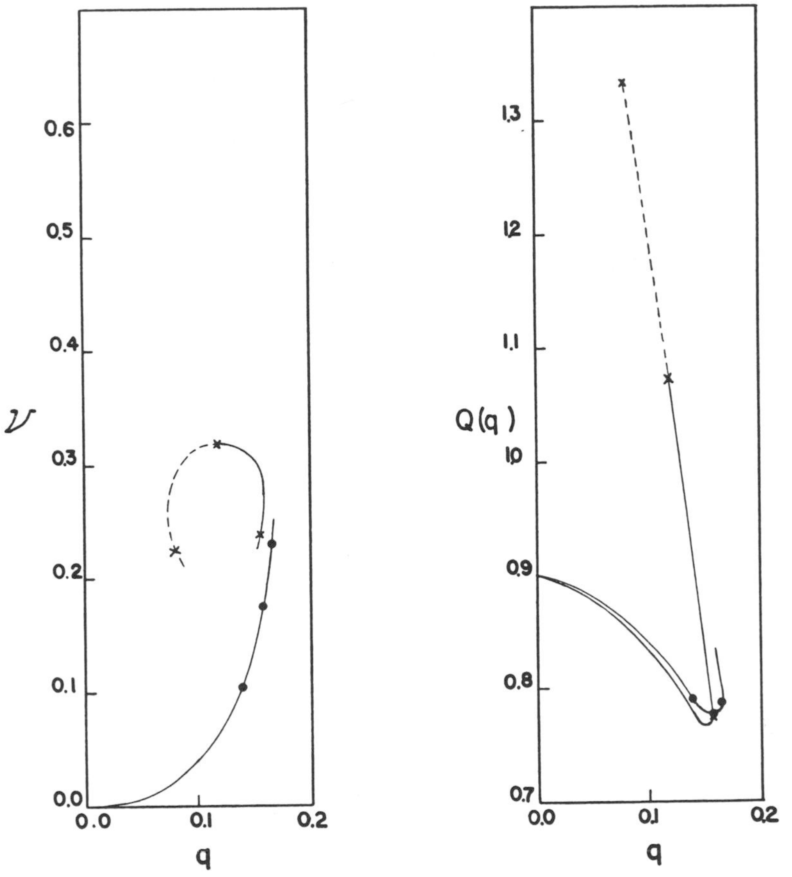

FIG. 2

A result of some importance in the present connection is the variation of the luminosity as q varies at constant central temperature. We shall discuss this variation of the luminosity on the basis of the models considered in § 4.

We may first recall that for a homologous family of stellar configurations derived on the basis of the law of opacity (29) there exists a luminosity formula of the form

$$L = \text{constant}\, \frac{M^{5.5}}{\kappa_0 R^{0.5}}\, \mu^{7.5}\,, \tag{39}$$

[15] This phenomenon of spiraling need not cause any particular surprise. It arises essentially from the oscillatory behavior of the solutions of the isothermal equation as $\xi \to \infty$ (see *ibid.*, pp. 163–66). Situations similar to these described in the text have also been encountered in other connections (see, e.g., S. Chandrasekhar, *Ap. J.*, **87**, 535, 1938; and *M.N.*, **99**, 673, 1939).

where the constant is a characteristic of the family. Consequently, for stellar models with isothermal cores and negligible radiation pressure, we must have a formula of the form

$$L = L_0(q)\,\frac{1}{\kappa_0}\,\frac{M^{5.5}}{R^{0.5}}\,\mu^{7.5}\,, \tag{40}$$

where, as the notation implies, L_0 depends on q only. Remembering that in obtaining the models I of Table 2 we have used integrations of the point-source envelope computed for the solar values of L, M, and R with $\mu = 1$, it follows that

$$L_0(q) = (\kappa_0)_{\text{integration}}\,, \tag{41}$$

if we suppose that in equation (40) L, M, and R are expressed in solar units. Similarly, for the solutions II (see Table 2)

$$L_0(q) = \frac{(\kappa_0)_{\text{integration}}}{(2.2)^{7.5}}\,. \tag{42}$$

Consider now the variation in L as q varies at constant T_c, μ, M, and κ_0. During such a change, R will alter according to (cf. eq. [38])

$$R(q) = Q(q)\,\frac{\mu H}{k}\,\frac{GM}{T_c}\,. \tag{43}$$

Eliminating R between equations (40) and (43),

$$L = \frac{L_0(q)}{[Q(q)]^{1/2}}\left(\frac{k}{GH}\right)^{1/2}\frac{1}{\kappa_0}\,M^5 T_c^{0.5}\mu^7\,. \tag{44}$$

Hence, the variation in the luminosity is governed by the factor

$$\frac{L_0(q)}{[Q(q)]^{1/2}}\,. \tag{45}$$

If we now consider the more general case in which the radiation pressure in equation (27) is taken into account, it is clear that we can still construct a homologous sequence of configurations. But a homologous family is now determined by two parameters: q and $M\mu^2/\odot$. However, as long as we are interested only in the changes in the luminosity occurring in a star of given M and μ, we can always write down a relation of the form (44). Moreover, any such relation will be valid for a sequence of configurations of constant $M\mu^2$.

The factor (45) governing the variation of L for constant M, T_c, κ_0, and μ is tabulated in Table 2. According to the values given in this table, the luminosity increases by a factor of about 3 from the stage where there is no isothermal core to the stage where the core contains the maximum possible mass.

The variation in the luminosity predicted by equation (44) implies a corresponding variation in the thickness of the energy-generating shell, for, according to equations (26) and (44),

$$4\pi r_i^2 \rho_i \Delta r_i \epsilon_0 = \frac{L_0(q)}{[Q(q)]^{1/2}} \frac{1}{\kappa_0} \left(\frac{k}{HG}\right)^{1/2} M^5 T_c^{0.5} \mu^7 . \tag{46}$$

The foregoing equation can be simplified by using equation (30). We find

$$\frac{\Delta r_i \epsilon_0}{r_i} = \frac{L_0(q)}{[Q(q)]^{1/2} u_\infty(\xi_i)\nu} \frac{1}{\kappa_0} \left(\frac{k}{HG}\right)^{1/2} M^4 T_c^{0.5} \mu^7 , \tag{47}$$

or

$$\frac{\Delta r_i}{R} \propto \frac{qL_0(q)}{[Q(q)]^{1/2} u_\infty(\xi_i)\nu} . \tag{48}$$

The quantity on the right-hand side (apart from a constant factor) is tabulated in Table 2. We notice that the variation in the thickness of the shell is not very marked.

6. *General remarks.*—We shall now consider briefly the bearing of the results summarized in § 5 on the physical problems outlined in § 1 and in particular the implication for the Gamow-Teller theory of the energy production in giants. Suppose that to begin with a star has a central temperature $T_c(\sim 10^{6°})$ at which the disintegration of lithium can provide for an adequate source of energy. Under these circumstances the star will approximate to the Cowling model which has a convective core occupying 17 per cent of the radius and containing 15 per cent of the mass of the star. Suppose now that the lithium in the central regions is exhausted and that the process of the diffusion of elements does not take place rapidly enough for the restoration of adequate amounts of lithium to the center. We shall then have a shell-source model. In the early stages ($\nu < 0.15$) the star will consist of an isothermal core, a convective fringe, and a point-source radiative envelope. However, very soon (i.e., when $\nu > 0.15$) the star will consist only of an isothermal core and a radiative envelope. It is now clear that energy production from the disintegration of lithium can continue only as long as the mass in the isothermal core increases. But we have seen that ν cannot increase beyond a certain maximum value $\nu_{max}(\sim 35$ per cent). When this happens the liberation of energy from the process considered will cease. The star must then readjust itself to a contractive model ($\epsilon \propto T$) and evolve according to the Helmholtz-Kelvin time scale. This will continue till the central temperature increases sufficiently for the liberation of nuclear energy from the disintegration of the next element, beryllium, to become effective. The whole cycle of changes will now be repeated.

In considering the course of changes we have described in the foregoing paragraph, it is of interest to trace the track of evolution in the Hertzsprung-Russell diagram. To illustrate this we have plotted

$$\log \frac{L_0(q)}{[Q(q)]^{1/2}} \tag{49}$$

against

$$\log \frac{L_0(q)}{[Q(q)]^{1/2}} - 2 \log Q(q) \tag{50}$$

in Figure 3. According to our earlier remarks, an evolution of the kind we are considering must cease when the luminosity has reached about its maximum value (cf. Table 2). We may note at this point that at no stage during such an evolution does the isothermal core occupy a large fraction of the radius; indeed, it is always less than about 17 per cent.

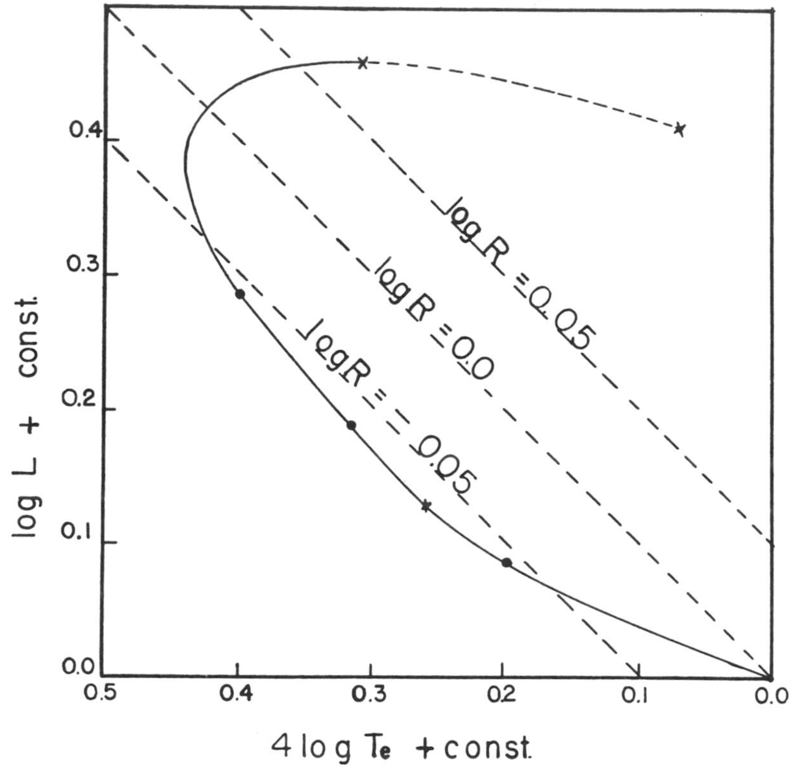

Fig. 3

Consequently, we cannot expect any significant changes in the stability of a star along such a course.

Finally, we may remark that with suitable modifications we may similarly follow the eventual course of evolution of a star as the hydrogen in the central regions becomes exhausted.

Yerkes Observatory
July 30, 1941

ON THE EVOLUTION OF THE MAIN-SEQUENCE STARS

M. SCHÖNBERG[1] AND S. CHANDRASEKHAR

ABSTRACT

The evolution of the stars on the main sequence consequent to the gradual burning of the hydrogen in the central regions is examined. It is shown that, as a result of the decrease in the hydrogen content in these regions, the convective core (normally present in a star) eventually gives place to an isothermal core. It is further shown that there is an upper limit ($\sim$ 10 per cent) to the fraction of the total mass of hydrogen which can thus be exhausted. Some further remarks on what is to be expected beyond this point are also made.

1. *General considerations.*—The problem of stellar evolution is intimately connected with that of energy production in the stars. Both Bethe and Weizsäcker[2] showed that the source of the energy radiated by the main-sequence stars is the transformation of hydrogen into helium through the so-called "carbon cycle." On the basis of the Bethe-Weizsäcker theory, G. Gamow[3] outlined a picture of stellar evolution.

The Gamow theory is based on three fundamental assumptions: (*a*) the stars evolve gradually through a sequence of equilibrium configurations; (*b*) the successive equilibrium configurations are homologous; and (*c*) the nuclear reaction continues to take place until the entire hydrogen in the star is exhausted.

Such a picture of stellar evolution presents certain difficulties. The assumption that the successive equilibrium configurations are homologous cannot be expected to be rigorously valid; for the nuclear reaction reduces the hydrogen content in the neighborhood of the center of the star, and therefore the molecular weight in this region becomes increasingly larger than that of the rest of the stellar material, unless we suppose that a diffusion process rapidly mixes the whole of the stellar mass. The only region in which the mixing can be supposed to take place is the Cowling convective core; but stellar configurations in which the ratios of the molecular weights of the convective core and envelope are different are not homologous, contrary to assumption (*b*).

Another difficulty of the Gamow theory is the uncertainty of the amount of hydrogen that can be burned. It is important to determine this quantity accurately, since its value affects essentially the final luminosity and effective temperature given by the homology formulae. Again it does not appear probable that the entire hydrogen content could be

[1] Fellow of the J. S. Guggenheim Foundation, at the Yerkes Observatory.

[2] H. Bethe, *Phys. Rev.*, **55**, 434, 1939; and C. F. von Weizsäcker, *Phys. Zs.*, **39**, 633, 1938.

[3] *Phys. Rev.*, **53**, 59, 1938; **55**, 718, 1939; *Nature*, **144**, 575, 1939.

exhausted, since it would imply a thorough mixing of the stellar material, in order that the entire content could reach the center, where the nuclear reactions principally take place. Sometimes it is assumed that only 14.5 per cent of the entire hydrogen content (which is the fraction of the total mass contained inside the Cowling convective core) participate in the carbon cycle. This hypothesis, though apparently more plausible, should be further amended, for it would be valid only if the turbulence in the convective core mixed the material rapidly enough to avoid the formation of an isothermal region at the center which would tend to stop convection. However, even if there is no formation of an isothermal core, the fraction of the stellar mass contained in the convective region would be expected to diminish as the molecular weight increases relatively to that of the rest of the stellar material. We shall obtain the precise amount of this diminution; but it is clear that only a small fraction of the stellar hydrogen could be burned if only the hydrogen in the convective core was available for the nuclear reaction. We should not conclude, however, that only that small amount of hydrogen could be burned, since there is the possibility of the formation of an isothermal core after the exhaustion of hydrogen has stopped the convection in the central regions. Such a possibility results, as we shall show, from the existence of equilibrium configurations formed by isothermal cores surrounded by point-source envelopes, the mass of the isothermal cores being larger than that of the limiting convective core of vanishing hydrogen content.

In the isothermal core-radiative envelope models the nuclear reaction takes place at the interface of core and envelope. The fraction of the mass contained in the isothermal core cannot exceed a fixed value, so that the nuclear reaction will finally cease when the mass in the core reaches its maximum value. A possibility which should not be overlooked is that, during the transition to gravitational energy production, larger cores could be formed that would not be equilibrium configurations; however, the lifetime of such configurations is presumably small, so that in a first approximation we can neglect such possibilities.

2. *Stellar models formed by convective cores and radiative envelopes with different molecular weights.*—The construction of these models can be done either by the method used by T. G. Cowling[4] or by that proposed by Chandrasekhar.[5]

At the interface the values of the temperature, pressure, and mass of the core should be identical:

$$P(r_i)_{\text{core}}=P(r_i)_e\,;\qquad T(r_i)_{\text{core}}=T(r_i)_e\,;\qquad M(r_i)_{\text{core}}=M(r_i)_e\,,\tag{1}$$

where P, M, and T denote the total pressure, the mass within the radius r, and the temperature, respectively. The index i indicates that the values refer to the interface and the index e that the quantities correspond to the envelope solution of the equilibrium equations.

Conditions (1) are not sufficient; it is further necessary that the effective polytropic index of the envelope be 1.5 at the interface

$$\left(\frac{d\ln P}{d\ln T}\right)_e=\left(\frac{d\ln P}{d\ln T}\right)_{\text{core}}\tag{2}$$

It is convenient to introduce the homology invariant quantities U and V for the core as follows:

$$U=\frac{4\pi\rho r^3}{M(r)}\,;\qquad V=\frac{2}{5}\,\frac{GM(r)\,\rho}{rP}\,.\tag{3}$$

Conditions (1) can now be written in terms of U_i and V_i as

$$U_i=\frac{4\pi\rho_e(r_i)\,r_i^3}{M_e(r_i)}\,\frac{\mu_c}{\mu_e}\,;\qquad V_i=\frac{2}{5}\,\frac{GM_e(r_i)\,\rho_e(r_i)}{r_iP_e(r_i)}\,\frac{\mu_c}{\mu_e}\,,\tag{4}$$

[4] *M.N.*, **91**, 92, 1931. [5] *An Introduction to the Study of Stellar Structure*, p. 352, Chicago, 1939.

where μ_c and μ_e denote the molecular weights of the core and envelope, respectively. The appearance of the ratio of the molecular weights in formulae (4) is due to the discontinuity of the density at the interface. If radiation pressure is negligible, the ratio of the values of the density on both sides of the interface is simply the ratio of the corresponding molecular weights. Indeed, in that case the total pressure may be identified with the gas pressure; and so

$$P = \frac{k}{\mu H} \rho T . \tag{5}$$

Since both P and T are continuous at the interface, equation (5) implies the continuity of ρ/μ.

For given values of the total mass M, luminosity L, radius R, and molecular weights μ_c and μ_e it is generally possible to find ∞^1 solutions satisfying conditions (4), each of these solutions corresponding to a value of the opacity constant κ_0 that appears in Kramer's law,

$$\kappa = \kappa_0 \rho T^{-3.5} . \tag{6}$$

Condition (2) eliminates the arbitrariness of κ_0.

In order to determine the effect of the increase of the ratio μ_c/μ_e on the size of the convective core, we have applied the method described to a star with the solar values $M\odot$, $L\odot$, $R\odot$, $\mu_e = 1$, and $\mu_c/\mu_e = 2$, neglecting radiation pressure. We found that such a convective core could be approximately fitted to an envelope described by the Inger Nielsen solution with $\log \kappa_0 = 24.992$. The characteristics of the Cowling model and those of the limiting convective core with $\mu_c/\mu_e = 2$ are given in Table 1. In this table we have tabulated the quantities

TABLE 1

$\frac{\mu_c}{\mu_e}$	ξ	η	S	$\frac{\rho_c}{\bar{\rho}}$	$\frac{L_0}{[\frac{2}{5}S]^{1/2}} \times 10^{-25}$	$(1-\beta_c)\left(\frac{M\odot}{M}\right)^2 \mu_e^{-4}$	$-\omega$
1.	0.169	0.145	2.250	37	0.573	0.0053	0.554
2.	0.083	0.076	1.853	179	1.140	0.0190	0.803

$$\xi = \frac{r_i}{R} ; \quad \eta = \frac{M_i}{M} ; \quad S = \frac{T_c}{T_i} \frac{\eta}{\xi V_i} ; \quad \omega = -\left[\frac{6}{7} \frac{\eta^2}{\xi V_i}\left(\tfrac{4}{5} U_i + V_i - 1\right) + \tfrac{6}{5} J\right] , \tag{7}$$

where T_c is the central temperature and J an integral that has to be evaluated numerically, as will be explained later. All the tabulated quantities are homology invariants, and it is possible to express in terms of them the radius R, the luminosity L, and the gravitational energy Ω of the star as follows:

$$R = \frac{2}{5} \frac{\mu_c H}{k} \frac{GM}{T_c} S\left(\frac{\mu_c}{\mu_e}\right) \tag{8}$$

$$L = \frac{L_0\left(\frac{\mu_c}{\mu_e}\right)}{\left[\frac{2}{5} S\left(\frac{\mu_c}{\mu_e}\right)\right]^{1/2}} \left(\frac{k}{GH}\right)^{1/2} \frac{1}{\kappa_0} M^5 T_c^{0.5} \mu_e^{7.5} \mu_c^{-0.5} , \tag{9}$$

$$\Omega = \omega \frac{GM^2}{R} , \tag{10}$$

where L_0 is a numerical constant depending only on the ratio of the molecular weights. Before proceeding further we shall indicate how formulae (8) ,(9) and (10) arise.

3. According to the second of equations (4), we have

$$r_i = \frac{2}{5}\frac{GM\eta\,\rho_{\text{core}}(r_i)}{V_i P_i},$$

and so

$$R = \tfrac{2}{5}GM\frac{\eta}{\xi V_i}\frac{\rho_{\text{core}}(r_i)}{P_i};$$

but in the core we have

$$\frac{\rho}{P} = \frac{\mu_c H}{kT}.$$

Hence,

$$R = \frac{2}{5}\frac{\mu_c H}{k}\frac{GM}{T_c}\frac{\eta}{\xi V_i}\frac{T_c}{T_i} = \frac{2}{5}\frac{\mu_c H}{k}\frac{GM}{T_c}S\left(\frac{\mu_c}{\mu_e}\right),$$

which is our formula (8).

For any homologous family of configurations, derived on the basis of Kramer's law of opacity, the luminosity is given by the formula

$$L = L_0 \cdot \frac{M^{5.5}}{\kappa_0 R^{0.5}}\mu_e^{7.5}, \tag{11}$$

L_0 being a characteristic constant of the family. In the present case L_0 can depend only on μ_c/μ_e. Applying equation (11) to the star of the family with the solar values $M\odot$, $L\odot$, $R\odot$, and $\mu_c = 1$ and using solar units, we get for L_0 the formula

$$L_0 = (\kappa_0)_{\text{int}}. \tag{9a}$$

Introducing into equation (11) expression (8) for R, we obtain formula (9).

4. We shall now evaluate the potential energy Ω of the composite models. Quite generally

$$\Omega = -\int_0^R \frac{GM(r)\,dM(r)}{r}, \tag{12}$$

or

$$\Omega = \frac{1}{2}\int_0^R W\,dM(r). \tag{13}$$

W being the gravitational potential.

Now, for a composite model, Ω is the sum of the potential energies of the core and envelope, i.e.,

$$\Omega = \Omega_{\text{core}} + \Omega_e, \tag{14}$$

$$\Omega_{\text{core}} = \frac{1}{2}\int_0^{r_i} W\,dM(r), \qquad \Omega_e = \frac{1}{2}\int_{r_i}^R W\,dM(r).$$

The core is a gaseous sphere of polytropic index 1.5; hence[6]

$$\frac{5}{2}\left(\frac{P}{\rho} - \frac{P_i}{\rho_i}\right) = W_i - W = -\frac{GM_i}{r_i} - W; \tag{15}$$

and therefore

$$\Omega_{\text{core}} = -\frac{5}{4}\int_{\text{core}} P\,d\tau + \frac{5}{4}\frac{P_i M_i}{\rho_i} - \frac{1}{2}\frac{GM_i^2}{r_i},$$

[6] Chandrasekhar, *op. cit.*, p. 100.

where $d\tau$ is an element of volume. Using a formula due to E. A. Milne,[7] we can express Ω_c as

$$\Omega_c = -3\int_{\text{core}} P d\tau + 4\pi P_i r_i^3 . \tag{16}$$

Introducing this value of $\int P d\tau$ in the expression for Ω_c, we get

$$\Omega_{\text{core}} = \tfrac{5}{12}\Omega_{\text{core}} - \frac{5\pi}{3} P_i r_i^3 + \frac{5}{4}\frac{P_i M_i}{\rho_i} - \frac{1}{2}\frac{GM_i^2}{r_i}$$

or

$$\Omega_{\text{core}} = -\frac{6}{7}\left[\frac{GM_i^2}{r_i} + \tfrac{5}{2}P_i\left(\frac{4\pi}{3} r_i^3 - \frac{M_i}{\rho_i}\right)\right] . \tag{17}$$

But

$$\frac{GM_i^2}{r_i} = \frac{GM^2}{R}\frac{\eta^2}{\xi} . \tag{17a}$$

On the other hand, according to the conditions (4) we have

$$P_i r_i^3 = \frac{2}{5}\frac{GM\eta P_{\text{core}}(r_i)}{V_i} r_i^2 = \frac{2}{5}\frac{GM^2}{4\pi R}\frac{U_i\eta^2}{\xi V_i} \tag{17b}$$

and

$$\frac{P_i}{\rho_i} = \frac{2}{5}\frac{GM}{R}\frac{\eta}{\xi V_i} . \tag{17c}$$

Taking into account these last three formulae, we get for Ω_{core}

$$\Omega_{\text{core}} = -\frac{6}{7}\frac{\eta^2}{\xi}\left[1 + \frac{1}{3}\frac{U_i}{V_i} - \frac{1}{V_i}\right]\frac{GM^2}{R} . \tag{18}$$

Turning next to the part Ω_e of Ω, we start with the formula (cf. eq. [16])

$$\Omega_e = -3\int_{\text{env}} P d\tau - 4\pi P_i r_i^3 . \tag{16a}$$

The integral in equation (16*a*) can be put in the form

$$\int_{\text{env}} P d\tau = \tfrac{2}{5} J \frac{GM^2}{R} , \tag{19}$$

where J is a homology invariant quantity that has to be evaluated numerically. Taking into account equations (17*b* and (19) we get for Ω_e

$$\Omega_e = -\left[\tfrac{6}{5}J + \frac{2}{5}\frac{U_i\eta^2}{\xi V_i}\right]\frac{GM^2}{R} . \tag{20}$$

Thus, combining equations (18) and (20), we finally obtain

$$\Omega = \Omega_e = \Omega_{\text{core}} = -\left[\frac{6}{7}\frac{\eta^2}{\xi V_i}(\tfrac{4}{5}U_i + V_i - 1) + \tfrac{6}{5}J\right]\frac{GM^2}{R} , \tag{21}$$

which is the last of the formulae (4). We can express equation (21) more simply as

$$\Omega = \omega\frac{GM^2}{R} . \tag{10}$$

It is this quantity ω which is tabulated in Table 1.

[7] *M.N.*, **89**, 739, 1929; **96**, 179, 1936.

It is now easy to derive the formulae for the internal energy H and the total energy E as follows:

$$H=\int_0^R c_v T dM(r)=\left[(c_v)_{\text{core}}\frac{\mu_c H}{k}\int_0^{r_i}\frac{P}{\rho}dM+(c_v)_e\frac{\mu_e H}{k}\int_{r_i}^{R}\frac{P}{\rho}dM\right],$$

or

$$H=\left[\left(\frac{c_v}{c_p-c_v}\right)_{\text{core}}\int_{\text{core}}P d\tau+\left(\frac{c_v}{c_p-c_v}\right)_e\int_{\text{env}}P d\tau\right]$$

$$=\left[\frac{1}{\gamma_{\text{core}}-1}\int_{\text{core}}P d\tau+\frac{1}{\gamma_e-1}\int_{\text{env}}P d\tau\right].$$

Taking into account equations (16) and (16*a*), we get

$$H=-\frac{1}{3}\left[\frac{\Omega_{\text{core}}}{\gamma_{\text{core}}-1}+\frac{\Omega_e}{\gamma_e-1}-4\pi P_i r_i^3\left(\frac{1}{\gamma_{\text{core}}-1}-\frac{1}{\gamma_e-1}\right)\right].$$

But $\gamma_{\text{core}}=\gamma_{\text{env}}=\gamma$; and so

$$H=-\frac{\Omega}{3(\gamma-1)}. \tag{22}$$

For the total energy E we have

$$E=\Omega+H=\Omega-\frac{\Omega}{3(\gamma-1)}=\frac{3\gamma-4}{3(\gamma-1)}\Omega. \tag{23}$$

Formulae (22) and (23) hold for any model in which γ is constant through the whole stellar mass. The preceding argumentation is, however, necessary to prove the validity of the Ritter-Perry formula (22) for composite configurations with different values of μ in the different regions.

5. *Stellar models with isothermal cores and radiative envelopes with different molecular weights.*—The method of constructing stellar models formed by an isothermal core surrounded by a radiative envelope was discussed by L. R. Henrich and S. Chandrasekhar[8] for the case in which the values of μ are the same in both regions. We will now examine the more general case of different molecular weights.

It is necessary to fit an E solution corresponding to an isothermal core with molecular weight μ_c to a radiative envelope with molecular weight μ_e. At the interface the values of the pressure, temperature, and mass of the core given by both solutions should be identical:

$$P(r_i)_{\text{core}}=P(r_i)_e;\quad T(r_i)_{\text{core}}=T(r_i)_e;\quad M(r_i)_{\text{core}}=M(r_i)_e. \tag{24}$$

The density has a discontinuity at the interface. Neglecting radiation pressure, we get, as in section 2, for the ratio of the densities on both sides of the interface the value of the ratio of the respective molecular weights. The conditions (24) are the only ones to be fulfilled, and so we get a family of ∞^1 configurations for any given set of values of M, L, and R.

It is convenient to introduce the homology invariant functions u and v to describe the isothermal core,

$$u=4\pi\frac{r^3\rho_{\text{core}}}{M r)},\qquad v=\frac{\mu_c H}{k}\frac{GM(r)}{rT(r)}. \tag{25}$$

The equations of fit in the new variables are

$$u_i=4\pi\left[\frac{r^3\rho(r_i)_e}{M(r_i)_e}\right]\frac{\mu_c}{\mu_e},\qquad v_i=\left[\frac{\mu_e H}{k}\frac{GM(r_i)_e}{r_i T(r_i)_e}\right]\frac{\mu_c}{\mu_e}. \tag{26}$$

[8] *Ap. J.*, **94**, 525, 1941.

The quantities q and ν,

$$q = \frac{r_i}{R} \quad \text{and} \quad \nu = \frac{M(r_i)}{M}, \tag{27}$$

are homology invariants, and each of them can be used to label the different configurations corresponding to the same stellar mass and the same central temperature.

From the second of equations (26) we readily obtain the following formulae for the radius R:

$$R(q) = Q(q)\frac{\mu_c H}{k}\frac{GM}{T_c}, \tag{28}$$

where

$$Q(q) = \frac{\nu}{v_i q}. \tag{29}$$

For the present case the luminosity formula takes the form

$$L = L_0(q, \mu_c/\mu_e)\frac{1}{\kappa_0}\frac{M^{5.5}}{R^{0.5}}\mu_e^{7.5}. \tag{30}$$

Equation (30) is analogous to equation (9) and can be derived in the same way. The quantity L_0 depends on both q and μ_c/μ_e. Our numerical integrations were done with the solar values $M_\odot$, $L_\odot$, $R_\odot$, and $\mu_e = 2.2$; hence we get for L_0

$$L_0 = \frac{(\kappa_0)_{\text{int}}}{(2.2)^{7.5}}. \tag{31}$$

Introducing the expression (28) of R in equation (30), we find

$$L = \frac{L_0}{\sqrt{Q}}\left(\frac{k}{GH}\right)^{1/2}\frac{1}{\kappa}M^5 T_c^{0.5}\mu_e^{7.5}\mu_c^{-0.5}. \tag{32}$$

The variation of the luminosity at constant central temperature is determined by the factor $L_0/\sqrt{Q}$.

The thickness Δr_i of the energy-producing shell is related to the luminosity by the equation

$$L = 4\pi r_i^2 \rho(r_i)_e \Delta r_i \epsilon_0, \tag{33}$$

where ϵ_0 is the energy production per gram per second.

Introducing in equation (33) the expression for L, we get

$$\frac{\epsilon_0 \Delta r_i}{R} = \frac{qL_0}{Q^{1/2}u_i\nu}\frac{1}{\kappa_0}\left(\frac{k}{GH}\right)^{1/2} M^4 T_c^{0.5}\mu_e^{6.5}\mu_c^{0.5}, \tag{34}$$

or

$$\frac{\Delta r_i}{R} \propto \frac{qL_0}{Q^{1/2}u_i\nu}. \tag{34a}$$

The potential energy Ω is the sum of two terms Ω_{core} and Ω_e. The quantity Ω_c is given by formula (16), since in the derivation of this formula the nature of the core does not matter. Introducing in equation (16) the value of P given by the gas equation and taking into account that the temperature is constant throughout the core, we obtain for Ω_c

$$\Omega_c = -3\frac{kT_c}{\mu_c H}\int_{\text{cor}} \rho\, d\tau + 4\pi P_i r_i^3. \tag{35}$$

Similarly, Ω_e is given by equation (16a), and therefore

$$\Omega = \Omega_e + \Omega_c = -3\,\frac{kT_c}{\mu_c H}\,\nu M - 3\int_{\text{env}} P\,d\tau\,. \tag{36}$$

Taking into account formulae (19) and (28), we can simplify the expression of

$$\Omega = -3\,(Q\nu + \tfrac{2}{5}J)\,\frac{GM^2}{R}\,. \tag{37}$$

The results of the numerical integration are given in Table 2 and illustrated graphically in Figures 1, 2, 3, 4, 5, and 6. The points represented by circles refer to the models with isothermal cores, the squares refer to models with convective cores.

TABLE 2*

q	ν	Q	$-\omega$	$\frac{L_0}{\sqrt{Q}} \times 10^{-25}$	$\frac{\rho_c}{\bar{\rho}} \times 10^{-2}$	$C\,\frac{\Delta r_i}{R}$	$(1-\beta_i)\left(\frac{M_\odot}{M}\right)^2 \mu_e^{-4}$
0.045	0.079	0.747	0.936	1.759	369	1.680	0.0004
.046	.071	.709	.896	1.610	628	1.750	.0005
.049	.093	.761	.988	1.956	124	1.520	.0005
.058	.101	.713	.976	2.020	32.1	1.306	.0005
.063	.094	.661	.918	1.869	14.5	1.176	.0004
.067	.089	.625	.879	1.714	9.21	1.116	.0004
.071	.072	.571	.813	1.425	4.49	1.024	.0004
0.073	0.065	0.555	0.786	1.318	3.46	1.000	0.0004

* C is a constant factor; $(1-\beta)$ is the ratio of radiation pressure to gas pressure; ρ_c is the central density; and $\bar{\rho}$ the mean density.

6. From Table 1 we derive the main features of the models with convective cores. As the ratio of molecular weights μ_c/μ_e increases from 1 to 2;

a) the fraction of the stellar mass in convective equilibrium decreases from 0.145 to 0.076.

b) the fraction of the radius occupied by the convective core decreases from 0.169 to 0.083.

c) the radius of the star increases by a factor $\sim$1.65, assuming that T_c remains constant.

d) the potential energy and the total energy decrease, the heat content, H, increases, and there is, therefore, liberation of gravitational energy when the star is still burning hydrogen.

e) The luminosity increases by a factor $\sim$ 1.41, assuming again the constancy of T_c.

f) The effective temperature decreases by a factor $\sim$ 1.18. The decrease of T_{eff} is due to the large increase of the radius. Actually, the situation is different, for the sharp reduction of hydrogen content in the core raises the central temperature and hence counteracts the tendency of the radius to grow.

For the models with isothermal cores we have the following properties:

a) There are no equilibrium configurations with cores containing less than 0.065 or more than 0.101 of the stellar mass. The lower limit is due to the appearance of convective instability at the interface, while the upper one is due to the impossibility of fitting a core to an envelope. The upper limit is a decreasing function of μ_c/μ_e, since it is $\sim$0.35 for the case of equal molecular weights, as was shown by Henrich and Chandrasekhar.[7]

b) Starting from its minimum value, ν increases rapidly as q grows, reaches its absolute maximum, and starts spiraling around a certain value.

c) The radius of the model starts increasing as ν grows, reaches an absolute maximum 1.7 $R_{\text{Cow.}}$, and afterward starts to spiral.

d) The potential energy of gravitation Ω and the total nonnuclear energy E start decreasing as ν increases, reach an absolute minimum, and then finally start increasing and spiraling. In other words, the gravitational binding increases until it reaches a maximum and afterward spirals. Since the nuclear binding is the same for two models with the same ν, it results that there can be two or more configurations with the same amount of burned hydrogen and with different total bindings.

e) The luminosity starts increasing as ν grows until an absolute maximum $\sim$2.5 $L_{\text{Cow.}}$ is reached and then spirals. The total change in the luminosity from the Cowling model to the model with the maximum isothermal core is only 1 mag., instead of 5 mag. as in the Gamow theory.

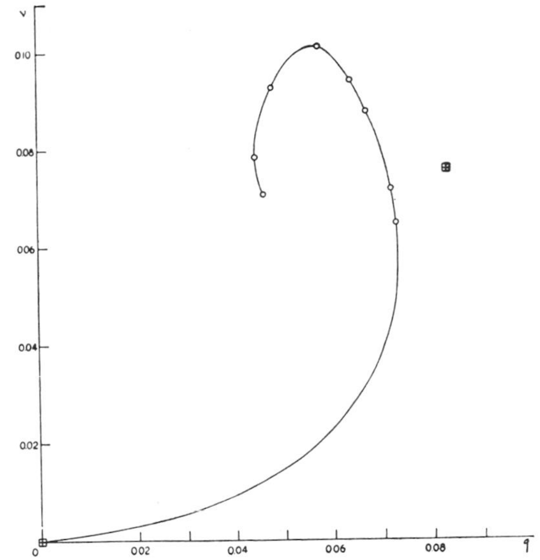

FIG. 1

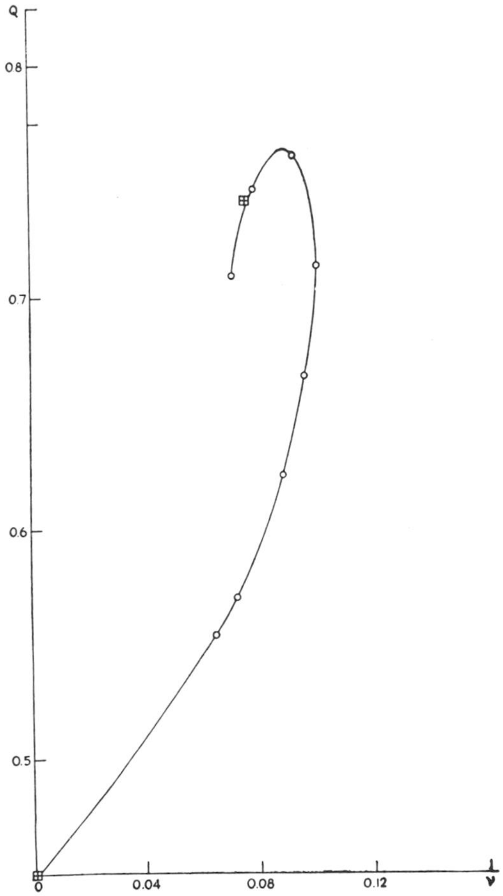

FIG. 2

f) The effective temperature remains practically constant as L grows, then decreases and starts spiraling.

g) The thickness of the energy-generating shell varies very slightly.

h) The central density increases continually along the spiral in Figure 1. The ratio of the central density of the isothermal core with maximum ν to that of the Cowling model is $\sim$22.

7. *Applications to stellar evolution.*—Now we can give a more detailed picture of the evolution of the main-sequence stars. The hydrogen combustion in the center may have either of the following effects: formation of an isothermal core at the center or shrinkage of the convective core. Which of the effects takes place depends on the rapidity of mix-

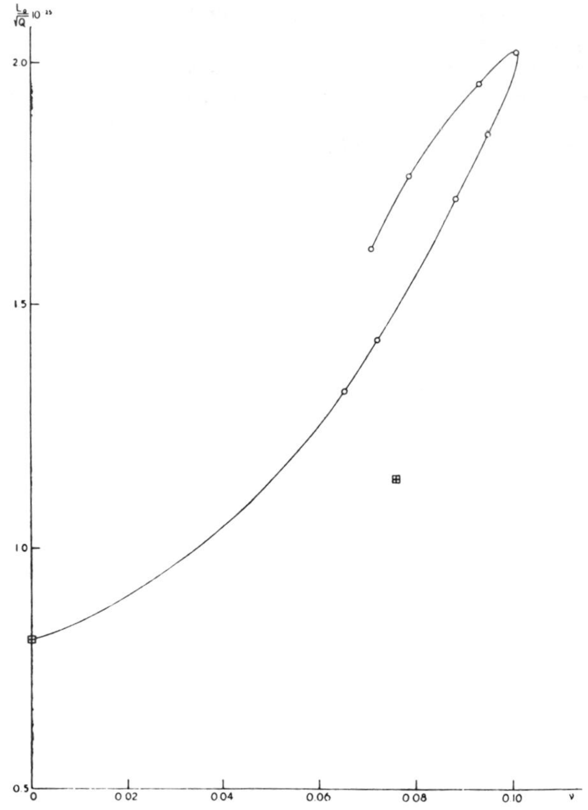
$\frac{L_o}{\sqrt{Q}}$ 10 25
2.0
1.5
1.0
0.5
0
0.02
0.04
0.06
0.08
0.10
v

FIG. 3

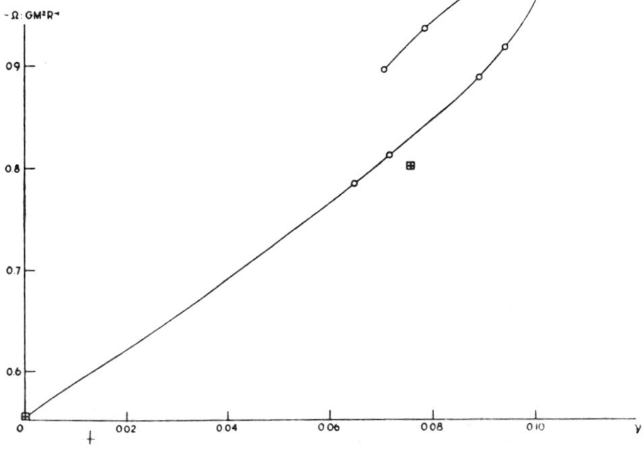
-Ω: GM²R⁻¹
0.9
0.8
0.7
0.6
0
0.02
0.04
0.06
0.08
0.10
y

FIG. 4

ing. The superadiabatic gradient being larger for very bright stars, the mixing will proceed faster and the convective core will shrink; in stars of low luminosity an isothermal core may be formed.

The formation of an isothermal core surrounded by a convective shell and a radiative envelope seems, at first sight, to present difficulties, because at the interface of the convective and isothermal regions the polytropic indices would be different: ∞ and 1.5. On the other hand, since the energy-producing layers would lie at the boundary of the isothermal core, it is reasonable to suppose that there should be convective instability at the outer boundary of those layers and stability at the inner one. It should be remembered, in this connection, that the nonexistence of sources of energy in the isothermal core would make convection impossible, since there would be no driving mechanism. Consequently, we need not require the equality of polytropic indices at the interface of the convective and isothermal regions. In any case it may be expected that, even if there are no equilibrium configurations of the type considered, there may still exist quasi-equilibrium configurations of that type.

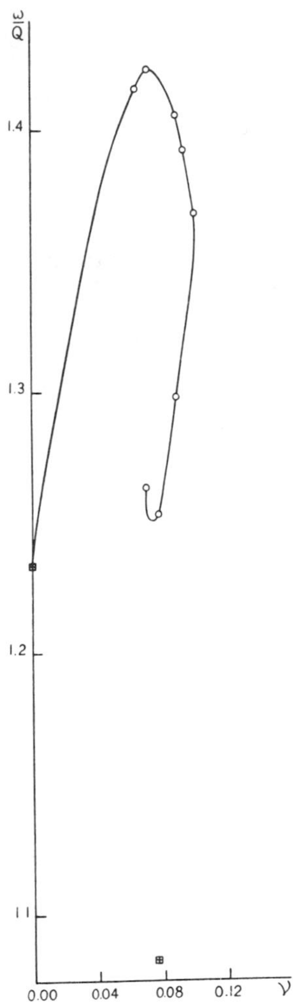

Fig. 5

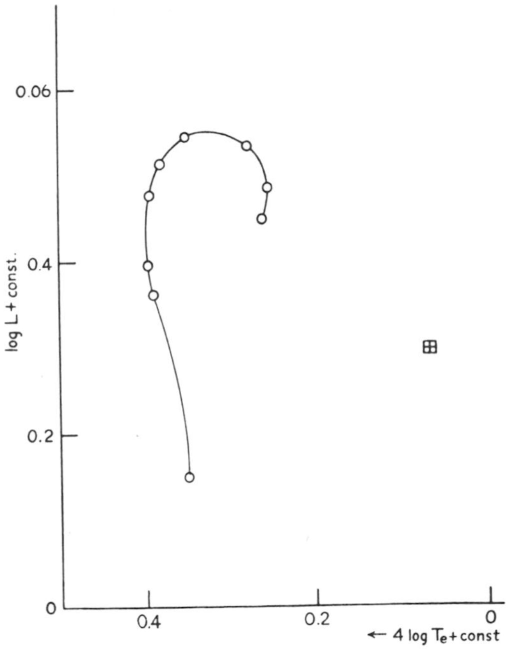

Fig. 6

The growth of an isothermal core at the center would slow down and finally stop convection when the fraction of the stellar mass inside it exceeded 0.065. The stellar model would then go over into the sequence formed by isothermal cores surrounded by radiative envelopes.

In the case where there is rapid mixing the convective core would start shrinking and the luminosity would increase somewhat, while the radius would expand rapidly. The star would not, however, expand so much as the results of Table 1 might suggest, for the scarcity of hydrogen would make the central temperature rise. As the central temperature rises, so will the temperature at the outer points, and the energy production will spread to outer regions. The convective currents must stop before the whole hydrogen is burned, and that will be taken care of by the spreading of the energy production and consequent leveling of the temperature gradient. Finally, the star will readjust itself to an isothermal core–radiative envelope model. The later stages of the evolution would be the same as in the case of slow mixing.

It is to be noted that the liberation of the gravitational energy has already started during the time when the nuclear reaction is going on and not merely after the reaction is over, as it is usually supposed. This liberation of gravitational energy is in turn responsible for the increase of the heat content of the star.

So far we have discussed the evolution of a star only during the relatively early stages of the exhaustion of hydrogen in its central regions. The question now arises as to what can be said concerning the evolution during the later stages, i.e., after the isothermal core has grown to include the maximum possible mass. When this stage has been reached, the liberation of energy from the carbon cycle must cease, and we should expect the star to adjust itself to a contractive model ($\epsilon a T$) and evolve according to the Helmholtz-Kelvin time scale. But this gravitational contraction cannot proceed indefinitely, for with continued contraction the temperature at the interface between the central regions exhausted of hydrogen and the outer envelope containing hydrogen will steadily increase. And, when the temperature exceeds a certain value, the nuclear reactions will start, and the carbon cycle will again become operative. At first, the energy liberated by the nuclear processes will be small compared to the gravitational liberation. But very soon, because of the high-termperature sensitiveness of the nuclear reactions, the energy liberation from the carbon cycle will exceed the liberation of energy by the gravitational contraction. In other words, the central regions must again tend to become isothermal. However, no equilibrium configuration is possible under these circumstances: for the isothermal core would have to contain a greater fraction of the total mass than is possible under equilibrium conditions. It therefore appears difficult to escape the conclusion that beyond this point the star must evolve through nonequilibrium configurations. It is difficult to visualize what form these nonequilibrium transformations will take; but, whatever their precise nature, they must depend critically on whether the mass of the star is greater or less than the upper limit M_3 ($=5.7\ \mu^{-2}\ \odot$) to the mass of degenerate configurations.[9] For masses less than M_3 the nonequilibrium transformations need not take particularly violent forms, as finite degenerate white-dwarf states exist for these stars. However, when $M > M_3$, the star must eject the excess mass first, before it can evolve througn a sequence of composite models consisting of degenerate cores and gaseous envelopes toward the completely degenerate state. Our present conclusions tend to confirm a suggestion made by one of us (S. C.) on different occasions that the supernova phenomenon may result from the inability of a star of mass greater than M_3 to settle down to the final state of complete degeneracy without getting rid of the excess mass.[10]

YERKES OBSERVATORY
WILLIAMS BAY, WISCONSIN
AND
UNIVERSITY OF SÃO PAULO
BRAZIL

[9] *An Introduction to the Study of Stellar Structure*, p. 423, Chicago, 1939.

[10] At a symposium held at the Yerkes Observatory in the fall of 1941, Dr. R. Minkowski stated that the analysis of his spectroscopic observations on the Crab nebula supports this suggestion.

AN ATTEMPT TO INTERPRET THE RELATIVE ABUNDANCES OF THE ELEMENTS AND THEIR ISOTOPES

S. Chandrasekhar and Louis R. Henrich

ABSTRACT

In this paper an attempt is made to derive some information concerning the prestellar stage at which the elements are supposed to have been formed. By using first the relative abundances of the isotopes of a single element (e.g., *O*, *Ne*, *Mg*, *Si*, and *S*), it is shown that a temperature of the order of a few billion degrees is indicated. The equilibrium between the fundamental nuclear particles (protons, neutrons, α-particles, electrons, and positrons) at temperatures ranging from 5 to 10 billion degrees is then studied to establish the relative concentrations of protons and neutrons as a function of the temperature. This relation is then used to compute theoretical mass-abundance-curves under different physical conditions. From such calculations it is concluded that under the physical conditions specified by $T = 8 \times 10^9$ degrees and $\rho = 10^7$ gm/cm^3 the theoretical mass-abundance-curve from oxygen to sulphur agrees fairly satisfactorily with the known abundance-curve according to V. M. Goldschmidt (Fig. 2). An important feature of the nuclear mixture considered is that hydrogen and helium are the two most abundant constituents, which is in agreement with known facts. However, the conditions indicated are seen to be quite insufficient to account for the existence of the heavy nuclei to any appreciable extent. It is, therefore, suggested that we should distinguish at least two epochs in the development of the prestellar stage. We imagine that at the earliest stages conditions of extreme temperatures and densities prevailed at which the heavier nuclei could have been formed. As the matter cooled to lower temperatures and densities, appreciable amounts (1 part in 10^6) of the heavy elements must have been "frozen" into the mixture. At temperatures of the order of from 5×10^9 to 8×10^9 degrees and densities of the order of from 10^4 to 10^7 gm/cm^3 the present known relative abundances of the elements from oxygen to sulphur may have been established.

1. *Introduction.*—It is now generally agreed that the chemical elements cannot be synthesized under conditions now believed to exist in stellar interiors. Consequently, the question of the origin of the elements is left open. On the other hand, the striking regularities which the relative abundances of the elements and their isotopes reveal (e.g., Harkins' rule) require some explanation. It has therefore been suggested that the elements were formed at an earlier, *prestellar*, stage of the universe. If this is accepted, we then have a tentative basis for deriving some information concerning the physical conditions which would have prevailed during this hypothetical prestellar stage. More particularly, we may attempt to interpret the relative abundances of the elements and their isotopes in terms of the physical conditions under which these observed abundances can be realized as a consequence of thermal equilibrium between nuclei of all sorts, neutrons, electrons, and positrons. We do not know a priori whether such an interpretation is possible in terms of a single density and temperature. Indeed, a preliminary discussion of this problem by von Weizsäcker[1] has indicated that we should distinguish at least two distinct epochs in the prestellar state: an initial epoch of extreme density and temperature, when the heaviest elements, like gold and lead, were formed; and a later epoch of relatively "moderate" conditions, during which the present relative abundances of the lighter elements beyond oxygen (to at least sulphur, as we shall see in § 4) came to be established. However, von Weizsäcker's discussion was largely qualitative and was based on very few comparisons with experimental data. Since that discussion our knowledge, both of relative abundances and masses, has advanced sufficiently to justify a more detailed examination of the problem. We therefore propose to rediscuss the problem of the relative abundances of the elements in the light of the increased information which is now available.

[1] *Phys. Zs.*, **39**, 633, 1938 (see particularly pp. 641–645).

2. *Determination of the neutron concentration and the temperature from the relative abundances of the isotopes of a single element.*—In some ways the simplest method of obtaining some information concerning the prestellar stage is from the relative abundances of the isotopes of a single element which has three (or more) isotopes.[2] An accurate knowledge of the masses and the relative abundances of the isotopes of a single such element will lead to a direct determination of the neutron concentration and the temperature.

The equilibrium between the successive isotopes of an element differing by one mass number is maintained according to the scheme

$$ {}^{A}_{Z}X + \nu \rightleftarrows {}^{A+1}_{Z}X \, . \tag{1} $$

Accordingly, we have

$$ \frac{n^{A}_{Z} n_{\nu}}{n^{A+1}_{Z}} = 2 \, \frac{G^{A}_{Z}}{G^{A+1}_{Z}} \left(\frac{A}{A+1} \right)^{3/2} \frac{(2\pi M k T)^{3/2}}{h^{3}} \, e^{-E_{A}/kT} \, , \tag{2} $$

where n^{A}_{Z}, n^{A+1}_{Z}, and n_{ν} denote, respectively, the number of nuclei of species (A, Z) and $(A + 1, Z)$ and the number of neutrons per unit volume; further G^{A}_{Z} and G^{A+1}_{Z} are the statistical weights of the ground states of the respective nuclei, M the unit of atomic mass,[3] T the temperature,

$$ E_{A} = c^{2}(M^{A}_{Z} + M_{\nu} - M^{A+1}_{Z}) \, , \tag{3} $$

where M^{A}_{Z} is the weight of the atom of nuclear charge Z containing A heavy particles and M_{ν} is the mass of the neutron, while the rest of the symbols have their usual meanings.

Similarly, the equilibrium equation

$$ {}^{A+1}_{Z}X + \nu \rightleftarrows {}^{A+2}_{Z}X \tag{4} $$

will provide another equation between n_{ν} and T. From these two equations we readily obtain the relation

$$ kT = \frac{E_{A+1} - E_{A}}{ln \left[\frac{n^{A}_{Z} n^{A+2}_{Z}}{(n^{A+1}_{Z})^{2}} \frac{(G^{A+1}_{Z})^{2}}{G^{A}_{Z} G^{A+2}_{Z}} \left(\frac{[A+1]^{2}}{A[A+2]} \right)^{3/2} \right]} \, , \tag{5} $$

where ln represents the logarithm to the base e. From the foregoing equation we can obtain T if we know the relative abundances of the three isotopes, their masses (*very* accurately), and their statistical weights. Once T has been determined, equation (2) will suffice to specify $\log n_{\nu}$.

We have used the foregoing method to make five independent determinations of $\log n_{\nu}$ and T from the data available for the isotopes of oxygen, neon, magnesium, silicon, and sulphur. The results are summarized in Table 1. An examination of this table shows that, while an average temperature of several billion degrees is indicated, the neutron concentration varies between limits too wide to draw any safe conclusion.[4] It

[2] This method was first suggested by von Weizsäcker (*ibid.*).

[3] The mass of the proton, the mass of the hydrogen atom, or the mass of the neutron may be used at this point in the equation without any noticeable difference in the result. Actually, the mass of the hydrogen atom was used in the computation.

[4] It should not, however, be concluded that these variations in the neutron concentration correspond to equally wide variations in the physically more important quantity, namely, the density (cf. §§ 3, 4, and 5).

therefore appears that a more detailed discussion of the known mass-abundance-curve for the elements is necessary before any trustworthy estimate of the density can be made. But we already have the suggestion that the principal uncertainty in the discussion will be the density.

3. *The equilibrium between protons, neutrons, α-particles, electrons, and positrons.*— Consider an assembly of nuclei of all sorts, protons, free neutrons, electrons, and positrons at some given temperature. Let n_p, n_ν, n^-, and n^+ denote the equilibrium concentrations of protons, neutrons, electrons, and positrons, respectively. Further, let n_Z^A

TABLE 1

DETERMINATION OF LOG n_ν AND T FROM THE RELATIVE ABUNDANCES OF THE ISOTOPES OF A SINGLE ELEMENT

Element		Mass (Atomic)*	Relative Abundance†	T	$\log n_\nu$
O	^{16}O	16	99.76		
	^{17}O	17.00450	0.04	4.2×10^9	26.5
	^{18}O	18.00490	0.20		
Ne	^{20}Ne	19.99881	90.00		
	^{21}Ne	21.00018	0.27	2.9×10^9	19.7
	^{22}Ne	21.99864	9.73		
Mg	^{24}Mg	23.99189	77.4		
	^{25}Mg	24.99277	11.5	10.0×10^9	30.7
	^{26}Mg	25.99062	11.1		
Si	^{28}Si	27.98639	89.6		
	^{29}Si	28.98685	6.2	12.9×10^9	31.2
	^{30}Si	29.98294	4.2		
S	^{32}S	31.98306	95.0		
	^{33}S	32.98260	0.74	3.3×10^9	19.1
	^{34}S	33.97974	4.2		

* Masses of the oxygen isotopes as listed by O. Hahn, S. Flügge, and J. Mattauch in *Phys. Z.*, **41**, 1, 1940. Other masses as listed by E. Pollard in *Phys. Rev.*, **57**, 1186, 1940.

† The relative abundances of the isotopes were taken from J. J. Livingood and G. T. Seaborg in *Rev. of Modern Phys.*, **12**, 30, 1940. Also, for nuclei of known spins the statistical weights were taken to be $(2S+1)$; when such information was lacking, a weight of 1 was adopted for nuclei of even mass numbers and a weight of 2 for the odd ones.

denote the concentration of a typical nucleus of charge Z and mass-number A. The fundamental reactions maintaining equilibrium are:

$$p+e^- \rightleftarrows \nu \, , \tag{6}$$

$$e^-+e^+ \rightleftarrows \gamma\text{-rays} \, , \tag{7}$$

and

$$^A_Z X + p\,\Delta p + \nu\,\Delta\nu \rightleftarrows {}^{A'}_{Z'} X \, , \tag{8}$$

where

$$A' = A+\Delta p + \Delta\nu \, ; \qquad Z' = Z + \Delta p \, . \tag{9}$$

In equations (8) and (9) Δp and $\Delta \nu$ are arbitrary integers. (It should be remarked that the product nucleus ${}^{A'}_{Z'}X$ need not, in general, be a stable nucleus.) Finally, we have the condition that the whole assembly is electrically neutral:

$$n^- = \Sigma Z n_Z^A + n^+ , \tag{10}$$

where the summation on the right-hand side is extended over nuclei of all charge and mass numbers.

Strictly speaking, the rigorous solution of the equations (6)–(10) under given physical conditions of density and temperature involves the consideration of a simultaneous system of equations of a very high order. In practice, however, the system of equations is effectively reduced very considerably, for, under prescribed conditions only a few of the nuclear particles occur with sufficient abundance to affect appreciably equation (10). As we shall see later, in § 4, under conditions of greatest interest in the present connection, the most abundant particles are protons, neutrons, α-particles, electrons, and positrons. We shall therefore begin our discussion of the theoretical mass-abundance-curves under given conditions by considering the equilibrium between these particles.

First, the equation governing the concentrations of free neutrons, protons, and electrons is *either*

$$\frac{n_p n^-}{n_\nu} = 2 \frac{(2\pi m_e kT)^{3/2}}{h^3} e^{-c^2(M_1^1 - M_\nu)/kT} \tag{11}$$

or

$$\frac{n_p n^-}{n_\nu} = 16\pi \left(\frac{kT}{hc}\right)^3 e^{-c^2(M_1^1 - M_\nu)/kT} , \tag{12}$$

depending on whether the electrons are assumed to obey the nonrelativistic or the relativistic equations of statistical mechanics.[5] In the foregoing equations m_e, M_ν, and M_1^1 denote the mass of the electron, neutron, and hydrogen atom, respectively. A comparison of equations (11) and (12) shows that in a first approximation we may use equation (11) for $T \leqslant 4 \times 10^9$ degrees, while equation (12) should be used for higher temperatures.

Second, the equation governing the concentrations of electrons and positrons at a given temperature is *either*

$$n^- n^+ = 4 \frac{(2\pi m_e kT)^3}{h^6} e^{-2m_e c^2/kT} \tag{13}$$

or

$$n^- n^+ = 256\pi^2 \left(\frac{kT}{hc}\right)^6 e^{-2m_e c^2/kT} , \tag{14}$$

again depending on whether T is less than or greater than 4×10^9 degrees.[6]

Third, the equation giving the concentration of α-particles in terms of the proton and neutron concentrations is

$$\frac{(n_p)^2 (n_\nu)^2}{n_2^4} = 2 \left(\frac{1}{4}\right)^{3/2} \left[2 \frac{(2\pi M kT)^{3/2}}{h^3} \right]^3 e^{-D_2^4/kT} , \tag{15}$$

[5] See, e.g., S. Chandrasekhar, *M.N.*, **91**, 446, 1931.

[6] See R. H. Fowler, *Statistical Mechanics*, 2d ed., p. 654, Cambridge, England, 1936.

where, quite generally D_Z^A is given by

$$D_Z^A = c^2[(A - Z)M_\nu + ZM_1^1 - M_Z^A] \, . \tag{16}$$

Finally, we have the equation insuring the electrical neutrality of the whole assembly:

$$n^- = n^+ + n_p + 2n_2^4 + \ldots \, . \tag{17}$$

For any assigned neutron concentration and temperature the foregoing equations can be solved (by trial and error), and Table 2 summarizes the results of such computations

TABLE 2

THE ELECTRON—POSITRON—NEUTRON—PROTON—α-PARTICLE EQUILIBRIUM

LOG n_ν	$T=5\times10^9$ DEGREES				$T=6.5\times10^9$ DEGREES				$T=8\times10^9$ DEGREES			
	log n_p	log $2n_2^4$	log n^-	log n^+	log n_p	log $2n_2^4$	log n^-	log n^+	log n_p	log $2n_2^4$	log n^-	log n^+
23.0	24.28	18.42	29.81	29.81								
23.5	24.78	20.42	29.81	29.81								
24.0	25.28	22.42	29.81	29.81								
24.5	25.78	24.42	29.81	29.81								
25.0	26.28	26.42	29.81	29.81								
25.5	26.78	28.42	29.81	29.81								
26.0	27.02	29.90	30.07	29.55	26.99	22.78	30.27	30.27				
26.5	26.91	30.68	30.68	28.94	27.49	24.77	30.27	30.27				
27.0	26.74	31.34	31.35	28.27	27.98	26.76	30.27	30.27	27.79	21.91	30.61	30.61
27.5					28.47	28.75	30.28	30.26	28.29	23.91	30.61	30.61
28.0					28.76	30.31	30.49	30.04	28.79	25.90	30.61	30.61
28.5					28.65	31.11	31.11	29.43	29.28	27.87	30.62	30.60
29.0					28.49	31.78	31.78	28.77	29.73	29.78	30.67	30.55
29.5									29.83	30.98	31.08	30.15
30.0									29.71	31.72	31.71	29.52
31.0									29.38	33.05	33.04	28.19

for three different temperatures. The resulting variations are illustrated graphically in Figure 1.

An examination of Figure 1 reveals that for low neutron concentrations the material density is almost entirely due to the electron-positron pairs which occur in a "vacuum" in thermal equilibrium. At higher neutron concentrations the protons and α-particles increase in numbers. Eventually the proton abundance reaches a maximum and decreases (asymptotically) according to the law

$$n_p \propto n_\nu^{-1/3} \, . \tag{18}$$

At the same time,

$$n_2^4 \propto n_\nu^{4/3} \, . \tag{19}$$

It should, however, be remarked that this increase of the α-particle concentration cannot continue indefinitely, for, as the neutron concentration increases appreciably beyond the maximum proton concentration, some of the heavier nuclei become abundant and the corresponding nuclear equilibrium must be included in setting up the primary neutron-proton equilibrium and satisfying the condition for electrical neutrality. However, as we have already remarked, in the following sections we shall not be interested in conditions when the abundant nuclei are different from protons or α-particles.

4. *The theoretical mass-abundance-curves for elements beyond oxygen and comparison with Goldschmidt's empirical curve.*—For the range of physical conditions considered in § 3, the relative concentrations of protons and neutrons being known as a function of temperature, it is now possible to compute the complete nuclear abundance-curve to be expected under equilibrium conditions at given temperatures and neutron concentra-

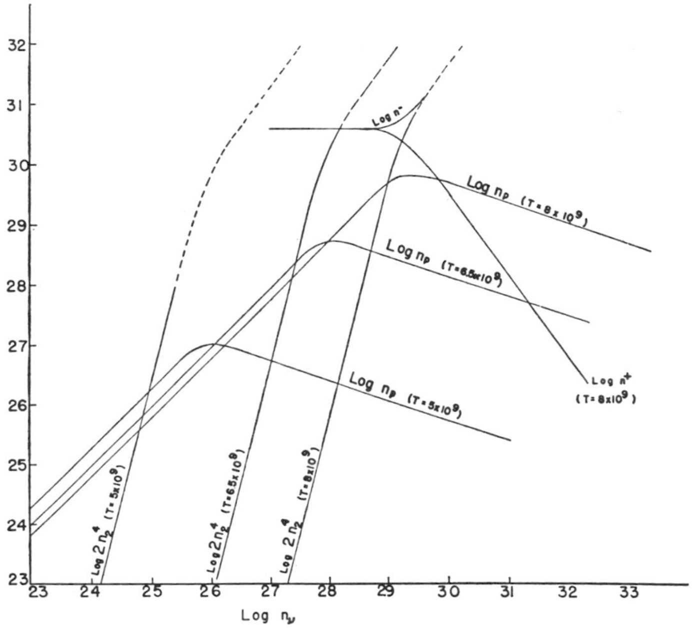

FIG. 1

tions. The relative concentration of the nuclei ${}^{A}_{Z}X$ and ${}^{A'}_{Z'}X(A' = A + \Delta p + \Delta\nu;$ $Z' = Z + \Delta p)$ is determined by the equation

$$\frac{n^A_Z (n_p)^{\Delta p} (n_\nu)^{\Delta\nu}}{n^{A'}_{Z'}} = \frac{G^A_Z}{G^{A'}_{Z'}} \left(\frac{A}{A'}\right)^{3/2} \left[2\,\frac{(2\pi MkT)^{3/2}}{h^3}\right]^{\Delta p + \Delta\nu} e^{-(D^{A'}_{Z'} - D^A_Z)/kT}\,, \tag{20}$$

where D^A_Z has the same meaning as in equation (16). We may note that the foregoing equation can be re-written in the following form, for computing:

$$\left.\begin{aligned}\log n^{A'}_{Z'} = {} & \log n^A_Z - \log (G^A_Z/G^{A'}_{Z'}) - \tfrac{3}{2}\log (A/A') \\ & + \Delta p(\log n_p - 34.08 - \tfrac{3}{2}\log T) \\ & + \Delta\nu(\log n_\nu - 34.08 - \tfrac{3}{2}\log T) + \frac{4.692}{T}\,(D^{A'}_{Z'} - D^A_Z)\,,\end{aligned}\right\} \tag{21}$$

where T is expressed in units of billions of degrees and D^A_Z in millimass units.

It should be noted that, according to equations (20) and (21), we can compute the equilibrium concentrations not only of the stable nuclei but also of the unstable nuclei if we have the necessary data for them. The table of atomic masses given by W. H. Barkas[7] gives theoretical masses for some of the unstable nuclei. Using these masses, we may compute theoretical abundance-curves. In doing this we can take the "net" abundance of a stable nuclei to be equal to the abundance of this nucleus *plus* the abundances of all unstable nuclei, which, after an appropriate number of β^- or β^+ disintegrations, transform into the particular nucleus under consideration. However, it was found, for the densities and temperatures for which computations were made, that the relative abundance of the unstable nuclei "bordering" on a given stable nucleus was generally quite small in the range of the periodic table considered. Consequently, the effect of these unstable nuclei was ignored.

Our object, then, is to compare the computed theoretical abundance-curves with the observed relative abundances of the stable nuclei, as given by V. M. Goldschmidt.[8] Remembering that our object is to obtain some information concerning the prestellar stage (as indicated by the known relative abundances of the nuclei), we adopted the following procedure.

For an assigned temperature the neutron concentration was so adjusted that the relative concentration of the nuclei ${}^{16}_{8}O$ and ${}^{36}_{18}A$ occurring in the equilibrium mixture was approximately the same as that known to occur in the "cosmos"—according to Goldschmidt, approximately in the ratio 15,000:1. After having adjusted the physical conditions in this manner, the complete theoretical mass-abundance-curve was computed according to equation (20) and the table of atomic masses given by Barkas.

Calculations of the kind outlined in the preceding paragraph have been made for different initially assigned temperatures, and the results are summarized in Table 3.

In Figure 2 we have compared the theoretical abundances of the elements beyond oxygen, according to Table 3, with the abundances given by Goldschmidt.[9] An examination of this figure shows that the better agreement with the computed and the observed abundances is obtained under the conditions:

$$\left.\begin{aligned} T &= 8 \times 10^9 \text{ degrees} , \\ \log n_p &= 29.83 , \\ \log n_\nu &= 29.30 , \\ \log n_2^4 &= 30.3 , \end{aligned}\right\} \qquad (22)$$

or

$$\rho = 10^7 \text{ gm/cm}^3 ; \qquad T = 8 \times 10^9 \text{ degrees} . \qquad (23)$$

It is further seen that the theoretical abundance-curve under these conditions agrees with Goldschmidt's curve quite satisfactorily for all elements from oxygen to sulphur. It is difficult to extend the theoretical calculations beyond argon, since the mass defects are not known to sufficient accuracy.

5. *Further discussion of the physical conditions for the prestellar stage derived in section 4.*—We shall now consider in some detail the special features of the physical conditions ($T \sim 8 \times 10^9$; $\rho \sim 10^7$ gm/cm³) we have derived for the prestellar stage on the

[7] *Phys. Rev.*, **55**, 691, 1939.

[8] V. M. Goldschmidt, *Geochemische Verteilungsgesetze der Elemente. IX. Die Mengenverhältnisse der Elemente und der Atom-Arten*, Oslo: J. Dybwad, 1938.

[9] *Ibid.* (see particularly the table on p. 120). In the case of neon and argon, approximate values have been read from the graph, p. 123.

basis of the relative abundances of the elements from $^{16}_{8}O$ to $^{32}_{16}S$, as known from a variety of terrestrial, meteoric, and stellar sources.

TABLE 3

THE COMPUTED RELATIVE ABUNDANCES OF THE NUCLEI FOR DIFFERENT INITIAL CONDITIONS

Element	Z	A	$T=6.5\times10^9$ $\log n_\nu=27.65$ $\log n_p=28.60$	$T=8\times10^9$ $\log n_\nu=29.30$ $\log n_p=29.83$
H	1	1	28.6	29.8
H	1	2	22.8	25.2
He	2	3	20.9	23.6
He	2	4	29.0	30.3
He	2	5	....	....
Li	3	6	17.8	21.2
Li	3	7	16.3	20.1
Li	3	8	....	....
Be	4	9	16.4	20.2
B	5	10	15.9	19.5
B	5	11	16.1	19.8
C	6	12	21.5	24.0
C	6	13	17.9	21.3
N	7	14	16.6	20.0
N	7	15	17.9	21.2
O	8	16	20.5	23.1
O	8	17	16.4	20.0
O	8	18	15.2	19.1
F	9	19	14.6	18.4
Ne	10	20	17.4	20.5
Ne	10	21	15.9	19.4
Ne	10	22	15.8	19.5
Na	11	23	16.4	19.8
Mg	12	24	18.3	21.1
Mg	12	25	17.0	20.2
Mg	12	26	17.5	20.7
Al	13	27	17.1	20.3
Si	14	28	18.0	20.7
Si	14	29	17.1	20.2
Si	14	30	18.1	21.1
P	15	31	17.1	20.1
S	16	32	17.3	20.1
S	16	33	16.9	19.9
S	16	34	18.2	21.1
Cl	17	35	16.6	19.6
A	18	36	16.1	19.0
Cl	17	37	16.4	19.6
A	18	38	17.8	20.6
K	19	39	15.9	18.9
A	18	40	14.5	18.2

From one point of view the most significant characteristic of our equilibrium mixture is the enormous abundance of hydrogen and helium. In itself, this is a very satisfactory feature, for one of the most striking facts known about the abundances of the elements is precisely the extreme abundance of the two lightest elements, compared to all the others. However, according to Table 3, hydrogen and helium are, together, as much as 10^7 times more abundant than oxygen, while it is currently estimated that this ratio should be more nearly 10^4 or 10^5. In considering this discrepancy it should be remembered that in our calculations of the abundances of the different nuclear species according to equation (20) we have ignored the partition-function factor, which should, strictly, be included.

It is known that in the corresponding formulae in the theory of ionization equilibrium these partition-function factors can, under favorable circumstances, amount to as much as 20 in favor of the species in the higher stages of excitation.[10] In the case of nuclear equilibrium, these factors are likely to be even more important, for the known increase

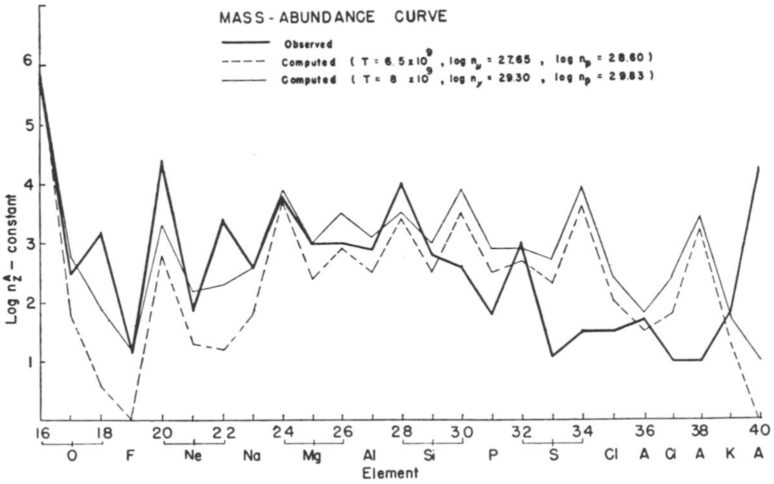

FIG. 2

in the density of excited nuclear levels with increasing mass number (particularly for the odd nuclei) would be expected to bring in large factors in the direction of increasing the oxygen-helium ratio. However, special circumstances may intervene in the cases of particular nuclei. On the whole, it appears to us that, by taking properly into account the partition-function factors in the equation of nuclear equilibrium, it should not be impossible to increase the values for the oxygen-helium ratio given in Table 3 by a factor of the order 100, while keeping the general agreement between the computed and the observed relative abundances of the nuclei from oxygen to sulphur at substantially the same level as that achieved in Figure 2 for $T = 8 \times 10^9$ degrees and $\rho = 10^7$ gm/cm^3.

The second feature to which attention should be drawn is the hydrogen-helium ratio. For the conditions derived for the pre-stellar stage in § 4 this ratio is 1:3 in favor of helium. This is contrary to the generally accepted view that hydrogen is actually the more abundant of the two. It should, however, be noted in this connection that at lower temperatures and densities the equilibrium ratio $H:He$ rapidly shifts in favor of hydrogen (see Fig. 1). It therefore seems likely that the observed high ratio between hydrogen and helium may be the result of a later stage in the prestellar development—a stage in which the relative abundances of the elements beyond oxygen had already been

[10] See, e.g., Fowler, *op. cit.*, chap. xiv, pp. 562 ff.

frozen; but the equilibrium between the fundamental nuclear particles, protons, neutrons, and α-particles still functioned sufficiently to shift the hydrogen-helium ratio in favor of the former.

And, finally, we should refer to the abundances of the heavy nuclei. It is found that the physical conditions under which we would predict anything like the observed relative abundances of the elements beyond oxygen and up to potassium (say) will be wholly inadequate to account for any appreciable amounts of the heavy nuclei. Thus, under the conditions indicated by our considerations in § 4 (namely, $T = 8 \times 10^9$ and $\rho = 10^7 \text{gm/cm}^3$) the equilibrium oxygen-iron ratio is $1:10^{-10}$; this is in complete disagreement with the known ratio. Of course, we might ignore iron as an exceptionally abundant element. But, if we should go to elements beyond iron, the amounts predicted decrease so very rapidly that the conclusion is inescapable that to predict anything like the observed relative abundances of the heaviest nuclei (e.g., an oxygen-lead ratio of 10^{-7}) we need distinctly different conditions from those indicated by the relative abundances of the isotopes of a single element, the general run of the mass-abundance-curve beyond oxygen, and particularly the extreme abundance of hydrogen and helium.[11] We shall return to this matter in § 6.

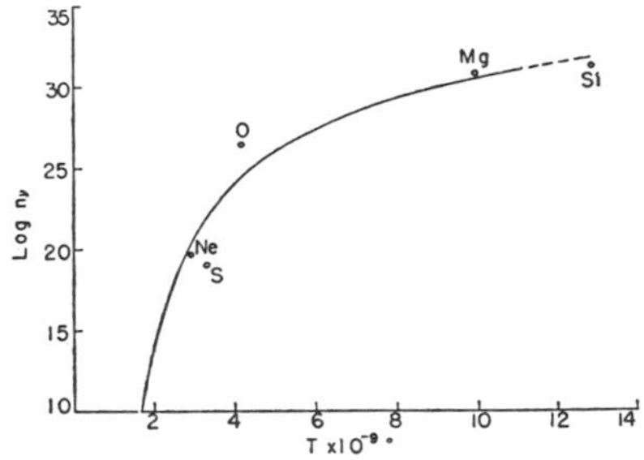

FIG. 3

6. *General remarks concerning the development of the prestellar stage.*—According to the considerations advanced in the preceding sections, we may picture the development of the prestellar stage in the following terms.

Originally, conditions of extreme temperatures ($T \sim 10^{10}$ to 10^{11} degrees) and densities (anywhere between the densities used in this paper and nuclear densities) should have prevailed. Under such conditions the heavy nuclei could presumably be formed. As the matter cooled to lower temperatures and densities, appreciable amounts of the heavy elements (about a few parts in a million) must have been "frozen" into the mixture. At temperatures of the order of from 6 to 8 billion degrees and densities ranging from 10^5 to 10^7 gm/cm^3 the present known relative abundances of the elements beyond and including oxygen must have been established; further, the extreme abundances of hydrogen and helium relative to all the other elements must also have been established during this same epoch in the development of the prestellar stage. At somewhat lower temperatures and densities the hydrogen-helium ratio must have been fixed at the known value. It may also be noted that nuclear equilibrium, as such, must have ceased completely to function at a temperature of the order of 4 billion degrees, as is evident from the discussion of § 3 and Figure 1. At still lower temperatures the abundances of the lighter nuclei—lithium, beryllium, boron, carbon, and nitrogen—must have suffered further changes from nonequilibrium processes of the character now occurring in stellar interiors. One peculiar circumstance should be mentioned at this point. In Figure 3 we have plotted the neutron concentration at which the proton maximum occurs as a function of the temperature. In the same diagram we have also plotted the neutron concentrations and temperatures derived in § 2 from the relative abundances of the isotopes of oxygen, neon, magnesium, silicon, and sulphur. It is remarkable how these points lie along the proton-maximum-curve. We are tempted to ask: "Did the original matter cool in such a way that the protons continued to exist with the maximum possible abundance?"

[11] It has already been noted by von Weizsäcker (*op. cit.*) that, to account for the relative abundances of isotopes of oxygen, conditions are needed which are very different, indeed, from that required to account, for example, for the observed oxygen-lead ratio. But our conclusions stated in the text are more general than this.

Finally, we may briefly refer to the cause of the original expansion and cooling. One suggestion is that it may be connected with the beginning of the expansion of the universe. Another (possibly related) suggestion is that it might have arisen from the loss of energy by neutrino emission in the manner contemplated by Gamow and Shoenberg in a different connection.[12]

In conclusion, it should perhaps be emphasized that the considerations of this paper should be regarded as of a purely exploratory nature and that such "agreements" as may have been obtained should not be overstressed. It should, indeed, be remembered that we are here dealing with a stage in the evolution of the universe in which conditions were utterly different from the present conditions. To emphasize this fact, it may be remarked that at an average density of the order of 10^7 gms/cm^3 the entire mass in the universe (*ca.* 10^{54} gm.) may be inclosed in a sphere of radius of the order of 10^{16} cm., i.e., somewhat less than a hundredth of a parsec.

Yerkes Observatory
December 19, 1941

[12] *Phys. Rev.*, **59**, 539, 1941.

NATURE

The Cosmological Constants

PROF. P. A. M. DIRAC's recent letter in NATURE[1] encourages me to direct attention to certain 'coincidences' which I had noticed some years ago, but which I have been hesitating to publish from the conviction that purely 'dimensional arguments' will not lead one very far.

If we consider the natural constants h (Planck's constant), c (velocity of light), H (mass of the proton), G (the constant of gravitation), we can form the following combination M_α which is of the dimension of mass:

$$M_\alpha = \left(\frac{hc}{G}\right)^{\alpha} \frac{1}{H^{2\alpha-1}}. \qquad (1)$$

where α is an arbitrary numerical constant. Now a particular case of the above occurs in the theory of stellar interiors, namely, when $\alpha = 3/2$. Then

$$M_{3/2} = \left(\frac{hc}{G}\right)^{3/2} \frac{1}{H^2} \doteqdot 5{\cdot}76 \times 10^{34} \text{ gm.}, \qquad (2)$$

which is about thirty times the mass of the sun. Now, the apparent success of steady state considerations in 'explaining' the observed order of stellar masses can be traced to the circumstance that the above combination (2) of the natural constants gives a mass of the correct order. It may be noticed that apart from numerical constants, (2) is the same as the upper limit to the mass of completely degenerate (degenerate in the sense of the Fermi-Dirac statistics) configurations[2]. The occurrence of (2) in stellar structure equations need not cause any surprise, since one can easily convince oneself by considerii two homologous stellar configurations that if formula for mass exists, it must contain the me molecular weight μH with an inverse power 2, ai this would, according to (1), fix the value of tl exponent α as 3/2.

It is of interest to see what (1) leads to for oth values of α. If $\alpha = 2$, then

$$M_2 = \left(\frac{hc}{G}\right)^{2} \frac{1}{H^3} \doteqdot 9{\cdot}5 \times 10^{20} \text{ mass of sun}$$

If we divide M_2 by H, then we get for the correspon ing 'number of protons or/and neutrons',

$$N = \left(\frac{hc}{G}\right)^{2} \frac{1}{H^4} \doteqdot 1{\cdot}1 \times 10^{78},$$

which is of the right order as the 'number of partic in the universe'. We may notice that if $G \sim t^{-1}$ (t Milne's cosmological time), then $N \sim t^2$, which agre with Dirac's speculation.

It may be further pointed out that if $\alpha =$ then

$$M_{1\frac{3}{4}} = 1{\cdot}7 \times 10^{11} \text{ mass of sun},$$

which is of the same order as the mass of our Mil Way system. If we 'identify' $M_{1\frac{3}{4}}$ as representi the mass of a galaxy (external or otherwise), th we should have, according to Dirac's ideas, that t 'number of particles in the galaxy' should vary $t^{1{\cdot}75}$ Similarly, the number of particles in a s should vary as $t^{1{\cdot}5}$.

S. CHANDRASEKHAR.

Yerkes Observatory,
Wisconsin.

[1] NATURE, 139, 323 (Feb. 20, 1937).

[2] Chandrasekhar, S., *Mon. Not. Roy. Ast. Soc.*, 91, 456 (1931).

PART FOUR

Integral Theorems on the Equilibrium of a Star

30

On the maximum possible Central Radiation Pressure in a Star of a given Mass.

GENTLEMEN,—

According to a theorem due to Eddington the total pressure P inside a star cannot anywhere exceed the value $P_{max.}$ given by

$$P_{max.} = \frac{1}{2}\left(\frac{4}{3}\pi\right)^{1/3} GM^{2/3}\rho_0^{4/3}, \quad . \quad . \quad . \quad . \quad (1)$$

where ρ_0 is the greatest density inside the star. $P_{max.}$ given by (1) is just equal to the central pressure in a configuration of mass M with a uniform density ρ_0. If it is further assumed that the density increases monotonically inwards then ρ_0 in (1) specifies the central density.

Let the radiation pressure be a fraction $(1-\beta_c)$ of the total pressure at the centre of a star. We shall simply refer to $(1-\beta_c)$ as the *central radiation pressure.* If the perfect gas laws are obeyed at the centre then we can express the central pressure P_c in terms of $(1-\beta_c)$ in the form

$$P_c=\left[\left(\frac{k}{\mu H}\right)^4 \frac{3}{a}\frac{1-\beta_c}{\beta_c^4}\right]^{1/3} \rho_0^{4/3}, \quad . \quad . \quad (2)$$

where k=Boltzmann's Constant, μ=molecular weight, H=mass of the proton, a=radiation constant.

On comparing (1) and (2) we have the inequality

$$\left[\left(\frac{k}{\mu H}\right)^4 \frac{3}{a}\frac{1-\beta_c}{\beta_c^4}\right]^{1/3} \leqslant \frac{1}{2}\left(\frac{4}{3}\pi\right)^{1/3} GM^{2/3}. \quad . \quad . \quad (3)$$

After some minor transformations (3) is found to be equivalent to

$$M \leqslant 0{\cdot}3035 M_3\left(\frac{960}{\pi^4}\frac{1-\beta_c}{\beta_c^4}\right)^{1/2}, \quad . \quad . \quad . \quad . \quad (4)$$

where M_3 is the limiting mass for completely degenerate configurations. Thus we have proved that *for a given mass* M *the central radiation pressure* $(1-\beta_c)$ *cannot exceed the value* $(1-\beta_m)$ *which satisfies the quartic equation*

$$M=0{\cdot}3035 M_3\left(\frac{960}{\pi^4}\frac{1-\beta_c}{\beta_c^4}\right)^{1/2}. \quad . \quad . \quad . \quad . \quad (5)$$

Eddington's quartic equation, which determines the "actual value" of $(1-\beta)$ (now assumed constant in the star), differs from equation (5) only by the factor 0·304 in front of M_3 being absent.

As an illustration of the use of (5) we see that a star which has a $(1-\beta_c)$ greater than $(1-\beta_\omega)$, where

$$\frac{960}{\pi^4}\frac{1-\beta_\omega}{\beta_\omega}=1,$$

must have a mass certainly greater than $0{\cdot}304\mathfrak{M}$, where $\mathfrak{M}$ is the mass on the standard model which has a "$1-\beta_1$"$=1-\beta_\omega$. (Clearly $\mathfrak{M}=M_3\beta_\omega^{3/2}$.) In other words, stars for which central degeneracy cannot set in at all on contraction must have masses at least greater than $0{\cdot}304\mathfrak{M}$ (which is about $2{\cdot}01\odot\mu^{-2}$).

I am Gentlemen,
Yours faithfully,
S. CHANDRASEKHAR.

Harvard College Observatory,
Cambridge, Mass, U.S.A.,
1935 December 31.

31

THE PRESSURE IN THE INTERIOR OF A STAR.

S. Chandrasekhar, Ph.D.

1. In a recent paper * Professor E. A. Milne has established by direct methods certain inequalities which should be true for stellar configurations in hydrostatic equilibrium. A fundamental theorem which is proved in Milne's paper is the following :—

If P_c denotes the central pressure in any equilibrium configuration, and P_1 the pressure at a conventionally assigned boundary where the radius is R and the mass enclosed is M, then

$$P_c > P + \frac{3}{8\pi}\frac{GM^2(r)}{r^4} \geqslant P_1 + \frac{3}{8\pi}\frac{GM^2}{R^4}. \quad (1)†$$

In the above formula $M(r)$ denotes the mass enclosed inside a sphere of radius r. The inequalities expressed in (1) are true, provided $\bar{\rho}(r)$, the mean density inside r, always exceeds the actual density $\rho(r)$ at r.

2. The outer members in the above inequality yield

$$P_c \geqslant \frac{3}{8\pi}\frac{GM^2}{R^4}. \quad (2)$$

Formula (2) provides the *minimum* pressure at the centre of an equilibrium configuration of assigned mass and radius. But in certain stellar applica-

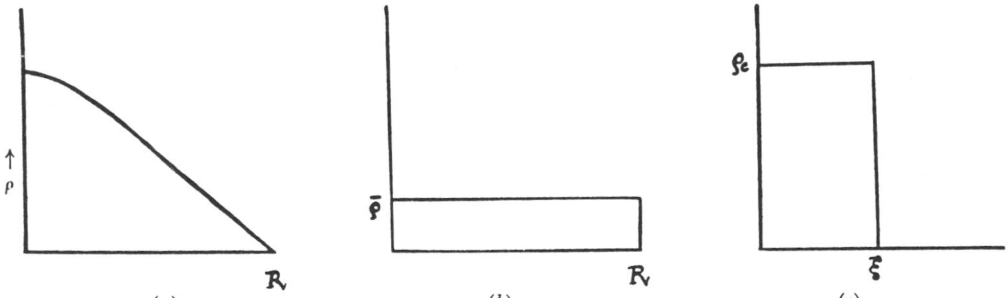

(*a*) (*b*) (*c*)

FIG. 1.

tions ‡ the inequality which is required is in the opposite direction to (2), *i.e.* one needs an inequality which would give an upper bound to the central pressure in an equilibrium configuration of assigned mass and central density.

Essentially, inequalities of the kind proved by Milne arise from a "comparison" § of the given equilibrium configuration with another configuration of the same mass and radius but at a uniform density equal to the

* E. A. Milne, *M.N.*, **96**, 179, 1936.

† The outer members of (1) form an inequality which is the statement of a theorem due to Eddington.

‡ E. C. Stoner, *M.N.*, **92**, 662, 1932 ; S. Chandrasekhar, *Observatory*, **59**, 47, 1936.

§ The word "comparison" is not to be taken too literally.

mean density of the original configuration (see fig. 1; compare (*a*) and (*b*)). One gets a different set of inequalities by a corresponding "comparison" of the given configuration with another homogeneous configuration of the same mass but at a uniform density now equal to the central density in the original configuration (see fig. 1; compare (*a*) and (*c*)).

3. *Theorem* 1.—*In any equilibrium configuration in which the mean density inside r decreases outwards we have the inequality*

$$\tfrac{1}{2}G(\tfrac{4}{3}\pi)^{1/3}\bar{\rho}^{4/3}(r)M^{2/3}(r) \leqslant P_c - P \leqslant \tfrac{1}{2}G(\tfrac{4}{3}\pi)^{1/3}\rho_c^{4/3}M^{2/3}(r), \tag{3}$$

where $\bar{\rho}(r)$ *denotes the mean density inside* r, *and* ρ_c *the central density, and* P_c *the central pressure.*

Proof.—The equation of hydrostatic equilibrium is

$$\frac{dP}{dr} = -\frac{GM(r)}{r^2}\rho. \tag{4}$$

Integrating this we have

$$P_c - P = G\int_0^r \frac{M(r)}{r^2}\rho dr, \tag{5}$$

or since

$$dM(r) = 4\pi r^2 \rho dr, \tag{6}$$

we have

$$P_c - P = \frac{G}{4\pi}\int_0^r \frac{M(r)dM(r)}{r^4}. \tag{7}$$

By definition

$$\tfrac{4}{3}\pi r^3 \bar{\rho}(r) = M(r). \tag{8}$$

Hence

$$r^4 = \left[\frac{M(r)}{\tfrac{4}{3}\pi\bar{\rho}(r)}\right]^{4/3}. \tag{9}$$

Substituting (9) in (7) we have

$$P_c - P = \frac{1}{4\pi}(\tfrac{4}{3}\pi)^{4/3}G\int_0^r \bar{\rho}^{4/3}(r)M^{-1/3}(r)dM(r). \tag{10}$$

Since by hypothesis $\bar{\rho}(r)$ decreases outwards we have

$$P_c - P \leqslant \frac{1}{4\pi}(\tfrac{4}{3}\pi)^{4/3}G\rho_c^{4/3}\int_0^r M^{-1/3}(r)dM(r), \tag{11}$$

$$\leqslant \tfrac{1}{2}G(\tfrac{4}{3}\pi)^{1/3}\rho_c^{4/3}M^{2/3}(r). \tag{12}$$

Again from (10) we have

$$P_c - P \geqslant \frac{1}{4\pi}(\tfrac{4}{3}\pi)^{4/3}G\bar{\rho}^{4/3}(r)\int_0^r M^{-1/3}(r)dM(r), \tag{13}$$

$$\geqslant \tfrac{1}{2}G(\tfrac{4}{3}\pi)^{1/3}\bar{\rho}^{4/3}(r)M^{2/3}(r). \tag{14}$$

Combining (12) and (14) we have the required result.

Corollary 1.—If we put $r = R$ in (12) and (14) we have

$$\tfrac{1}{2}G(\tfrac{4}{3}\pi)^{1/3}\bar{\rho}^{4/3}M^{2/3} \leqslant P_c - P_1 \leqslant \tfrac{1}{2}G(\tfrac{4}{3}\pi)^{1/3}\rho_c^{4/3}M^{2/3}, \tag{15}$$

where P_1 is the boundary pressure.

Corollary 2.—If $(1-\beta_c)$ is the ratio of the radiation pressure to the total pressure at the centre of a wholly gaseous configuration satisfying the conditions of theorem 1, then

$$1-\beta_c \leqslant 1-\beta^*, \tag{16}$$

where $(1-\beta^*)$ satisfies the quartic equation

$$M=\left(\frac{6}{\pi}\right)^{1/2}\left[\left(\frac{k}{\mu H}\right)^4 \frac{3}{a}\frac{1-\beta^*}{\beta^{*4}}\right]^{1/2}\frac{1}{G^{3/2}}. \tag{17}$$

In terms of $(1-\beta_c)$ the central pressure P_c is given by

$$P_c=\left[\left(\frac{k}{\mu H}\right)^4 \frac{3}{a}\frac{1-\beta_c}{\beta_c^4}\right]^{1/3}\rho_c^{4/3}. \tag{18}$$

Comparing (18) with the inequality on the right-hand side of (15) we have

$$M^{2/3} \geqslant \left(\frac{6}{\pi}\right)^{1/3}\left[\left(\frac{k}{\mu H}\right)^4 \frac{3}{a}\frac{1-\beta_c}{\beta_c^4}\right]^{1/3}\frac{1}{G}. \tag{19}$$

From (19) the required inequality (16) follows, since $(1-\beta)/\beta^4$ is a montonically increasing function of $1-\beta$.

In the same way one can prove that

$$\frac{1}{G^{3/2}}\left(\frac{6}{\pi}\right)^{1/2}\left[\left(\frac{k}{\mu H}\right)^4 \frac{3}{a}\frac{1-\beta_c}{\beta_c^4}\right]^{1/2} \geqslant M\cdot\left(\frac{\bar{\rho}}{\rho_c}\right)^2. \tag{19$'$}$$

4. We shall now establish certain inequalities similar to those in §§ 5, 6 in Milne's paper.

Theorem 2.—*If I_ν is the integral defined by*

$$I_\nu=\int_0^R \frac{GM(r)dM(r)}{r^\nu}, \qquad (\nu<6), \tag{20}$$

then under the conditions of Theorem 1

$$\frac{3}{6-\nu}\frac{GM^2}{R^\nu} \leqslant I_\nu \leqslant \frac{3}{6-\nu}\frac{GM^2}{\xi^\nu}, \tag{21}$$

where ξ is defined by the equation

$$\tfrac{4}{3}\pi\xi^3\rho_c=M. \tag{22}$$

(The first part of the inequality (21) is proved by a different method by Milne (*loc. cit.*, § 5).)

Proof.—We have

$$\begin{aligned} I_\nu &= G\int_0^R \frac{M(r)dM(r)}{r^\nu}, \\ &= G(\tfrac{4}{3}\pi)^{\nu/3}\int_0^R \bar{\rho}^{\nu/3}(r)M^{(3-\nu)/3}(r)dM(r), \qquad (23) \\ &\leqslant G(\tfrac{4}{3}\pi\rho_c)^{\nu/3}\int_0^R M^{(3-\nu)/3}(r)dM(r), \\ &\leqslant \frac{3}{6-\nu}G(\tfrac{4}{3}\pi\rho_c)^{\nu/3}M^{(6-\nu)/3}, \end{aligned}$$

or by (22)

$$I_\nu \leqslant \frac{3}{6-\nu} G \frac{M^2}{\xi^\nu}. \tag{24}$$

From (23) we also have

$$I_\nu \geqslant G(\tfrac{4}{3}\pi\bar{\rho})^{\nu/3} \int_0^R M^{(3-\nu)/3}(r) dM(r) \tag{25}$$

$$\geqslant \frac{3}{6-\nu} G \frac{M^2}{R^\nu}. \tag{26}$$

Combining (25) and (26) we have the required inequality (21).

Corollary 1.—If $\nu = 1$, then I_1 as defined in (20) is precisely the negative potential energy Ω, and (21) now shows that

$$\frac{3}{5} \frac{GM^2}{R} \leqslant \Omega \leqslant \frac{3}{5} \frac{GM^2}{\xi}. \tag{27}$$

Corollary 2.—If $\nu = 2$, then I_2 defines $M\bar{g}$ where $\bar{g}$ is the mean value of gravity in the configuration. By (21) we now have

$$\frac{3}{4} \frac{GM^2}{R^2} \leqslant M\bar{g} \leqslant \frac{3}{4} \frac{GM^2}{\xi^2}. \tag{28}$$

Trinity College, Cambridge:
1936 *April* 28.

THE PRESSURE IN THE INTERIOR OF A STAR

S. CHANDRASEKHAR

ABSTRACT

In this paper certain integral theorems on the equilibrium of a star are proved.

I

In a recent paper[1] the author proved the following theorems:

THEOREM 1.—*In any equilibrium configuration in which the mean density inside* r *decreases outward, we have the inequality*

$$\tfrac{1}{2}G(\tfrac{4}{3}\pi)^{1/3}\bar{\rho}^{4/3}(r)M^{2/3}(r) \leqslant P_c - P \leqslant \tfrac{1}{2}G(\tfrac{4}{3}\pi)^{1/3}\rho_c^{4/3}M^{2/3}(r)\ , \tag{1}$$

where $\bar{\rho}$ (r) *denotes the mean density inside* r, ρ_c *the central density*, P_c *the central pressure, and* M(r) *the mass inclosed inside* r.

THEOREM 2.—*If* $(1 - \beta_c)$ *is the ratio of the radiation pressure to the total pressure at the center of a wholly gaseous configuration, then under the conditions of Theorem 1*

$$1 - \beta_c \leqslant 1 - \beta^*\ , \tag{2}$$

where $(1-\beta^*)$ *satisfies the quartic equation*

$$M = \left(\frac{6}{\pi}\right)^{1/2}\left[\left(\frac{k}{\mu H}\right)^4 \frac{3}{a}\,\frac{1-\beta^*}{\beta^{*4}}\right]^{1/2}\frac{1}{G^{3/2}}\ . \tag{3}$$

THEOREM 3.—*If* I_ν *is the integral defined by*

$$I_\nu = \int_0^R \frac{GM(r)dM(r)}{r^\nu}\ , \qquad (\nu < 6)\ , \tag{4}$$

where R *denotes the radius of the configuration, then*

$$\frac{3}{6-\nu}\,G(\tfrac{4}{3}\pi\bar{\rho})^{\nu/3}M^{(6-\nu)/3} \leqslant I_\nu \leqslant \frac{3}{6-\nu}\,G(\tfrac{4}{3}\pi\rho_c)^{\nu/3}M^{(6-\nu)/3}\ . \tag{5}$$

[1] *M.N.*, **96**, 644, 1936. See also E. A. Milne, *ibid.*, 179.

II

In this paper we shall prove certain additional theorems on the equilibrium of a star.

THEOREM 4.—*Under the conditions of Theorem 1 we have*

$$\frac{P_c}{\rho_c^{\nu/3}} \leqslant \frac{1}{6-\nu} (\tfrac{4}{3}\pi)^{(\nu-3)/3} G R^{\nu-4} M^{(6-\nu)/3} , \tag{6}$$

provided

$$6 > \nu \geqslant 4 , \tag{7}$$

where (6) *is a strict inequality for* $\nu > 4$.

Proof: We have equation 7 (*loc. cit.*)

$$dP = -\frac{G}{4\pi} \frac{M(r)dM(r)}{r^4} . \tag{8}$$

Introducing equation (8) in equation (4), we have

$$I_\nu = -4\pi \int_0^R \frac{dP}{r^{\nu-4}} . \tag{9}$$

Since $\nu \geqslant 4$, we clearly have

$$I_\nu \geqslant -\frac{4\pi}{R^{\nu-4}} \int_0^R dP = \frac{4\pi}{R^{\nu-4}} P_c . \tag{10}$$

In equation (10) we have the equality sign only for the case $\nu = 4$. For $\nu > 4$ we have a strict inequality.

Combining equation (10) with the inequality of Theorem 3, we have

$$\frac{4\pi}{R^{\nu-4}} P_c \leqslant I_\nu \leqslant \frac{3}{6-\nu} G (\tfrac{4}{3}\pi\rho_c)^{\nu/3} M^{(6-\nu)/3} \tag{11}$$

or

$$\frac{P_c}{\rho_c^{\nu/3}} \leqslant \frac{1}{6-\nu} (\tfrac{4}{3}\pi)^{(\nu-3)/3} G R^{\nu-4} M^{(6-\nu)/3} . \tag{12}$$

Again, equation (12) is a strict inequality for $\nu > 4$. This proves the theorem.

III

Comments on Theorem 4.—When $\nu = 4$, equation (12) reduces to an inequality from which Theorem 2 would immediately follow. If we write

$$\nu = 3\left(1 + \frac{1}{n}\right), \tag{13}$$

then we can re-write equation (12) as

$$\frac{P_c}{\rho_c^{(n+1)/n}} \leqslant S_n G\, R^{(3-n)/n} M^{(n-1)/n}, \qquad (1 < n \leqslant 3)\,, \tag{14}$$

where S_n stands for the numerical coefficient

$$S_n = (\tfrac{4}{3}\pi)^{1/n} \frac{n}{3(n-1)}\,. \tag{15}$$

Equations (14) and (15) bring out the very general "critical" nature of $n = 1$ and $n = 3$, a circumstance only very partially disclosed in the theory of polytropes. For, if we consider a polytrope of index n, then of course

$$\frac{P}{\rho^{(n+1)/n}} = \text{Constant} = \frac{P_c}{\rho_c^{(n+1)/n}}\,. \tag{16}$$

For the polytropic case (16) we have

$$\frac{P_c}{\rho_c^{(n+1)/n}} = T_n G\; R^{(3-n)/n} M^{(n-1)/n}\,, \tag{17}$$

where

$$T_n = \frac{1}{(n+1)} \left(\frac{4\pi}{\omega_n^{n-1}}\right)^{1/n}, \tag{18}$$

$$\omega_n = -\left(\xi^{\frac{n+1}{n-1}} \frac{d\theta_n}{d\xi}\right)_{\xi=\xi_1}\,. \tag{19}$$

The symbols ξ and θ stand for the Emden variables, θ_n is *any* Emden solution of index n, and the quantity in brackets in equation (19) has to be taken at the first zero $\xi = \xi_1$ of the Emden solution.[2]

[2] An Emden solution of index n is a function which satisfies Emden's equation of index n and is finite at the origin. Equation (19) can be evaluated at the boundary of any Emden solution because it is homology invariant.

Table I gives the values of S_n and T_n for different values of n.

TABLE I

n	S_n	T_n	S_n/T_n
3.0	0.806	0.364	2.214
2.5	0.985	.351	2.803
2.0	1.364	.365	3.741
1.5	2.599	.424	6.125
1.0	∞	0.637	∞

IV

THEOREM 5.—*If* $I_{\sigma,\nu}$ *stands for the integral*

$$I_{\sigma,\nu} = \int_0^R \frac{GM^\sigma(r)dM(r)}{r^\nu}, \qquad [3(\sigma + 1) > \nu], \tag{20}$$

then under the conditions of Theorem 1

$$\frac{3}{3\sigma + 3 - \nu}\frac{GM^{\sigma+1}}{\xi^\nu} \geqslant I_{\sigma,\nu} \geqslant \frac{3}{3\sigma + 3 - \nu}\frac{GM^{\sigma+1}}{R^\nu}, \tag{21}$$

where ξ *is defined by the relation*

$$\tfrac{4}{3}\pi\rho_c\xi^3 = M. \tag{22}$$

Proof: Since

$$r^\nu = \left[\frac{M(r)}{\frac{4}{3}\pi\bar{\rho}(r)}\right]^{\nu/3}, \tag{23}$$

we have from equation (20) that

$$I_{\sigma,\nu} = G(\tfrac{4}{3}\pi)^{\nu/3}\int_0^R \bar{\rho}^{\nu/3}(r)M^{(3\sigma-\nu)/3}dM(r). \tag{24}$$

Since we have assumed that $\bar{\rho}(r)$ decreases outward, it is clear that the minimum value of equation (24) is obtained by replacing $\bar{\rho}(r)$ by its minimum $\bar{\rho}$ (the mean density for the whole configuration) and taking it out of the integral sign. In the same way the maximum value of equation (24) is obtained by replacing $\bar{\rho}(r)$ by its maximum value ρ_c and taking it out of the integral sign. One thus finds that

$$\left.\begin{aligned} &\frac{3}{3\sigma+3-\nu}\, G(\tfrac{4}{3}\pi\bar{\rho})^{\nu/3} M^{(3\sigma+3-\nu)/3} \leqslant I_{\sigma,\nu} \\ &\qquad \leqslant \frac{3}{3\sigma+3-\nu}\, G(\tfrac{4}{3}\pi\rho_c)^{\nu/3} \cdot M^{(3\sigma+3-\nu)/3}\,, \end{aligned}\right\} \tag{25}$$

which is easily seen to be equivalent to equation (21).

V

Theorem 6.—*If $\bar{P}_{p,q}$ is the mean pressure defined by*

$$M^p R^q \bar{P}_{p,q} = \int_0^R P d(M^p(r) r^q)\,, \qquad (p,\, q > 0)\,, \tag{26}$$

then under the conditions of Theorem 1

$$\bar{P}_{p,q} \geqslant \frac{3}{4\pi}\, \frac{1}{3p+q+2}\, \frac{GM^2}{R^4}\,. \tag{27}$$

Proof: Integrating equation (26) by parts, we have

$$M^p R^q \bar{P}_{p,q} = -\int_0^R M^p(r) r^q dP\,. \tag{28}$$

By equation (8) we have

$$M^p R^q \bar{P}_{p,q} = \frac{1}{4\pi}\int_0^R \frac{GM^{p+1}(r) dM(r)}{r^{4-q}}\,, \tag{29}$$

$$= \frac{1}{4\pi}\, I_{p+1,\, 4-q}\,. \tag{30}$$

Hence, by Theorem 5 we easily find that

$$\bar{P}_{p,q} \geqslant \frac{3}{4\pi}\, \frac{1}{3p+q+2}\, \frac{GM^2}{R^4}\,, \tag{31}$$

which proves the theorem.

VI

THEOREM 7.—*In a wholly gaseous configuration in which the mean density inside* r, *the temperature, and the ratio of the radiation pressure to the total pressure decrease outward, we have*

$$T_c > \tfrac{1}{6}\frac{\mu H}{k}\frac{GM}{R}\beta^* , \tag{32}$$

where T_c *is the central temperature,* β^* *has the same meaning as in Theorem 2 and* μ, H, *and* k *are, respectively, the mean molecular weight (assumed constant in the whole configuration), the mass of the proton, and the Boltzmann constant.*

Proof: Let $\bar{P}$ denote the mean pressure defined by

$$R^3\bar{P} = \int_0^R Pd(r^3) . \tag{33}$$

By equation (31)

$$\bar{P} \geqslant \frac{3}{20\pi}\frac{GM^2}{R^4} = \tfrac{1}{5}(\tfrac{4}{3}\pi)^{1/3}G\bar{\rho}^{4/3}M^{2/3} . \tag{34}$$

In a wholly gaseous configuration

$$P = \frac{k}{\mu H}\beta^{-1}\rho T . \tag{35}$$

Hence

$$\bar{P} = \frac{k}{\mu H}\overline{(\beta^{-1}\rho T)} . \tag{36}$$

Since $(1 - \beta)$ is assumed to decrease outward, β must increase outward, and hence

$$\bar{P} < \frac{k}{\mu H}\beta_c^{-1}T_c\bar{\rho} , \tag{37}$$

where $\bar{\rho}$ is now the mean density defined in the usual way, since the means we are now taking are weighted according to the volume element; compare equation (33). Hence

$$T_c > \tfrac{1}{5}\frac{\mu H}{k}G(\tfrac{4}{3}\pi\bar{\rho})^{1/3}M^{2/3}\beta_c , \tag{38}$$

or

$$T_c > \tfrac{1}{5} \frac{\mu H}{k} \frac{GM}{R} \beta_c . \tag{39}$$

But by Theorem 2

$$\beta_c \geqslant \beta^* , \tag{40}$$

where β^* satisfies the quartic equation (3). Hence, combining equations (39) and (40), we have

$$T_c > \tfrac{1}{5} \frac{\mu H}{k} \frac{GM}{R} \beta^* , \tag{41}$$

which proves the theorem.

The inequality equation (41) is not a "best possible" one, but it has the advantage of not neglecting the radiation pressure and is, in fact, the first of the kind to be established.

We notice that the *mean* temperature $\overline{T}$, defined by

$$M\overline{T} = \int_0^R T dM(r) ,$$

satisfies the same inequality as T_c. For

$$M\overline{T} = \int_0^R T dM(r) = \frac{\mu H}{k} \int_0^R \beta P \rho^{-1} dM(r) .$$

$$= \frac{\mu H}{k} \int_0^R \beta P dv ,$$

where dv is the volume element. But since β is assumed to increase outward,

$$M\overline{T} \geqslant \frac{\mu H}{k} \beta_c \int_0^R P dv = \tfrac{1}{3} \frac{\mu H}{k} \beta_c \Omega ,$$

where Ω is the negative potential energy of the configuration. But by Theorem 3,

$$\Omega = I_1 \geqslant \tfrac{3}{5} \frac{GM^2}{R} .$$

Hence

$$\bar{T} \geqslant \tfrac{1}{5}\frac{\mu H}{k}\frac{GM}{R}\beta_c \geqslant \tfrac{1}{5}\frac{\mu H}{k}\frac{GM}{R}\beta^* .$$

VII

Corollary to Theorem 7.—In a wholly gaseous configuration in which the mean density inside r *and the ratio of the radiation pressure to the total pressure decreases outward, we have*

$$M < \left(2.5\,\frac{\overline{(\rho^{4/3})}}{(\bar{\rho})^{4/3}}\right)^{3/2}\left(\frac{6}{\pi}\right)^{1/2}\left[\left(\frac{k}{\mu H}\right)^4\frac{3}{a}\frac{1-\beta_c}{\beta_c^4}\right]^{1/2}\frac{1}{G^{3/2}} . \tag{42}$$

Proof: Equation (35) can also be written

$$P = \left[\left(\frac{k}{\mu H}\right)^4\frac{3}{a}\frac{1-\beta}{\beta^4}\right]^{1/3}\rho^{4/3} , \tag{43}$$

so that

$$\bar{P} = \overline{\left[\left(\frac{k}{\mu H}\right)^4\frac{3}{a}\frac{1-\beta}{\beta^4}\right]^{1/3}\rho^{4/3}} , \tag{44}$$

or, since $(1-\beta)$ decreases outward,

$$\bar{P} < \left[\left(\frac{k}{\mu H}\right)^4\frac{3}{a}\frac{1-\beta_c}{\beta_c^4}\right]^{1/3}\overline{(\rho^{4/3})} . \tag{45}$$

Combining equations (45) and (34), we have

$$\left[\left(\frac{k}{\mu H}\right)^4\frac{3}{a}\frac{1-\beta_c}{\beta_c^4}\right]^{1/3}\overline{(\rho^{4/3})} > \tfrac{1}{5}(\tfrac{4}{3}\pi)^{1/3}G\bar{\rho}^{4/3}M^{2/3} , \tag{46}$$

which, after some minor transformation, goes over into equation (42).

It may be noticed that in equation (42) we cannot replace $\overline{(\rho^{4/3})}/(\bar{\rho})^{4/3}$ by unity, since

$$\overline{(\rho^{4/3})} > (\bar{\rho})^{4/3} . \tag{47}$$

Equation (42) is not a "best possible" inequality, but it is much "sharper" than the inequality established previously.[3]

YERKES OBSERVATORY
April 12, 1937

[3] Equation (19′) (*loc. cit.*).

33

THE OPACITY IN THE INTERIOR OF A STAR

S. CHANDRASEKHAR

ABSTRACT

In this paper two integral theorems on the radiative equilibrium of a gaseous star are proved.

I

In two recent papers[1] the author has proved some general theorems on the equilibrium of a star. These theorems are of some importance in the theory of stellar structures, in so far as they provide inequalities for the physical variables, e.g., central pressure, mean pressure, central radiation pressure $(1 - \beta_c)$, etc., which should be valid under very general circumstances. The method consists in obtaining inequalities for the physical variables which are direct consequences of the equation of hydrostatic equilibrium:

$$\frac{dP}{dr} = -\frac{GM(r)}{r^2}\rho . \tag{1}$$

In obtaining inequalities based on equation (1), one generally restricts one's self to such equilibrium configurations as are characterized by the mean density $\bar{\rho}(r)$, inside r decreasing outward. Further, in obtaining inequalities for the mean temperature a further restriction, namely, that $(1 - \beta)$ decreases outward, is introduced (cf. II, Theorem 7).

In this paper we shall obtain certain inequalities for equilibrium configurations in *radiative equilibrium*. We shall then have an additional differential equation for the radiation pressure $p_r(= \frac{1}{3}aT^4)$, namely,

$$\frac{dp_r}{dr} = -\frac{\kappa L(r)}{4\pi c r^2}\rho , \tag{2}$$

where κ is the opacity coefficient, c is the velocity of light, and $L(r)$ is the amount of energy crossing the spherical surface of radius r.

The numbering of the theorems is continued from II, and references to the equations of that paper are inclosed in square brackets.

[1] *M.N.*, **96**, 644, 1935; *Ap. J.*, **85**, 372, 1937. These papers will be referred to as "I" and "II," respectively.

[78]

II

In the following, L, M, and R refer to the luminosity, the mass, and the radius of a star; and the auxiliary variable η is defined by

$$\eta = \frac{L(r)}{M(r)} \Big/ \frac{L}{M} . \tag{3}$$

THEOREM 8.—*In a wholly gaseous configuration in radiative equilibrium, in which the mean density* $\bar{\rho}(r)$ *inside* r *decreases outward, we have*

$$L \leqslant \frac{4\pi c G M(1 - \beta^*)}{\overline{\kappa\eta}} , \tag{4}$$

where $(1 - \beta^*)$ *has the same meaning as in Theorem 2 and* $\overline{\kappa\eta}$ *is defined by*

$$P_c\overline{\kappa\eta} = \int_R^0 \kappa\eta dP . \tag{5}$$

Part of the analysis leading up to this theorem is originally due to B. Strömgren.[2]

Proof: The equation of radiative equilibrium (2) can be written as

$$dp_r = - \frac{\kappa}{4\pi^2 c} \frac{L(r)dM(r)}{r^4} . \tag{6}$$

Since

$$dP = - \frac{G}{4\pi} \frac{M(r)dM(r)}{r^4} , \tag{7}$$

we have, using equation (3), that

$$dp_r = \frac{L}{4\pi c G M} \kappa\eta dP . \tag{8}$$

Integrating equation (8) and using the boundary condition that $p_r = 0$ at $r = R$, we have

$$p_r = \frac{L}{4\pi c G M} \int_R^r \kappa\eta dP . \tag{9}$$

[2] *Handbuch der Astrophysik*, **8**, 159, 1936.

The foregoing equation has been given before by Strömgren. From equation (9) we can easily show that *if* $\kappa\eta$ *decreases outward,* $(1 - \beta)$ *would also decrease outward.* Extending the integral from 0 to R and using the definition (5) for the average value for $\kappa\eta$, we clearly have that

$$1 - \beta_c = \frac{L}{4\pi cGM}\,\overline{\kappa\eta}\,, \tag{10}$$

or

$$L = \frac{4\pi cGM(1 - \beta_c)}{\overline{\kappa\eta}}\,. \tag{11}$$

Since we have assumed that the mean density decreases outward, we can apply Theorem 2, which states that

$$1 - \beta_c \leq 1 - \beta^*\,, \tag{12}$$

where β^* satisfies a certain quartic equation (equation [3]) and is determined by the mass M, uniquely. Combining equations (11) and (12), we have

$$L \leqslant \frac{4\pi cGM(1 - \beta^*)}{\overline{\kappa\eta}}\,, \tag{13}$$

which proves the theorem.

III

Theorem 9.—*In a wholly gaseous configuration in which the mean density* $\bar{\rho}(r)$ *inside* r *and the rate of generation of energy* ϵ *decrease outward, we have*

$$\bar{\kappa} \leqslant \frac{4\pi cGM(1 - \beta^*)}{L}\,, \tag{14}$$

where $\bar{\kappa}$ *is the mean opacity coefficient defined by*

$$P_c\bar{\kappa} = \int_R^0 \kappa dP\,, \tag{15}$$

and the equality sign in equation (14) is possible only when ϵ *is constant.*

Proof: This is an immediate consequence of Theorem 8. For, if ϵ decreases outward, η must also decrease outward, and consequently the ***minimum*** value of η is unity. Hence

$$\overline{\kappa\eta} \geqslant \bar{\kappa}\,, \tag{16}$$

the equality sign in equation (16) being possible only when $\eta =$ constant $= 1$, i.e., when ϵ is constant. By Theorem 8

$$\overline{\kappa\eta} \leqslant \frac{4\pi c G M(1-\beta^*)}{L}\,. \tag{17}$$

Combining equations (16) and (17), we have the required result.

IV

We will apply equation (14) to certain practical cases of interest. Numerically, equation (14) reduces to

$$\bar{\kappa} \leqslant 1.318 \times 10^4 \frac{M}{\odot} \cdot \frac{L_\odot}{L} (1-\beta^*)\,, \tag{18}$$

where $L_\odot$ refers to the luminosity of the sun.

For Capella we have $M = 4.18\odot$ and $L = 126\ L_\odot$. Assuming $\mu = 1$, the solution of the quartic equation for β^* yields $1-\beta^* = 0.22$. Hence

$$\bar{\kappa}_{\text{Capella}} < 96.1\ gm^{-1}\ cm^2\,. \tag{19}$$

In the same way for the sun, we find ($\mu = 1$, $1-\beta^* = 0.03$)

$$\bar{\kappa}_\odot < 395\ gm^{-1}\ cm^2\,. \tag{20}$$

V

There is one interesting application of equation (14) to stellar models in which the opacity coefficient κ is assumed to be constant. For equation (14) can then be written as

$$L \leqslant L^* = \frac{4\pi c G M(1-\beta^*)}{\kappa}\,. \tag{21}$$

Equation (21) has to be interpreted in the following sense: If $L > L^*$, then the configuration *must* be characterized by the occurrence of negative density gradients (i.e., $\bar{\rho}(r)$ increases outward in some finite regions of the interior), no matter what the law of energy generation is, provided only the rate of generation of energy ϵ decreases outward. On the other hand, if $L < L^*$, it does not *necessarily* follow that the configuration is characterized by positive density gradients throughout its interior. But if $L < L^*$ we can always find a "mild"-enough law for the rate of generation of energy such that the configuration is characterized by a positive density gradient throughout its interior. Further, it should be noticed that if

$$L > L_1 = \frac{4\pi cGM}{\kappa}, \tag{22}$$

then no equilibrium configuration is possible.[3] The inequality (22) is interpreted by the statement that if $L > L_1$ then the configuration would "blow up." We now see that this tendency to "blow up" must set in at lower values for the luminosity, in the event of negative density gradients in its interior. Negative density gradients must certainly exist for configurations with $L > L^*$. Depending on the concentration of the energy sources toward the center, the negative density gradients will set in for some $L < L^*$.

VI

Corollary to Theorem 9.—If, in addition to the conditions of Theorem 9, κ is assumed to increase outward, then

$$\kappa_c \leqslant \frac{4\pi cGM(1-\beta^*)}{L}, \tag{23}$$

where κ_c is the opacity at the center. This is, of course, obvious.

If we assume any definite law for opacity, then equation (23) can be converted into an inequality for the central temperature, for a star of known mass and luminosity. Thus, if we assumed that

$$\kappa = \kappa_1 \frac{\rho}{T^{3+S}}, \qquad (S > 0), \tag{24}$$

[3] E. A. Milne, *M.N.*, **91**, 4, 1930. See esp. pp. 12, 13, and 53 of this paper.

then κ would *increase* outward if $(1 - \beta)$, and T would *decrease* outward; for we can write equation (24) as

$$\kappa = \kappa_1 \frac{\mu H}{k} \frac{a}{3} \frac{\beta}{1-\beta} T^{-s} . \tag{25}$$

Hence, by Theorem 2,

$$\kappa_c \geqslant \kappa_1 \frac{\mu H}{k} \frac{a}{3} \frac{\beta^*}{1-\beta^*} T_c^{-s} . \tag{26}$$

Combining this with (23), we have

$$T_c^s \geqslant \frac{L\kappa_1}{4\pi c G M} \frac{\mu H}{k} \frac{a}{3} \frac{\beta^*}{(1-\beta^*)^2} . \tag{27}$$

The foregoing inequality giving the minimum central temperature for a star of known M and L is generally not as good as the minimum central temperature set by Theorem 7 for a star of known M and R.

APPENDIX

We have seen that $(1 - \beta^*)$, giving the maximum possible $(1 - \beta_c)$ in a wholly gaseous configuration in which the mean density $\bar{\rho}(r)$, inside r, is assumed to decrease outward, plays an important role in Theorems 7, 8, and 9. It is therefore convenient to have a table giving M for different values of $(1 - \beta^*)$. Table I should be sufficient for most purposes.

TABLE 1

SOLUTIONS OF THE EQUATION

$$M = \left(\frac{6}{\pi}\right)^{1/2} \left[\left(\frac{k}{\mu H}\right)^4 \frac{3}{a} \frac{1-\beta^*}{\beta^{*4}}\right]^{1/2} \frac{1}{G^{3/2}}$$

$1-\beta^*$	$\left(\frac{M}{\odot}\right)\mu^2$	$1-\beta^*$	$\left(\frac{M}{\odot}\right)\mu^2$
0.025	0.908	0.5	15.432
.05	1.352	0.6	26.41
.1	2.130	0.7	50.72
.2	3.812	0.8	122.0
.3	6.099	0.9	517.6
0.4	9.585	1.0	....

YERKES OBSERVATORY
June 8, 1937

AN INTEGRAL THEOREM ON THE EQUILIBRIUM OF A STAR

S. CHANDRASEKHAR

ABSTRACT

In this paper an integral theorem on the equilibrium of a star is proved which gives the lower limit to the value of $P_c/\rho_c^{(n+1)/n}$, assuming that both ρ and $P/\rho^{(n+1)/n}$ do not increase outward. As a special case of the theorem ($n = 3$) it is shown that for a gaseous star of a given mass in radiative equilibrium, in which ρ and $[\overline{\kappa\eta}]_R^r$ do not increase outward, the minimum value of $1 - \beta_c$ is the constant value of $(1 - \beta)$ ascribed to a standard model configuration of the same mass. For $n = \infty$ the theorem gives the minimum central temperature for a gaseous star with negligible radiation pressure.

In some recent papers[1] the author has proved a number of integral theorems on the equilibrium of a star. In particular it was shown that for any equilibrium configuration (of prescribed mass and radius) in which the mean density $\bar{\rho}(r)$ inside r decreases outward it is possible to set an *upper limit* to the value of $P_c/\rho_c^{(n+1)/n}$ for $1 < n \leqslant 3$. The inequality in question is (II, Eqs. [14] and [15])

$$\frac{P_c}{\rho_c^{(n+1)/n}} \leqslant S_n G M^{(n-1)/n} R^{(3-n)/n} \,, \qquad (1 < n \leqslant 3) \,, \quad (1)$$

where

$$S_n = (\tfrac{4}{3}\pi)^{1/n} \frac{n}{3(n-1)} \,. \qquad (2)$$

Furthermore, (1) is a *strict* inequality for $n < 3$, and for $n = 3$ is equivalent to setting an upper limit to $(1 - \beta_c)$ for gaseous stars (cf. Theorem 2, II).

The problem of finding a *lower limit* to the ratio $P_c/\rho_c^{(n+1)/n}$ has proved to be rather an elaborate one. In this paper we prove theorems in this direction.

The numbering of the theorems is continued from II and III.

I

THEOREM 10.—*In any equilibrium configuration of prescribed mass and radius in which both* ρ *and* $K = P/\rho^{(n+1)/n}$, $(n > 1)$ *do not increase*

[1] *M.N.*, **96**, 644, 1936; *Ap. J.*, **85**, 372, 1937, and **86**, 78, 1937. These papers will be referred to as "I," "II," and "III," respectively.

outward, the minimum value of K_c *is attained in the sequence of equilibrium configurations which consist of polytropic cores of index* n *and homogeneous envelopes.*

More explicitly, wc consider a *composite* configuration in which the polytropic core extends to a fraction A of the radius R of the star. Inside the polytropic core, K is constant and equal to K_c. For such composite configurations, K_c will be a function $K_c(A)$ of A only. The theorem states that the minimum value of the function $K_c(A)$ is the *absolute* minimum of K_c for any equilibrium configuration in which ρ and K are restricted not to increase outward.

Proof: We shall first prove the following lemma:

Lemma: The configuration in which K_c *attains the minimum, either* $d\rho/dr = 0$ *or* $dK/dr = 0$ *for all* $0 \leqslant r \leqslant R$.

For, if not, in the configuration in which K attains its minimum there must exist a finite interval

$$0 < r_1 \leqslant r \leqslant r_2 < R \,, \tag{3}$$

in which

$$\frac{d\rho}{dr} < 0 \,; \qquad \frac{dK}{dr} < 0 \,. \tag{4}$$

Let P and ρ refer to the configuration we are considering, namely, the one in which K_c attains its minimum.

By means of the following transformation we construct the pressure and density distributions defined by

$$0 \leqslant r \leqslant r_1: \qquad P^* = (1-\epsilon)^2 P \,; \qquad \rho^* = (1-\epsilon)\rho \,, \tag{5}$$

$$r_1 < r \leqslant r_2: \qquad P^* = P + \epsilon P_1 \,; \qquad \rho^* = \rho + \epsilon\rho_1 \,, \tag{6}$$

$$r_2 < r \leqslant R: \qquad P^* = P \,; \qquad \rho^* = \rho \,, \tag{7}$$

where P^* and ρ^* refer to the new distributions of pressure and density, ϵ is a sufficiently small *positive* constant, and P_1 and ρ_1 (which are functions of r in the interval $r_1 \leqslant r \leqslant r_2$) are, for the present, unspecified.

If P^* and ρ^* should refer to an *equilibrium* configuration of the same mass M as the original configuration, then the following conditions would be fulfilled.

i) *Continuity of* P^* *and* ρ^*.—From (5), (6), and (7) we see that to insure continuity at $r = r_1$ and $r = r_2$ we should have

$$\rho_1 = -\rho \,, \qquad (r = r_1) \; ; \qquad \rho_1 = 0 \,, \qquad (r = r_2) \; ; \tag{8}$$

$$P_1 = -2P \,, \qquad (r = r_1) \; ; \qquad P_1 = 0 \,, \qquad (r = r_2) \, . \tag{9}$$

ii) *Constancy of mass.*—This requires

$$4\pi \int_0^R \rho^* r^2 dr = 4\pi \int_0^R \rho r^2 dr \, . \tag{10}$$

By (5), (6), and (7) we find that (10) reduces to

$$M(r_1) = 4\pi \int_{r_1}^{r_2} \rho_1 r^2 dr \, . \tag{11}$$

The left-hand side of (11) is a known quantity. We can clearly choose a function ρ_1 in the interval $r_1 \leqslant r \leqslant r_2$ such that the equation (11) and the boundary conditions at r_1 and r_2 (Eq. [8]) are all satisfied. We assume that ρ_1 has been chosen to satisfy these conditions.

iii) *The distributions* P^* *and* ρ^* *satisfy the equation of hydrostatic equilibrium.*—The pressure-density distributions in any configuration of equilibrium must satisfy the equation

$$\frac{1}{r^2}\frac{d}{dr}\left(\frac{r^2}{\rho}\frac{dP}{dr}\right) = -4\pi G\rho \, . \tag{12}$$

Given that P and ρ satisfy (12), we have to show that P^* and ρ^* distributions (with a suitable choice of P_1) satisfy (12).

It is immediately obvious that in the interval $0 \leqslant r \leqslant r_1$ and

$r_2 \leqslant r \leqslant R$, equation (12) is satisfied. For $r_1 \leqslant r \leqslant r_2$ we have, according to (5), (6), and (7),

$$\frac{1}{r^2}\frac{d}{dr}\left(\frac{r^2}{\rho}\frac{dP_1}{dr} - \frac{r^2}{\rho^2}\rho_1\frac{dP}{dr}\right) = -4\pi G\rho_1 . \tag{13}$$

From (13) we derive

$$\frac{1}{r^2}\frac{d}{dr}\left(\frac{r^2}{\rho}\frac{dP_1}{dr}\right) = F(r) , \tag{14}$$

where

$$F(r) = -4\pi G\rho_1 + \frac{1}{r^2}\frac{d}{dr}\left(\frac{r^2}{\rho^2}\rho_1\frac{dP}{dr}\right) . \tag{15}$$

Since P and ρ are assumed to be known functions of r, and ρ_1 has been chosen according to (ii) above, we can regard $F(r)$ as a known function of r. From (14) we easily derive

$$P_1 = \int_{r_1}^{r}\frac{\rho}{r^2}\left\{\int_{r_1}^{r}\xi^2F(\xi)d\xi + c_1\right\}dr + c_2 , \tag{16}$$

where c_1 and c_2 are two integration constants. We now choose c_1 and c_2 such that P_1 defined by (16) satisfies the boundary conditions (9).

We thus see that with P_1 and ρ_1 chosen as specified in (ii) and (iii) above, P^* and ρ^* refer to an equilibrium configuration of the same mass and radius as the original configuration.

Finally, since in the interval $r_1 \leqslant r \leqslant r_2$, ρ and K are strictly decreasing (Eq. [4]), it is clear that we can choose a positive (*nonzero*) ϵ sufficiently small that ρ^* and K^* (defined with respect to P^* and ρ^*) are decreasing functions of r.

We have thus shown that from the given equilibrium configuration we can construct another satisfying the restrictions on ρ and K. But the configuration specified by the functions P^* and ρ^* defines

$$K_c^* = (1 - \epsilon)^{1-(1/n)} K_c . \tag{17}$$

Since we have assumed $n > 1$, we see that (17) implies

$$K_c^* < K_c \, , \tag{18}$$

which contradicts our hypothesis that in the configuration in which K_c attains its minimum there exists an interval (3) in which (4) holds. This proves the lemma.

The theorem now follows almost immediately. It is only necessary to exclude the types of density distributions shown by the full-line curves in Figures 1 and 2. In Figure 1, the regions *1* and *3* are regions of constant K, while *2* is a region of constant ρ; in Figure 2, *1* and *3* are regions of constant K, while *2* and *4* are regions of constant ρ.

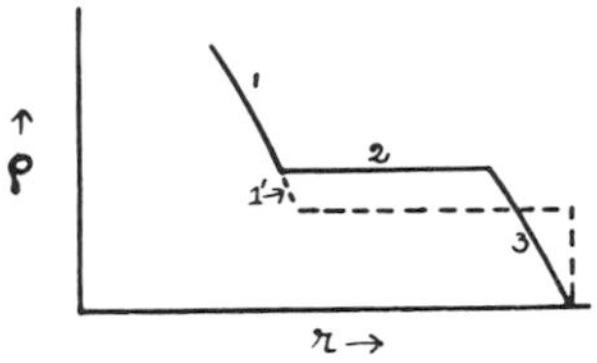

FIG. 1

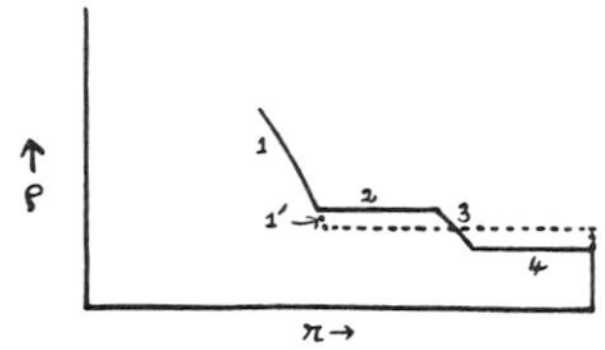

FIG. 2

But it is clear that by the constructions indicated by the dotted curves (in both cases *1′* corresponds to the analytic continuation of the density distribution specified by *1*) we are led to configurations with a smaller P_c, and hence a smaller K_c. This proves the theorem.

It has to be noticed that we have only proved that the minimum of K_c along the sequence of composite configurations (consisting of polytropic cores and homogeneous envelopes) is the absolute minimum of K_c under the restrictions $d\rho/dr \leqslant 0$, $dK/dr \leqslant 0$. But we have not yet specified the *particular* composite configuration in which the minimum of K_c is attained; to be able to do so, we shall have to study the function $K_c(A)$ where A is the fraction of the radius occupied by the polytropic core. We now proceed to study this function.

II

The composite configurations.—We consider a composite configuration in which the polytropic core extends to a fraction A of the radius R. Hence, if $r = r_1$ defines the place at which we have the *in-*

terface between the polytropic and the homogeneous regions, we have

$$r_1 = AR\,. \tag{19}$$

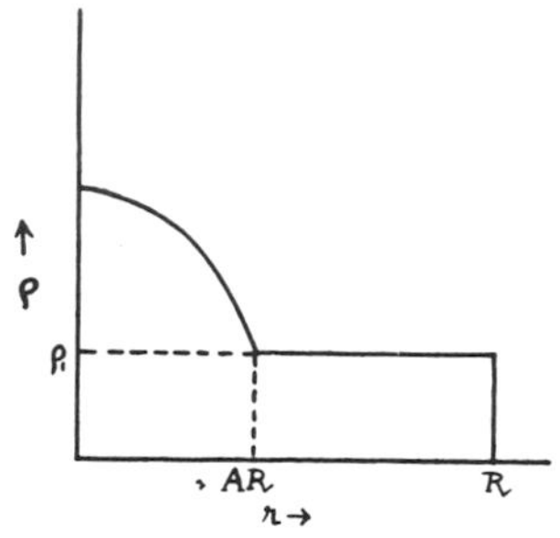

FIG. 3

Let $\bar{\rho}_1$ be the mean density of the polytropic core and ρ_1 the constant density in the homogeneous part. Let B denote the ratio

$$B = \frac{\bar{\rho}_1 - \rho_1}{\rho_1}\,. \tag{20}$$

Finally, let P_1 be the pressure at the interface.

Consider first the equilibrium of the homogeneous envelope: We have

$$M(r) = \tfrac{4}{3}\pi r_1^3\bar{\rho}_1 + \tfrac{4}{3}\pi(r^3 - r_1^3)\rho_1\,, \tag{21}$$

or, by (19) and (20),

$$M(r) = \tfrac{4}{3}\pi R^3\rho_1 A^3 B + \tfrac{4}{3}\pi r^3\rho_1\,. \tag{22}$$

The mass M of the configuration is therefore given by

$$M = \tfrac{4}{3}\pi R^3\rho_1\,(1 + A^3 B)\,. \tag{23}$$

The equation of hydrostatic equilibrium is

$$\frac{dP}{dr} = -\frac{GM(r)}{r^2}\,\rho_1\,, \tag{24}$$

or, by (22),

$$\frac{dP}{dr} = -G\,\tfrac{4}{3}\pi\rho_1^2\left(\frac{R^3A^3B}{r^2} + r\right), \tag{25}$$

or, integrating,

$$P = G\,\tfrac{4}{3}\pi\rho_1^2\left[\frac{R^3A^3B}{r} - \tfrac{1}{2}r^2\right]_R^r. \tag{26}$$

The pressure at the interface is obtained by putting $r = AR$ in (26). After some reductions we find that

$$P_1 = \tfrac{2}{3}\pi G\rho_1^2R^2[1 - A^2 + 2A^2B(1 - A)]\,. \tag{27}$$

Consider, now, the equilibrium of the polytropic core: in the core we can write

$$P = K\rho^{(n+1)/n}\,, \tag{28}$$

where K is a constant. The reduction to Emden's equation of index n is made by the substitutions

$$\rho = \lambda\theta^n\,; \qquad P = K\,\lambda^{(n+1)/n}\,\theta^{n+1}\,, \tag{29}$$

$$r = \left[\frac{(n+1)K}{4\pi G}\right]^{1/2}\lambda^{(1-n)/2n}\xi\,. \tag{30}$$

Let θ and ξ refer to the interface. Then P_1, ρ_1, and r_1 are given by the foregoing formulae. By (27), (29), and (30) we have

$$K\lambda^{(n+1)/n}\theta^{n+1} = \tfrac{2}{3}\pi G(\lambda\theta^n)^2\left[\frac{(n+1)K}{4\pi G}\right]\lambda^{(1-n)/n}\xi^2\,\frac{1 - A^2 + 2A^2B(1 - A)}{A^2}\,. \tag{31}$$

After some reductions the foregoing equation reduces to

$$1 = \xi^2\theta^{n-1}\cdot\frac{n+1}{6}\cdot\frac{1 - A^2 + 2A^2B(1 - A)}{A^2}\,. \tag{32}$$

Now introduce the homology invariant functions u and v defined by

$$u = -\frac{\xi\theta^n}{\theta'}\,; \qquad v = -\frac{\xi\theta'}{\theta}\,, \tag{33}$$

where θ' refers to the derivative of θ with respect to ξ. In the terms of u and v (32) can be re-written as

$$\frac{n+1}{6}\,uv = \frac{A^2}{1 - A^2 + 2A^2B(1-A)}\,. \tag{34}$$

Furthermore,

$$B = \frac{\bar{\rho}_1 - \rho_1}{\rho_1} = \frac{\bar{\rho}_1}{\rho_1} - 1\,, \tag{35}$$

or, using the well-known relation between the mean and the central densities for polytropic configurations, we have

$$B = -\frac{\lambda \dfrac{3}{\xi}\dfrac{d\theta}{d\xi}}{\lambda\theta^n} - 1 = -3\,\frac{\theta'}{\xi\theta^n} - 1\,, \tag{36}$$

or, by (33),

$$B = \frac{3}{u} - 1\,. \tag{37}$$

Equations (34) and (37) are our *equations of fit*. If the Emden function $\theta_n(\xi)$ is known, then for a given ξ, u and v are known and (34) and (37) determine A as the solution of a cubic equation. The configuration thus becomes determinate.

We have next to determine K in terms of A, R, and M. Using (34), we can re-write (27) as

$$P_1 = \tfrac{2}{3}\pi G\rho_1^2 R^2 A^2\,\frac{6}{(n+1)uv}\,. \tag{38}$$

By (28) and (38) we now have

$$K = \tfrac{2}{3}\pi G\rho_1^{(n-1)/n} R^2 A^2\,\frac{6}{(n+1)uv}\,. \tag{39}$$

We now eliminate ρ_1 between (39) and the mass relation (23). We thus have

$$K = \tfrac{2}{3}\pi G R^2 A^2\,\frac{6}{(n+1)uv}\left[\frac{M}{\tfrac{4}{3}\pi R^3(1 + A^3B)}\right]^{(n-1)/n}, \tag{40}$$

which, after some reductions, can be expressed as

$$K = \tfrac{1}{2}(\tfrac{4}{3}\pi)^{1/n} G M^{(n-1)/n} R^{(3-n)/n} \cdot Q_n \,, \tag{41}$$

where

$$Q_n = \frac{6}{(n+1)uv} \frac{A^2}{(1+A^3B)^{(n-1)/n}} \,. \tag{42}$$

We verify the following: Q_n *measures* K *in units of the value of* K_c *for the configuration of uniform density of mass* M *and radius* R. *The minimum of* $Q_n(A)$ *defines, according to our theorem, the minimum value of* K_c *in the specified units.*

Equation (34) is a cubic equation for A. Eliminating B between (34) and (37), we find that the equation for A can be written more conveniently as

$$2(n+1)v(3-u)A^3 + 3A^2[(n+1)v(u-2)+2] - (n+1)uv = 0 \,. \tag{43}$$

Furthermore,

$$B = \frac{3}{u} - 1 \,. \tag{44}$$

Equations (42), (43), and (44) define, then, the function $Q_n(A)$.

We notice that, as $u \to 0$,

$$A \to 1 \, ; \qquad B \to \frac{3}{u} \,, \tag{45}$$

$$Q_n \to \frac{6}{n+1}\left(\frac{1}{3\omega_n}\right)^{(n-1)/n} , \tag{46}$$

where

$$\omega_n = -\left(\xi^{(n+1)/(n-1)} \frac{d\theta_n}{d\xi}\right)_{\xi=\xi_1} \,. \tag{47}$$

Inserting (46) in (41), we obtain equation (17), II.

III

The case n = 5.—The case $n = 5$ presents some interesting features. It is found that the Schuster-Emden integral for the case $n = 5$ reduces to

$$3v + u = 3 \tag{48}$$

in the (u, v)-plane. The cubic equation for A (Eq. [43]) now becomes

$$3(1 + 2A)(1 - A)^2v^2 - 3(1 - A^2)v + A^2 = 0 , \tag{49}$$

or, solving for v,

$$v = \frac{3(1 + A) - \sqrt{3(1 - A)(3 + 9A + 8A^2)}}{6(1 + 2A)(1 - A)} . \qquad (50)^2$$

Also,

$$B = \frac{v}{1 - v} ; \qquad Q_5 = \frac{1}{3v(1 - v)} \frac{A^2}{(1 + A^3B)^{4/5}} . \tag{51}$$

Equations (50) and (51) present the explicit solution for the problem.

Furthermore, we notice that if $v = 1$, (49) reduces to

$$6A^3 - 5A^2 = 0 , \qquad \text{or} \qquad A = \tfrac{5}{6} . \tag{52}$$

Hence, for $A = \frac{5}{6}$, $v = 1$, $B = \infty$, $Q_5 = \infty$. Hence,

$$Q_5(A) \to \infty , \qquad A \to \tfrac{5}{6} . \tag{53}$$

The details of the solution are given in Table 1.

IV

The case $1 < n < 5$.—The solutions of the equations of fit have been effected for $n = 4.5, 4, 3,$ and 2. The details of the solution are given in Table 1. The respective $Q_n(A)$ curves are shown in Figure 4. From the figure we infer the following theorems.

THEOREM 11.—*In any equilibrium configuration of prescribed mass and radius in which both* ρ *and* $K = P/\rho^{(n+1)/n}$, $(1 < n \leqslant 3)$ *do not increase outward, the minimum value of* K_c *is the constant value of* K *which must be ascribed to a complete polytrope of index* n *having the given mass and radius.*

For the case $n = 3$ the foregoing theorem can be stated in the following alternative form:

THEOREM 12.—*In a gaseous stellar configuration in which both* ρ *and* $(1 - \beta)$ *do not increase outward,* $(1 - \beta_c)$ *must be greater than the*

[2] The positive sign before the square root in (50) is easily seen to correspond to a physically impossible solution.

constant value of (1 − β) *ascribed to a standard model configuration of the same mass.*

We shall comment on the implications of Theorem 12 in the studies of stellar structure in Section VIII.

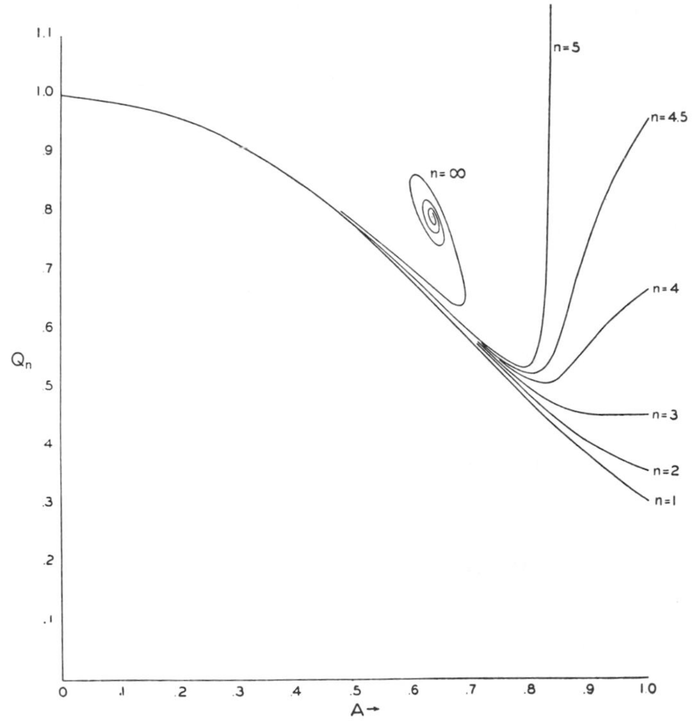

FIG. 4

V

The case n = 1.—The theorem has been proved only for the case $n > 1$. The arguments of Section I fail for this case (cf. Eqs. [17] and [18]). However, it seems likely that the theorem is also true for $n = 1$. If it is true, Theorem 11 is also valid for this case.

For the case $n = 1$ the Emden function is known:

$$\theta_1 = \frac{\sin \xi}{\xi}. \tag{54}$$

The appropriate equations are

$$\tfrac{1}{3}\xi^2 = \frac{A^2}{1 - A^2 + 2A^2B(1 - A)}, \tag{55}$$

$$B = 3\left[\frac{1}{\xi^2} - \frac{\cot \xi}{\xi}\right] - 1, \tag{56}$$

$$Q_1 = \frac{3A^2}{\xi^2}. \tag{57}$$

As $\xi \to \pi$, $A \to 1$ and $Q_1 = 3/\pi^2$. The details of the solution are given in Table 1.

VI

The case n $= \infty$.—For this case the conditions of the theorem are that both ρ and P/ρ do not increase outward. For gaseous stars with negligible radiation pressure $P/\rho = (k/\mu H)\,T$, and the theorem therefore sets a lower limit to the central temperature of such configurations.[3]

To determine the minimum value of Q_∞, we have to consider composite configurations consisting of *isothermal* cores and homogeneous envelopes. The analysis of these configurations is quite similar to that given in Section II.

If ψ is the Emden isothermal function, we define the functions u and v by

$$u = \frac{\xi e^{-\psi}}{\psi'}; \qquad v = \xi\frac{d\psi}{d\xi}. \tag{58}$$

The equations of fit are

$$\tfrac{1}{6}uv = \frac{A^2}{1 - A^2 + 2A^2B(1 - A)}, \tag{59}$$

$$B = \frac{3}{u} - 1, \tag{60}$$

[3] The problem of determining the minimum central temperature of a gaseous star with negligible radiation pressure has been considered earlier by Eddington (*Internal Constitution of the Stars* [Cambridge, 1924], pp. 91–93). However, his treatment of the problem is different from our approach to it, as it is a special case of a more general problem.

and

$$\left(\frac{P_c}{\rho_c}\right)_{\min} = \frac{1}{2}\frac{GM}{R}Q_\infty\,, \tag{61}$$

where

$$Q_\infty = \frac{6}{uv}\frac{A^2}{1 + A^3B}\,. \tag{62}$$

Equation (59) can be written alternatively as

$$2v(3 - u)A^3 + 3A^2[v(u - 2) + 2] - uv = 0\,. \tag{63}$$

Now, it is well known that the isothermal function oscillates about the singular solution

$$e^{-\psi} = \frac{2}{\xi^2} \tag{64}$$

as $\xi \to \infty$. In the (u, v)-plane this corresponds to the appropriate curve spiraling around

$$u = 1\,, \qquad v = 2\,. \tag{65}$$

Introducing (65) into (60) and (64) we find that

$$B = 2\,; \qquad 4A^3 = 1 \qquad \text{or} \qquad A = \sqrt[3]{0.25}\,. \tag{66}$$

With these values of A and B, (62) gives

$$Q_\infty = \sqrt[3]{0.5}\,. \tag{67}$$

Hence, the $Q_\infty(A)$ curve spirals around the point

$$Q_\infty^{(s)} = \sqrt[3]{0.5} = 0.7937\,; \qquad A^{(s)} = 0.62996\,. \tag{68}$$

The details of the solution are given in Table 1*a*. See also Figure 4.

VII

Numerical results.—In Tables 1 and 1*a* we give the details of the solution for the cases

$$n = 1,\ 2,\ 3,\ 4,\ 4.5,\ 5,\ \text{and}\ \infty\ . \tag{69}$$

TABLE 1

VALUES OF Q_n

$n=1$		$n=2$		$n=3$		$n=4$		$n=4.5$		$n=5$	
A	Q_1	A	Q_2	A	Q_3	A	Q_4	A	$Q_{4.5}$	A	Q_5
0.	1.	0.	1.	0.	1.0	0.	1.	0.	1.0	0.	1.
				0.366	0.8710					0.15	0.9768
				0.423	0.8339			0.410	0.841	0.30	0.914
0.685	0.592	0.557	0.722	0.590	0.6946			......	......	0.40	0.8505
0.737	0.537	......	......	0.708	0.5787	0.662	0.626	0.716	0.576	0.50	0.7738
0.784	0.489	0.785	0.499	0.818	0.4833	0.775	0.526	0.764	0.538	0.60	0.6859
0.866	0.410	......	......	0.872	0.4578	0.819	0.508	0.807	0.524	0.70	0.5931
0.919	0.363	0.893	0.408	0.912	0.4523	0.845	0.516	0.840	0.564	0.75	0.5521
......	......	......	......	0.949	0.4519	0.879	0.551	0.875	0.681	0.78	0.5372
......	......	......	......	0.964	0.4517	0.907	0.585	0.933	0.837	0.80	0.5410
......	......	......	......	0.968	0.4516	0.931	0.613	0.961	0.893	0.832	1.0361
0.968	0.325	0.934	0.384	0.971	0.4516	0.974	0.648	......	......	0.833	1.1500
0.984	0.314	0.996	0.358	0.986	0.4515	......	......	0.990	0.940	0.8333	2.2054
1.000	0.30396	1.000	0.3564	1.000	0.45154	1.000	0.6671	1.000	0.9572	$\frac{5}{6}$	∞

TABLE 1*a*

VALUES OF Q_∞

ξ	A	Q_∞	ξ	A	Q_∞
0.	0.	1.	70	0.608	0.849
2.0	.542	0.743	100	.594	.833
4.0	.663	0.646	150	.616	.808
5.0	.680	0.640	200	.615	.786
7.0	.689	0.655	300	.629	.783
10.0	.683	0.695	400	.633	.778
12	.675	0.721	500	.635	.776
16	.661	0.766	700	.636	.778
25	.636	0.826	1000	.636	.784
30	.626	0.840	1500	.627	.821*
35	.620	0.850	2000	.632	.796
50	0.609	0.859	∞	0.62996	0.7937

* There seems to be an error in Emden's table of the isothermal function at this point.

In Table 2 the minimum values of $P_c/\rho_c^{(n+1)/n}$ under the conditions of Theorem 10 are given. From (41) and (42) it is clear that we can express the theorem as

$$\frac{P_c}{\rho_c^{(n+1)/n}} \geqslant \Sigma_n G M^{(n-1)/n} R^{(n-3)/n} , \tag{70}$$

where

$$\Sigma_n = \tfrac{1}{2}(\tfrac{4}{3}\pi)^{1/n} \text{ Minimum } Q_n(A) . \tag{71}$$

The values of Σ_n are also given in Table 2.

TABLE 2

n	A	Min Q_n	Σ_n
1.0	1.0	0.30396	0.63662
1.5	1.0	.3265	.42422
2.0	1.0	.3564	.36475
2.5	1.0	.3964	.35150
3.0	1.0	.45154	.36394
4.0	0.821	.508	.363
4.5	0.801	.524	.360
5.0	0.788	.535	.356
∞	0.678	0.640	0.320

VIII

Some remarks on Theorem 12.—We have shown that if ρ and $(1-\beta)$ decrease outward, then it is possible to set a lower limit to $(1-\beta_c)$ which depends on the mass only. We now examine the physical meaning of the assumption that "$(1-\beta)$ does not increase outward" for the case of radiative equilibrium.

We then have Strömgren's relation (cf. III, Eq. [9])

$$1-\beta = \frac{L}{4\pi c G M}\,\overline{\kappa\eta}(r) , \tag{72}$$

where

$$\overline{\kappa\eta}(r) = \frac{1}{P}\int_R^r \kappa\eta\, dP . \tag{73}$$

Hence, $(1-\beta)$ *will decrease outward if* $\overline{\kappa\eta}(\mathrm{r})$ *decreases outward.* Analytically, the condition is

$$\frac{d}{dr}\,\overline{\kappa\eta}(r) \leqslant 0\,, \tag{74}$$

or, since dP/dr is negative,

$$\frac{d}{dP}\,\overline{\kappa\eta}(r) \geqslant 0\,. \tag{75}$$

From (73) and (75) we derive that (74) is equivalent to

$$\kappa\eta(r) \geqslant [\overline{\kappa\eta}]_R^r\,. \tag{76}$$

In words: *The necessary and sufficient condition for* $(1-\beta)$ *decreasing outward is that* $\kappa\eta$ *at any point inside the star must be greater than the average value of* $\kappa\eta$ *for material exterior to* r. It should be noticed that the condition stated is less restrictive than the requirement that $\kappa\eta$ decreases outward. It is clear from (76) that we can actually allow a decrease of $\kappa\eta$ (within limits) as we approach the center. In actual stellar configurations η might be expected to decrease outward, but this will not generally be true of κ. For this reason it is important to realize that (76) does not require $\kappa\eta$ to decrease outward. We can now express Theorem 12 in the following alternative way. *In a wholly gaseous configuration in radiative equilibrium in which the density and* $\overline{\kappa\eta}(\mathrm{r})$ *as defined by* (73) *both decrease outward, the central value* $(1-\beta_c)$ *of the ratio of the radiation pressure to the total pressure must satisfy the inequality.*

$$1-\beta_c \geqslant 1-\beta_s\,, \tag{77}$$

where $(1-\beta_s)$ *satisfies the quartic equation*

$$M = -4\pi\,\frac{1}{(\pi G)^{3/2}}\left[\left(\frac{k}{\mu H}\right)^4 \frac{3}{a}\,\frac{1-\beta_s}{\beta_s^4}\right]^{1/2}\left(\xi^2\,\frac{d\theta_3}{d\xi}\right)_{\xi=\xi_1}. \tag{78}$$

It might be recalled that under very much less restrictive circumstances than in the foregoing theorem we have shown that

$$1-\beta_c \leqslant 1-\beta^*\,, \tag{79}$$

where $(1 - \beta^*)$ satisfies a similar quartic equation (cf. II, Theorem 2). It is thus seen that we can solve the problem of finding both the upper and the lower limits of $1 - \beta_c$. We have (cf. III, Eq. [11])

$$L = \frac{4\pi cGM(1 - \beta_c)}{\overline{\kappa\eta}}, \tag{80}$$

where $\overline{\kappa\eta}$ refers now to the average over the whole star. Hence, under the conditions of Theorem 12 by (77),

$$L \geqslant \frac{4\pi cGM(1 - \beta_s)}{\overline{\kappa\eta}}. \tag{81}$$

Let

$$\overline{\kappa\eta} = \kappa_c \bar{\eta}_c, \tag{82}$$

where

$$\bar{\eta}_c = \frac{1}{P_c} \int_0^{P_c} \left(\frac{\kappa}{\kappa_c} \right) \eta dP. \tag{83}$$

If, further, we assume a law of opacity of the form

$$\kappa = \kappa_1 \rho T^{-3-S}, \tag{84}$$

then

$$\kappa_c = \kappa_1 \frac{\mu H}{k} \frac{a}{3} \frac{\beta_c}{1 - \beta_c} T_c^{-S}, \tag{85}$$

or, again by (77),

$$\kappa_c \leqslant \kappa_1 \frac{\mu H}{k} \frac{a}{3} \frac{\beta_s}{1 - \beta_s} T_c^{-S}. \tag{86}$$

Combining (81), (82), and (86), we have

$$L \geqslant \frac{4\pi cGM}{\kappa_1 \bar{\eta}_c} \frac{k}{\mu H} \frac{3}{a} \frac{(1 - \beta_s)^2}{\beta_s} T_c^S. \tag{87}$$

Comparing (77), (81), and (87) with the standard formulae in Eddington's theory, we see that the equations in that theory now become inequalities. This makes the conclusions drawn on the basis of the standard model have a "minimal" character which is of considerable physical importance.

If in addition to the conditions of Theorem 12 we assume that T decreases outward, then, according to Theorem 7 (II),

$$T_c > \frac{1}{5}\frac{\mu H}{k}\frac{GM}{R}\beta^* . \tag{88}$$

For the case of vanishing radiation pressure we can improve (88). For then (cf. the last row of Table 2)

$$T_c \geqslant 0.32\frac{\mu H}{k}\frac{GM}{R} . \tag{89}$$

We can now eliminate T_c between (87) and (88) or (89) and obtain an inequality of the same form as the luminosity formula used in current studies on gaseous stars.

The main Theorem 10 was conjectured by the writer over a year ago, but the fundamental idea in the proof as given in the text suggested itself only during a discussion with Professor J. von Neumann. It is a pleasure to record here my appreciation of the kind interest which Professor J. von Neumann has shown in this and other problems of the stellar interior. I am also indebted to Mr. E. Ebbighausen for his assistance in the numerical work connected with Tables 1 and 1*a*.

YERKES OBSERVATORY
March 8, 1938

THE MINIMUM CENTRAL TEMPERATURE OF A GASEOUS STAR

S. CHANDRASEKHAR

1. This paper is devoted to the consideration of the following problem :—

What is the minimum central temperature T_c for a gaseous star of a given mass M and radius R in which the density ρ and the temperature T are both assumed not to increase outward?

In solving this problem we shall further assume that the mean molecular weight μ is a constant throughout the entire configuration. It is clear from dimensional considerations that we should have a relation of the form

$$(T_c)_{\min} = \tfrac{1}{2} Q(M\mu^2) \frac{\mu H}{k} \frac{GM}{R}, \tag{1}$$

where Q is a dimensionless quantity depending on $M\mu^2$ only. The problem we are to consider is essentially one of specifying the function $Q(M\mu^2)$.

2. The mathematical problem that is presented can be formulated as follows :—

Since the configuration is one of equilibrium we have, in a standard notation,

$$\frac{dP}{dr} = -\frac{GM(r)}{r^2}\rho, \tag{2}$$

and

$$\frac{dM(r)}{dr} = 4\pi r^2 \rho. \tag{3}$$

For a gaseous star, which is the case under consideration,

$$P = \frac{k}{\mu H}\rho T + \tfrac{1}{3} a T^4, \tag{4}$$

where H is the mass of the proton and the other symbols have their usual meanings. We regard equation (4) as giving T implicitly as a function of P and ρ:

$$T \equiv T(P, \rho). \tag{5}$$

According to our assumption, $T(P, \rho)$ and ρ both do not increase outward. The problem is to find the minimum value of $T_c \equiv T(P_c, \rho_c)$ for an equilibrium configuration of assigned mass and radius. The problem, when formulated in this manner is seen to be a special case of the following more general problem :—

Consider an arbitrary continuous function $K(P, \rho)$, $(P, \rho \geqslant 0)$. Let us assume that K and ρ both do not increase outward. What is the minimum value of $K_c \equiv K(P_c, \rho_c)$ for an equilibrium configuration of a given mass M and radius R? With suitable restrictions on $K(P, \rho)$ we can solve this problem. We shall prove:

Theorem.—If $K(P, \rho)$ be an arbitrary continuous function in the variables P and ρ such that

$$2P\frac{\partial K}{\partial P} + \rho\frac{\partial K}{\partial \rho} > 0, \tag{6}$$

and

$$\frac{\partial K}{\partial P} > 0, \tag{7}$$

then in any equilibrium configuration in which both $K(P, \rho)$ and ρ do not increase outward, the minimum value of $K_c \equiv K(P_c, \rho_c)$ is attained in the sequence of equilibrium configurations which consist of cores of constant K and homogeneous envelopes.

It will be noticed that the special case of the above theorem which arises when K has the form

$$K = P/\rho^{(n+1)/n} \tag{8}$$

has already been proved by the author.* The condition (6) on K is now readily seen to be equivalent to the restriction $n > 1$ which was imposed in proving the special case (cf., *loc. cit.*). We shall verify in paragraph 3 that the function $T(P, \rho)$ [implicitly defined by equation (4)] satisfies the conditions of the theorem.

The proof of the theorem depends on the following lemma:—

Lemma.—In the configuration in which K_c attains the minimum, either $d\rho/dr = 0$ or $dK/dr = 0$ for all r, $0 \leqslant r \leqslant R$.

For, if not, in the configuration in which K_c attains its minimum there must exist a finite interval

$$0 < r_1 \leqslant r \leqslant r_2 < R, \tag{9}$$

in which

$$\frac{d\rho}{dr} < 0; \qquad \frac{dK}{dr} < 0. \tag{10}$$

Let P and ρ refer to the configuration we are considering, namely, the one in which K_c attains its minimum.

By means of the following transformation we construct the pressure and density distributions defined by

$$0 \leqslant r \leqslant r_1 \qquad P^* = (1-\epsilon)^2 P; \qquad \rho^* = (1-\epsilon)\rho, \tag{11}$$

$$r_1 < r \leqslant r_2 \qquad P^* = P + \epsilon P_1; \qquad \rho^* = \rho + \epsilon\rho_1, \tag{12}$$

$$r_2 < r \leqslant R \qquad P^* = P; \qquad \rho^* = \rho, \tag{13}$$

where P^* and ρ^* refer to the new distributions of pressure and density,

* S. Chandrasekhar, *Ap. J.*, **87**, 535, 1938. This paper will be referred to as *loc. cit.*

ϵ is a small *positive* constant, and P_1 and ρ_1 (which are functions of r in the interval $r_1 \leqslant r \leqslant r_2$) are for the present unspecified.

If P^* and ρ^* should refer to an equilibrium configuration of the same mass M as the original configuration the following conditions have to be fulfilled :—

(i) *Continuity of P^* and ρ^*.* From (11), (12) and (13) we see that to ensure the continuity at $r=r_1$ and $r=r_2$ we should require

$$\rho_1 = -\rho, \qquad (r=r_1); \qquad \rho_1 = 0, \qquad (r=r_2); \tag{14}$$

$$P_1 = -2P, \qquad (r=r_1); \qquad P_1 = 0, \qquad (r=r_2). \tag{15}$$

Equation (15) is correct to the first order in ϵ.

(ii) *Constancy of Mass.*—This requires

$$4\pi\int_0^R \rho^* r^2 dr = 4\pi\int_0^R \rho r^2 dr. \tag{16}$$

By (11), (12) and (13) we find that (16) reduces to

$$M(r_1) = 4\pi\int_{r_1}^{r_2} \rho_1 r^2 dr. \tag{17}$$

The left-hand side of (17) is a known quantity. We can clearly choose a function ρ_1 in the interval $r_1 \leqslant r \leqslant r_2$ such that equation (17) and the boundary conditions (14) at r_1 and r_2 are all satisfied. We shall assume that ρ_1 has been chosen to satisfy these conditions.

(iii) *The distributions P^* and ρ^* satisfy the equation of hydrostatic equilibrium.*—The pressure density distributions in any configuration of equilibrium must satisfy the equation

$$\frac{1}{r^2}\frac{d}{dr}\left(\frac{r^2}{\rho}\frac{dP}{dr}\right) = -4\pi G\rho. \tag{18}$$

Given that P and ρ satisfy (18), we have to show that P^* and ρ^* distributions (with a suitable choice of P_1) satisfy (18).

It is immediately obvious that in the interval $0 \leqslant r \leqslant r_1$ and $r_2 < r \leqslant R$ equation (18) is satisfied. For $r_1 < r \leqslant r_2$ we have according to (12) and (18)

$$\frac{1}{r^2}\frac{d}{dr}\left(\frac{r^2}{\rho}\frac{dP_1}{dr} - \frac{r^2}{\rho^2}\rho_1\frac{dP}{dr}\right) = -4\pi G\rho_1. \tag{19}$$

From (19) we derive

$$\frac{1}{r^2}\frac{d}{dr}\left(\frac{r^2}{\rho}\frac{dP_1}{dr}\right) = F(r), \tag{20}$$

where

$$F(r) = -4\pi G\rho_1 + \frac{1}{r^2}\frac{d}{dr}\left(\frac{r^2}{\rho^2}\rho_1\frac{dP}{dr}\right). \tag{21}$$

Since P and ρ are assumed to be known functions of r and ρ_1 has been chosen according to (ii) above, we can regard $F(r)$ as a known function of r. From (20) we easily derive

$$P_1 = \int_{r_2}^{r} \frac{\rho}{r^2}\left\{\int_{r_1}^{r} \xi^2 F(\xi)d\xi + c_1\right\}dr + c_2, \tag{22}$$

where c_1 and c_2 are two constants of integration. We now choose c_1 and c_2 such that P_1 defined by (22) satisfies the boundary conditions (15).

We thus see that with P_1 and ρ_1 chosen as specified in (ii) and (iii) above, P^* and ρ^* refer to an equilibrium configuration of the same mass and radius as the original configuration.

Finally, since in the interval $r_1 \leqslant r \leqslant r_2$, ρ and K are strictly decreasing [inequality (10)] we can choose a positive (*non zero*) ϵ sufficiently small that ρ^* and K^* (defined with respect to P^* and ρ^*) are decreasing functions of r.

We have thus shown that from the given equilibrium configuration we can construct another satisfying the restrictions on ρ and K. But the configuration specified by the functions P^* and ρ^* defines

$$K_c^* = K(P_c - 2\epsilon P_c, \rho_c - \epsilon\rho_c), \tag{23}$$

or, to the first order in ϵ,

$$K_c^* = K(P_c, \rho_c) - \epsilon\left(2P\frac{\partial K}{\partial P} + \rho\frac{\partial K}{\partial \rho}\right)_{P=P_c;\ \rho=\rho_c}. \tag{24}$$

From (6) and (24) it now follows that

$$K_c^* < K_c, \tag{25}$$

which contradicts our hypothesis that in the configuration in which K_c attains its minimum there exists an interval (9) in which (10) holds. This proves the lemma.

The theorem now follows almost immediately. The argument proceeds as in *loc. cit.* (p. 539), appeal now being made to our second restriction (7) on K.

3. The theorem that we have proved in paragraph 2 enables us to solve the problem of the minimum central temperatures of gaseous stars. For, according to equation (4),

$$\delta P = \frac{k}{\mu H}(\rho\delta T + T\delta\rho) + \tfrac{4}{3}aT^3\delta T \tag{26}$$

or

$$\frac{\delta T}{T} = \frac{1}{\frac{k}{\mu H}\rho T + \frac{4}{3}aT^4}\left(\delta P - \frac{k}{\mu H}T\delta\rho\right). \tag{27}$$

Hence

$$\frac{P}{T}\frac{\partial T}{\partial P} = \frac{\frac{k}{\mu H}\rho T + \frac{1}{3}aT^4}{\frac{k}{\mu H}\rho T + \frac{4}{3}aT^4} > 0, \tag{28}$$

and

$$\frac{\rho}{T}\frac{\partial T}{\partial \rho} = -\frac{\dfrac{k}{\mu H}\rho T}{\dfrac{k}{\mu H}\rho T + \frac{4}{3}aT^4}. \tag{29}$$

Hence

$$2\frac{P}{T}\frac{\partial T}{\partial P} + \frac{\rho}{T}\frac{\partial \rho}{\partial T} = \frac{\dfrac{k}{\mu H}\rho T + \frac{2}{3}aT^4}{\dfrac{k}{\mu H}\rho T + \frac{4}{3}aT^4} > 0. \tag{30}$$

Hence $T(P, \rho)$ satisfies the restrictions required for the validity of the theorem proved in paragraph 2. As stated for this special case the theorem will read:

In any gaseous equilibrium configuration of prescribed mass and radius and of constant mean molecular weight μ in which both ρ and T do not increase outward, the minimum value of T_c is attained in the sequence of equilibrium configurations which consist of isothermal cores and homogeneous envelopes.

It has to be noticed that we have only proved that the minimum value of T_c along the sequence of composite configurations (consisting of isothermal cores and homogeneous envelopes) is the absolute minimum of T_c under the restrictions $d\rho/dr \leqslant 0$, $dT/dr \leqslant 0$. But we have not yet specified the *particular* composite configuration in which the minimum value of T_c is attained; to be able to do so, we shall have to study the function $T_c(A)$, where A is the fraction of the radius occupied by the isothermal core. We now proceed to study this function.

4. We consider a composite configuration in which the isothermal core extends to a fraction A of the radius R. Hence if $r = r_1$ defines the place at which we have the interface between the isothermal and the homogeneous regions we have

$$r_1 = AR. \tag{31}$$

Let $\bar{\rho}_1$ be the mean density of the isothermal core and ρ_1 the constant density in the homogeneous part. Let B denote the ratio

$$B = \frac{\bar{\rho}_1 - \rho_1}{\rho_1}. \tag{32}$$

Finally, let P_1 be the pressure at the interface.

By considering the equilibrium of the homogeneous envelope we obtain [cf., *loc. cit.*, equations (23) and (27)]

$$M = \tfrac{4}{3}\pi R^3 \rho_1 (1 + A^3 B), \tag{33}$$

$$P_1 = \tfrac{2}{3}\pi G \rho_1^2 R^2 [1 - A^2 + 2A^2 B(1 - A)]. \tag{34}$$

Consider next the equilibrium of the isothermal core: in the core we can write

$$P = K\rho + D, \tag{35}$$

where

$$K=\frac{k}{\mu H}T_c\,; \qquad D=\tfrac{1}{3}aT_c^{\,4}, \tag{35'}$$

where K and D are constants. The reduction to the Emden isothermal equation is made by the substitutions *

$$\rho=\lambda e^{-\psi}\,; \qquad P=K\lambda e^{-\psi}+D, \tag{36}$$

$$r=\left(\frac{K}{4\pi G\lambda}\right)^{1/2}\xi. \tag{37}$$

Let ψ and ξ refer to the interface. Then P_1, ρ_1 and r_1 are given by the above formulæ. Further, let the radiation pressure be a fraction $(1-\beta_1)$ of the total pressure P_1 at the interface, so that

$$\beta_1 P_1=\frac{k}{\mu H}\rho_1 T_c\,; \qquad (1-\beta_1)P_1=\tfrac{1}{3}aT_c^{\,4}. \tag{38}$$

From (36) and (38) we have

$$P_1=\frac{1}{\beta_1}\frac{k}{\mu H}\rho_1 T_c=\frac{K}{\beta_1}\lambda e^{-\psi}. \tag{39}$$

Equating (34) and (39) and using equations (36) and (37) we have

$$\frac{K}{\beta_1}\lambda e^{-\psi}=\tfrac{2}{3}\pi G(\lambda e^{-\psi})^2\left(\frac{K}{4\pi G\lambda}\right)\xi^2\frac{1-A^2+2A^2B(1-A)}{A^2}. \tag{40}$$

After some reductions the above equation simplifies to

$$\frac{6}{\beta_1}=\xi^2 e^{-\psi}\,.\,\frac{1-A^2+2A^2B(1-A)}{A^2}. \tag{41}$$

Introduce the homology-invariant functions u and v defined by

$$u=\frac{\xi e^{-\psi}}{\psi'}\,; \qquad v=\xi\psi', \tag{42}$$

where ψ' refers to the derivative of ψ with respect to ξ. In terms of u and v equation (41) can be rewritten as

$$\tfrac{1}{6}\beta_1 uv=\frac{A^2}{1-A^2+2A^2B(1-A)}. \tag{43}$$

Further,

$$B=\frac{\bar\rho_1}{\rho_1}-1, \tag{44}$$

or, using the relation between the mean and the central densities for isothermal configurations, we have

$$B=3\frac{\psi'}{\xi e^{-\psi}}-1, \tag{45}$$

* See S. Chandrasekhar, *An Introduction to the Study of Stellar Structure* (Chicago), p. 155.

or, according to (42),

$$B = \frac{3}{u} - 1. \tag{46}$$

Equations (43) and (46) are our equations of fit. If the isothermal function $\psi(\xi)$ is known, then for a given ξ, u and v are known, and (43) and (46) determine A as the solution of a cubic equation.

We have next to determine T_c in terms of A, R, M and β_1. In terms of β_1 we can write for the interfacial pressure P_1, the expression

$$P_1 = \left[\left(\frac{k}{\mu H}\right)^4 \frac{3}{a} \frac{1-\beta_1}{\beta_1^4}\right]^{1/3} \rho_1^{4/3}. \tag{47}$$

Hence, according to equations (34) and (47),

$$\left[\left(\frac{k}{\mu H}\right)^4 \frac{3}{a} \frac{1-\beta_1}{\beta_1^4}\right]^{1/3} = \tfrac{2}{3}\pi G \rho_1^{2/3} R^2[1 - A^2 + 2A^2 B(1-A)]. \tag{48}$$

From (33) we have, on the other hand,

$$\rho_1^{2/3} = \frac{M^{2/3}}{(\frac{4}{3}\pi)^{2/3}} \frac{1}{R^2} \frac{1}{(1+A^3B)^{2/3}}. \tag{49}$$

Eliminating $\rho_1^{2/3}$ between the equations (48) and (49) we obtain

$$\left[\left(\frac{k}{\mu H}\right)^4 \frac{3}{a} \frac{1-\beta_1}{\beta_1^4}\right]^{1/3} = \tfrac{1}{2}(\tfrac{4}{3}\pi)^{1/3} G M^{2/3} \frac{1 - A^2 + 2A^2B(1-A)}{(1+A^3B)^{2/3}}, \tag{50}$$

or alternatively

$$M = \frac{1 + A^3 B}{[1 - A^2 + 2A^2B(1-A)]^{3/2}} M^*(\beta_1), \tag{51}$$

where

$$M^*(\beta_1) = \left(\frac{6}{\pi}\right)^{1/2} \left[\left(\frac{k}{\mu H}\right)^4 \frac{3}{a} \frac{1-\beta_1}{\beta_1^4}\right]^{1/2} \frac{1}{G^{3/2}}. \tag{52}$$

Again, from equations (34) and (38) we have

$$\frac{k}{\beta_1 \mu H} \rho_1 T_c = \tfrac{2}{3}\pi G \rho_1^2 R^2 [1 - A^2 + 2A^2B(1-A)], \tag{53}$$

or

$$\frac{k}{\mu H} T_c = \beta_1 \,.\, \tfrac{2}{3}\pi G \,.\, \rho_1 R^2 [1 - A^2 + 2A^2 B(1-A)]. \tag{54}$$

Eliminating ρ_1 between the equations (33) and (54) we have

$$\frac{k}{\mu H} T_c = \tfrac{1}{2}\beta_1 \frac{1 - A^2 + 2A^2B(1-A)}{1 + A^3 B} \frac{GM}{R}. \tag{55}$$

We can therefore write

$$T_c = \tfrac{1}{2} Q \frac{\mu H}{k} \frac{GM}{R}, \tag{56}$$

where

$$Q=\beta_1\frac{1-A^2+2A^2B(1-A)}{1+A^3B}. \tag{57}$$

Using (43) we can rewrite equation (57) as

$$Q=\frac{6}{uv}\,\frac{A^2}{1+A^3B}. \tag{58}$$

This solves the equilibrium problem of the composite configurations. For an assumed value for ξ and β_1, determine A and B according to equations (43) and (46); equations (51) and (52), and (56) and (58) determine the mass of the configuration and the temperature of the isothermal core respectively.

Equation (43) is a cubic equation for A. Eliminating B between (43) and (46) we find that the equation for A can be written more conveniently as

$$2v(3-u)A^3+3\left[v(u-2)+\frac{2}{\beta_1}\right]A^2-uv=0. \tag{59}$$

Also, collecting our other results,

$$B=\frac{3}{u}-1, \tag{60}$$

$$M=\frac{1+A^3B}{[1-A^2+2A^2B(1-A)]^{3/2}}M^*(\beta_1), \tag{61}$$

$$T_c=\tfrac{1}{2}Q\frac{\mu H}{k}\,\frac{GM}{R}, \tag{62}$$

where

$$M^*(\beta_1)=\left(\frac{6}{\pi}\right)^{1/2}\left[\left(\frac{k}{\mu H}\right)^4\frac{3}{a}\,\frac{1-\beta_1}{\beta_1{}^4}\right]^{1/2}\frac{1}{G^{3/2}}, \tag{63}$$

and

$$Q=\frac{6}{uv}\,\frac{A^2}{1+A^3B}. \tag{64}$$

5. Our next problem is to determine the minimum value of T_c as given by equations (59)–(64) for prescribed values of M and R; in other words, we need the minimum value of Q for an assigned value of M. According to the equations (59), (60) and (61)

$$M=M(\xi,\beta_1); \tag{65}$$

while according to (59), (60) and (64)

$$Q=Q(\xi,\beta_1). \tag{66}$$

For a given M and an assumed value of ξ, equation (65) determines β_1. Equation (66) then specifies the appropriate value of Q. In this manner, we can effect the elimination of β_1 between the relations (65) and (66). We then repeat this calculation for a series of values of ξ and then determine the minimum value of the derived $Q(\xi)$'s. This is the procedure that should be adopted to determine the minimum central temperature for a gaseous star of a *given* mass. The actual labour that will be involved in effecting the elimination of β_1 between the equations (65) and (66) for different values of ξ and for different initially prescribed values of M will be enormous. There is, however, an indirect but a more convenient method which can be adopted in practice. In this indirect method we first prescribe a certain value for β_1. We then determine M and Q according to equations (59)–(64) for various values of ξ. The calculations are then repeated for a series of initially assigned values of β_1. We obtain in this way a family of $(M\mu^2, Q)$ curves. We next draw the envelope to these curves. It is clear that this envelope to the $(M\mu^2, Q)$ curves will give the relation between $M\mu^2$ and the corresponding minimum value of Q, which we shall denote by Q_{min}.

Table I summarizes the results of the calculations that were undertaken for determining the minimum central temperatures for gaseous stars. Only such parts of the calculations as are relevant to the drawing of the envelope of the $(M\mu^2, Q)$ curves are reproduced. Further, in this table the unit of mass used is M_3 defined by

$$M_3 = -4\pi\left(\frac{K_2}{\pi G}\right)^{3/2}\left(\xi^2\frac{d\theta_3}{d\xi}\right)_{\xi=\xi_1}, \tag{67}^*$$

where K_2 is the relativistic-degenerate constant and θ_3 is the Lane-Emden function of index 3. Numerically, we have

$$M_3 = 5{\cdot}75\odot\mu^{-2}. \tag{68}$$

In terms of (67) we can rewrite equation (63) as

$$M^*(\beta_1) = 0{\cdot}3034 M_3\left(\frac{960}{\pi^4}\frac{1-\beta_1}{\beta_1{}^4}\right)^{1/2}. \tag{69}$$

It will be noticed that equation (63) is identical with the quartic equation † which determines the maximum value $(1-\beta^*)$ of the radiation pressure $(1-\beta_c)$, at the centre of a gaseous configuration of mass M^*.

The computations summarized in Table I were made possible through the courtesy of Mr. Gordon Wares who kindly provided the writer with a table of the isothermal function more accurate and more extensive than any that has been published so far.

The $(M/M_3, Q)$ curves for the different values of β_1 given in Table I were plotted and the envelope to these curves was drawn. From this envelope the values of Q_{min} for different values of M/M_3 were directly read off. The results are given in Table II. For any given value of the mass (expressed in units of M_3) we can now obtain by interpolation the

* See S. Chandrasekhar, *An Introduction to the Study of Stellar Structure* (Chicago), p. 422.

† S. Chandrasekhar, *Observatory*, **59**, 47, 1936. Also *M.N.*, **96**, 644, 1936. An alternative form for (69) is

$$M^*(\beta_1)/M_3 = 0{\cdot}9525(1-\beta_1)^{1/2}\beta_1{}^{-2}. \tag{69'}$$

TABLE I

β_1	ξ	A	M	Q
	4·6	0·6745	0	0·6407
	4·8	·6773	0	·64014
1·00	5·0	·6798	0	·64010
	5·2	·6817	0	·64044
	3·2	0·6244	0·3946	0·6465
	3·6	·6402	·3971	·6363
	4·0	·6516	·3958	·6302
	4·4	·6601	·3920	·6272
	4·6	·6634	·3894	·6267
0·95	4·8	·6662	·3863	·6266
	5·0	·6685	·3831	·6268
	5·2	·6704	·3796	·6275
	5·6	·6734	·3722	·6298
	6·0	·6754	·3646	·6331
	3·6	0·6285	0·6047	0·6183
	4·0	·6400	·6017	·6137
	4·4	·6482	·5947	·6116
0·90	4·6	·6515	·5905	·6115
	4·8	·6542	·5854	·6119
	5·2	·6584	·5746	·6134
	5·6	·6613	·5630	·6163
	6·0	·6631	·5509	·6200
	3·2	0·6006	0·8012	0·6062
	3·6	·6162	·8023	·5991
	4·0	·6274	·7966	·5955
	4·2	·6318	·7919	·5949
0·85	4·4	·6356	·7862	·5947
	4·6	·6388	·7798	·5950
	4·8	·6414	·7728	·5956
	5·2	·6455	·7575	·5980
	5·6	·6482	·7415	·6014
	3·2	0·5876	1·0112	0·5842
	3·6	·6030	1·0106	·5787
	4·0	·6140	1·0011	·5764
	4·2	·6184	0·9947	·5761
0·80	4·4	·6220	·9867	·5763
	4·6	·6252	·9778	·5771
	4·8	·6277	·9682	·5781
	5·2	·6317	·9482	·5812
	5·6	·6343	·9272	·5852
	3·2	0·5738	1·246	0·5608
	3·6	·5889	1·242	·5569
	3·8	·5948	1·236	·5559
	4·0	·5998	1·228	·5557
0·75	4·2	·6040	1·218	·5559
	4·4	·6075	1·207	·5566
	4·8	·6130	1·184	·5591
	5·2	·6168	1·158	·5627

β_1	ξ	A	M	Q
	2·8	0·5386	1·510	0·5423
	3·2	·5590	1·517	·5362
	3·6	·5738	1·508	·5336
0·70	3·8	·5795	1·499	·5331
	4·0	·5844	1·488	·5334
	4·2	·5885	1·476	·5340
	4·4	·5920	1·462	·5352
	4·8	·5972	1·431	·5383
	2·8	0·5232	1·836	0·5144
	3·2	·5432	1·839	·5100
	3·4	·5510	1·834	·5089
0·65	3·6	·5576	1·825	·5085
	3·8	·5632	1·812	·5087
	4·0	·5679	1·798	·5094
	4·4	·5752	1·763	·5121
	2·8	0·5068	2·237	0·4853
	3·2	·5262	2·234	·4822
	3·4	·5338	2·226	·4816
0·60	3·6	·5402	2·212	·4818
	3·8	·5456	2·195	·4825
	4·0	·5501	2·175	·4835
	2·8	0·4890	2·742	0·4544
	3·0	·4992	2·741	·4532
	3·2	·5079	2·732	·4526
0·55	3·4	·5152	2·718	·4527
	3·6	·5214	2·699	·4532
	4·0	·5309	2·650	·4557
	2·8	0·4700	3·397	0·4221
	3·0	·4798	3·391	·4213
	3·2	·4882	3·377	·4213
0·50	3·4	·4952	3·356	·4217
	3·6	·5010	3·330	·4226
	4·0	·5100	3·265	·4256
	2·8	0·4493	4·274	0·3880
	3·0	·4588	4·262	·3877
	3·2	·4667	4·239	·3880
0·45	3·4	·4734	4·209	·3888
	3·6	·4790	4·173	·3900
	4·0	·4874	4·086	·3935
	2·4	0·4036	5·494	0·3533
	2·6	·4162	5·500	·3524
	2·8	·4268	5·492	·3520
0·40	3·0	·4358	5·471	·3521
	3·2	·4434	5·437	·3529
	3·6	·4548	5·345	·3553
	4·0	·4626	5·229	·3587

β_1	ξ	A	M	Q
	2·4	0·3802	7·281	0·3148
	2·6	·3922	7·282	·3143
	2·8	·4022	7·262	·3143
0·35	3·0	·4106	7·227	·3147
	3·2	·4177	7·179	·3156
	3·6	·4284	7·051	·3182
	2·2	0·3413	10·020	0·2752
	2·4	·3545	10·034	·2747
	2·6	·3657	10·024	·2744
0·30	2·8	·3750	9·988	·2748
	3·2	·3894	9·863	·2762
	3·6	·3990	9·686	·2786
	2·2	0·3137	14·60	0·2331
	2·4	·3259	14·60	·2328
0·25	2·6	·3362	14·57	·2329
	2·8	·3448	14·51	·2333
	3·2	·3577	14·32	·2348

β_1	ξ	A	M	Q
	2·2	0·2824	23·03	0·1894
	2·4	·2934	23·02	·1894
0·20	2·6	·3027	22·95	·1896
	2·8	·3104	22·85	·1900
	3·2	·3218	22·56	·1914
	2·0	0·2348	41·28	0·1443
	2·2	·2462	41·29	·1443
0·15	2·4	·2558	41·25	·1443
	2·6	·2638	41·13	·1445
	2·8	·2704	40·95	·1449
	3·2	·2801	40·47	·1459
	2·0	0·1930	93·66	0·0976
	2·2	·2023	93·66	·0976
0·10	2·4	·2102	93·51	·0976
	2·8	·2220	92·92	·0980
	3·2	·2296	92·06	·0986

TABLE II

M/M_3	$Q_{\min}$	$1-\beta^*$	M/M_3	$Q_{\min}$	$1-\beta^*$
0	0·640	0	4·0	0·396	0·575
0·4	·624	0·110	5·0	·364	·614
0·6	·610	·180	6·0	·337	·643
0·8	·592	·238	7·0	·318	·667
1·0	·574	·286	8·0	·300	·686
1·2	·557	·326	9·0	·287	·702
1·4	·541	·361	10·0	·273	·717
1·6	·524	·390	12·5	·246	·744
1·8	·508	·416	15·0	·230	·764
2·0	·494	·440	23·0	·189	·807
2·5	·463	·485	41·2	·144	·854
3·0	·438	·521	93·7	·0976	0·902
3·5	0·414	·551	∞	0	1·000

appropriate value of $Q_{\min}$. The minimum central temperature for a gaseous star having the prescribed mass is then given by

$$(T_c)_{\min} = \tfrac{1}{2}Q_{\min}\frac{\mu H}{k}\frac{GM}{R}, \tag{70}$$

or numerically

$$(T_c)_{\min} = 11{\cdot}54\,Q_{\min}\mu\left(\frac{M}{\odot}\right)\left(\frac{R_\odot}{R}\right)\times 10^6 \text{ degrees.} \tag{71}$$

Before we proceed to numerical applications of Table II to practical cases, we shall first consider the limiting forms for $Q_{\min}$ for the cases $M\to 0$ and $M\to\infty$.

For $M\to 0$, it is clear that the minimum central temperature can be obtained by neglecting the radiation pressure in the equation of hydrostatic equilibrium. In that case, the problem effectively reduces to to determining the minimum value of P/ρ. The solution to this problem is contained in the author's earlier paper (*loc. cit.*, § VI). We therefore have

$$Q_{\min} \to 0{\cdot}640, \qquad M \to 0. \tag{72}$$

On the other hand, as $M \to \infty$, it is readily verified that

$$\beta_1 \to 0, \qquad A \to 0, \tag{73}$$

and that

$$Q_{\min} \to \beta^*, \tag{74}$$

where

$$M = \left(\frac{6}{\pi}\right)^{\frac{1}{2}} \left[\left(\frac{k}{\mu H}\right)^4 \frac{3}{a} \frac{1-\beta^*}{\beta^{*4}}\right]^{\frac{1}{2}} \frac{1}{G^{3/2}}. \tag{75}$$

Fig. 1 illustrates the general $(1-\beta^*, Q_{\min})$ relation. We notice that this curve intersects the $(1-\beta^*)$ axis at an angle of 45°; this is of course in agreement with (74).

6. Finally, Table III illustrates the method of using the results of § 5 to practical cases. In this table we have given the minimum central temperatures for a few typical cases.

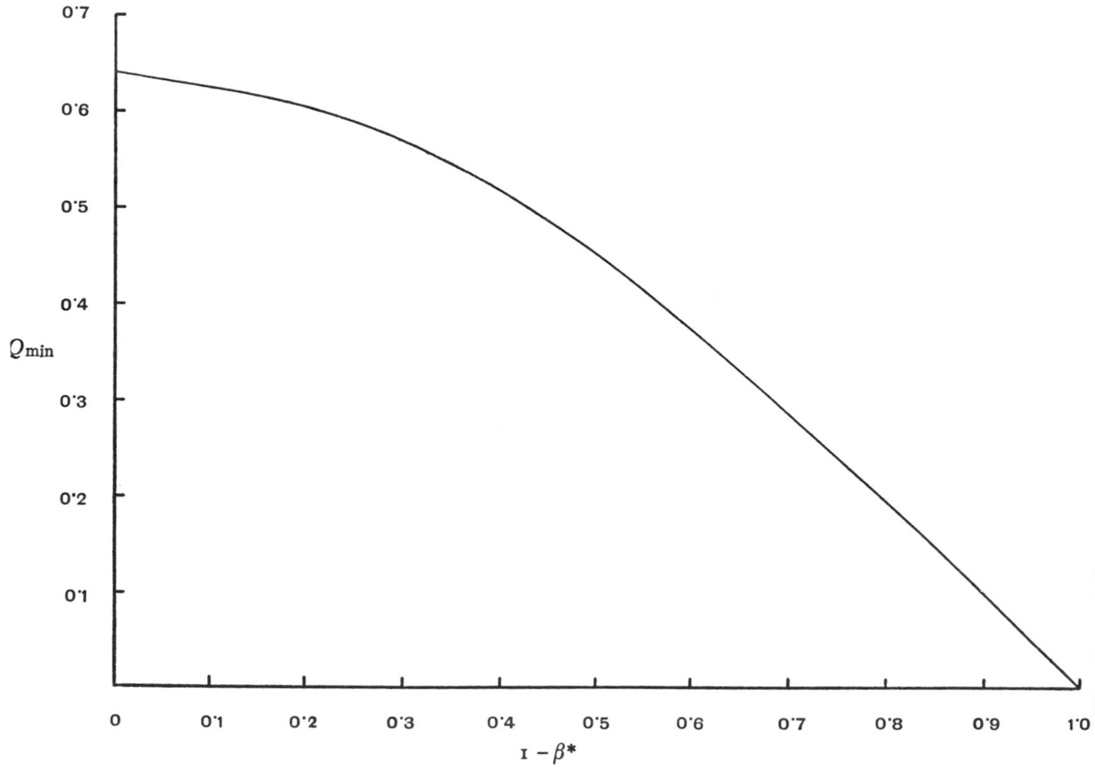

FIG. 1.—*The* $[(1-\beta^*), Q_{\min}]$ *Curve.*

TABLE III

Star	$M/\odot$	$R/R_\odot$	μ	M/M_3	$1-\beta^*$	Q_{min}	T_c min millions of degrees
Sun	1·00	1·00	1·00	0·174	0·030	0·636	7·3
Sirius A	2·34	1·78	1·00	0·41	·114	·622	9·4
Capella A	4·18	15·9	1·00	0·73	·22	·60	1·82
ζ Aur (K star)	14·8	200	1·00	2·57	·49	·46	0·39
Trumpler's Stars T_2	100	7·2	0·50	4·3	·59	·38	30
Trumpler's Stars T_3	140	6·6	0·50	6·1	·645	·34	42
Trumpler's Stars T_6	400	17	0·50	17	0·78	0·21	29

Summary

In this paper the problem of determining the minimum central temperatures of gaseous stars is solved. It is shown that in any equilibrium configuration in which the temperature and the density do not increase outward the minimum central temperature is attained in the sequence of equilibrium configurations consisting of isothermal cores and homogeneous envelopes. By a study of the appropriate sequence of composite configurations it is found that we can write

$$(T_c)_{\min} = \tfrac{1}{2} Q_{\min} \frac{\mu H}{k} \frac{GM}{R},$$

where $Q_{\min}$ is a dimensionless quantity depending on $M\mu^2$ only. In Table II, $Q_{\min}$ is tabulated as a function of the mass of the star expressed in a suitable unit (M_3). Numerical applications (Table III) are also given.

In conclusion I wish to express my indebtedness to Mrs. T. Belland who undertook the rather tiring numerical work connected with the preparation of Table I; but for Mrs. Belland's assistance this paper would not have been written. My thanks are also due to Mr. G. W. Wares who generously placed at my disposal his unpublished integration of the isothermal function.

Yerkes Observatory:
1939 *March* 24.

THE INTERNAL CONSTITUTION OF THE STARS

S. CHANDRASEKHAR

Yerkes Observatory, University of Chicago

(*Read February 17, 1939, in Symposium on Progress in Astrophysics*)

IN THIS paper an attempt is made to describe the general methods that have been developed to determine the physical conditions in stellar interiors. In view of the complexity of the problem, it is of value to consider, in the first instance, only those methods which involve the minimum of assumptions. Three such methods have been developed in recent years. They are

I. The method of the integral theorems.
II. The method of the homologous transformations.
III. The method of stellar envelopes.

I. THE METHOD OF THE INTEGRAL THEOREMS

In this method the fundamental assumption is made that stars are in hydrostatic equilibrium.[1] We should then have

$$\frac{dP}{dr} = -\frac{GM(r)}{r^2}\rho, \tag{1}$$

where P is the total pressure and ρ the density at a distance r from the center and $M(r)$ is the mass enclosed inside r. Since P is the total pressure we can write

$$P = p_r + p_g, \tag{2}$$

where p_g is the gas pressure and p_r the radiation pressure; *i.e.*,

$$p_r = \tfrac{1}{3}aT^4; \qquad p_g = f(\rho, T), \tag{3}$$

where a is the Stefan-Boltzmann constant, and $f(\rho, T)$ specifies the equation of state. If the assumption is made that stellar material behaves like a perfect gas then

$$p_g = \frac{k}{\mu H}\rho T, \tag{4}$$

where k is the Boltzmann constant, μ the mean molecular mass, and H the mass of the hydrogen atom.

[1] We thus exclude from our considerations rotating and variable stars.

From the geometry of the case we have in addition to (1)

$$\frac{dM(r)}{dr} = 4\pi r^2 \rho. \tag{5}$$

Using (5) we can rewrite (1) as

$$dP = -\frac{1}{4\pi}\frac{GM(r)dM(r)}{r^4}. \tag{6}$$

A direct integration of (6) yields [2]

$$P_c = \frac{1}{4\pi}\int_0^R \frac{GM(r)dM(r)}{r^4}, \tag{7}$$

where R is the radius of the star and P_c is the central pressure. From (7) we immediately infer that

$$P_c > \frac{G}{4\pi}\int_0^R \frac{M(r)dM(r)}{R^4} = \frac{1}{8\pi}\frac{GM^2}{R^4}. \tag{8}$$

Hence with no assumption except that stars are in hydrostatic equilibrium we can set a lower limit to the central pressure. Numerically (8) reduces to

$$P_c > 4.50 \times 10^8 \left(\frac{M}{\odot}\right)^2 \left(\frac{R_\odot}{R}\right)^4 \text{ atmospheres,} \tag{9}$$

where $\odot$ and $R_\odot$ refer to the mass and the radius of the sun.

We can similarly obtain a lower limit to the mean pressure defined by

$$M\bar{P} = \int_0^R P dM(r). \tag{10}$$

After an integration by parts (10) reduces to

$$M\bar{P} = -\int_0^R M(r)dP, \tag{11}$$

or using (6)

$$M\bar{P} = \frac{G}{4\pi}\int_0^R \frac{M^2(r)dM(r)}{r^4}. \tag{12}$$

From (12) we obtain without difficulty [3] that

$$\bar{P} > \frac{1}{12\pi}\frac{GM^2}{R^4} \tag{13}$$

[2] Using the boundary condition that $P = 0$ at $r = R$ the radius of the star.
[3] By replacing r by R and taking it outside the integral sign.

or numerically

$$\bar{P} > 3.0 \times 10^8 \left(\frac{M}{\odot}\right)^2 \left(\frac{R_\odot}{R}\right)^4 \text{ atmospheres.} \tag{14}$$

In other words, we can expect pressures of the order of 10^9 atmospheres in the stellar interiors.

We can get somewhat sharper inequalities if we supplement our assumption of hydrostatic equilibrium by another one, namely, that the mean density $\bar{\rho}(r)$ interior to r does not increase outward. This assumption implies that

$$\bar{\rho}(r) \geqslant \rho(r). \tag{15}$$

It will be noticed that we do not exclude completely the possibility of negative density gradients. We only insist that the actual density $\rho(r)$ at any point does not exceed the mean density $\bar{\rho}(r)$ interior to the point considered.

To establish inequalities for P_c etc., we consider the following expression:

$$I_{\sigma,\,\nu}(r) = \int_0^r \frac{GM^\sigma(r)dM(r)}{r^\nu}. \tag{16}$$

We further assume that

$$3(\sigma + 1) - \nu > 0. \tag{17}$$

By the definition of the mean density $\bar{\rho}(r)$ we have

$$\bar{\rho}(r) = M(r)/\tfrac{4}{3}\pi r^3. \tag{18}$$

From (18) we derive that

$$r^\nu = \left\{\left(\frac{3}{4\pi}\right)\frac{M(r)}{\bar{\rho}(r)}\right\}^{\nu/3}. \tag{19}$$

Substituting (19) in (16) we have

$$I_{\sigma,\,\nu}(r) = G\left(\frac{4\pi}{3}\right)^{\nu/3}\int_0^r \bar{\rho}^{\nu/3}(r)M^{(3\sigma-\nu)/3}(r)dM(r). \tag{20}$$

Since according to our assumption $\bar{\rho}(r)$ does not increase outward we get an upper bound for the integral on the right hand side of (20) by replacing $\bar{\rho}(r)$ by ρ_c and taking it outside the integral sign. In the same we get a lower bound by replacing $\bar{\rho}(r)$ by its value at r[4] and taking it outside the integral sign. We obtain in this way

$$\frac{3G}{3(\sigma + 1) - \nu}\left(\frac{4\pi}{3}\right)^{\nu/3} M^{(3\sigma+3-\nu)/3}(r)\bar{\rho}^{\nu/3}(r) \leqslant I_{\sigma,\,\nu}(r)$$
$$\leqslant \frac{3G}{3(\sigma + 1) - \nu}\left(\frac{4\pi}{3}\right)^{\nu/3} M^{(3\sigma+3-\nu)/3}(r)\rho_c^{\nu/3}. \tag{21}$$

[4] r here refers to the upper bound of the integral defining $I_{\sigma,\,\nu}(r)$.

(21) is a fundamental inequality from which several results of importance can be derived.

From (6) we derive that

$$P_c - P = \frac{1}{4\pi} I_{1,\,4}(r). \tag{22}$$

From (21) we now infer that

$$\frac{1}{2}\left(\frac{4\pi}{3}\right)^{1/3} GM^{2/3}(r)\bar{\rho}^{4/3}(r) \leqslant P_c - P \leqslant \frac{1}{2}\left(\frac{4\pi}{3}\right)^{1/3} GM^{2/3}(r)\rho_c^{4/3}. \tag{23}$$

If we put $r = R$ in the above inequality, we find

$$\frac{1}{2}\left(\frac{4\pi}{3}\right)^{1/3} GM^{2/3}\bar{\rho}^{4/3} \leqslant P_c \leqslant \frac{1}{2}\left(\frac{4\pi}{3}\right)^{1/3} GM^{2/3}\rho_c^{4/3}. \tag{24}$$

The left hand part of the above inequality can be rewritten as

$$P_c \geqslant \frac{3}{8\pi}\frac{GM^2}{R^4} \tag{25}$$

or numerically

$$P_c > 1.35 \times 10^9 \left(\frac{M}{\odot}\right)^2 \left(\frac{R_\odot}{R}\right)^4 \quad \text{atmospheres.} \tag{26}$$

(26) improves the earlier inequality (9) by a factor 3.

If we put $r = R$ in (21) we obtain

$$\frac{3}{3\sigma + 3 - \nu}\frac{GM^{\sigma+1}}{R^\nu} \leqslant I_{\sigma,\,\nu} \leqslant \frac{3}{3\sigma + 3 - \nu}\frac{GM^{\sigma+1}}{r_c^\nu}, \tag{27}$$

where r_c is defined by

$$\frac{4}{3}\pi r_c^3 \rho_c = M. \tag{28}$$

From (12) we see that

$$M\bar{P} = \frac{1}{4\pi} I_{2,\,4}. \tag{29}$$

Hence from (27) we derive that

$$\frac{3}{20\pi}\frac{GM^2}{R^4} \geqslant \bar{P} \geqslant \frac{3}{20\pi}\frac{GM^2}{r_c^4}. \tag{30}$$

The right hand part of the inequality (30) reduces to

$$\bar{P} > 5.4 \times 10^8 \left(\frac{M}{\odot}\right)^2 \left(\frac{R_\odot}{R}\right)^4 \quad \text{atmospheres.} \tag{31}$$

Remembering that the potential energy Ω and the mean value of gravity

$\bar{g}$ are given by

$$-\Omega = G\int_0^R \frac{M(r)dM(r)}{r} = I_{1,1}, \tag{32}$$

and

$$M\bar{g} = G\int \frac{M(r)dM(r)}{r^2} = I_{1,2}, \tag{33}$$

we have [again using (27)] that

$$\frac{3}{5}\frac{GM^2}{R} \leqslant -\Omega \leqslant \frac{3}{5}\frac{GM^2}{r_c}, \tag{34}$$

and

$$\frac{3}{4}\frac{GM}{R^2} \leqslant \bar{g} \leqslant \frac{3}{4}\frac{GM}{r_c^2}. \tag{35}$$

The inequality for Ω further enables us to set a lower limit to the mean temperatures of gaseous stars in which radiation pressure can be neglected.[5] We define the mean temperature $\bar{T}$ by

$$M\bar{T} = \int_0^R TdM(r). \tag{36}$$

If the radiation pressure can be neglected

$$P = p_g = \frac{k}{\mu H}\rho T. \tag{37}$$

Hence from (36) and (37) we have

$$M\bar{T} = \frac{\mu H}{k}\int_0^R \frac{P}{\rho}dM(r) \tag{38}$$

$$= \frac{\mu H}{k}\int_0^R PdV, \tag{38'}$$

where dV is the volume element. However, according to a well known theorem in potential theory

$$-\Omega = 3\int_0^R PdV. \tag{39}$$

Hence combining (34), (38′) and (39) we find

$$\frac{1}{5}\frac{\mu H}{k}\frac{GM}{R} \leqslant \bar{T} \leqslant \frac{1}{5}\frac{\mu H}{k}\frac{GM}{r_c}. \tag{40}$$

The left hand side of the above inequality reduces to

$$\bar{T} \geq 4.6 \times 10^6 \, \mu \frac{M}{\odot}\frac{R_\odot}{R} \text{ degrees.} \tag{41}$$

[5] As we shall see below this is justifiable for the stars of normal mass.

In other words we can expect temperatures of the order of a few million degrees in stellar interiors.

The physical content of the results (24), (27), (34), (35) and (40) is the following: We are given a certain equilibrium configuration of mass M and radius R with an arbitrary density distribution, arbitrary except for the conditions that $\bar{\rho}(r)$ does not increase outward. From the given configuration we can construct two other configurations of uniform density—one with a constant density equal to $\bar{\rho}$ and the other with a constant density equal to ρ_c (see Fig. 1). The radii of these two

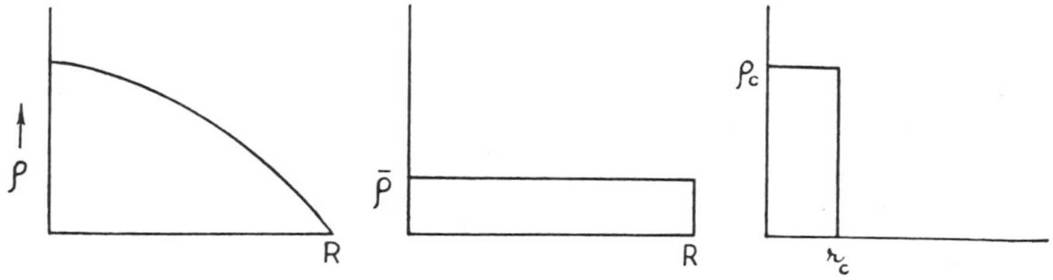

FIG. 1.

configurations are clearly R and r_c, respectively. What we have shown is that the physical variables characterizing the given equilibrium configuration, namely, P_c, $\bar{P}$, $-\Omega$, $\bar{g}$ and $\bar{T}$ (for the case of negligible radiation pressure) have values respectively less than those for the configuration of uniform density with $\rho = \rho_c$ and respectively greater than those for the configuration of uniform density with $\rho = \bar{\rho}$. Thus the given configuration is, in this sense, intermediate between the two configurations of uniform density with $\rho = \rho_c$ and $\rho = \bar{\rho}$ respectively.

Now one of the quantities which is of considerable importance in the discussion of the physical conditions in stellar interiors is the relative importance of the radiation pressure. This can be conveniently measured by the fraction $(1 - \beta)$ defined according to the relations:

$$\left.\begin{aligned}(1-\beta)P &= p_r = \frac{1}{3}aT^4,\\ \beta P &= p_g = \frac{k}{\mu H}\rho T.\end{aligned}\right\} \tag{42}$$

By a simple elimination of T between the equations (42) we obtain

$$P = \left[\left(\frac{k}{\mu H}\right)^4 \frac{3}{a}\frac{1-\beta}{\beta^4}\right]^{1/3} \rho^{4/3}. \tag{43}$$

Hence

$$P_c = \left[\left(\frac{k}{\mu_c H}\right)^4 \frac{3}{a}\frac{1-\beta_c}{\beta_c{}^4}\right]^{1/3} \rho_c{}^{4/3}. \tag{44}$$

From (24) and (44) we now have

$$\left[\left(\frac{k}{\mu_c H}\right)^4 \frac{3}{a} \frac{1-\beta_c}{\beta_c^4}\right]^{1/3} \leqslant \left(\frac{\pi}{6}\right)^{1/3} GM^{2/3}, \tag{45}$$

or

$$M \geqslant \left(\frac{6}{\pi}\right)^{1/2} \left[\left(\frac{k}{\mu_c H}\right)^4 \frac{3}{a} \frac{1-\beta_c}{\beta_c^4}\right]^{1/2} \frac{1}{G^{3/2}}. \tag{47}$$

Define $(1 - \beta^*)$ by the equation

$$M = \left(\frac{6}{\pi}\right)^{1/2} \left[\left(\frac{k}{\mu_c H}\right)^4 \frac{3}{a} \frac{1-\beta^*}{\beta^{*4}}\right]^{1/2} \frac{1}{G^{3/2}}. \tag{48}$$

Combining (47) and (48) we obtain

$$\frac{1-\beta^*}{\beta^{*4}} \geqslant \frac{1-\beta_c}{\beta_c^4}. \tag{49}$$

Hence we should have

$$1 - \beta_c \leqslant 1 - \beta^*. \tag{50}$$

The inequality (50) shows that for a gaseous star the value of $(1 - \beta)$ at the center cannot exceed an amount depending on the mass of the star only. Table 1 gives the value of $(1 - \beta^*)$ for different values of the mass M. As an example of the application of Table 1, we see that for the sun, $1 - \beta_c < 0.03$ while for Capella ($M = 4.18 \odot$), $1 - \beta_c < 0.2$ assuming in both cases that $\mu_c = 1$. Table 1 clearly illustrates that for the normal stars the radiation pressure as a factor in the equation of hydrostatic equilibrium can be neglected.

TABLE 1

VALUES OF $(1 - \beta^*)$

$1-\beta^*$	$\frac{M}{\odot}\mu_c^2$	$1-\beta^*$	$\frac{M}{\odot}\mu_c^2$	$1-\beta^*$	$\frac{M}{\odot}\mu_c^2$
0.01	0.56	0.14	2.77	0.35	7.67
.02	0.81	.15	2.94	.40	9.62
.03	1.01	.16	3.11	.45	12.15
.04	1.19	.17	3.28	.50	15.49
.05	1.36	.18	3.46	.55	20.06
.06	1.52	.19	3.64	.60	26.52
.07	1.68	.20	3.83	.65	36.05
.08	1.83	.21	4.02	.70	50.92
.09	1.98	.22	4.22	.75	75.89
.10	2.14	.23	4.43	.80	122.5
.11	2.29	.24	4.65	.85	224.4
.12	2.45	.25	4.87	.90	519.6
0.13	2.61	0.30	6.12	1.00	∞

A problem of importance that has not been considered so far is the question of the minimum central temperature of stars. In solving this problem we assume (1) that the density decreases outward and (2) that the temperature decreases outward. In considering this problem we regard T as a known function of P and ρ. This problem of the minimum central temperature leads to the consideration of the following more general problem.

Given that ρ and a certain function $F(P, \rho)$ both do not increase outward, what is the minimum value of $F_c \equiv F(P_c, \rho_c)$ for an equilibrium configuration of known mass M and radius R? This problem can be solved if suitable restrictions are imposed on F. We can, in fact, prove the following.[6]

Let $F(P, \rho)$ be such that

$$2\frac{\partial \log F}{\partial \log P} + \frac{\partial \log F}{\partial \log \rho} > 0 \qquad (P, \rho > 0), \tag{51}$$

and

$$\frac{\partial F}{\partial P} > 0. \tag{52}$$

Then in any equilibrium configuration of prescribed mass and radius in which both F and ρ do not increase outward the minimum value of F_c is attained in the sequence of equilibrium configurations which consist of cores of constant F and homogeneous envelopes.

If we choose for F the form

$$F = P\rho^{\delta}, \tag{53}$$

then the condition (51) implies that

$$\delta > -2. \tag{54}$$

Hence an immediate consequence of the theorem stated is the following: In any equilibrium configuration of prescribed mass and radius in which both ρ and $K = P/\rho^{(n+1)/n}$, $(n > 1)$, do not increase outward the minimum value of K_c is attained in the sequence of equilibrium configurations which consist of polytropic cores of index n and homogeneous envelopes.[7] It will be noticed that if we put $n = \infty$ in this special theorem we have a means of finding the lower limit to the central temperatures of stars with negligible radiation pressure. For the general problem however, we should use the theorem in its more general form. We have

$$P = \frac{k}{\mu H}\rho T + \frac{1}{3}aT^4. \tag{55}$$

[6] The details of the proof the theorem stated will be published by the author elsewhere.
[7] This special theorem has been proved by the author, see *Ap. J.*, **87**, 535, 1938.

Equation (55) when regarded as implicitly determining T as a function of P and ρ specifies the function F. It is easily verified that $T(P, \rho)$ satisfies the restrictions (51) and (52). Consequently, for a star of given mass and radius we have to construct the sequence of equilibrium configurations consisting of isothermal cores and homogeneous envelopes and determine the minimum central temperature along this sequence. The theorem now asserts that this would then give the absolute minimum for T_c under the restrictions imposed. The numerical work required to determine the minimum central temperatures for stars of different masses is rather long and tedious. Only the results will be quoted:

It is found that we can write

$$T_c \geqslant \frac{1}{2} Q(M) \frac{\mu H}{k} \frac{GM}{R}, \tag{56}$$

where Q is factor which depends in a complicated way on M and μ. The following table gives the values of $Q(M)$ for certain values of M.

TABLE 2

$\frac{M}{\odot}\mu^2$	$Q(M)$
3.4	0.61
4.5	0.595
5.75	0.575
7.1	0.554
19.4	0.421

(56) can be written also as:

$$T_c \geqslant 11.5\mu Q(M) \frac{M}{\odot} \frac{R_\odot}{R} \times 10^6 \text{ degrees.} \tag{57}$$

It is found that

$$Q \to 0.64 \quad \text{as} \quad M \to 0, \tag{58}$$

and

$$Q \to \beta^* \quad \text{as} \quad M \to \infty. \tag{59}$$

From Table 2 and the inequality (57) we see that for the sun $T_c > 7.4 \times 10^6$ degrees, while for Capella $T_c > 1.8 \times 10^6$ degrees. These results again show that we can expect temperatures of at least a few million degrees in stellar interiors. Though we have found only the minimum values it is clear that the actual values must be of the same order. They may differ from the minimum values by a factor of the order 10 but we cannot expect values of an entirely different order of magnitude.

There is one other application of the theorem we are considering which is of some importance. If we put $\delta = -4/3$ in (53) then the theorem will enable us to set a *lower* limit to $(1 - \beta_c)$ for stars in which ρ

and $(1 - \beta)$ do not increase outward. The analysis of the appropriate composite configurations leads to the surprising result that the minimum value of $(1 - \beta_c)$ is the constant value which we would ascribe to a star of the same mass if it were a complete polytrope of index $n = 3$. Now a stellar model which has played a conspicuous role in the development of the subject of stellar interiors is the so-called *standard model* in which $(1 - \beta)$ is *assumed* to be a constant. The use of this model has been criticized from several directions. For this reason it is important to realize that apart from other considerations the standard model has a definite value in so far as it has a maximal (or minimal as the case may be) characteristic. We shall return later to the physical meaning of the assumption that $(1 - \beta)$ does not increase outward.

If we now assume that stars are in radiative equilibrium then the *radiative temperature gradient* is determined by

$$\frac{dp_r}{dr} = -\frac{\kappa L(r)}{4\pi c r^2}\rho, \tag{60}$$

where κ is the stellar opacity coefficient and $L(r)$ is the amount of energy in ergs per second crossing the spherical surface of radius r. Further in (60) c is the velocity of light and the other symbols have their usual meanings. From (1) and (60) we derive that

$$\frac{dp_r}{dP} = \frac{L}{4\pi c G M}\kappa\eta, \tag{61}$$

where

$$\eta = \frac{L(r)/M(r)}{L/M} = \frac{\bar{\epsilon}(r)}{\bar{\epsilon}}. \tag{62}$$

The physical meaning of η is that it is the ratio of the average rate of liberation of energy $\bar{\epsilon}(r)$ interior to the point r to the corresponding average $\bar{\epsilon}$ for the whole star.

Integrating (61) from $r = r$ to $r = R$ and using the boundary condition $p_r = 0$ at $r = R$ we have

$$p_r = \frac{L}{4\pi c G M}\overline{\kappa\eta}(r)P, \tag{63}$$

where

$$\overline{\kappa\eta}(r) = \frac{1}{P}\int_R^r \kappa\eta\, dP. \tag{64}$$

Equation (63) can be written alternatively in the form

$$L = \frac{4\pi c G M(1 - \beta)}{\overline{\kappa\eta}(r)}. \tag{65}$$

If we put $r = R$ in the above equation we get a fundamental relation which enables the evaluation of L in terms of an average value of $\kappa\eta$:

$$L = \frac{4\pi cGM(1 - \beta_c)}{\overline{\kappa\eta}}. \tag{66}$$

From equation (65) we conclude that the condition $(1 - \beta)$ not increasing outward (which we used earlier in the discussion) is equivalent to the assumption of $\overline{\kappa\eta}(r)$ not increasing outward. Alternatively we should have

$$\kappa\eta(r) \geqslant [\overline{\kappa\eta}]_R{}^r. \tag{67}$$

It is clear from (67) that we can actually allow a decrease of $\kappa\eta$ (within limits) as we approach the center. In actual stellar configurations η might be expected to decrease outward, but this will not be generally true of κ. For this reason it is important to realize that the minimal characteristic of the standard model was proved under restrictions which do not exclude the possibility of $\kappa\eta$ actually decreasing outward [within the limits set by (67)].

Equation (66) enables us to set an upper limit to a mean value of the stellar opacity. For if we combine (66) with the inequality (50) we obtain

$$\overline{\kappa\eta} \leqslant \frac{4\pi cGM(1 - \beta^*)}{L}. \tag{68}$$

If we assume that $\epsilon(r)$ does not increase outward then η will not increase outward and the minimum value of η is unity. Hence

$$\overline{\kappa\eta} \geqslant \bar{\kappa}, \tag{69}$$

where

$$\bar{\kappa} = \frac{1}{P_c}\int_0^{P_c} \kappa dP. \tag{70}$$

Hence we have

$$\bar{\kappa} < \frac{4\pi cGM(1 - \beta^*)}{L}. \tag{71}$$

If we apply (71) to the case of Capella ($M = 4.18\odot$, $L = 120L_\odot$) we find that

$$\bar{\kappa}_{\text{Capella}} < 100 \text{ gm}^{-1} \text{ cm}^2. \tag{72}$$

It should be noticed that our method of averaging weights the central regions of the configuration very heavily and hence the upper limit (71) is essentially an upper limit to the opacity at the center of the configuration. The physical meaning of the inequality (71) is this: If for a star of given mass M and luminosity L, $\bar{\kappa}$ should be greater than the limit set by (71) then either the density or the rate of generation of energy ϵ or both must increase outward in some finite regions of the interior of a star.

We thus see that no special assumptions are required to establish the orders of magnitude of the physical variables in stellar interiors.[8]

II. The Method of the Homologous Transformations

We have already seen that if we restrict ourselves to the considerations of stars of masses less than five to six times the solar mass then we can neglect the radiation pressure in the equation of hydrostatic equilibrium, *i.e.*,

$$P \backsimeq \frac{k}{\mu H} \rho T, \tag{73}$$

or in a somewhat better approximation we can write

$$P = \frac{k}{\bar{\beta} \mu H} \rho T, \tag{74}$$

where $\bar{\beta}$ is a certain average value of β inside the star. We assume that $\bar{\beta}$ is very nearly unity. Our two equations of equilibrium are:

$$\frac{k}{\bar{\beta} \mu H} \frac{d}{dr} (\rho T) = - \frac{GM(r)}{r^2} \rho, \tag{75}$$

$$\frac{dM(r)}{dr} = 4\pi r^2 \rho. \tag{76}$$

To make the system of equations governing the equilibrium of the star complete we need an additional equation to determine the temperature gradient. If radiative equilibrium obtains then [*cf.* equation (60)]

$$\frac{d}{dr} \left(\frac{1}{3} a T^4 \right) = - \frac{\kappa L(r)}{4\pi c r^2} \rho, \tag{77}$$

where it is clear from definition that

$$L(r) = \int_0^r 4\pi r^2 \rho \epsilon dr. \tag{78}$$

In (78) ϵ is the rate of generation of energy in ergs per second per gram of the substance. An alternative form for (77) can be obtained by combining equations (61) and (65). We have

$$\frac{dp_r}{dP} = (1 - \beta) \frac{\kappa \eta}{\overline{\kappa \eta}(r)} = \frac{p_r}{P} \frac{\kappa \eta}{\overline{\kappa \eta}(r)}, \tag{79}$$

[8] The references to the literature will be found in the author's monograph *An Introduction to the Study of Stellar Structure* (Chicago, 1939).

or in a somewhat different form:

$$\frac{dT}{T} = \frac{1}{4} \frac{\kappa\eta}{\overline{\kappa\eta}(r)} \frac{dP}{P}. \tag{80}$$

Now, according to certain well known methods of argument [9] an existing temperature is stable or unstable according as it is less than or greater than the corresponding adiabatic gradient. For an enclosure containing matter and radiation the condition of adiabacy is given by

$$dQ = d(aVT^4 + c_V T) + \left(\frac{1}{3} aT^4 + \frac{k}{\mu H} \rho T\right) dV = 0, \tag{81}$$

where c_V is the specific heat of the gas at constant volume. Equation (81) can be rewritten in the form [10]

$$\frac{dT}{T} = \frac{\Gamma_2 - 1}{\Gamma_2} \frac{dP}{P}, \tag{82}$$

where

$$\Gamma_2 = 1 + \frac{(4 - 3\beta)(\gamma - 1)}{\beta^2 + 3(\gamma - 1)(1 - \beta)(4 + \beta)}, \tag{83}$$

where γ is the ratio of the specific heats ($= 5/3$) for the gas and β has the same meaning as hitherto. The condition for the stability of the radiative gradient is therefore

$$4 \frac{\Gamma_2 - 1}{\Gamma_2} \geqslant \frac{\kappa\eta}{\overline{\kappa\eta}(r)}. \tag{84}$$

The radiative gradient becomes "unstable" if $\kappa\eta/\overline{\kappa\eta}(r)$ exceeds the value on the right hand side of the above inequality. The following table (Table 3) gives the values of $4(\Gamma_2 - 1)/\Gamma_2$ for different values of $1 - \beta$.

TABLE 3

$1-\beta$	0	0.1	0.2	0.3	0.4	0.5	0.6	0.7	0.8	0.9	1.0
$4(\Gamma_2 - 1)/\Gamma_2$ . . .	1.6	1.304	1.177	1.108	1.065	1.039	1.022	1.010	1.004	1.000	1.000

It follows from Table 3 that if the energy sources are concentrated towards the center then as we approach the central regions the radiative gradient will become unstable. For the case of vanishing radiation pressure, the nature of the steady state that will be set up if the radiative gradient becomes unstable has been investigated in detail

[9] See, for instance, the author's monograph on *An Introduction to the Study of Stellar Structure* (Chicago, 1939), pp. 222–227.

[10] See reference 9, pp. 55–59.

by Cowling and Biermann. These authors come to the conclusion that if

$$\frac{\kappa\eta}{\overline{\kappa\eta(r)}} \geqslant 1.6, \tag{85}$$

then the adiabatic gradient obtains:

$$p \propto T^{\gamma/(\gamma-1)}. \tag{86}$$

Now a rigorous attack on the equations of equilibrium (75), (76), (77) and (78) [11] is possible only if the dependence of ϵ and κ on ρ and T are known. Until recently we had no information concerning the nature of the dependence of ϵ on ρ and T. At the present time we have some notions on this [12] but even now our knowledge is by no means entirely satisfactory. It is for this reason that the analysis of stellar structure has depended so largely on the study of stellar models. It would appear at first sight that the uncertainty in the law of energy generation is a serious matter. We can however show that *for stars in which the radiation pressure is negligible throughout the configuration*

$$L = \text{Constant}\,\frac{1}{\kappa_0}\,\frac{M^{5+s}}{R^s}\,(\mu\bar{\beta})^{7+s}, \tag{87}$$

if the rate of generation of energy ϵ *and the coefficient of opacity* κ *follow the laws*

$$\epsilon = \epsilon_0\rho^\alpha T^\nu, \qquad \kappa = \kappa_0\rho T^{-3-s}, \tag{88}$$

where α, ν *and* s *are arbitrary. The constant in* (87) *depends on the exponents* α, ν and s.[13]

To prove this let us first consider stars in radiative equilibrium. Then the equations of equilibrium can be written as

$$\frac{dP}{dr} = -\frac{GM(r)}{r^2}\rho, \tag{89}$$

$$\frac{dM(r)}{dr} = 4\pi r^2\rho, \tag{90}$$

$$P = \frac{k}{\bar{\beta}\mu H}\rho T, \tag{91}$$

$$\frac{dT}{dr} = -\frac{3}{4ac}\,\kappa_0\rho^2 T^{-6-s}\,\frac{\int_0^r \rho^{\alpha+1}T^\nu r^2 dr}{r^2}. \tag{92}$$

[11] Or (86) depending upon whether (84) is satisfied or not.
[12] See H. N. Russell's article in this volume..
[13] This theorem is due to B. Strömgren.

The system of equations (89)–(92) has to be solved with the boundary conditions

$$M(r) = M, \qquad \rho = 0, \qquad T = 0 \qquad \text{at} \qquad r = R, \tag{93}$$

and

$$M(r) = 0 \qquad \text{at } r = 0. \tag{94}$$

These provide four boundary conditions and since the system of equations (89)–(92) is equivalent to a single differential equation of the fourth order, it follows that there is just exactly one solution which will satisfy the boundary conditions (93) and (94). We shall now show that from such a solution we can construct another solution such that it will describe another configuration with a different M, R, and $\bar{\beta}\mu$; we shall see, in fact, that the transformations required to go over from one set of values M, R and μ to another set M_1, R_1 and μ_1 is the successive application of three elementary homologous transformations. To show this we proceed as follows:

Let the physical variables, after a general homologous transformation has been applied be denoted by attaching a suffix "1." for a general homologous transformation we should have

$$\begin{aligned} r_1 &= y^{n_1} r, & (\bar{\beta}\mu)_1 &= y^{n_5}\bar{\beta}\mu \\ P_1 &= y^{n_2} P, & T_1 &= y^{n_6} T \\ M(r_1)_1 &= y^{n_3} M(r), & (\kappa_0\epsilon_0)_1 &= y^{n_7}(\kappa_0\epsilon_0) \\ \rho_1 &= y^{n_4}\rho, \end{aligned} \tag{95}$$

where $n_1, \cdots, n_7$ are, for the present, arbitrary constants and y is the transformation constant. The exponents $n_1, \cdots, n_7$ should satisfy certain conditions, namely, those which are necessary for the continued validity of equations (89)–(92) in the suffixed variables. Substituting (95) in (89) we find that we should have

$$y^{n_2 - n_1} = y^{n_3 + n_4 - 2n_1}, \tag{96}$$

or

$$n_2 - n_1 = n_3 + n_4 - 2n_1. \tag{97}$$

In the same way equations (90), (91) and (92) yield:

$$n_3 - n_1 = 2n_1 + n_4, \tag{98}$$

$$n_2 = n_4 + n_6 - n_5, \tag{99}$$

$$n_6 - n_1 = n_7 + (\alpha + 3)n_4 - (6 + s - \nu)n_6 + n_1. \tag{100}$$

We have thus four equations between the seven unknowns. Hence, we should be able to express any four of the n's in terms of the other three. We shall choose n_1, n_3 and n_5 as the independent quantities.

Solving for n_2, n_4, n_6 and n_7 in terms of n_1, n_3 and n_5 we find that

$$n_2 = -4n_1 + 2n_3, \tag{101}$$

$$n_4 = -3n_1 + n_3, \tag{102}$$

$$n_6 = -n_1 + n_3 + n_5, \tag{103}$$

$$n_7 = -(s - \nu - 3\alpha)n_1 + (4 + s - \nu - \alpha)n_3 + (7 + s - \nu)n_5. \tag{104}$$

If we choose $n_1 = 1$, $n_3 = 0$ and $n_5 = 0$ we have a homologous transformation in which a star of a given mass M and mean molecular mass $\mu\bar{\beta}$ is expanded or contracted. In thc same way, the choice $n_1 = 0$, $n_3 = 1$ and $n_5 = 0$ corresponds to an alteration of M while R and $\mu\bar{\beta}$ are kept unchanged. Finally, the choice $n_1 = 0$, $n_3 = 0$ and $n_5 = 1$ corresponds to an alteration of $\mu\bar{\beta}$ while M and R are kept unchanged.

We have now to consider how the luminosity is changed by a homologous transformation. Since

$$L = 4\pi \int_0^R r^2 \rho \epsilon dr, \tag{105}$$

we have according to our law (88) for ϵ

$$\kappa_0 L = 4\pi \kappa_0 \epsilon_0 \int_0^R r^2 \rho^{\alpha+1} T^\nu dr. \tag{106}$$

Hence, by a general homologous transformation $\kappa_0 L$ alters to $(\kappa_0 L)_1$ where

$$(\kappa_0 L)_1 = y^{n_7 + 3n_1 + (\alpha+1)n_4 + \nu n_6}(\kappa_0 L), \tag{107}$$

or by (101), (102), (103) and (104)

$$(\kappa_0 L)_1 = y^{-sn_1 + (5+s)n_3 + (7+s)n_5}(\kappa_0 L). \tag{108}$$

In other words

$$L = \text{constant} \frac{M^{5+s}}{\kappa_0 R^s} (\mu\bar{\beta})^{7+s}. \tag{109}$$

Another relation of importance is that which is equivalent to (104):

$$\kappa_0 \epsilon_0 = \text{constant } R^{(3\alpha+\nu-s)} M^{(4+s-\nu-\alpha)} (\mu\bar{\beta})^{7+s-\nu}. \tag{110}$$

It is clear that the constants in (109) and (110) can depend only on the exponents α, ν and s. We have thus proved the invariance of the luminosity formula for stars in radiative equilibrium. If, however, the law of energy generation is such that it leads to a sufficiently strong concentration of the energy sources towards the center then we should reach a stage when

$$\frac{\kappa\eta}{\overline{\kappa\eta}(r)} > 4 \frac{\Gamma_2 - 1}{\Gamma_2}. \tag{111}$$

In other words going inward toward the interior of a star the radiative gradient will become unstable at some definite point $r = r_i$ (say). For stars with negligible radiation pressure we have

$$\frac{\kappa\eta}{\overline{\kappa\eta}(r_i)} = 1.6 \qquad (r = r_i). \tag{112}$$

For $r < r_i$, $\kappa\eta/\overline{\kappa\eta}(r) > 1.6$. Now the right hand side of (112) is a pure number, while the quantity on the left hand side is homology invariant. Hence the fraction $q = r_i/R$ of the radius at which the instability of the radiative gradient sets in, is the same for all stars with vanishing radiation pressure. The fraction q depends only on the exponents α, ν and s which occur in the laws for ϵ and κ. Further the material interior to r_i will be in convective equilibrium and we should have

$$\frac{p}{p_i} = \left(\frac{\rho}{\rho_i}\right)^{\gamma}, \tag{113}$$

where p_i and ρ_i refer to the pressure and the density at the interface, *i.e.*, at $r_i = qR$. Equation (113) is clearly homology invariant. We have thus proved the invariance of the form of the mass-luminosity-radius relation quite generally.

The next thing that we shall have to examine is the range of variation in the constant of proportionality in the relation (109) for the possible range of stellar models. In these discussions we shall restrict ourselves to the case $s = 1/2$, *i.e.*, assume for the law of opacity the form

$$\kappa = \kappa_0 \rho T^{-3.5}, \tag{114}$$

where κ_0 will depend upon the chemical composition.

We shall consider two models: (*a*) the model ϵ = constant and (*b*) the point source model $\epsilon = 0$, $r \neq 0$. The model ϵ = constant corresponds to a uniform distribution of the energy sources while the point source model corresponds to the complete concentration of the energy sources at the center of the star. We may expect that in the actual stars an energy source distribution is realized which is intermediate to these two limiting cases.

Consider first the model ϵ = constant. Remembering that we are restricting ourselves to stars with negligible radiation pressure we can rewrite (61) in the form:

$$\frac{k}{\mu H}\frac{3}{a}\frac{d(\rho T)}{d(T^4)} = \frac{4\pi c G M}{L\kappa\eta}. \tag{115}$$

For the model under consideration $\eta = 1$. Using (88) as our law of opacity we have

$$\frac{k}{\mu H}\frac{3}{a}\frac{d(\rho T)}{d(T^4)} = \frac{4\pi cGM}{\kappa_0 L}\frac{T^{3+s}}{\rho}. \tag{116}$$

Let

$$\rho = \frac{\mu H}{k}\frac{a}{3}T^3 y. \tag{117}$$

Then we have

$$\frac{d(yT^4)}{d(T^4)} = \frac{4\pi cGM}{\kappa_0 L}\frac{k}{\mu H}\frac{3}{a}\frac{T^s}{y}, \tag{118}$$

or after some minor transformations:

$$\frac{1}{4}T\frac{dy}{dT} = \frac{4\pi cGM}{\kappa_0 L}\frac{k}{\mu H}\frac{3}{a}\frac{T^s}{y} - y. \tag{119}$$

Introduce the new variable x defined by

$$x = \frac{4\pi cGM}{\kappa_0 L}\frac{k}{\mu H}\frac{3}{a}T^s. \tag{120}$$

(119) now takes the form

$$\frac{1}{4}sxy\frac{dy}{dx} = x - y^2. \tag{121}$$

The general solution of (121) is easily seen to be

$$y^2 = \frac{8}{8+s}x + Bx^{-8/s}, \tag{122}$$

where B is a constant of integration. From (122) we see that if $s > 0$ the second term in (122) becomes rapidly small compared to the first term as we descend into the deeper layers of the star. Hence for layers not immediately near the boundary

$$y^2 \backsim \frac{8}{8+s}x. \tag{123}$$

From (117), (120) and (123) we now have

$$\rho = \text{constant } T^{3+\frac{1}{2}s}; \tag{124}$$

or again

$$P = \frac{k}{\mu H}\rho T = \text{constant } T^{4+\frac{1}{2}s}. \tag{125}$$

From (124) and (125) we see that

$$P = \text{constant } \rho^{(1+[1/(3+\frac{1}{2}s)])}. \tag{126}$$

In other words, the configurations are polytropes of index $n = 3 + \frac{1}{2}s$. For the physically interesting case $s = 1/2$ so that the stars on this model are polytropes of index $n = 3.25$. The march of density and temperature in such configurations is given in Table 4 (see Figs. 2 and 3).

TABLE 4

DENSITY AND TEMPERATURE DISTRIBUTIONS FOR THE MODEL ϵ = CONSTANT

ξ	ρ/ρ_c	T/T_c
0	1.000	1.000
0.4	0.918	0.974
0.8	0.719	0.903
1.2	0.495	0.805
1.6	0.311	0.698
2.0	0.184	0.594
2.4	0.105	0.500
2.8	0.0588	0.418
3.2	0.0326	0.349
3.6	0.0179	0.290
4.0	0.00970	0.240
4.4	0.00518	0.198
4.8	0.00271	0.162
5.2	0.00137	0.131
5.6	6.57×10^{-4}	0.105
6.0	2.92×10^{-4}	0.0818
6.4	1.16×10^{-4}	0.0615
6.8	3.78×10^{-5}	0.0436
7.2	8.63×10^{-6}	0.0277
7.6	8.19×10^{-7}	0.0134
8.0	2.96×10^{-11}	0.0006
8.0189	0	0

Further from the integration appropriate for the polytrope $n = 3.25$ we find that [14]

$$\left.\begin{aligned} \rho_c &= 88.15\,\bar{\rho}, \\ T_c &= 0.968\,\frac{\mu H}{k}\frac{GM}{R}, \\ P_c &= 20.37\,\frac{GM^2}{R^4}. \end{aligned}\right\} \qquad (127)$$

Further by a simple transformation of the luminosity formula (66) we obtain [15]

$$L = 1.43 \times 10^{25}\,\frac{1}{\kappa_0}\,\frac{M^{5.5}}{R^{0.5}}\,\mu^{7.5}, \qquad (128)$$

where L, M and R are expressed in the corresponding solar units.

[14] The integration for the polytrope $n = 3.25$ is given in *Ap. J.*, **89**, 116, 1939.
[15] For the details of the derivation see the author's monograph (Ref. 9, pp. 322–327).

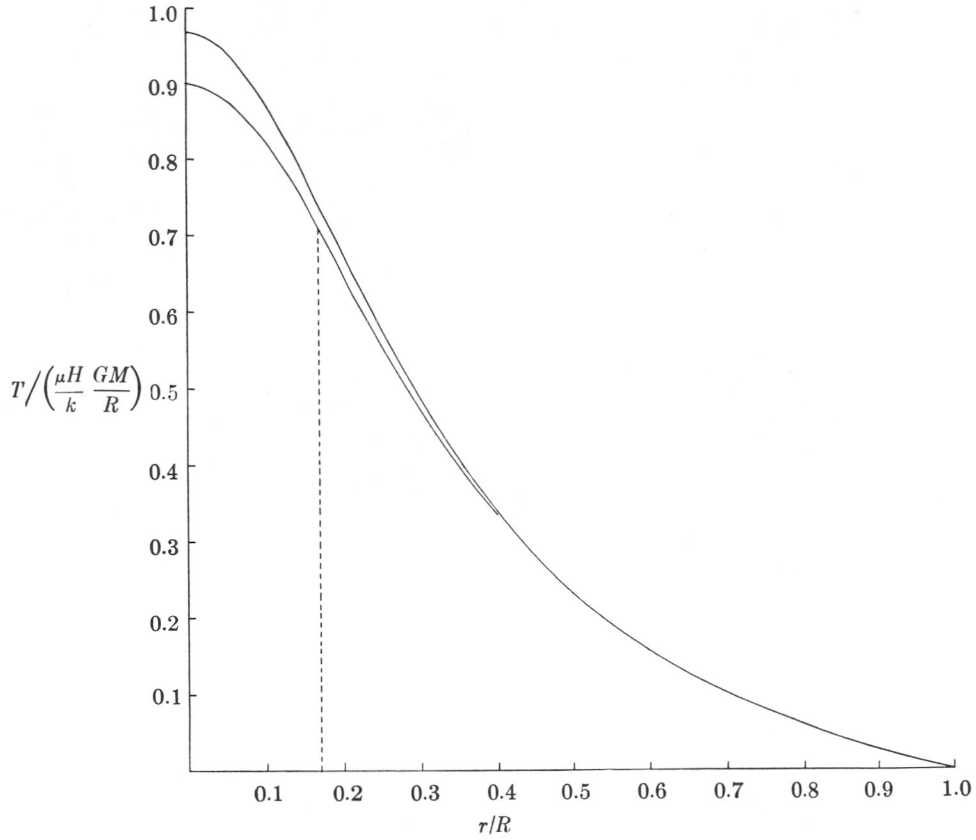

FIG. 2. The temperature distributions in the models (a) ϵ = constant and (b) the point source model.

Let us next consider the point source model. For this model it is clear that the central regions must be in convective equilibrium and therefore consist of polytropic cores of indcx $n = 1.5$ surrounded by " point-source envelopes," *i.e.*, regions governed by the equations

$$\frac{k}{\mu H}\frac{d}{dr}(\rho T) = -\frac{GM(r)}{r^2}\rho, \tag{129}$$

$$\frac{dp_r}{dr} = -\frac{\kappa_0 L}{4\pi c r^2}\frac{\rho^2}{T^{3.5}}, \tag{130}$$

$$\frac{dM(r)}{dr} = 4\pi r^2 \rho. \tag{131}$$

The integration has to be effected numerically. To start the integration we assume that the polytropic core extends to a fraction q of the

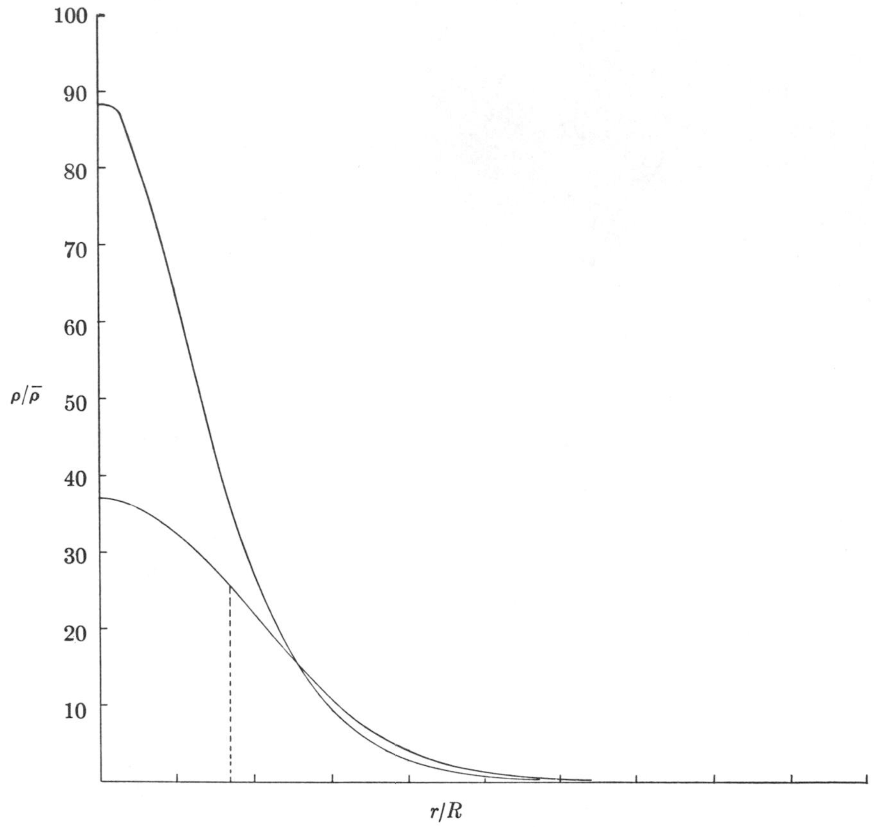

FIG. 3. The density distributions in the models (a) ϵ = constant and (b) the point source model.

radius of the star. From this point onward the integration is continued by means of equations (129)–(131). For an arbitrarily assigned fraction q, the density ρ and the temperature T will not tend to zero simultaneously. We can, however, adjust q by trial and error until ρ and T do tend to zero simultaneously. In this way we can construct a configuration of assigned mass M, radius R and mean molecular mass μ. From the homology argument it follows that the relative distributions of density, temperature, etc., will be the same for all stars with negligible radiation pressure built on this model. It is found [16] that the convective core extends to a fraction 0.17 of the radius of the star and encloses 14.5 per cent of the mass. The march of ρ and T for this model is shown in Table 5 (see Figs. 2 and 3). Further, it is found that

[16] The integration was first effected by Cowling. For greater details see the author's monograph (Ref. 9), pp. 351–355.

TABLE 5

DENSITY AND TEMPERATURE DISTRIBUTIONS FOR THE POINT-SOURCE MODEL

ξ	ρ/ρ_c	T/T_c
0	1.000	1.000
0.4	0.961	0.974
0.8	0.852	0.898
1.2	0.693	0.784
1.6	0.508	0.661
2.0	0.330	0.551
2.4	0.197	0.455
2.8	0.110	0.372
3.2	5.90×10^{-2}	0.303
3.6	3.05×10^{-2}	0.245
4.0	1.51×10^{-2}	0.196
4.4	7.16×10^{-3}	0.156
4.8	3.20×10^{-3}	0.121
5.2	1.30×10^{-3}	0.0919
5.6	4.60×10^{-4}	0.0666
6.0	1.26×10^{-4}	0.0448
6.4	2.06×10^{-5}	0.0256
6.8	6.22×10^{-7}	0.00873
7.0	5.4×10^{-10}	0.00100
7.027	0	0

$$\left.\begin{aligned} \rho_c &= 37.0\,\bar{\rho}, \\ T_c &= 0.900\,\frac{\mu H}{k}\frac{GM}{R}, \\ P_c &= 7.95\,\frac{GM^2}{R^4}. \end{aligned}\right\} \tag{132}$$

The luminosity formula is found to be

$$L = 5.43\times10^{24}\,\frac{1}{\kappa_0}\frac{M^{5.5}}{R^{0.5}}\,\mu^{7.5}, \tag{133}$$

where L, M and R are expressed in the corresponding solar units.

We now see that the mass-luminosity-radius relations for the two models are of the same form (in agreement with our general discussion). In addition, we notice that the constants of proportionality differ only by a factor 2.6 for the extreme range in the possible distributions of the energy sources.

For a comparison with these two models we may note that on the standard model in which stars are polytropes of index $n = 3$ we have

$$\left.\begin{aligned} \rho_c &= 54.2\,\bar{\rho}, \\ T_c &= 0.854\,\frac{\mu H}{k}\frac{GM}{R}, \\ P_c &= 11.05\,\frac{GM^2}{R^4}. \end{aligned}\right\} \tag{134}$$

We see that this model is "intermediate" to the two other models considered.

A fundamental result of importance which has come out of the present discussion is that there exists a relation of the type

$$L = \text{constant}\frac{1}{\kappa_0}\frac{M^{5.5}}{R^{0.5}}\mu^{7.5}, \tag{135}$$

in which the uncertainty in the constant of proportionality due to possible range of stellar models is less than the other uncertainties inherent in the problem.

As we have already pointed out κ_0 will depend on the chemical composition. The discussion of the stellar opacity is beyond the scope of the present report [17] except to mention that it is found that we can write

$$\kappa_0 = \frac{3.89 \times 10^{25}(1 - X_0^2)}{t}, \tag{136}$$

where X_0 is the hydrogen content by weight and t is a factor which is slowly varying function of ρ and T. For individual stars t can be replaced by a certain appropriate average value of $t = \bar{t}$. Thus for the sun it is found that $\bar{t} = 5$ while for Capella it is practically unity.

It is now obvious that an application of the relation (135) to the observational material on the masses, luminosities and the radii of the stars enables the determination of the mean molecular weight of individual stars. This problem has been investigated in great detail by Strömgren who finds a *systematic* variation of X_0 in the plane of the Hertzsprung-Russell diagram (more clearly however in a mass-radius diagram). As a result of his investigation Strömgren finds that within the limits of uncertainty of the observational material the stars can be satisfactorily arranged as a two parametric family, the two parameters being the mass M and the hydrogen content X_0. The interpretation of the Hertzsprung-Russell diagram which Strömgren arrives at is the following:

The main series up to spectral class A is the locus of stars of hydrogen content varying between 25 to 45 per cent—*i.e.*, about a mean of 35 per cent—and masses running up to 2.5 ⊙. Stars of small mass and low hydrogen content are relatively rare, they occur as subgiants of spectral classes G to K. The gap between M giants and the corresponding dwarfs (on the main series) arises from the circumstance that not even stars of low hydrogen content "scatter" in this region. The massive stars ($M > 5\odot$) occurring in the region of the B-stars which are rich in hydrogen (X_0 sometimes going up to 95 per cent)

[17] For details see the author's monograph, pp. 261–272.

form the continuation of the main series, the continuation arising from the circumstance that massive stars with " medium " hydrogen content ($0.4 < X_0 < 0.8$) which are on the main series occur in a very small region of the H.R diagram. (We shall obtain evidence in Section III for the breakdown of the model underlying these computations for the very massive stars. Further, along the main series the breakdown probably sets in at about $M = 10\odot$. The investigations of the hydrogen contents of the B-stars are therefore somewhat inconclusive.) The giant branch is characterized by stars having about the same hydrogen content as (or somewhat less than) the main series stars. The giant branch is limited on the side of low luminosity, since stars of low luminosity are relatively rare. On the side of high luminosity it is limited again, because for X_0 a little greater than 0.35 the characteristic bend of the curves of constant X_0 (see Fig. 4) disappears and also because the stars of large mass with hydrogen content greater than about 40 per cent scatter over a large area in the H.R diagram, which must, therefore, be sparsely populated. The gap in the giant branch in the region of the spectral class F is probably due to a real scarcity of stars with masses between 2.5 and 4.5 $\odot$. The supergiants then are interpreted as massive stars with medium hydrogen content (see Fig. 4).

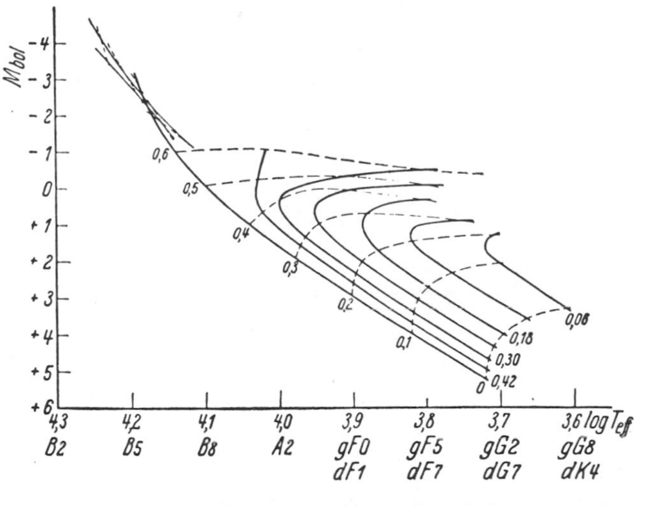

FIG. 4. Curves of constant X_0 (the full line curves) and the curves of constant M (the dotted curves) in the plane of the Hertzsprung-Russell diagram.

An important application of the method described has been recently made by Kuiper. By determining μ and X_0 for a few stars in the Hyades cluster for which he had derived the L, M and R values, he was able to show that the stars in this cluster are relatively poor in hydrogen as compared to the normal main series stars (*i.e.*, sun, α Centauri, etc.).

So far we have considered only the relation (109). But from the homology argument we established another relation, namely (110). Now, κ_0 and ϵ_0 will depend upon the chemical composition, *i.e.*, on μ and X_0. Hence if we consider stars with a given X_0 then we can eliminate R between (109) and (110) and obtain a *pure* mass-luminosity relation. Hence for stars with constant X_0 we have:

$$L = \text{constant}\ M^{5+s+\frac{(4+s-\nu-\alpha)s}{3\alpha+\nu-s}}. \tag{137}$$

If we assume $s = 1/2$ and $\alpha = 2$, (137) reduces to

$$L = \text{constant}\ M^{\frac{63+10\nu}{11+2\nu}}. \tag{138}$$

Hence by selecting stars of a given X_0 we can determine ν by a comparison of (138) with the result of observations. This in turn will give some indications of the type of nuclear reactions that are responsible as the source of stellar energy.[18]

III. The Method of Stellar Envelopes

In this method the equilibrium of the stellar envelopes is studied. By a stellar envelope we shall mean the outer parts of a star which though containing only a small fraction (for definiteness, we shall assume this fraction to be 10 per cent) of the total mass M nevertheless occupy a good fraction of the radius R. A study of stellar envelopes has a twofold application to astrophysical theories: first, it extends the study of the conventional stellar atmospheres into the far interior and, secondly, it has also a very definite bearing on the studies of the deep interiors which are the main concern in this report. Thus the central condensation of a star, defined as the fraction ξ^* of the radius R which encloses the inner 90 per cent of the mass M, must give some indication of the concentration of mass toward the center of the star under consideration. It is clear that $(1 - \xi^*)$ is a measure of the *extent* of the stellar envelope.

In writing down the equations of equilibrium of the stellar envelope we introduce two simplifications, (*a*) that there are no sources of energy in the stellar envelope and (*b*) that the mass contained in the envelope can be neglected in comparison with the mass of the star as a whole. Indeed, these two assumptions can be regarded as defining the stellar

[18] G. Gamow, *Ap. J.*, **89**, 130, 1939.

envelope. The equations of equilibrium then are

$$\frac{d}{dr}\left(\frac{k}{\mu H}\rho T + \frac{1}{3}aT^4\right) = -\frac{GM}{r^2}\rho, \tag{139}$$

$$\frac{dp_r}{dr} = -\frac{\kappa_0 L}{4\pi c r^2}\frac{\rho^2}{T^{3.5}}, \tag{140}$$

and

$$\frac{dM(r)}{dr} = 4\pi r^2 \rho. \tag{141}$$

The equations (139) and (140) can be solved explicitly [19] to give the distribution of density and temperature in the stellar envelope. Finally, equation (141) enables us to determine how far inwards we have to go to cover the first 10 per cent of the mass. The equations which determine the central condensation are found to be [20]

$$\alpha = 6.25 \times 10^{-3}\left[\frac{L^2 R\mu(1 - X_0{}^2)^2}{M^3}\right]^{1/4}, \tag{142}$$

$$f(\alpha; w^*) = 0.0618\frac{(1 - X_0{}^2)^{0.5}}{\mu^{3.75}}\left(\frac{LR^{0.5}}{M^{5.5}}\right)^{0.5}, \tag{143}$$

and

$$\xi^* = \frac{1}{(w^* + \alpha)^3\left(w^* + \frac{19}{51}\alpha\right) + 1 - \frac{19}{51}\alpha^4}. \tag{144}$$

In the above equations L, M and R are expressed in solar units. $f(\alpha; w)$ is a function defined by means of a definite integral. Tables of this function have been provided.[21]

The method of evaluating ξ^* for a star of given L, M and R and an assumed value of μ proceeds as follows:

We first determine α. Then by interpolation in the tables of the function $f(\alpha; w)$ we find the value w^* such that $f(\alpha; w^*)$ has the value given by the right hand side of (143). (144) then determines ξ^*.

Figs. 5, 6 and 7 illustrate the dependence of ξ^* on X_0 for different stars.

Without going into too much detail it is clear that the evaluation of ξ^* for the normal stars (sun, Capella, ζ Herculis) confirms the conclusions drawn by Strömgren on the basis of the standard model. To consider an example: compare the sun and ζ Herculis. Both are stars of small mass and hence of negligible radiation pressure. According to our

19 For the solutions see the author's monograph, Chapter VIII.

20 See reference 19.

21 Tables of the function $f(\alpha; w)$ will be found in the author's monograph, p. 361.

discussion of the homologous transformations in Section II, it is clear that these two stars must be homologous—and hence must be characterized by the same value of ξ^*. Fig. 5 shows that if ζ Herculis and the sun should have the same value, then the former must be poorer in hydrogen than the sun. This confirms the conclusion of Strömgren who has derived for the sun and ζ Herculis the values $X_0 = 0.37$ and 0.11 respectively.

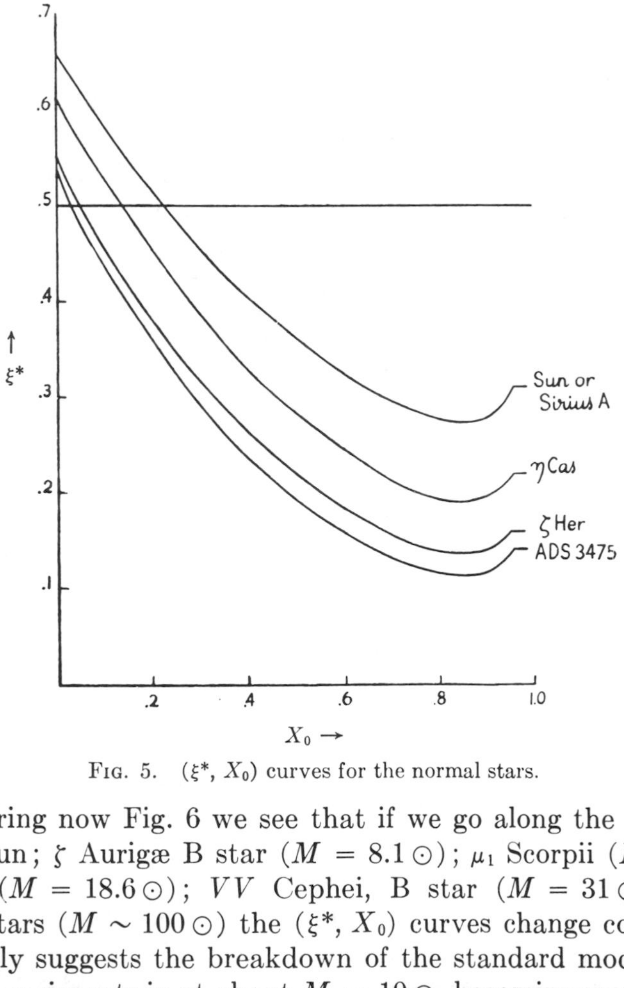

FIG. 5. (ξ^*, X_0) curves for the normal stars.

Considering now Fig. 6 we see that if we go along the sequence of stars, the sun; ζ Aurigæ B star ($M = 8.1\odot$); μ_1 Scorpii ($M = 12\odot$); V Puppis ($M = 18.6\odot$); VV Cephei, B star ($M = 31\odot$) and the Trumpler stars ($M \sim 100\odot$) the (ξ^*, X_0) curves change continuously. This strongly suggests the breakdown of the standard model for stars on the main series sets in at about $M = 10\odot$, becoming more and more pronounced on passing toward the larger masses. This breakdown is most clearly shown by the Trumpler stars where no adjustment of the mean molecular mass can make them homologous to the normal stars.

Turning next to Fig. 7 we notice that if we go along the sequence of stars, the sun ($M = \odot, R = R_\odot$); ζ Herculis ($M = .98\odot$, $R = 1.9R_\odot$);

Capella ($M = 4.2\odot$, $R = 15.8R_\odot$); ζ Aurigæ, K-star ($M = 14.8\odot$, $R = 200R_\odot$); ϵ Aurigæ, I-star ($M = 24.6\odot$, $R = 2140R_\odot$); *VV* Cephei, M-star ($M = 49\odot$, $R = 2130R_\odot$) we infer again the possibility of a breakdown of the standard model also in the region of the massive supergiants (stars of high luminosity and large radius). The

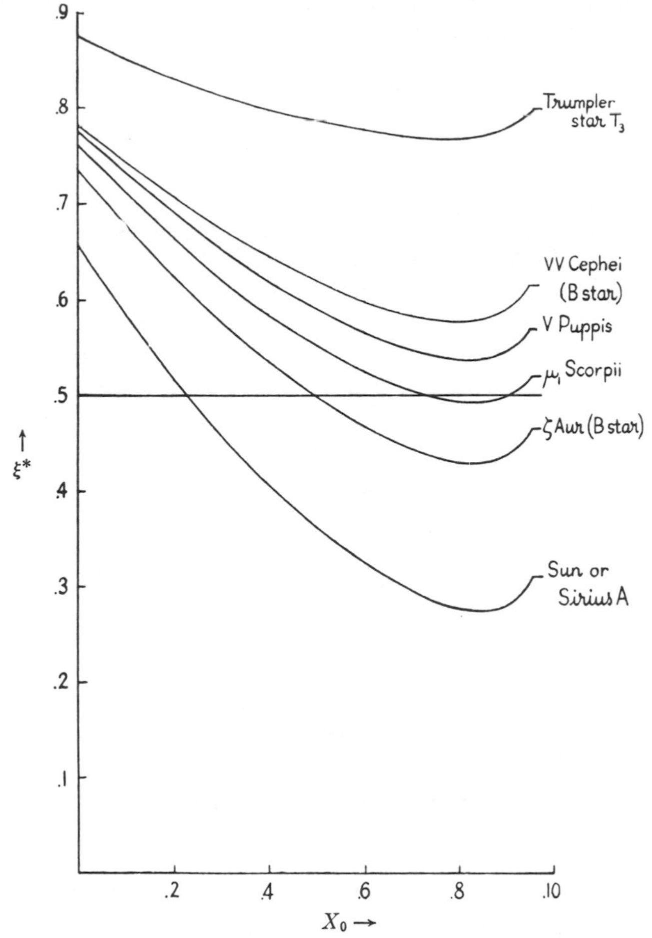

FIG. 6. (ξ^*, X_0) curves for the stars on the main series.

breakdown is now, however, in the sense of becoming more centrally condensed; this differs from the behavior of the massive stars which form an extension of the main series; the latter are certainly more homogeneous than the normal stars. Among the supergiants the possibility of finding stars with ξ^* as small as 0.05 (*e.g.*, *VV* Cephei, *M*-component) cannot be excluded.

The main results which emerge from the discussion of the central condensations of stars can be summarized as follows:

(*a*) The general way in which the theory of stellar envelopes supports the essential conclusions reached in Section II concerning the structures and the hydrogen contents of the normal stars.

(*b*) The increasing homogeneity of the massive stars on the main series, the breakdown of the standard model setting in probably at values of the mass of about 10 ⊙.

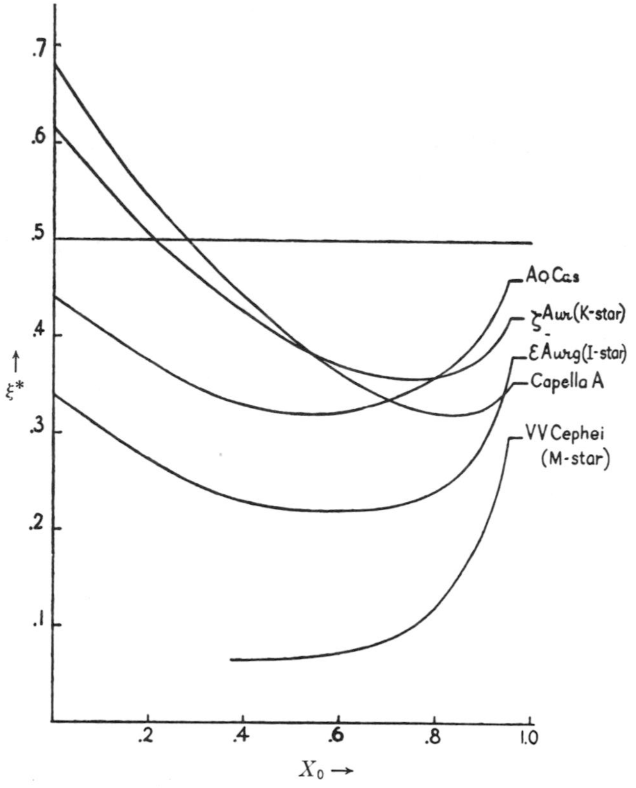

FIG. 7. (ξ^*, X_0) curves for the giants and supergiants.

(*c*) The centrally condensed nature of the massive supergiants.

We may therefore infer from the examples discussed that a certain systematic variation of the stellar model in the (M, R) plane exists.

To avoid misunderstanding, it should be pointed out that the discussion of the mass-luminosity-radius relation is satisfactory only for stars with negligible radiation pressure and consequently can be expected to be valid only for stars of normal masses ($M < 5$ ⊙). On the other hand, in the theory of stellar envelopes the effect of the radiation pressure is taken into account " exactly " and it is for this reason that we are able to draw unambiguous conclusions concerning the structures of stars to which the standard theory cannot be applied. There is thus no contradiction involved in our discovering the " break-down " nature of the massive stars by the method of stellar envelopes.

IV. The Theory of White Dwarfs

So far we have restricted ourselves to a consideration of gaseous stars. However, as is well known, R. H. Fowler showed for the first time that the electrons assembling in the interior of the white dwarfs must be highly degenerate in the sense of the Fermi-Dirac statistics.

The theory of degeneracy in the form required for application in the theory of white dwarfs can be derived in an entirely elementary way. Now, a given number N of electrons can be confined in a given volume V by one of two methods, either by means of potential walls such that electrons inside the " potential hole " cannot escape, or by means of imposing certain periodicity conditions. We shall not consider these restrictions but the essential result of such discussions is that we can label the possible energy states for an electron inside a given volume V by means of quantum numbers in somewhat the same manner as the quantum states of an electron in an atom. If we assume that volume V is large, then it follows from the general theory that the number of quantum states with momenta between p and $p + dp$ is given by

$$V\frac{8\pi p^2 dp}{h^3}. \tag{145}$$

The meaning of (145) is simply that the accessible six-dimensional phase space can be divided into " cells " of volume h^3 and that in each cell there are two possible states. Now, the Pauli principle states that no two electrons can occupy the same quantum state. This implies that if $N(p)dp$ denotes the number of electrons with momenta between p and $p + dp$ then

$$N(p)dp \leqslant V\frac{8\pi p^2 dp}{h^3}. \tag{146}$$

A completely degenerate electron gas is one in which the lowest quantum states are all occupied. In other words, we should have

$$N(p) = V\frac{8\pi p^2}{h^3}. \tag{147}$$

It is clear that if there is only a finite number N of electrons in the specified volume, then all the electrons must have momenta less than a certain maximum value p_0 such that

$$N = V\int_0^{p_0} \frac{8\pi p^2}{h^3}dp = \frac{8\pi V}{3h^3}p_0^3. \tag{148}$$

This " threshold value " p_0 of p is related to the electron concentration n by

$$n = \frac{N}{V} = \frac{8\pi}{3h^3}p_0^3. \tag{149}$$

To calculate the pressure, we recall that by definition the pressure P exerted by a gas is simply the mean rate of transfer of momentum across an ideal surface of unit area in the gas. From this definition it follows quite generally that

$$PV = \frac{1}{3}\int_0^\infty N(p)pv_pdp, \tag{150}$$

where v_p is the velocity associated with the momentum p. According to (147) we have for the case under consideration

$$P = \frac{8\pi}{3h^3}\int_0^{p_0} p^3\frac{\partial E}{\partial p}dp, \tag{151}$$

where E is the kinetic energy of the electron having a momentum p. According to the special theory of relativity

$$E = mc^2\left\{\left(1 + \frac{p^2}{m^2c^2}\right)^{1/2} - 1\right\}, \tag{152}$$

which gives

$$\frac{\partial E}{\partial p} = \frac{1}{m}\left(1 + \frac{p^2}{m^2c^2}\right)^{-1/2}p. \tag{153}$$

Substituting (153) in (151) we have

$$P = \frac{8\pi}{3mh^3}\int_0^{p_0}\frac{p^4dp}{\left(1 + \frac{p^2}{m^2c^2}\right)^{1/2}}. \tag{154}$$

The integral occurring in (154) can be evaluated and we find that we can express P as

$$P = \frac{\pi m^4c^5}{3h^3}f(x), \tag{155}$$

where

$$f(x) = x(2x^2 - 3)(x^2 + 1)^{1/2} + 3\sinh^{-1}x, \tag{156}$$

and

$$x = p_0/mc. \tag{157}$$

We can now write (149) in the form

$$n = \frac{8\pi m^3 c^3}{3h^3} x^3. \tag{158}$$

Equations (155), (156) and (158) represent parametrically the equation of state of a completely degenerate electron gas. (158) can be written alternatively as

$$\rho = n\mu_e H = Bx^3, \tag{159}$$

where

$$B = \frac{8\pi m^3 c^3 \mu_e H}{3h^3} = 9.82 \times 10^5 \mu_e. \tag{160}$$

Similarly we can write (155) as

$$P = Af(x), \tag{161}$$

where

$$A = \frac{\pi m^4 c^5}{3h^3} = 6.01 \times 10^{22}. \tag{162}$$

Completely degenerate stellar configurations are then those which are in hydrostatic equilibrium and in which P and ρ are related according to (159) and (161). We should therefore introduce equations (159) and (161) in the equation of equilibrium,

$$\frac{1}{r^2}\frac{d}{dr}\left(\frac{r^2}{\rho}\frac{dP}{dr}\right) = -4\pi G\rho. \tag{163}$$

By the transformations

$$r = \left(\frac{2A}{\pi G}\right)^{1/2} \frac{1}{By_0}\eta; \qquad y = y_0\phi, \tag{164}$$

where

$$y_0^2 = x_0^2 + 1, \tag{165}$$

equation (163) reduces to

$$\frac{1}{\eta^2}\frac{d}{d\eta}\left(\eta^2\frac{d\phi}{d\eta}\right) = -\left(\phi^2 - \frac{1}{y_0^2}\right)^{3/2}. \tag{166}$$

Equation (166) has to be solved with the boundary conditions

$$\phi = 1; \qquad \frac{d\phi}{d\eta} = 0 \qquad \text{at} \qquad \eta = 0. \tag{167}$$

For each specified value of y_0 we have one such solution. The boundary is defined at the point where the density vanishes, and this by (159) and (165) means that if η_1 specified the boundary

$$\phi(\eta_1) = 1/y_0. \tag{168}$$

The integrations for the function ϕ have been carried out for ten different values of y_0 and the physical characteristics of the resulting configuration are shown in Table 6. The mass radius relation is shown in Fig. 8.

TABLE 6*

THE PHYSICAL CHARACTERISTICS OF COMPLETELY DEGENERATE CONFIGURATIONS

$1/y_0^2$	$M/\odot$	ρ_0 in Grams per Cubic Centimeter	ρ_{mean} in Grams per Cubic Centimeter	Radius in Centimeters
0	5.75	∞	∞	0
0.01	5.51	9.85×10^8	3.70×10^7	4.13×10^8
0.02	5.32	3.37×10^8	1.57×10^7	5.44×10^8
0.05	4.87	8.13×10^7	5.08×10^6	7.69×10^8
0.1	4.33	2.65×10^7	2.10×10^6	9.92×10^8
0.2	3.54	7.85×10^6	7.9×10^5	1.29×10^9
0.3	2.95	3.50×10^6	4.04×10^5	1.51×10^9
0.4	2.45	1.80×10^6	2.29×10^5	1.72×10^9
0.5	2.02	9.82×10^5	1.34×10^5	1.93×10^9
0.6	1.62	5.34×10^5	7.7×10^4	2.15×10^9
0.8	0.88	1.23×10^5	1.92×10^4	2.79×10^9
1.0	0	0	0	∞

* The values given in this table differ slightly from the published values (S. Chandrasekhar, *M. N.*, **95**, 208, 1935, Table III). The difference is due to the change in the accepted values of the fundamental physical constants.

The calculations are for $\mu_e = 1$. For other values of μ_e, M should be multiplied by μ_e^{-2}, R by μ_e^{-1} and ρ_0 by μ_e.

The most important characteristic of these configurations is that they possess a natural limit, *i.e.* as

$$y_0 \to \infty, \qquad \phi \to \theta_3 \tag{169}$$

(where θ_3 is the Lane-Emden function of index 3), and the mass tends to a finite limit M_3. Numerically it is found that

$$M_3 = 5.75\,\mu_e^{-2}\,\odot. \tag{170}$$

A glance at Table 6 shows that the mean density, the mass and the radius of these degenerate configurations are all of the right order of magnitude to provide the basis for the theoretical discussion of the white dwarfs. However, a really satisfactory test of the theory will consist in providing an observational basis for the existence of a mass such that as we approach it the mean density increases several times, even for a slight increase in mass. At the present time there is just one case which seems to support this aspect of the theoretical prediction. The case in question is Kuiper's white dwarf (AC 70° 8247) which is from several points of view a very remarkable star. According to Kuiper, the most probable values of L and R are

$$\text{Log } L = -1.76; \qquad \text{Log } R = -2.38, \tag{171}$$

L and R being expressed in solar units. If we assume that $\mu = 1.5$, the mass-radius relation leads to a mass of 2.5 ⊙. On the other hand if we neglect relativity effects then the unrelativistic mass-radius relation (the

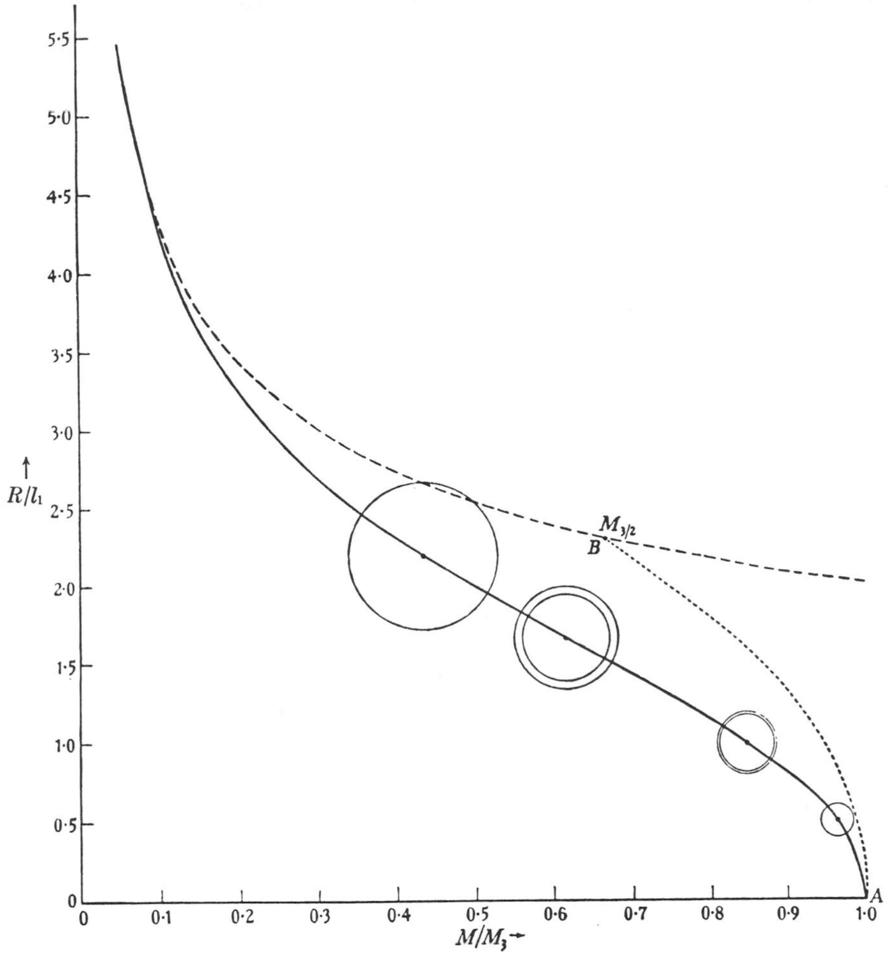

FIG. 8. The mass-radius relation for the white dwarfs.

dotted curve in Fig. 6) leads to a minimum mass of 28 ⊙. The "probable" value predicted on the unrelativistic theory will be of the order of 100 solar masses. We conclude then that the relativistic effects are confirmed by observations.[22]

[22] There is a possibility that Wolf 219, another white dwarf discovered by Kuiper, may be comparable to AC 70° 8247). If confirmed this star would be even more extraordinary than AC 70° 8247) since it is of lower luminosity.

From this brief survey of a few of the methods that have been developed in the study of stellar constitution, it should be clear that there is no reason to be sceptical of the value of such studies. Indeed, the progress that has been made in the last years, while it has been largely due to the realization of the peculiar limitations which a rational approach to the subject necessarily imposes on the investigator, justifies also the hope for still greater advances in the future.

PART FIVE

Theory of Stellar Atmospheres

MODEL STELLAR PHOTOSPHERES.

S. Chandrasekhar.

(Communicated by Professor Milne)

1. The physical structure of the photospheric layers has been studied by Woltjer, Jeans and Milne among others.* For our purposes here the treatment given by Milne in his Bakerian Lecture is the most convenient. In the analysis Milne assumes a law $\kappa \propto P/T^{\frac{9}{2}}$, and in his numerical work to evaluate the temperature, temperature gradient, pressure, density, etc., he uses for the constant of proportionality in the law $\kappa \propto P/T^{\frac{9}{2}}$, that derived by him " astrophysically " from the maxima of zinc lines.

But McCrea † has recently pointed out that another profitable approach to the problems associated with the stellar atmospheres is to work with simplified " models " and to work out fully the consequences of the *physical theory*. It is proposed therefore to apply here this general scheme of advance to the problem of the physical structure of the photosphere on the basis of the author's recent work on the stellar absorption coefficient.‡

2. *Statement of the Problem.*—The problem is to determine the distribution of pressure, density, temperature, temperature gradient and opacity through the photospheric layers (in terms of the optical thickness τ and depth h) of a one-constituent atmosphere where the first stage ionization has set in. In the solution of this problem we make (following Milne) the simplifying assumption of using a mean value for the ionization, instead of treating this as a variable with the " depth." The formula for the absorption coefficient is (a special case of equation (63) of my paper)

$$\kappa = \frac{40}{\pi^4\sqrt{3}}\frac{e^2h^4}{cm(2\pi m)^{\frac{3}{2}}}\chi_1\left(1+\frac{2\cdot 43\chi_1}{kT}\right)\frac{P\bar{x}}{(kT)^{\frac{9}{2}}\text{ atomic mass}}, \tag{1}$$

* Woltjer, *B.A.N.*, **2**, 171 (No. 66), 1924 ; Jeans, *M.N.*, **85**, 199, 1925 ; Milne, *M.N.*, **85**, 768, 1925 ; Milne, *M.N.*, **90**, 17, 1929, §§ 20, 21 and 22 ; Milne, *Phil. Trans.*, A, **228**, 421, 1929, § 16 (Bakerian Lecture).

† W. H. McCrea, " Model Stellar Atmospheres," *M.N.*, **91**, 836, 1931. This kind of approach to stellar problems in *general* has been very often urged by Professor Milne on various occasions. See, for instance, his remarks in *Zeits. für Astrophysik*, **1**, 98, 1930, where he compares stellar problems to those in geometry.

‡ S. Chandrasekhar, *Proc. Roy. Soc.*, A (in press).

where P = electron pressure,
χ_1 = the first " ionization potential " in ergs,
$\bar{x}$ = the fraction of the atoms ionized.

Since for the temperatures usually met with in the photospheric layers ($\sim 10^4$ degrees) the second term in the brackets in equation (1) is by far the most important, we take as our basic formula in the analysis

$$\kappa = \alpha P\bar{x}/T^{\frac{11}{2}}, \tag{2}$$

where

$$\alpha = \frac{40}{\pi^4\sqrt{3}} \frac{e^2h^4}{cm(2\pi m)^{\frac{3}{2}}(k)^{\frac{11}{2}}} \frac{2{\cdot}43\chi_1{}^2}{\text{atomic mass}}$$

$$= 5{\cdot}62 \times 10^{19} \times \bar{\chi}^2/a, \tag{2'}$$

where $\bar{\chi}$ = is the ionization-potential in *electron-volts* ;
a = the atomic weight (not *mass*).

3. *Solution of the Problem.*—Let p' denote the radiation-pressure at depth h in the photospheric layers measured from some convenient reference level. A net flux πF traversing a layer ρdh, of absorption coefficient κ, communicates a momentum $\pi F\kappa\rho dh/c$. Hence

$$\frac{dp'}{dh} = \frac{\kappa\rho\pi F}{c}. \tag{3}$$

Milne has proved that the relation $p' = \frac{1}{3}aT^4$ holds to the extent to which $T^4 = \frac{1}{2}T_1{}^4(1 + \frac{3}{2}\tau)$ is an accurate solution of Schwarzschild's problem. The equation of mechanical equilibrium is

$$\frac{dp}{dh} + \frac{dp'}{dh} = g\rho. \tag{4}$$

Dividing (4) by (3),

$$1 + \frac{dp}{dp'} = \frac{cg}{\pi\kappa F}. \tag{5}$$

Let M be the mass of the star, L its bolometric magnitude ; then

$$g = GM/r_1{}^2 \quad \text{and} \quad \pi F = L/4\pi r_1{}^2, \quad \textit{i.e.} \quad g/\pi F = 4\pi GM/L.$$

Further,

$$\kappa = \frac{\alpha P\bar{x}}{T^{\frac{11}{2}}} = \alpha P\bar{x}\left(\frac{\frac{1}{3}a}{p'}\right)^{\frac{11}{8}}, \tag{6}$$

where $\bar{x}$ is the average number of free electrons per atom in the photospheric layers, *i.e.*

$$\frac{P}{p} = \frac{\bar{x}}{1 + \bar{x}}. \tag{7}$$

Hence (5) becomes

$$1 + \frac{dp}{dp'} = \frac{4\pi cGM}{L} \frac{(1 + \bar{x})}{\alpha\bar{x}^2(\frac{1}{3}a)^{\frac{11}{8}}} \frac{p'^{\frac{11}{8}}}{p}. \tag{8}$$

Make the substitution

$$p = u\frac{p'^{\frac{19}{16}}}{p_1{}'^{\frac{3}{16}}}, \tag{9}$$

where
$$p_1' = \tfrac{1}{3}aT_1^4. \tag{10}$$
(8) now becomes
$$\left(\frac{p_1'}{p'}\right)^{\frac{3}{16}} + p'\frac{du}{dp'} + \frac{19}{16}u = \frac{4\pi cGM}{L}\,\frac{(1+\bar{x})}{\bar{x}^2\alpha(\frac{1}{3}a)^{\frac{11}{8}}}p_1'^{\frac{3}{8}}\cdot\frac{1}{u}$$
$$= \left[\frac{4\pi cGM}{L}\,\frac{(1+\bar{x})}{\bar{x}^2\alpha\frac{1}{3}a}T_1^{\frac{3}{2}}\right]\cdot\frac{1}{u}$$
$$= \frac{A}{u}\ \text{(say)}. \tag{11}$$

Now $(p_1'/p')^{\frac{3}{16}}$ deviates from unity only very little, and we replace it by unity. Hence we have
$$\frac{dp'}{p'} = \frac{u\,du}{A - u - \frac{19}{16}u^2}. \tag{12}$$
Write
$$A - u - \tfrac{19}{16}u^2 = -\tfrac{19}{16}(u-u_1)(u+u_2). \tag{13}$$
From (13) we see immediately that
$$\left.\begin{aligned} A &= (\tfrac{19}{16})u_1u_2\,;\\ \tfrac{16}{19} &= u_2 - u_1.\end{aligned}\right\} \tag{13'}$$
Integration of (12) gives us now
$$\log\frac{p'}{p_0'} = -\frac{\frac{16}{19}}{u_1+u_2}\left[u_1\log\left(1-\frac{u}{u_1}\right) + u_2\log\left(1+\frac{u}{u_2}\right)\right], \tag{14}$$
remembering the " boundary condition " that, when $u=0$,
$$p' = p_0' = \tfrac{1}{3}aT_0^4.$$
Since as the solution of Schwarzschild's problem we have
$$\frac{p'}{p_0'} = 1 + \tfrac{3}{2}\tau, \tag{15}$$
we see that as τ increases, p'/p_0' increases and u rapidly increases and approaches the limit u_1. Hence by (9) the ratio of gas-pressure to radiation-pressure increases slowly according to $u_1(p'/p_1')^{\frac{3}{16}}$. Thus
$$\frac{p}{p'}\ \text{ultimately increases as } T^{\frac{3}{4}}. \tag{16}$$

It may be remarked here that the corresponding result on the Jeans-Milne analysis is that " p/p' ultimately increases as $T^{\frac{1}{4}}$." Now equations (14), (15) and (9) determine p', τ and p for a given u, p' in turn determining the temperature T. It now remains to determine the heights to which these values of u refer. In equation (3) substitute $\rho = p/(R/\mu)T$ and insert also the formula (6) for κ. We find
$$\frac{dp'}{dh} = \frac{\alpha\bar{x}^2}{1+\bar{x}}\left(\frac{\frac{1}{3}a}{p'}\right)^{\frac{11}{8}}\frac{p^2}{(R/\mu)T}\,\frac{gL}{4\pi cGM}, \tag{17}$$

or by (9),

$$\frac{1}{T}\frac{dT}{dh}=\frac{1}{4}\frac{1}{p'}\frac{dp'}{dh}=\frac{\frac{1}{4}\alpha\bar{x}^2}{1+\bar{x}}\frac{(\frac{1}{3}a)^{\frac{11}{8}}}{p'^{\frac{19}{8}}}\frac{p^2}{(R/\mu)T}\frac{gL}{4\pi cGM}=\frac{\frac{1}{4}g}{A(R/\mu)T}u^2\,; \tag{17'}$$

or by (13′),

$$\frac{dT}{dh}=\frac{\frac{1}{4}g}{R/\mu}\frac{u^2}{(\frac{19}{16})u_1u_2}. \tag{18}$$

We observe that as $u \to u_1$,

$$\frac{dT}{dh}\to\frac{\frac{1}{4}g}{R/\mu}\frac{u_1}{(\frac{19}{16})u_2}, \tag{19}$$

or again by (13′),

$$\left(\frac{dT}{dh}\right)_{\infty}=\frac{\frac{1}{4}g}{R/\mu}\frac{1}{(\frac{19}{16})+u_1^{-1}}. \tag{19'}$$

From (17′) we get, after some minor reductions,

$$\frac{gdh}{(R/\mu)T}=-\frac{u_1u_2du}{u(u-u_1)(u+u_2)}. \tag{20}$$

(20) *is identical* with Milne's equation (80) of his Bakerian Lecture. (Of course our u_1 and $-u_2$ are the roots of a slightly different quadratic equation in u and our variable u itself is differently defined.)

Denoting a mean value of T by $\bar{T}$ we find that (Bakerian Lecture, equation (81))

$$h-h_0=\frac{(R/\mu)\bar{T}}{g}\left[\log u-\frac{u_2}{u_1+u_2}\log\left(1-\frac{u}{u_1}\right)-\frac{u_1}{u_1+u_2}\log\left(1+\frac{u}{u_2}\right)\right]. \tag{21}$$

We have now obtained the complete solution to our problem, with, of course, certain simplifying assumptions.

4. *Numerical Applications.*—We use the above formulæ to derive the physical characteristics of the photospheric layers of the " Sun," *i.e.* of a " model " whose mass, luminosity, surface gravity, T_1 etc., are the same as those of the Sun. Now Russell * estimates that the " level of ionization " in the solar atmosphere is such that the atoms of ionization-potential 8·3 volts are 50 per cent. ionized. Hence in our numerical work we take $\bar{\chi}=8{\cdot}3$ volts and $\bar{x}=\frac{1}{2}$. Also, to determine the value of α in (2), we need in addition a *mean* atomic weight. We assume this to be 1·5, to be in conformity with the estimates of Russell and Unsöld regarding the hydrogen-abundance in stellar atmospheres. Also, for " μ " in equation (18) we take 0·75 m_H. The results of the calculation are given below:

$$\alpha=2{\cdot}58\times10^{21}\,;$$
$$\bar{x}=\tfrac{1}{2}\,;$$
$$\mu=0{\cdot}75m_H\,;$$
$$\text{atomic weight}=1{\cdot}5\,;$$
$$\bar{\chi}=8{\cdot}3\text{ volts}\,;$$
$$g=2{\cdot}74\times10^4\text{ cm. sec.}^{-2}\,;$$
$$T_1=5740\text{ degrees.}$$

* H. N. Russell, *Ap. J.*, **70**, 11, 1929.

Using these values,
$$A=\frac{4\pi c G_1 M}{L}\frac{T_1^{\frac{3}{2}}}{\frac{1}{3}a\alpha}\frac{1+\bar{x}}{\bar{x}^2}=6582\ ;$$

$$u_1=\ 74\cdot 02\ ;$$
$$u_2=\ 74\cdot 86\ ;$$
$$u_1+u_2=148\cdot 88.$$

The physical structure of the photospheric layers calculated with the above data and the formulæ of the previous section is tabulated in Table I.

It is seen that the physical characteristics thus calculated offer a highly plausible description of the state of the Sun's photospheric layers, the one exception being that the pressures appear to be perhaps rather large. This is a consequence of the low absorption coefficient consequent on the 50 per cent. ionization assumed in the calculations. To illustrate the effect of ionization, therefore, we repeat the calculations with the following data:—

$$\bar{\chi}=7\cdot 812 \text{ volts}\ ;$$
$$\bar{x}=1\ ;$$
$$\mu=0\cdot 5 m_H\ ;$$
$$\text{atomic weight}=1\ ;$$
$$a=3\cdot 43\times 10^{21}\ ;$$
$$A=1650\ ;$$
$$u_1=36\cdot 86\ ;$$
$$u_2=37\cdot 70\ ;$$
$$u_1+u_2=74\cdot 56.$$

The results of the calculations are contained in Table II. We note the general reduction in pressure and the increased absorption coefficient.

TABLE I

u	Optical Thickness	Radiation-pressure, p', in Dynes cm.$^{-2}$	Gas-pressure, p, in Dynes cm.$^{-2}$	Temperature T Degrees	Temperature Gradient dT/dh Degrees Km.$^{-1}$	Depth h, Km.	Density ρ Gms. cm.$^{-3}$	Absorption Coefficient κ
0	0	1·386	0	4830	0	$-\infty$	0	0
10	0·0058	1·399	12·3	4841	0·0945	447·6	$2\cdot 309\times 10^{-11}$	28·43
20	0·0215	1·432	25·27	4868	0·3781	591·6	$4\cdot 716\times 10^{-11}$	57·07
30	0·0521	1·494	39·92	4921	0·8506	681·6	$7\cdot 367\times 10^{-11}$	86·08
40	0·1038	1·602	57·84	5008	1·513	764·5	$1\cdot 049\times 10^{-10}$	111·1
50	0·1933	1·788	82·38	5147	2·364	846·6	$1\cdot 453\times 10^{-10}$	134·8
60	0·3764	2·169	124·4	5402	3·404	935·2	$2\cdot 090\times 10^{-10}$	157·9
65	0·5688	2·568	164·6	5635	3·994	1034	$2\cdot 654\times 10^{-10}$	166·1
70	1·0410	3·550	260·4	6109	4·632	1185	$3\cdot 872\times 10^{-10}$	169·5
72	1·5930	4·699	373·7	6554	4·900	1315	$5\cdot 178\times 10^{-10}$	164·5
73	2·335	6·240	530·7	7036	5·037	1455	$6\cdot 850\times 10^{-10}$	157·9
73·5	3·334	8·320	752·1	7736	5·107	1631	$9\cdot 016\times 10^{-10}$	147·9
74	8·75	19·57	2095	9363	5·176	2137	$20\cdot 32\ \times 10^{-10}$	127·3

On comparing Tables I and II (particularly Table II) with table IX of Milne's Bakerian Lecture, it will be observed that there is practically no difference between the *predicted* distribution of temperature, density, pressure, etc., in the photospheric layers as τ varies, and that deduced by Milne on the basis of his empirical law for the absorption coefficient. The physical structure of the photospheric layers as deduced by Milne has been generally regarded as representing what is "likely," and it is satisfactory that our results, based on a purely physical theory of the absorption coefficient, should be in general agreement.

TABLE II

u	τ	Radiation-pressure, p', in Dynes cm.$^{-2}$	Gas-pressure, p, in Dynes cm.$^{-2}$	Temperature T Degrees	Temperature Gradient dT/dh Degrees Km.$^{-1}$	Depth h, Km.	Density $\rho \times 10^{12}$ Gms. cm.$^{-3}$	Absorption Coefficient κ
0	0	1·386	0	4830	0	0	0	0
5	0·0052	1·397	6·146	4839	0·0629	468·2	7·693	57·15
10	0·213	1·432	12·64	4865	0·2514	607·5	15·71	113·2
15	0·522	1·494	19·96	4921	0·567	844·3	24·57	169·0
20	0·1037	1·602	28·93	5008	1·006	943·2	34·97	222·3
25	0·1937	1·789	41·20	5147	1·571	1065	48·48	268·6
30	0·3805	2·180	62·36	5409	2·263	1223	69·98	314·2
33	0·6380	2·712	89·15	5712	2·738	1374	94·51	332·7
35	1·0861	3·644	134·3	6137	3·080	1568	132·2	333·5
36	1·731	4·985	200·4	6651	3·26	1783	182·5	323·7
36·5	2·794	7·196	314·3	7292	3·35	2046	261·0	306·3
36·8	6·741	15·40	781·2	8819	3·40	2677	536·5	274·2

But our results differ in one point from those of Milne, in that *our linear depths* (in kms.) *are uniformly about* 40 *times larger*. This is due to the fact that we have assumed in our calculations a very low mean molecular weight, $0·75m_H$ as against Milne's $20m_H$.

The predicted pressures and the absorption coefficients are of the expected order, though it should be remembered that this is largely connected with the low mean atomic weight used in our calculations.

5. *Alternative Treatment.*—It will be noticed that for the observed order of magnitudes for L, M, T, etc., A is very large—of order 1000. In fact we have

$$A = 1097\left(\frac{M}{M_\odot}\right)\left(\frac{L_\odot}{L}\right)\left(\frac{T_1}{5740}\right)^{\frac{3}{2}}\left(\frac{\alpha'}{\alpha}\right)\frac{1+\bar{x}}{\bar{x}^2}, \tag{22}$$

where α' is the coefficient in the expression (2′) for κ, calculated with $\bar{\chi}=8·3$ volts and $a=1·5$. From (13′) we easily obtain for u_1 and u_2,

$$u_1 = \sqrt{\beta^2+0·18}-0·42\,;$$

$$u_2 = \sqrt{\beta^2+0·18}+0·42\,;$$

where

$$\beta^2 = \tfrac{16}{19}A. \tag{22′}$$

Since β is very large we see that u_1 and u_2 are both practically equal to β. Hence we can rewrite (12) as

$$\frac{19}{16}\frac{dp'}{p'} = \frac{u\,du}{\beta^2-u^2}. \tag{22″}$$

(22″) integrates immediately, and we obtain, remembering that when $u=0$, $p'=p_0'$,

$$p'=p_0'\left(1-\frac{u^2}{\beta^2}\right)^{-\frac{8}{19}} \tag{23}$$

By (9) we have for the gas-pressure,*

$$p=up_0'\left(1-\frac{u^2}{\beta^2}\right)^{-\frac{1}{2}}2^{-\frac{3}{16}}. \tag{23'}$$

In obtaining (23′) we have used the relation, $p_0'=\frac{1}{2}p_1'$.

To determine the temperature gradient we proceed as before, and by (17′) we have

$$\frac{dT}{dh}=\frac{\frac{1}{4}g}{A(R/\mu)}u^2. \tag{24}$$

By (23), on the other hand,

$$1-\frac{u^2}{\beta^2}=\left(\frac{T_0}{T}\right)^{\frac{19}{2}}; \tag{25}$$

or

$$u^2=\beta^2\left[1-\left(\frac{T_0}{T}\right)^{\frac{19}{2}}\right]. \tag{25'}$$

Inserting (25′) in (24) and using the relation (22′) we obtain

$$\frac{dT}{dh}=\frac{4}{19}\frac{g}{R/\mu}\left[1-\left(\frac{T_0}{T}\right)^{\frac{19}{2}}\right]. \tag{26}$$

Now as we descend into the photospheric layers the second term in brackets in the above equation very soon becomes negligibly small because of the high power to which a quantity less than unity is raised. Thus, if $T=1{\cdot}18T_0$, $(T_0/T)^{\frac{19}{2}}=0{\cdot}2$, and if $T=1{\cdot}275T_0$, $(T_0/T)^{\frac{19}{2}}=0{\cdot}1$. Hence we have, to a good approximation, for layers not immediately near the surface and for which $T \geqslant 1{\cdot}25T_0$,

$$\frac{dT}{dh}=\frac{4}{19}\frac{g}{R/\mu}, \tag{27}$$

or

$$h-h_0=\frac{19}{4}\frac{R/\mu}{g}T. \tag{28}$$

Comparing (27) with (19′), and remembering that $u_1^{-1} << 1$, we find that what we have called previously " $\left(\frac{dT}{dh}\right)_\infty$ " is in fact a very good approximation for layers not immediately near the boundary. Equation (26) in fact shows the extent to which we have to descend in the photospheric layers

* From (23) and (23′) we have

$$p/p' \propto u\left(1-\frac{u^2}{\beta^2}\right)^{-\frac{3}{38}} \propto up'^{\frac{3}{16}} \propto uT^{\frac{3}{4}}.$$

But as T increases, $u \rightarrow \beta$ and is practically constant. Hence that " p/p' ultimately varies as $T^{\frac{3}{4}}$ " comes out from this treatment also.

before our approximation becomes valid. The temperature will have to increase by about 25 per cent. over the boundary temperature. In fig. 1 the results of Tables I and II are plotted—the temperature against the depth—and we see at once how soon the linear relation between h and T sets in—

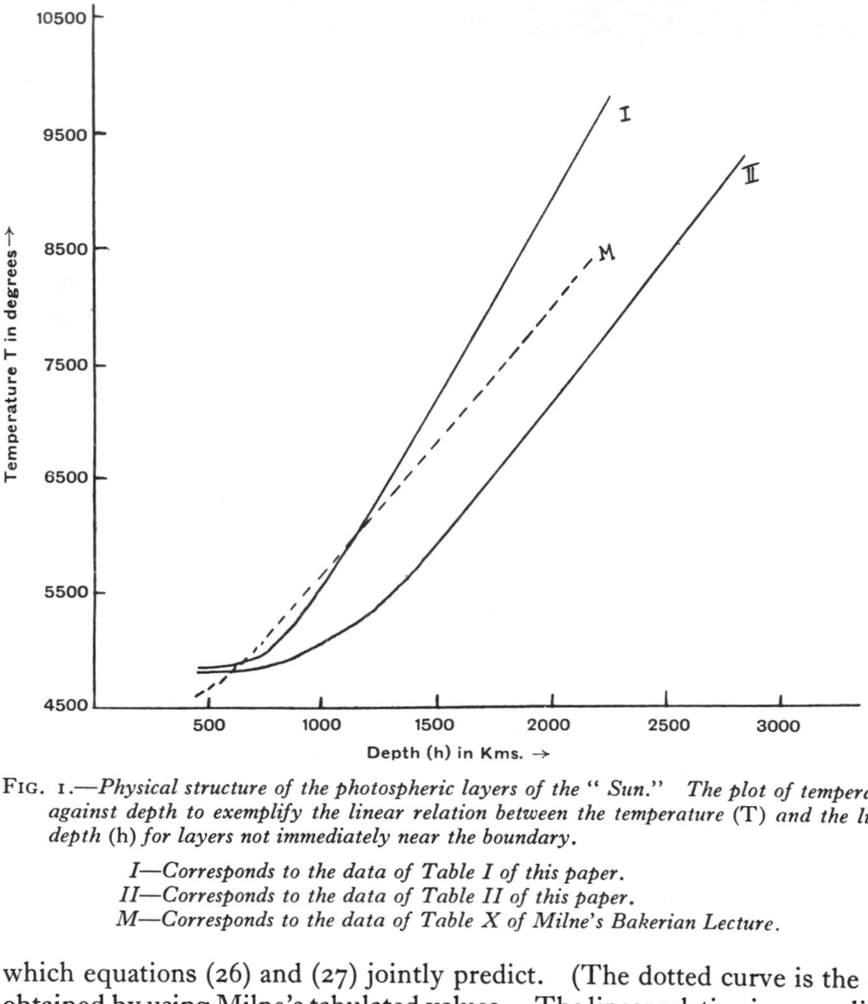

FIG. 1.—*Physical structure of the photospheric layers of the "Sun." The plot of temperature against depth to exemplify the linear relation between the temperature* (T) *and the linear depth* (h) *for layers not immediately near the boundary.*

I—Corresponds to the data of Table I of this paper.
II—Corresponds to the data of Table II of this paper.
M—Corresponds to the data of Table X of Milne's Bakerian Lecture.

which equations (26) and (27) jointly predict. (The dotted curve is the one obtained by using Milne's tabulated values. The linear relation is exemplified here also.*)

Again, (28) can be rewritten, with the help of (23), in the form

$$h-h_0=\frac{19}{4}\frac{(R/\mu)}{g}T_0\left(1-\frac{u^2}{\beta^2}\right)^{-\frac{2}{15}}. \tag{29}$$

* If $\kappa \propto P/T^{\frac{9}{2}}$ then the relation (26) will be replaced by

$$\frac{dT}{dh}=\frac{4}{17}\frac{g}{(R/\mu)}\left[1-\left(\frac{T_0}{T}\right)^{\frac{17}{2}}\right], \tag{26'}$$

and instead of (28) we have

$$h-h_0=\frac{17}{4}\frac{(R/\mu)}{g}T. \tag{28'}$$

Equation (29) is of course true only after we have descended some distance into the photospheric layers.

The solution for h for layers near the boundary ($T \leqslant 1{\cdot}25 T_0$) can be obtained by a procedure analogous to that employed in § 3. Now since T deviates only a little (less than 25 per cent.) from the boundary value, the use of a " mean " temperature $\overline{T}$ is amply justified. The solution is (*cf.* equation (21))

$$h - h_0 = \frac{(R/\mu)\overline{T}}{g}\left[\log u - \tfrac{1}{2}\log\left(1 - \frac{u_2}{\beta^2}\right)\right], \tag{30}$$

where $\overline{T}$, as before, represents some mean value of T for the layers traversed. (30) can be rewritten in a more convenient form. By (23′),

$$h - h_0 = \frac{(R/\mu)\overline{T}}{g}\left[\log\left\{\frac{p}{p_0'} \cdot 2^{\frac{3}{16}}\right\}\right], \tag{31}$$

or

$$h - h_0 = \frac{(R/\mu)\overline{T}}{g}[\log p - \log p_0' + 0{\cdot}130]. \tag{32}$$

Equations (23), (23′), (32), and (29) give the complete solution to our problem.

From (28) we can obtain one other useful piece of information. If we consider two stars with the same μ and boundary temperature (or T_1), then, since

$$\frac{p}{p_0'} = 1 + \tfrac{3}{2}\tau,$$

it is clear that for both these stars equality of τ signifies equality of p' and therefore of T at the corresponding points. Then relation (28) shows that if we neglect the arbitrary additive constant, the h's for the two stars should be inversely as the surface gravities, or, to put it more precisely, for *two stellar atmospheres of the same molecular weight* μ *and boundary temperature* $\mathrm{T_0}$, *the thicknesses of the photospheric layers between places of equal temperature* $\mathrm{T'}$ *and* $\mathrm{T''}$ ($\mathrm{T'} < \mathrm{T''}$) *in the two stars are inversely proportional to the surface gravities if* $\mathrm{T'} > 1{\cdot}25\ \mathrm{T_0}$.

It is clear that even if T' is not greater than $1{\cdot}25\ T_0$, the proportionality of the " thicknesses " to g^{-1} will still be approximately valid if T'' is sufficiently large. Thus if (following Milne) we adopt $\tau = 4$ * as defining the greatest depth to which we can see at the centre of the disc, it follows that the thicknesses between $\tau = 0{\cdot}01$ (say) and $\tau = 4$ should be roughly proportional to g^{-1}. This result has previously been noted by Milne from his numerical calculations contained in his Bakerian Lecture, but the above gives the result a precise analytical form and shows that the proportionality $(h'' - h') \propto g^{-1}$ should strictly hold good if the boundary temperatures are the same and the temperature T' (corresponding to h', the upper level of the two) is appreciably greater than T_0.

6. *General Remarks.*—It will be noted that the above calculations are not

* Actually $\tau = 4$ corresponds to $T'' = 1{\cdot}626\ T_0$ and $\left(\frac{T_0}{T}\right)^{\frac{19}{2}} = 0{\cdot}01$.

strictly deduced from the physical theory alone. Thus we have used in our numerical calculations values for $\bar{x}$ near that deduced by Russell for the level of ionization from different considerations. This is not, however, an intrinsic defect in the method, and is essentially due to the fact that to secure simplicity in the analysis we had to introduce a mean value $\bar{x}$ for the degree of ionization instead of treating this as a variable as we ought to have done ; and the penalty to be paid for this is that we have to submit to our method still having a semi-empirical air about it. For, the fact that we assume *beforehand* a mean value for $\bar{x}$ presupposes that we have already a knowledge of this from other sources. Thus for temperatures of the order of 6000°, if we take $\chi = 13{\cdot}54$ volts (say) then it is clear *a priori* that $\bar{x}$ can only be too small and the results of the calculations cannot surely be regarded as " representative." For, if $\bar{x} \rightarrow 0$ then $\kappa \rightarrow 0$, $A \rightarrow \infty$, and $dT/dh \rightarrow 0$, and the photospheric layers tend to have an infinite extension. Hence to be " reasonable " we must have values of x not too small and which should not be inconsistent with our choice of $\bar{\chi}$ and the physical conditions—a knowledge of which is presupposed when we assume $\bar{x}$.*

7. *Summary*.—(1) The physical structure of the photospheric layers is studied on the basis of an accurate theory of the absorption coefficient.

(2) It is found that ultimately the ratio of the gas-pressure to the radiation-pressure increases as $T^{\frac{3}{4}}$.

(3) Numerical applications are made to two typical cases, and though the values are derived purely on the basis of the physical theory and certain plausible assumptions regarding the " level of ionization," there seems nothing " impossible " about the predicted physical characteristics of the photospheric layers. The pressures and the absorption coefficients are of the expected order.

(4) An alternative treatment of the photospheric problem is given which shows that the linear depths are proportional to the temperatures for layers not immediately near the boundary.

(5) It is also pointed out that for two stars with the same μ and T_0 at points of equal τ (not too small) the linear depths are inversely as the surface gravities.

In conclusion I wish to record my thanks to Professor E. A. Milne, at whose suggestion the calculations presented here were undertaken.

* The consistency of the calculations can be tested as follows : To start the calculation we assume a mean value for $\bar{x}$. Then the method of the paper provides a complete march of T and p down the photospheric layers, and with these calculated values of T and p we could calculate x at each level. The mean of these x's should be equal to the initial value of $\bar{x}$ assumed. If it is different the calculation should be repeated with a new starting value of $\bar{x}$.

THE SOLAR CHROMOSPHERE.

S. Chandrasekhar, Ph.D.

§ 1. *Introduction.*—The problem of the chromosphere is, like the problem associated with any atmosphere, to determine its physical structure, *i.e.* to explain on an adequate physical basis its observed density distribution. The chromosphere, however, presents a great difficulty at a much earlier stage, and the fundamental problem had been for long merely the question relating to its enormous extensions, as it was recognised quite early that on the basis of our ordinary ideas of statistical equilibrium we cannot have on the Sun more than a hundred kilometres of visible atmosphere while the actual heights of the chromospheres are nearer ten thousand kilometres.

But this difficulty was overcome when Milne * developed in connection with the calcium chromosphere the idea of support by monochromatic radiative equilibrium. The underlying physical ideas are well known, but it may be recalled that the law of the density distribution predicted by the theory depended very much on whether the atmosphere is a "fully supported" one or only a "partially supported" one, which simply means as to whether the *very high-level* chromospheric atoms are completely supported by radiation-pressure or by the joint working of the radiation-pressure together with a small residual limiting pressure gradient. Actually it was found that

* E. A. Milne, (1) *M.N.*, **84**, 354, 1924; (2) **85**, 111, 1924; (3), (4) **86**, 459, 578, 1926. For convenient summaries see R. H. Fowler, *Statistical Mechanics* (Cambridge), pp. 397–408; E. A. Milne, *Handbuch der Astrophysik*, Band iii/1, pp. 173–182; and A. S. Eddington, *Internal Constitution of Stars*, pp. 362–369.

in a fully supported atmosphere we have a very slow decrease of density, but if we removed from the radiation-pressure even so much as 1 per cent. of the burden of supporting the high-level atoms, then the structure would simply collapse on to the lowest levels. The chromospheric equilibrium then appeared to be an extremely delicate one, the least departure from exact balance being followed by catastrophic consequences. This delicate poise of the chromosphere led to many interesting consequences, the two most important being (i) the possibility of the emission of high-speed particles by being continually accelerated away as a consequence of the Doppler effect and the increasing flux of radiation to which the atom gets exposed, and (ii) the possibility of a general explanation of the forms and motions of the "spot" and "erruptive prominences," the lack of radiation-pressure near sunspots leading to "attraction" and the excess of it near faculæ leading to "repulsion." The first of the above effects was discovered by Milne, and the second has been the subject-matter of the investigations of the late S. R. Pike.*

§ 2. *Difficulties of the Theory and a Criticism of the Alternative Theory based on Ideas of "Turbulence."*—As the theory based on monochromatic radiative equilibrium was successful, at least to the extent of providing an adequate physical mechanism which explains the very large extension of the calcium chromosphere, it was necessary to examine further as to whether the theory and the observations agreed in the finer details regarding the march of the density distribution, and also as to whether Milne's ideas, developed in connection with the calcium chromosphere, could be applied to other cases as well—to the hydrogen chromosphere, for instance. Menzel,† in his work on the *Solar Chromosphere*, has come to the conclusion that Milne's theory, in the form in which it stands, fails under these more general tests. The work of Pannekoek and Minnaert ‡ on the hydrogen chromosphere and the analysis based on their results by McCrea § also lend evidence in favour of Menzel's conclusions.

The work of Pannekoek, Minnaert and Menzel has established that to a first very rough approximation the chromosphere in its density distribution mimics an isothermal atmosphere at a temperature at which the chromospheric atoms would be moving about with thermal velocities of the order of 15 km./sec.$^{-1}$. Menzel and McCrea further adduce evidence based on Doppler broadening measurements that these velocities are *real*.

These facts appear at first sight to be so much in contradiction with the consequences of support by monochromatic radiative equilibrium that Menzel and McCrea have abandoned this theory altogether and wish to introduce fundamentally "new ideas." These attempts (at least in the form in which they have been presented) are not at all logical deductions from well-defined premises, but are merely methods to define in some *ad hoc*

* S. R. Pike, (1) *M.N.*, **87**, 56, 1926 ; (2) **88**, 3, 1927 ; (3) **88**, 635, 1928.

† D. H. Menzel, (1) *Lick O.P.*, **17** ; also (2) *M.N.*, **91**, 628, 1931.

‡ A. Pannekoek and M. Minnaert, *Verh. d. Kon. Akad. Amsterdam*, xiii, No. 5, 1928 ; also Davidson, Minnaert, Ornstein and Stratton, *M.N.*, **88**, 536, 1928.

§ W. H. McCrea, (1) *M.N.*, **89**, 483, 1929 ; (2) **89**, 718, 1929.

manner a temperature *parameter* T^* of the order of 20,000 degrees or more for the chromospheric atoms and then to assert that the density distribution must be the same as an isothermal atmosphere at the *thermodynamic* temperature T^*. They argue in the following manner :—

In an isothermal atmosphere the density law is

$$\rho = \rho_0 e^{-\frac{mgH}{kT}},$$

where m is the mass of the atom, g the value of gravity and H the height. The numerator mgH in the exponent of e represents the potential energy of an atom of mass m in a gravitational field of intensity g at height H above the surface. The denominator kT is equal to 2/3 the average kinetic energy of the moving atom. "If we now consider *groups* of atoms in *turbulent motion*, the individual velocities of the groups approximating to a Maxwellian distribution with an average velocity C, then since the potential energy would be unaffected we should expect the density distribution of such an atmosphere to follow the law" (Menzel (1)) :

$$\rho = \rho_0 e^{-\frac{mgH}{kT+\frac{1}{3}mC^2}} = e^{-\frac{mgH}{kT^*}}.$$

By *defining* sufficiently large velocities C we can define arbitrarily large values for the temperature parameter T^*. Menzel and McCrea *deduce* values for C to fit the observed law of density distribution.

There are three main objections against the theory based on these ideas of turbulence. *Firstly*, the assumption that the groups of atoms can be distributed according to Maxwell's law appears to be physically inadmissible, for a Maxwellian distribution implies that the various groups of atoms exchange momenta ; how then do groups exchange momenta without the individual atoms taking part in the "collision processes"? * *Secondly*, even if the type of equilibrium suggested was actually realised it would not attribute to the chromosphere that degree of delicate balance which is necessary for us to attribute to it on other grounds. *Thirdly*, no theoretical justification is provided as to why C should precisely be of the order of 15 km./sec.$^{-1}$, but a really adequate theory must certainly explain as to why kinetic velocities of just this order are encountered in the chromosphere.

These objections are enough to show the untenability of these ideas based on "turbulence," and it would appear more profitable to examine in greater detail the possibilities of the support by radiation-pressure, at least for this reason that it does not seem to be generally recognised that there are in fact other possibilities.

§ 3. *Monochromatic Radiative Equilibrium over a Non-uniformly Radiating Surface.*—Now, the Schuster idealisation of the existence of a definite radiating surface naturally underlies any theory of the chromosphere,† and

* I am indebted to Professor Milne for this remark.

† The same Schuster idealisation underlies also the determination of the "number of atoms per cm.^{2} above the photosphere" (by the method of Stewart and Unsöld by a simple application of Schuster's formula for the residual intensity), but here the radiating surface is placed *below* the reversing layers. In the theory

if necessary we can also attribute to the radiating surface additional characteristics such as the periodic variations in the emergent flux of radiation as are revealed by direct observation. It is a fact of observation that the emergent flux of radiation *in* the chromospheric lines is not constant but varies from point to point on the surface of the Sun and in fact oscillates about a certain mean value specified by the residual intensity in the line. Examinations of the spectroheliograms in monochromatic light indicate that an allowance of a maximum variation of the emergent flux by a factor 2 to 3 * over an average distance of about 10,000 kilometres is a fair estimate. It would be worth while, then, to examine if the non-uniformity of the radiating surface when incorporated into the theory of support by monochromatic radiation would be sufficient to remove the difficulties which the theory at present encounters.

If we assume that the mean flux just corresponds to "full support," then the variation of the flux over the surface would accelerate the chromospheric atoms, and in a hydrodynamically steady state the velocities of the chromospheric atoms must, from dimensional considerations alone, be of the order of $\sqrt{\lambda g}$, where λ is "wave-length" of the flux periodicity over the surface and g the solar value of gravity. If $\lambda = 20{,}000$ kilometres, then $\sqrt{\lambda g}$ is 73·6 km./sec.$^{-1}$. There will be other numerical factors coming in, but it is clear that it is reasonable to hope that the chromospheric atoms will have on this theory such kinetic velocities as are demanded by observation.

§ 4. *General Outline of the Method.*—It is convenient to regard the flux emitted by the radiating surface as being periodic in the X and Y directions (the Z direction is the vertical) with certain specified wave-lengths λ_x and λ_y. That is, we assume that

$$I(x, y) = I_0 + I_1(x, y), \tag{1}$$

and

$$I_1(x + n\lambda_x, y + m\lambda_y) = I_1(x, y), \tag{1'}$$

where $I(x, y)$ is the intensity of the emitted radiation at the point (x, y), which is assumed to be locally isotropic in the hemisphere; † I_0 is a constant and m and n are arbitrary integers. We shall further assume that the atom would be in equilibrium when the flux of radiation is that corresponding to I_0. Then

$$\tfrac{1}{4} B_{1\to 2} I_0 = mg, \tag{2}$$

of the chromosphere, on the other hand, the radiating surface is placed *above* the reversing layers, and it is the flux of radiation corresponding to the residual intensity which is responsible for supporting the chromospheric atoms. Thus the "number of atoms" determined by Stewart's method has nothing to do with the number of atoms in a column of the chromosphere *above* the reversing layers. The non-recognition of this fact has led to much confused statements in the literature.

* The allowance of a factor 2 appears to be a safe *lower* limit. See in this connection G. E. Hale and F. Ellerman, *Proc. Nat. Acad. Sci.*, 1, 102, 1916. I am indebted to Professor H. H. Plaskett for this reference and also for valuable information on the observational side of these questions.

† "Darkening effects" can be neglected in a first survey.

where $B_{1\to2}$ is the appropriate Einstein *absorption* coefficient,* and g the solar value of gravity. Hence it is only the component $I_1(x, y)$ which would give rise to a "field of force," † and we need the components of the force at an arbitrary point x, y, z. If we denote these by H_x, H_y and H_z, then clearly

$$H_x = \frac{zB_{1\to2}}{4\pi}\int_{-\infty}^{\infty}\int_{-\infty}^{\infty} I_1(x'+x, y'+y)\frac{x'dx'dy'}{(x'^2+y'^2+z^2)^2}, \tag{3}$$

$$H_y = \frac{zB_{1\to2}}{4\pi}\int_{-\infty}^{\infty}\int_{-\infty}^{\infty} I_1(x'+x, y'+y)\frac{y'dx'dy'}{(x'^2+y'^2+z^2)^2}, \tag{4}$$

$$H_z = \frac{z^2B_{1\to2}}{4\pi}\int_{-\infty}^{\infty}\int_{-\infty}^{\infty} I_1(x'+x, y'+y)\frac{dx'dy'}{(x'^2+y'^2+z^2)^2}. \tag{5}$$

This field of force (H_x, H_y, H_z) will yield a system of dynamical equations of motion which will have, in general, periodic as well as aperiodic solutions. It is clear that the permanence of the chromosphere requires the atoms in the chromosphere to describe *periodic* trajectories.‡ We should therefore have the components of the velocity u, v, w to be all periodic functions in x and y. In determining the density distribution associated with this derived field of velocities we must remember that we are treating an atmosphere which is not in thermodynamical equilibrium and that as a consequence we cannot introduce any conception of temperature. The problem is essentially a *hydrodynamical* one and we are provided with only one partial differential equation for ρ (the equation of continuity) as the condition for the maintenance of a *steady state*. We should have in fact

$$u\frac{\partial\rho}{\partial x}+v\frac{\partial\rho}{\partial y}+w\frac{\partial\rho}{\partial z}+\rho\left(\frac{\partial u}{\partial x}+\frac{\partial v}{\partial y}+\frac{\partial w}{\partial z}\right)=0. \tag{6}$$

This equation will not specify the density distribution uniquely, but only to the extent of an arbitrary function. But the boundary conditions which a density distribution function of an atmosphere will have to satisfy will remove this indeterminateness.

§ 5. In this paper I shall confine myself to working out the details of the equilibrium for the case

$$I_1(x, y) = I_1 \sin\frac{2\pi}{\lambda}x, \tag{7}$$

where I_1 is a constant. With this assumption the problem admits of exact mathematical solution and we can, with a start made this way, hope to obtain insight into the more general problem.§

* $B_{1\to2}$ as here defined is $h\nu_0/c$ times the usually defined Einstein *probability* coefficient.

† It may be mentioned that this field-force cannot be derived from a "potential."

‡ The aperiodic trajectories are likely to be associated with some types of prominences (see §§ 12, 17).

§ It is easy to write down the equations of motion for more general forms for $I_1(x, y)$ than (7), but as a full mathematical treatment even for the case

$$I_1(x, y) = I_1 \sin\frac{2\pi}{\lambda}x \sin\frac{2\pi}{\lambda}y, \tag{7'}$$

requires considerable numerical work, I shall reserve these refinements for a future occasion.

§ 6. *The Dynamical Equations of Motion.*—H_y is naturally zero. By (3), (5) and (7) we have

$$H_x = \frac{zB_{1\to 2}I_1}{\pi}\cos\frac{2\pi}{\lambda}x\int_0^\infty x'\sin\frac{2\pi}{\lambda}x' \,.\, dx'\int_0^\infty \frac{dy'}{(x'^2+y'^2+z^2)^2}, \tag{8}$$

$$H_z = \frac{z^2B_{1\to 2}I_1}{\pi}\sin\frac{2\pi}{\lambda}x\int_0^\infty \cos\frac{2\pi}{\lambda}x' \,.\, dx'\int_0^\infty \frac{dy'}{(x'^2+y'^2+z^2)^2}, \tag{8'}$$

which after some reductions can be expressed in the forms

$$\left.\begin{aligned} H_x &= \frac{B_{1\to 2}I_1}{4}\cos\xi \,.\, \zeta K_0(\zeta), \\ H_z &= \frac{B_{1\to 2}I_1}{4}\sin\xi \,.\, \zeta K_1(\zeta), \end{aligned}\right\} \tag{9}$$

where

$$\xi = \frac{2\pi}{\lambda}x\,; \qquad \zeta = \frac{2\pi}{\lambda}z, \tag{10}$$

and

$$K_\nu(\zeta) = \frac{\Gamma(\nu+\frac{1}{2})(2\zeta)^\nu}{\Gamma(\frac{1}{2})}\int_0^\infty \frac{\cos\xi d\xi}{(\xi^2+\zeta^2)^{\nu+\frac{1}{2}}}. \tag{11}$$

From (9), using the relation (2) to eliminate $B_{1\to 2}$, we derive our *fundamental set of dynamical equations* :

$$\left.\begin{aligned} \frac{d^2\xi}{dt^2} &= a\cos\xi \,.\, \zeta K_0(\zeta), \\ \frac{d^2\zeta}{dt^2} &= a\sin\xi \,.\, \zeta K_1(\zeta), \end{aligned}\right\} \tag{12}$$

where

$$a = \frac{2\pi}{\lambda} \,.\, rg\,; \qquad r = \frac{I_1}{I_0}. \tag{12'}$$

The functions $\zeta K_0(\zeta)$ and $\zeta K_1(\zeta)$ are tabulated in Table III of the Appendix.

§ 7. The functions K_ν introduced in the previous section are related to the Hankel functions with imaginary arguments : *

$$K_\nu(\zeta) = \tfrac{1}{2}\pi i e^{\frac{1}{2}\nu\pi i}H_\nu^{(1)}(i\zeta). \tag{13}$$

There are a number of recurrence relations which these K_ν satisfy (see Watson, *Bessel Functions*, p. 79), but the relations we most often need are the following (easily verified from the definition (11)) :—

$$K_0'(\zeta) = -K_1(\zeta), \tag{14}$$

$$\zeta K_1'(\zeta) + K_1(\zeta) = -\zeta K_0(\zeta), \tag{15}$$

where a dash denotes differentiation with respect to ζ. Further, for ζ large we have the asymptotic relation :

$$K_\nu(\zeta) = \left(\frac{\pi}{2\zeta}\right)^{\frac{1}{2}} e^{-\zeta}\left[1 + \frac{4\nu^2-1^2}{1\,!\,8\zeta} + \frac{(4\nu^2-1^2)(4\nu^2-3^2)}{2\,!\,(8\zeta)^2} + \ldots\right]. \tag{16}$$

* See G. N. Watson, *Bessel Functions* (Cambridge), 1922, p. 78.

Because of this exponential factor we see that the accelerations produced by the radiating surface tend to zero for ζ large, and one immediate consequence of this result is that the " height " of the chromosphere must be of the same order as the wave-length λ. This deduction is confirmed by observation.

Now the vertical acceleration is proportional to $\zeta K_1(\zeta)$, which is monotonic decreasing, taking the value unity for $\zeta = 0$. On the other hand, the horizontal acceleration which is proportional to $\zeta K_0(\zeta)$ passes through a *maximum*, $\zeta K_0(\zeta)$ taking the value 0 for both $\zeta = 0$ and $\zeta = \infty$. The maximum occurs at a value for ζ which is the root of the equation

$$\zeta K_0'(\zeta) + K_0(\zeta) = 0,$$

or

$$\zeta K_1(\zeta) = K_0(\zeta). \tag{17}$$

The solution of (16) is found numerically to be

$$\zeta = 0{\cdot}595. \tag{17'}$$

The occurrence of this maximum confirms an earlier result of Pike (2). He states that "a spot has influence high up above the chromosphere whilst affecting the lower regions very little."

§ 8. *An Integral of the Equations of Motion.*—The equations of motion (12) possess an integral which can be obtained as follows: Using (15), we derive

$$\frac{d}{dt}\left(\frac{d\xi}{dt}\frac{d\zeta}{dt}\right) = a\cos\xi \,.\, \zeta K_0(\zeta)\frac{d\zeta}{dt} + a\sin\xi \,.\, \zeta K_1(\zeta)\frac{d\xi}{dt} = -\frac{d}{dt}[a\cos\xi \,.\, \zeta K_1(\zeta)]. \tag{18}$$

Integrating (18) we have that

$$\frac{d\xi}{dt}\cdot\frac{d\zeta}{dt} + a\cos\xi \,.\, \zeta K_1(\zeta) = \text{constant}. \tag{19}$$

§ 9. *A Necessary and Sufficient Condition for the Existence of Periodic Solutions.*—As stated in § 4, we have to seek solutions of (12) which yield for $\dot\xi$, $\dot\zeta$ * and ζ (and therefore also for $\zeta K_1(\zeta)$) periodic functions in ξ, *i.e.* they should each be capable of being expanded in a Fourier series. We shall assume that

$$\dot\xi = \Sigma a_n e^{in\xi}\ ; \qquad \dot\zeta = \Sigma b_n e^{in\xi}, \tag{20}$$

and that further

$$\zeta K_1(\zeta) = \Sigma c_n e^{in\xi}. \tag{21}$$

In (20) and (21) (and also in the equations which follow) the summation over n extends from $+\infty$ to $-\infty$. By (20) we have

$$\ddot\xi = i[\Sigma a_n n e^{in\xi}]\dot\xi, \tag{22}$$

$$\ddot\zeta = i[\Sigma b_n n e^{in\xi}]\dot\xi, \tag{23}$$

and by (15)

$$\zeta K_0(\zeta)\dot\zeta = -i[\Sigma c_n n e^{in\xi}]\dot\xi. \tag{24}$$

* A dot denotes differentiation with respect to the time t.

Substituting (20), (22) and (24) in the first of the equations (12) and equating the coefficients of $e^{in\xi}$ we have

$$\sum_{r=0}^{n} a_r r b_{n-r} = -\frac{a}{2}[(n-1)c_{n-1}+(n+1)c_{n+1}]. \tag{25}$$

Similarly from (20), (21), (23) and the second of the equations (12) we obtain

$$\sum_{r=0}^{n} a_r(n-r)b_{n-r} = -\frac{a}{2}[c_{n-1}-c_{n+1}]. \tag{26}$$

Adding (25) and (26) we have

$$\sum_{r=0}^{n} a_r b_{n-r} = -\frac{a}{2}[c_{n-1}+c_{n+1}], \tag{27}$$

which by (20) and (21) is equivalent to the relation

$$\dot{\xi}\dot{\zeta} = -a\cos\xi\ .\ \zeta K_1(\zeta). \tag{28}$$

From (28) we deduce that *the necessary and sufficient condition that an atom starting at a point* (0, ζ_0) *on the ζ-axis should describe a periodic trajectory is that*

$$\dot{\xi}_0\dot{\zeta}_0 = -a\zeta_0 K_1(\zeta_0). \tag{29}$$

§ 10. *The Solution of the Equations of Motion.*—It is seen at once that the equations (12) combined with the condition (28) possess the following *first integrals* :—

$$\frac{d\xi}{dt} = \alpha\zeta K_1(\zeta), \tag{30}$$

$$\frac{d\zeta}{dt} = -\beta\cos\xi, \tag{31}$$

where α and β are positive constants such that

$$\alpha\beta = a. \tag{32}$$

It is easy to prove (by an analysis similar to that in § 9) that (30) and (31) are the only solutions which satisfy the further *necessary* condition that $\dot{\xi}$, $\dot{\zeta}$ and ζ are also periodic functions in time. We shall, however, omit the details of this proof.

From (30) and (31) we obtain that

$$-\beta\cos\xi d\xi = \alpha\zeta K_1(\zeta)d\zeta, \tag{33}$$

which on integration yields the following *equation for the trajectory* :—

$$\sin\xi = \frac{\alpha}{\beta}\int_{\zeta}^{\zeta_0}\zeta K_1(\zeta)d\zeta, \tag{34}$$

where ζ_0 is the point at which the trajectory intersects the ζ-axis. Equations (30), (31) and (34) express the complete solution of the dynamical problem.

§ 11. *A Discussion of the Form of the Trajectories.*—We shall introduce the quantity $Q(\zeta_1, \zeta_2)$ defined by

$$Q(\zeta_1, \zeta_2)=\int_{\zeta_2}^{\zeta_1} \zeta K_1(\zeta)d\zeta. \tag{35}$$

The equation of the trajectory can then be expressed by

$$\sin \xi = \frac{\alpha}{\beta} Q(\zeta_0, \zeta). \tag{36}$$

The values of $Q(\zeta, 0)$ for different values of ζ are tabulated in the last column of Table III in the Appendix. Further, we have the result easily obtained (see equation (56)):

$$Q(\infty, 0)=\frac{\pi}{2}. \tag{37}$$

Now, it is clear that *the atoms* (that are *in* the chromosphere) *should describe only such trajectories that they do not either strike the radiating surface or go off to infinity.* This will impose certain restrictions on the values of α and β.

Let ζ_1 be such that

$$Q(\zeta_1, 0)=Q(\infty, \zeta_1)=\tfrac{1}{2}Q(\infty, 0)=\frac{\pi}{4}. \tag{38}$$

Numerically ζ_1 is found to be 0·942. Consider firstly trajectories such that $\zeta_0 \leqslant \zeta_1$. ($\zeta_0$ is the point at which the trajectory intersects the ζ-axis.) The condition that the trajectory should have no intersections with the ξ-axis yields the inequality

$$\frac{\alpha}{\beta} Q(\zeta_0, 0) \geqslant 1, \tag{39}$$

or since $\alpha\beta = a$ we have

$$\alpha_{\min} = \sqrt{\frac{a}{Q(\zeta_0, 0)}}; \qquad \beta_{\max} = \sqrt{Q(\zeta_0, 0)\,.\,a}. \tag{40}$$

Actually the fact that $\alpha_{\min} \to \infty$ as $\zeta_0 \to 0$ need not trouble us since physically the condition that a chromospheric atom should not hit the radiating surface merely means that in the physical problem a chromospheric atom descending into the reversing layers gets into regions where approximations valid for the greater part of the chromosphere are no longer true and the mathematical singularity at $\zeta_0 = 0$ is merely a consequence of a rigorously physically inadmissible idealisation of the existence of a definite radiating surface which underlies the mathematical formulation.

Again since $\zeta_0 \leqslant \zeta_1$ the condition (39) automatically restricts the atom from not going off to infinity. Actually when $\zeta_0 = \zeta_1$ the atom which just grazes the radiating surface goes off to infinity but with a zero limiting velocity.

Now, let $\zeta_0 > \zeta_1$. The condition now that the atom does not go off to infinity yields instead of (39) that

$$\frac{\alpha}{\beta} Q(\infty, \zeta_0) > 1, \tag{41}$$

or

$$\alpha_{\min} = \sqrt{\frac{a}{Q(\infty, \zeta_0)}}; \qquad \beta_{\max} = \sqrt{Q(\infty, \zeta_0) \cdot a}. \tag{42}$$

The inequality (41) also restrains the atom from hitting the radiating surface.

If, however, the condition (41) is not satisfied, then the atom goes off to infinity with a certain *limiting vertical velocity* * given by

$$[\dot{\zeta}]_\infty = \beta \cos\left[\sin^{-1}\frac{\alpha}{\beta}Q(\infty, \zeta_0)\right]. \tag{43}$$

If $\alpha Q(\infty, \zeta_0)/\beta$ is unity, then the atom is at rest at infinity.

The values of $\alpha_{\min}$, $\beta_{\max}$ and the actual *minimum* horizontal velocities † specified by $\alpha_{\min} \cdot \zeta_0 K_1(\zeta_0)$ are tabulated in Table I.

TABLE I

ζ_0	$\alpha_{\min} \cdot a^{-1/2}$	$\beta_{\max} \cdot a^{-1/2}$	$[\alpha\zeta_0 K_0(\zeta_0)]_{\min} a^{-1/2}$
0	∞	0	∞
0·5	1·4667	0·6818	1·2148
1·0	1·1552	0·8656	0·6953
1·5	1·4186	0·7049	0·5903
2·0	1·7543	0·5700	0·4907
2·5	2·1807	0·4586	0·4028
3·0	2·721	0·3675	0·3278
3·5	3·405	0·2937	0·2651
4·0	4·272	0·2341	0·2134
4·5	5·371	0·1862	0·1711
∞	∞	0	0

To see precisely what these values correspond to in actual velocity measures we shall take $\lambda = 20{,}000$ km. and $r = 1/3$ (this corresponds to a maximum variation of flux by a factor 2 which, as has been pointed out, is a lower limit. $r = 1/2$ would appear to be more appropriate, but we shall assume that $r = 1/3$ to be on the safer side). On these assumptions ξ or ζ equal to $\sqrt{a}$ corresponds to the velocity components u or w given by the *fundamental unit of velocity* V specified by

$$V = \sqrt{rg \cdot \frac{\lambda}{2\pi}} = 17{\cdot}0 \text{ km./sec.}^{-1}. \tag{44}$$

We therefore see that the theory predicts just the order of kinetic velocities which observations indicate and which the advocates of turbulence have been looking for. We, however, notice that in contrast to McCrea and Menzel, who *deduced* values for their "*C*," we are now able to *predict* the

* The horizontal velocity is zero.

† The actual velocities are much greater. See § 14, equation (59).

correct order for the kinetic velocities on reasonable assumptions regarding the surface of the Sun and on the basis of a logically worked-out theory.

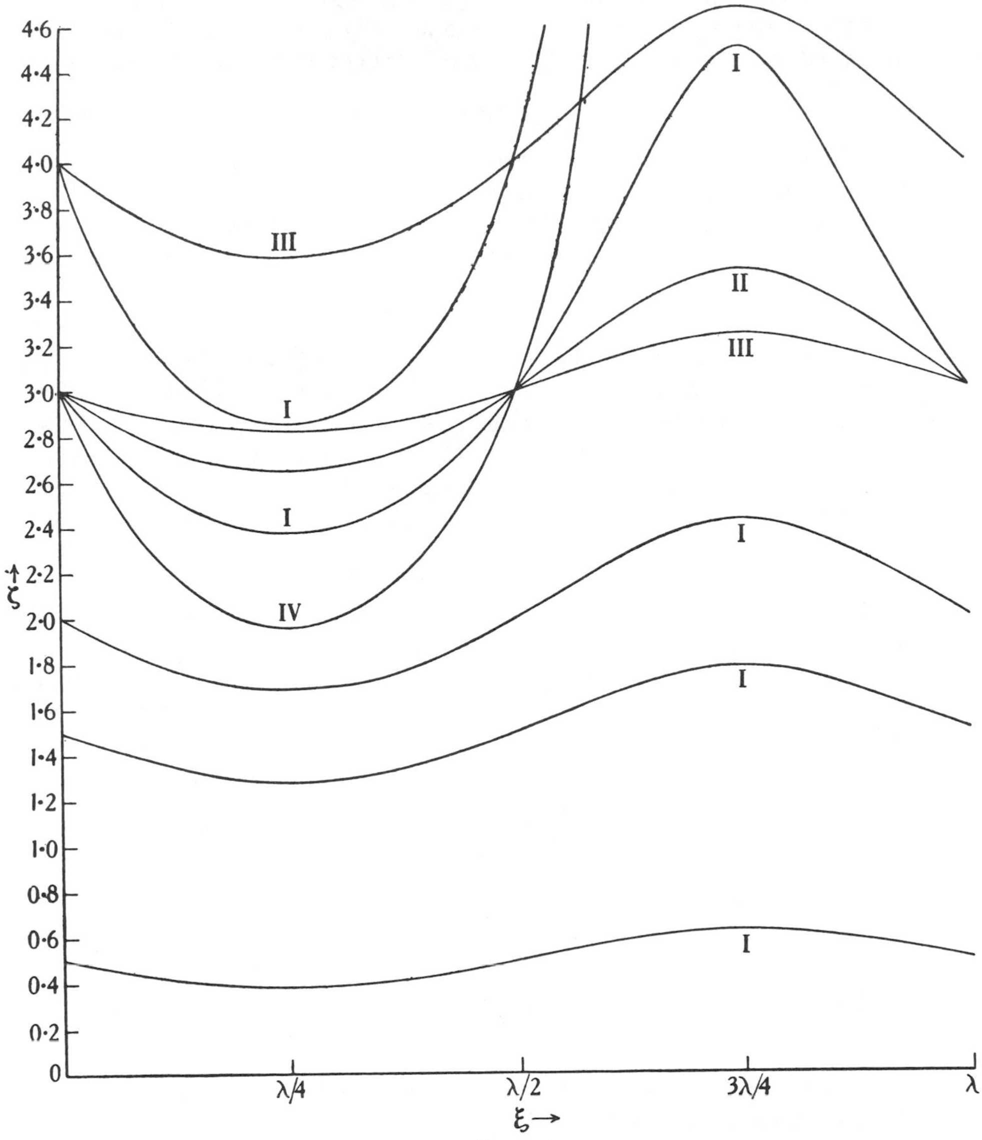

Figure 1.

Trajectories of the Chromospheric Atoms

I → $\alpha/\beta=10$; II → $\alpha/\beta=20$; III → $\alpha/\beta=40$; IV → $\alpha/\beta=5$

In fig. 1 we have drawn as illustration two different sets of trajectories.* (1) Trajectories with different ζ_0 but with a given value for $\alpha/\beta(=10)$.

* Different pairs of values (ξ, ζ) on the trajectories were calculated, using the table of values $Q(\zeta, 0)$ tabulated in the Appendix. It has not been thought worth while to give the details of the solutions.

The atoms describing these trajectories all have the same *downward* vertical velocity when crossing the ζ-axis. We notice that as we go higher up the trajectories become more and more "fluctuating," and in fact the atom starting with this value for α/β at $\zeta_0 = 4{\cdot}0$ actually goes off to infinity. The limiting value for $\dot{\zeta}$ as given by (43) for this trajectory is found to be

$$[\dot{\zeta}]_\infty = 0{\cdot}8365\beta = 0{\cdot}2645\sqrt{a}. \qquad (43')$$

(2) Trajectories of atoms all starting at the same height ($\zeta_0 = 3{\cdot}0$) but with decreasing downward vertical velocities corresponding to $\alpha/\beta = 5$, 10, 20 and 40. The atom characterised by $\alpha/\beta = 5$ actually goes off to infinity, and the limiting value in this case for $\dot{\zeta}$ is found to be

$$[\dot{\zeta}]_\infty = 0{\cdot}7375\beta = 0{\cdot}3298\sqrt{a}. \qquad (43'')$$

We have also drawn a trajectory starting at $\zeta_0 = 4$ with $\alpha/\beta = 40$.

§ 12. *An Aperiodic Solution of the Equations of Motion and the Possibility of the Emission of moderately High-speed Atoms.*—Before proceeding to the discussion of the density distribution in the chromosphere we shall now, while discussing the equations of motion, show that our theory predicts the possibility of a continual emission of moderately high-speed particles. We have already noticed a class of trajectories which go to infinity with certain limiting velocities, but these velocities are not very large. On the other hand there exist *aperiodic* solutions of the equations of motion, along the trajectories corresponding to which the atom will have quite large velocities. Thus consider an atom starting from the point (0, $\lambda/4$) with zero horizontal velocity. By (12) the trajectory is clearly a straight line parallel to the ζ-axis. The equations of motion (12) reduce to the one equation

$$\frac{d^2\zeta}{dt^2} = a\zeta K_1(\zeta), \qquad (45)$$

from which we deduce that

$$\dot{\zeta}^2 = \dot{\zeta}_0^2 + 2a\int_0^\zeta \zeta K_1(\zeta)d\zeta. \qquad (46)$$

By (37) we have

$$\dot{\zeta}_\infty^2 = \dot{\zeta}_0^2 + a\pi. \qquad (47)$$

An atom starting with even a zero vertical velocity will end with a limiting vertical velocity specified by

$$\dot{\zeta}_\infty = \sqrt{rg \cdot \frac{\lambda}{2}} = 30{\cdot}2 \text{ km./sec.}^{-1}. \qquad (48)$$

This possibility of this emission of *moderately high* speed atoms is of entirely different origin from the possibility of the emission of *very high* speed atoms pointed out by Milne (3).

§ 13. *The Density Distribution in the Chromosphere.*—In §§ 9, 10 and 11 the complete solution to the dynamical problem was obtained. It was found that an atom starting at a point (ζ_0, 0) can, consistent with the equations of motion and with the additional restriction that it should describe a periodic trajectory, have a range of values for α. We shall now impose the further condition that *the time derivatives* $\dot{\xi}$ *and* $\dot{\zeta}$ *are to be unique functions in the whole* ξ-ζ *plane.* We should not expect that this restriction would be rigorously satisfied in an actual chromosphere. Far more is it likely that the uniqueness of the velocity components is only very approximately realised in an actual chromosphere, but in assuming a rigorous uniqueness we shall be avoiding all the difficulties associated with the hydrodynamics of a fluid in which the velocity components are many valued functions of position.

It is easy to see that the condition for the uniqueness of $\dot{\xi}$ and $\dot{\zeta}$ restricts α/β *and therefore also* α *and* β *separately to be constants of the atmosphere.* With α/β constant the hydrodynamical problem is quite elementary. The equation of continuity (6) in the ξ, ζ variables takes the form

$$\dot{\xi}\frac{\partial\rho}{\partial\xi}+\dot{\zeta}\frac{\partial\rho}{\partial\zeta}+\rho\left(\frac{\partial\dot{\xi}}{\partial\xi}+\frac{\partial\dot{\zeta}}{\partial\zeta}\right)=0, \tag{49}$$

which by (30) and (31) reduces to

$$\alpha\zeta K_1(\zeta)\frac{\partial\rho}{\partial\xi}=\beta\cos\xi\frac{\partial\rho}{\partial\zeta}. \tag{50}$$

The solution of (50) is

$$\rho=\Phi\left\{\sin\xi+\frac{\alpha}{\beta}\int_0^\zeta\zeta K_1(\zeta)d\zeta\right\}, \tag{51}$$

where $\Phi(x)$ is an arbitrary function in the argument x.

§ 14. From (51) we see that the existence of a definite radiating surface emitting radiation with a flux which varies periodically over the surface, leads to a density distribution which necessitates the density at the base of the chromosphere to be a periodic function as well. In a first approximation it would be legitimate to assume that the density at the base of the chromosphere varies simply harmonically. This would mean that the function Φ is merely a multiple of the argument specified plus another constant. (A justification for this assumption is provided in § 15, where the conception of a "natural boundary" of a chromosphere is introduced.)

We do not, however, know what value we are to assign to α/β. We shall show in the following section that the law of density variation at the base of the chromosphere will itself control to some extent the precise value of α/β which will characterise the *normal* chromosphere. Since, however, the chromosphere extends (as a fact of observation) up to at least $4{\cdot}5\zeta$, *we shall assume provisionally that* $\alpha/\beta(=\alpha^2/a)$ *is* 40, for then the trajectories at least up to $\zeta_0=4{\cdot}0$ are not wildly fluctuating, and it is only when ζ_0 approaches 5 that the atoms characterised with this value for α/β describe trajectories which go off to infinity.* If we assign a value 40 we see that the second term in {} in (51) will be very large compared with sin ξ even for values of

* This gives a definite clue regarding the observed *instability* of the outer layers of the chromosphere (see § 16).

ζ only very appreciably different from zero. We can therefore in a *first approximation* neglect sin ξ. We should then have

$$\rho = \rho_0 \left[1 - \epsilon \frac{\alpha}{\beta} \int_0^\zeta \zeta K_1(\zeta) d\zeta \right], \tag{52}$$

where ρ_0 is the density at the base of the chromosphere and ϵ is a constant. Let us further impose the condition that $\rho \to 0$ as $\zeta \to \infty$. This condition would determine ϵ. We have

$$\epsilon = \frac{2}{\pi} \frac{\beta}{\alpha}. \tag{52'}$$

The *density distribution* then is specified by the relation

$$\rho = \rho_0 \cdot \frac{2}{\pi} \int_\zeta^\infty \zeta K_1(\zeta) d\zeta. \tag{53}$$

We notice that in (53) α/β does not appear explicitly. (53) predicts for the march of the *density-gradient* the law

$$\frac{\partial \rho}{\partial \zeta} = -\rho_0 \cdot \frac{2}{\pi} \zeta K_1(\zeta). \tag{54*}$$

In the following table the values of $2\zeta K_1(\zeta)/\pi$ for some values of ζ are tabulated and compared with the purely exponential function which the observations indicate as being very roughly the law realised in the actual chromosphere:—

TABLE II

ζ	$\frac{2}{\pi}\zeta K_1(\zeta)$	$e^{-\zeta}$	ζ	$\frac{2}{\pi}\zeta K_1(\zeta)$	$e^{-\zeta}$
0	0·6366	1·000	2·5	0·1176	0·0821
0·5	0·5273	0·6065	3·0	0·0767	0·0498
1·0	0·3832	0·3679	3·5	0·0496	0·0302
1·5	0·2649	0·2231	4·0	0·0318	0·0183
2·0	0·1781	0·1353	4·5	0·0203	0·0111

Considering the fact that the law (54) is only a first approximation, and considering also the very provisional nature of the observational material, the agreement is very satisfactory (*cf.* fig. 2). The *total mass M* supported in a unit column of the chromosphere is, according to (53),

$$M = \frac{\lambda}{2\pi} \int_0^\infty \rho d\zeta = \frac{\lambda}{\pi^2} \rho_0 \int_0^\infty d\zeta \int_\zeta^\infty x K_1(x) dx$$

$$= \frac{\lambda \rho_0}{\pi^2} \int_0^\infty \zeta^2 K_1(\zeta) d\zeta. \tag{55}$$

* It is easy to convince oneself by quite general arguments that *we should always have the law* (54) *as a first approximation whenever there exists a proportionality of* ξ *to* $\zeta K_1(\zeta)$. The law (54) is therefore quite general, and is not merely the feature of the two-dimensional case we are treating.

Now there is a general formula for the type of integrals which occur in (55):

$$\int_0^\infty K_\nu(\zeta)\zeta^{\mu-1}d\zeta = 2^{\mu-2}\Gamma\left(\frac{\mu-\nu}{2}\right)\Gamma\left(\frac{\mu+\nu}{2}\right). \qquad (56)^*$$

We have therefore

$$M = \frac{2\lambda\rho_0}{\pi^2}, \qquad (57)$$

or if N is *the number of atoms* † *in the unit column*, and n_0 the number per unit volume at the base, then

$$N = n_0 \cdot \frac{2\lambda}{\pi^2} = 0{\cdot}2026 n_0 \lambda. \qquad (58)$$

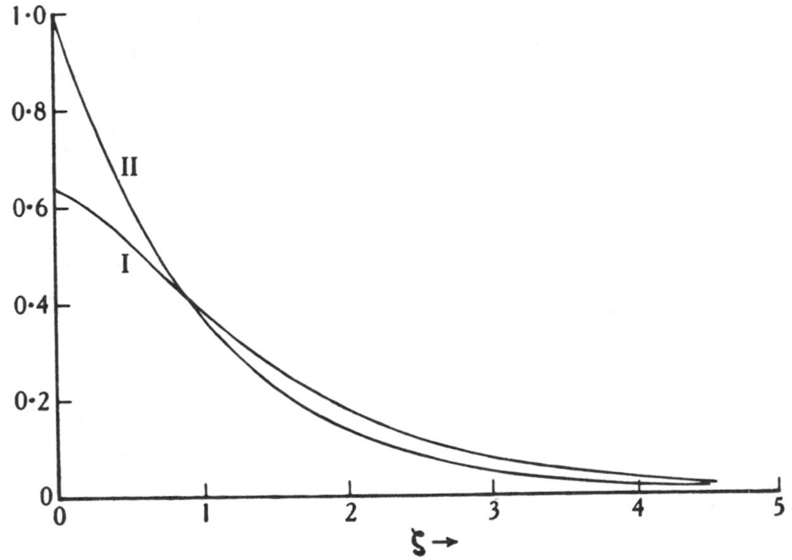

Figure 2.

Comparison of the Predicted March of the Density-Gradient (I) with the Exponential Law (II)

(58) states that *the number of atoms in a unit column of the chromosphere is equal to the number of atoms in a unit column of about a fifth of the wave-length λ filled uniformly at the concentration occurring at the base of the chromosphere.* This is a rather unexpected result and removes practically the whole of the difficulty associated with the hydrogen chromosphere.‡ In any case it suggests a valuable hint towards a solution.

Another quantity which is of interest in this connection is the *mean horizontal velocity* of the chromospheric atoms. Denoting the mean value of ξ by $\bar{\xi}$ we have

* See G. N. Watson, *Bessel Functions*, p. 388.

† *In that stage of ionization and excitation in which it is directly supported by radiation-pressure.*

‡ *Cf.* Menzel (1), McCrea (2).

$$\bar{\xi} = \int_0^\infty \xi\rho(\zeta)d\zeta \Big/ \int_0^\infty \rho(\zeta)d\zeta = \frac{\pi^2}{16}\alpha, \tag{59}$$

$$= 3\cdot901\sqrt{a}. \tag{59'}$$

By (44) this corresponds to about 66·3 km./sec.$^{-1}$, again consistent with the demands of observation.* Actually the velocities obtained appear to be about twice as large as those observed. To get a better fit λ should be nearer 10,000 km. It appears that the scale of length ($\lambda/2\pi$) provided by values of λ of the order of 8000 (to 6000) km. give on the whole a better agreement with the details of observation in addition to effecting a closer relation between the chromosphere and the "*granulations*." † I hope to deal with these questions in greater detail in a separate paper.

§ 15. In § 14, in discussing the density distribution and the structure of the chromosphere, we neglected the term in $\sin\xi$ in (51). We will now retain it and examine the consequent modifications in the structure of the chromosphere.

Let us assume that at the base of the chromosphere the variation of density is given by

$$\rho = \rho_0(1 - \epsilon \sin \xi), \tag{60}$$

where ϵ is a quantity unspecified for the present. (51) combined with (60) yields for the density distribution the following law :—

$$\rho = \rho_0\left\{(1 - \epsilon \sin \xi) - \epsilon\frac{\alpha}{\beta}\int_0^\zeta \zeta K_1(\zeta)d\zeta\right\}. \tag{61}$$

Now for a physically admissible density distribution we may expect the density ρ to vanish for some value of ζ finite or infinite. This imposes in the first instance a restriction on the possible values for ϵ. By (61) we should have

$$1 + \epsilon \leqslant \epsilon\frac{\alpha}{\beta}\int_0^\infty \zeta K_1(\zeta)d\zeta = \epsilon\frac{\pi}{2}\frac{\alpha}{\beta},$$

or

$$\epsilon\left(\frac{\pi}{2}\frac{\alpha}{\beta} - 1\right) \geqslant 1. \tag{62}$$

In § 14 we neglected the term in $\sin\xi$ and determined the value for ϵ (given by (52′)) from the condition that $\rho \to 0$ as $\zeta \to \infty$. We need now a similar condition to determine ϵ. We argue as follows :—

The atmosphere with a law of density distribution given by (61) has a *sharp boundary*. The equation of the boundary ζ_B is specified by (*cf.* equation (35))

$$Q(\zeta_B, 0) = \frac{\beta}{\epsilon\alpha} - \frac{\beta}{\alpha}\sin\xi. \tag{63}$$

* See A. Unsöld, *Ap. J.*, **69**, 209, 1929.

† Smaller values of λ would require larger values for α/β to obtain equal extensions for the chromosphere, and would thus provide large *mean* velocities (*cf.* equation (59)) in spite of the "unit" of velocity V being smaller.

Consider now the case when the relation (62) is an *equality*. Then equation (63) for this value of ϵ reduces to

$$Q(\infty, \zeta_B) = \frac{\beta}{\alpha}(1 + \sin \xi). \tag{64}$$

Now let ζ_0 be such that (*cf.* equation (41))

$$\frac{\alpha}{\beta} Q(\infty, \zeta_0) = 1. \tag{65}$$

Equation (64) then takes the form

$$\sin \xi = \frac{\alpha}{\beta} Q(\zeta_0, \zeta_B), \tag{66}$$

which is the equation of the trajectory of an atom starting at the point ζ_0 (defined in (65)) and which goes off to infinity but with a final zero limiting velocity. An atom starting at a point on the ζ-axis higher than ζ_0 would go off to infinity but with a finite limiting (vertical) velocity. Hence equations (66) and (65) specify the "*natural boundary*" of a chromosphere for a specified value of α/β. *We now assert that the natural boundary must coincide with the actual boundary.* We should then have

$$\epsilon\left(\frac{\pi}{2}\frac{\alpha}{\beta} - 1\right) = 1. \tag{67}^*$$

From (61) and (67) we now obtain

$$\frac{\partial \rho}{\partial \zeta} = -\rho_0 \cdot \frac{2}{\pi}\left(1 - \frac{2}{\pi}\frac{\beta}{\alpha}\right)^{-1} \cdot \zeta K_1(\zeta). \tag{68}$$

For $\alpha/\beta = 40$, $2\beta/\pi\alpha = 0{\cdot}0159$, which can be neglected in comparison with unity. To this order of accuracy then (68) and (54) predict the same law for the march of the density gradient. It is also easily verified that formulæ (55), (57), (58) and (59) all continue to be true to this order of accuracy, *i.e.* all correct to within 2 per cent.

In a *normal* chromosphere (with a specified α/β) it is probable that ϵ will be *slightly greater* than that given by (67), in which case the actual boundary defined by equation (63) will be *lower* than the natural boundary. Conversely, if we are given ϵ governing the actual variation of density at the base, then (67) will determine not only the natural boundary of the chromosphere but also the magnitude of the kinetic velocities that will be encountered among the chromosphere atoms.

An examination of the direct eclipse photographs of the chromosphere does indicate that it is in fact characterised by a sharp boundary, though I have not been able to find any explicit mention of this in the literature. Further, the boundary of the chromosphere has a wavy appearance (see

* Conversely if we had started with the "boundary condition" that the natural boundary of the chromosphere must coincide with the actual boundary, then we should have *derived* the law (60) for the variation of density at the base of the chromosphere with ϵ given by (67).

plate ix in Menzel's memoir), and a measurement made on this plate indicates that the actual distance between consecutive "crests" is of the order of 10,000 kilometres—a valuable confirmation of the general ideas on which we are working. Actually if we accept the theory we can from the form of the chromosphere and the known kinetic velocities of the chromospheric atoms (as deduced from the Doppler effect) determine the minute variations in density occurring at the base of the chromosphere.

§ 16. The accurate treatment for the derivation of the density distribution given in the previous section shows that the first approximation of § 14 is in fact a very good one. Also, as has been pointed out, the density law (53) is very much more general than the simplified idealisation on which it has been derived would lead us to expect (see footnote on p. 27). We therefore provisionally accept the law (53) as being valid for a "curved atmosphere" as well, and examine the consequences a step further.

Now, a quantity which is directly related to the actual observations that are made during eclipses is the number of atoms per cm.[2] column of the chromosphere viewed *edgewise*. (See Menzel (1), §§ 10, 16, especially pp. 242, 243, 266. For a clear understanding of what follows an explicit reference to these parts of Menzel's memoir will be found useful.) We have to compute, then, the number of atoms N_{ζ_1} in the column $-\infty < \eta < +\infty$ located at ζ_1.* Let R be the radius of the star in the *same scale* as ζ measures the vertical. Then

$$N_{\zeta_1} = \int_{-\infty}^{\infty} d\eta \int_{\frac{\eta^2}{2R}+\zeta_1}^{\infty} \frac{2}{\pi} n_0 \,.\, \zeta K_1(\zeta) d\zeta, \tag{69}$$

where n_0 has the same meaning as in equation (58). After some reduction (69) becomes

$$N_{\zeta_1} = \frac{2^{3/2}}{\pi} R^{\frac{1}{2}} \,.\, n_0 \int_{\zeta_1}^{\infty} \sqrt{\zeta - \zeta_1} \,.\, \zeta K_1(\zeta) d\zeta. \tag{70}$$

It is easily verified that the above equation is equivalent to equation (16.15) of Menzel (1) and hence shows that N_{ζ_1} is proportional to the ***intensity gradient*** along the chromospheric lines.

When $\zeta_1 = 0$ the integral in (70) can be evaluated (see equation (56)). We have

$$N_0 = \frac{32}{3\pi} [\Gamma(7/4)]^2 R^{\frac{1}{2}} \,.\, n_0 = 2{\cdot}868 R^{\frac{1}{2}} \,.\, n_0. \tag{71}$$

If the density law had been

$$n(\zeta) = n_0 \,.\, e^{-\zeta}, \tag{72}$$

we should have had instead of (70),

$$N_{\zeta_1} = 2{\cdot}507 R^{\frac{1}{2}} \,.\, n_0 e^{-\zeta_1}. \tag{73}$$

When $\zeta_1 = 0$, (73) and (70) give practically the same result.

* η is the y-axis measured in the same scale as ζ measures the vertical.

From (70) we have on differentiating

$$-\frac{d\log N_{\zeta_1}}{d\zeta_1}=\frac{\int_{\zeta_1}^{\infty}\sqrt{\zeta-\zeta_1}\,.\,\zeta K_0(\zeta)d\zeta}{\int_{\zeta_1}^{\infty}\sqrt{\zeta_1-\zeta_1}\,.\,\zeta K_1(\zeta)d\zeta}, \qquad (74)$$

while for the law (72) we have simply unity on the right-hand side. Though the formula (74) does not give unity for all ζ, it does tend to unity as $\zeta_1 \to \infty$, and the greatest deviation from unity occurs at $\zeta_1=0$, where it takes the value

$$-\frac{d\log N_0}{d\zeta_1}=\frac{3}{4}\left\{\frac{\Gamma(5/4)}{\Gamma(7/4)}\right\}^2=0{\cdot}7296. \qquad (75)$$

Hence

$$0{\cdot}7296 \leqslant -\frac{d\log N_{\zeta_1}}{d\zeta_1}<1. \qquad (76)$$

Thus the density law (53) cannot seriously be discordant with observations since an exponential law is found roughly to fit them.

§ 17. *Relation of the Chromosphere to the Prominences.*—It was first suggested by Milne and subsequently emphasised by Menzel that the difference between the chromosphere and the prominences is merely one of degree, and that the two phenomena apparently merge into one another directly. Indeed it is precisely from this observation that Menzel deduces the *necessity* for "turbulence," though it has to be mentioned that "turbulence" in the form he advocates will hardly explain this. To secure a close relation between the chromosphere and the prominences we need essentially a hydrodynamical treatment.

The prominences which are definitely associated with the chromosphere are the *Tornado* and the *quiescent* prominences.*

The Tornado prominences are quite small objects, and according to Pettit they generally start off from the outer layers of the chromosphere. The similarity of the appearance of these Tornados with the trajectories for which the inequality (41) is not satisfied (compare plate i in Pettit (*loc. cit.*) and the two trajectories in fig. 1 which start at $\zeta=3{\cdot}0$ and $4{\cdot}0$ with $\alpha/\beta=5$ and 10 respectively) suggests at once a possible origin for these. *Firstly*, that they should start off from the topmost levels is merely a consequence of the fact that in the lower levels α/β has some prescribed (mean) value, and that the atoms starting higher up do not satisfy the inequality (41) and in consequence go off to infinity. *Secondly*, we see that if locally the value of λ (or r) *decreased*, then a trajectory originally belonging to a *complete* periodic solution will now belong to the class of trajectories which go to infinity, and this may be the probable origin of the Tornados. If $\sqrt{r\lambda}$ is roughly the same for all the different chromospheric elements, then there would be *no appreciable "sorting out" effect.* This fact has to be especially emphasised,

* We shall use the nomenclature due to E. Pettit. For a summary see his report in the *Publications of the Astronomical Society of the Pacific*, **45**, 15, 1933.

because it has been frequently asserted that a separation of the different elements in the prominences is a necessary consequence of a theory based on the support by radiation-pressure; but this is not always true. It may be added that one may expect velocities of about 10 km./sec.$^{-1}$ for the atoms in the Tornados.

The second type of prominences which are closely related to the chromosphere are the quiescent prominences. These may merely be a temporary lifting up of the *normal boundary* of the chromosphere due to *variations* in *density* occurring at the base. Normally, as we have seen, ϵ in equation (61) will be slightly greater than the value given by (67). If for some reason the value of ϵ slightly *decreased* and became closer to the value given by (67), then the normal boundary of the chromosphere (which is sensitive to small variations of ϵ, near the value (67)) will get enormously lifted up and give rise to a quiescent prominence. That quiescent prominences exhibit considerable "internal motion of a turbulent kind" is nothing else than the exhibition in some relief of the normal conditions obtaining in the chromosphere itself. There should be no more separating out of the different elements in these quiescent prominences than what occurs in the normal chromosphere.

It is of interest also to notice the general similarity of the "fountains" associated with the spot-type prominences and our periodic trajectories (*cf.* fig. 1 with the "fountains" in plate i of Pettit (*loc. cit.*)). Though these "arched prominences" are not especially related to the chromosphere, yet since near the sunspots the conditions obtaining at the base of the normal chromosphere may be realised in an extreme form, the arched prominences may in fact correspond on a much larger scale to the periodic trajectories which on our ideas should be a normal feature in the chromosphere.

Finally, it may be mentioned that the "spikes" which are of such frequent occurrence in a normal chromosphere are likely to correspond to the emission of the moderately high-speed particles, which possibility has already been pointed out in § 12. The limiting velocities there obtained are of the same order as the velocities actually noticed in these "spikes."

Concluding Remarks.—We have not gone into any great length in a full analysis of the observational material, and it would in fact be hardly worth while at this stage of development in the theory; but on the basis of the evidence presented here it should be clear that the study of the monochromatic radiative equilibrium over a non-uniformly radiating surface may well give valuable hints regarding the general structure of the chromosphere and the prominences.

APPENDIX

As we have seen, the functions $\zeta K_{0,1}(\zeta)$ play an important rôle in the theory, and these functions make their appearance not only in the two-dimensional problem here treated, but also in the equations of motion associated with a radiating surface characterised by an intensity function of quite general types of periodicity. To facilitate the analysis of the more

TABLE III

ζ	$\zeta K_0(\zeta)$	$\zeta K_1(\zeta)$	$\int_0^{\zeta} \zeta K_1(\zeta) d\zeta$
0	0·00 000	1·00 000	0·00 000
0·1	·24 271	·98 538	·09 946
0·2	·35 054	·95 520	·19 658
0·3	·41 174	·91 680	·29 023
0·4	·44 581	·87 374	·37 978
0·5	·46 221	·82 823	·46 489
0·6	·46 651	·78 170	·54 539
0·7	·46 236	·73 520	·62 124
0·8	·45 228	·68 942	·69 246
0·9	·43 806	·64 488	·75 916
1·0	·42 102	·60 191	·82 149
1·1	·40 216	·56 073	·87 960
1·2	·38 221	·52 151	·93 370
1·3	·36 172	·48 431	·98 397
1·4	·34 112	·44 917	1·03 063
1·5	·32 071	·41 608	1·07 383
1·6	·30 073	·38 501	1·11 391
1·7	·28 135	·35 592	1·15 094
1·8	·26 268	·32 872	1·18 516
1·9	·24 481	·30 335	1·21 675
2·0	·22 779	·27 973	1·24 589
2·1	·21 165	·25 777	1·27 275
2·2	·19 639	·23 737	1·29 750
2·3	·18 202	·21 846	1·32 028
2·4	·16 852	·20 094	1·34 123
2·5	·15 587	·18 473	1·36 051
2·6	·14 404	·16 974	1·37 822
2·7	·13 299	·15 589	1·39 449
2·8	·12 270	·14 312	1·40 943
2·9	·11 312	·13 133	1·42 315
3·0	·10 422	·12 047	1·43 573
3·1	·09 596	·11 046	1·44 727
3·2	·08 830	·10 126	1·45 785
3·3	·08 122	·09 279	1·46 755
3·4	·07 466	·08 500	1·47 643
3·5	·06 860	·07 784	1·48 457
3·6	·06 300	·07 126	1·49 202
3·7	·05 783	·06 522	1·49 884
3·8	·05 307	·05 968	1·50 508
3·9	·04 868	·05 460	1·51 079
4·0	·04 464	·04 994	1·51 601
4·1	·04 092	·04 566	1·52 079
4·2	·03 738	·04 175	1·52 516
4·3	·03 435	·03 815	1·52 915
4·4	·03 146	·03 486	1·53 280
4·5	·02 880	·03 185	1·53 613
∞	0	0	1·57 080

general problem it has been thought useful to tabulate here the values of $\zeta K_{0,1}(\zeta)$, though the functions $e^{\zeta} K_{0,1}(\zeta)$ are available in Watson's *Bessel Functions.*

Another quantity of frequent occurrence is $Q(\zeta, 0) = \int_0^{\zeta} \zeta K_1(\zeta) d\zeta$. In the last column of Table III the values of this function are given. The integral was computed making use of the successive "differences" of the function, but the first two values had to be computed from the series expansion for $\zeta K_1(\zeta)$ because of the singularity in $K_1(\zeta)$ at $\zeta = 0$.

Trinity College, Cambridge :
1933 *October* 19.

THE RADIATIVE EQUILIBRIUM OF EXTENDED STELLAR ATMOSPHERES.

S. Chandrasekhar, Ph.D.

§ 1. The problems of radiative equilibrium which arise in the theory of stellar atmospheres have all been solved by idealising the outer layers of a star to be stratified in parallel planes. This last idealisation is of course a very justifiable one in all the commoner problems of the stellar atmospheres, but the recent work of Menzel, Miss Payne, Beals and others on the Wolf-Rayet stars and Novæ appear to show that we are here dealing with cases where the curvature of the outer layers must enter the abstract formulation of the respective problems in a very fundamental way. It is therefore convenient to have the solutions of the standard problems in the theory of radiative equilibrium with the curvature of the outer layers taken properly into account, and this paper, which is of a preliminary character, is a first step in that direction.

The plan of the paper is as follows: Section I, where the first integral of the equation of transfer is obtained, deals with the methods of approximation. Section II is devoted to the Schuster-Schwarzschild problems, and Section III to the formation of absorption lines. Lastly, Section IV treats the radiative equilibrium of a planetary nebula.

I. *The Fundamental Equations and the Methods of Approximation*

§ 2. *The Flux-Integral.*—Let I be the specific intensity of the radiation at a point P and at a distance r from the origin O and in a direction making an angle θ with OP in the positive direction of the radius vector. Let ρ and T be the density and temperature at any point and κ the coefficient of mass-absorption supposed independent of wave-length. Lastly, let B (a function of r only) be the intensity of the black body radiation corresponding to the temperature of the matter at any point r. The equation of transfer in polar co-ordinates is

$$\cos\theta\frac{\partial I}{\partial r}-\frac{\sin\theta}{r}\frac{\partial I}{\partial\theta}=\kappa\rho(B-I). \tag{1}$$

The equation of radiative equilibrium is, as usual,

$$2B=\int_0^\pi I\sin\theta d\theta. \tag{2}$$

Multiplying (1) by $\sin\theta d\theta$ and integrating between 0 and π we get after some transformations that

$$\frac{d}{dr}r^2\int_0^\pi I\cos\theta\sin\theta d\theta=0. \tag{3}$$

If we write
$$F=2\int_0^\pi I\sin\theta\cos\theta d\theta, \tag{4}$$
then πF is merely the net flux of radiation at the point r. Hence by (3) the *first integral* of the equation of transfer is
$$Fr^2=\text{constant}=F_0 \text{ (say)}. \tag{5}$$

§ 3. *The Methods of Approximation.*—The problem is to obtain the solution of (1) subject to (2) or its equivalent (5). Even in the case where the material is regarded as being stratified in parallel planes it has not been possible to integrate the equation of transfer exactly, and the two main methods of approximation that have been used in the "plane case" are the following :—

(*a*) The Schuster-Schwarzschild method of approximation in which we simply integrate the equation of transfer between $0 \leqslant \theta < \frac{1}{2}\pi$ and $\frac{1}{2}\pi \leqslant \theta \leqslant \pi$ and replace I in the two respective hemispheres by their average values I_1 and I_2.

(*b*) A second type of approximation which is due to Milne * is to seek a solution of the equation of transfer consistent with the flux integral. The approximations obtained by this method are much more accurate than those obtained by the method (*a*). A less rigorous but a much more rapid method of obtaining solutions (identical with those obtained by Milne's method) is the Eddington † type of approximation, which consists in the following two steps. If
$$J=\frac{1}{2}\int_0^\pi I\sin\theta d\theta, \tag{6}$$
and
$$K=\frac{1}{2}\int_0^\pi I\cos^2\theta\sin\theta d\theta, \tag{7}$$
then we put
$$K=\frac{1}{3}J, \tag{8}$$
suggested by the mean value of $\cos^2\theta$ over the sphere. Secondly, one uses the boundary condition that at the boundary
$$F(0)=2J(0). \tag{9}$$

§ 4. In treating the problems of radiative equilibrium in polar co-ordinates it is however not clear that one could use the same methods of approximation. Indeed, as McCrea ‡ first showed, the Schuster-Schwarzschild method of approximation applied to our case without any modification leads to inconsistencies. But, as will be shown subsequently, if one uses the Milne-Eddington type of approximation without any modification then one is not led to such inconsistencies. This, however, does not mean that this method of approximation is the proper one to work with when treating the problem in polar co-ordinates.

* E. A. Milne, *M.N.*, **81**, 382, 1921, which is the standard paper on the subject.
† A. S. Eddington, *Internal Constitution of Stars* (Cambridge), p. 322.
‡ W. H. McCrea, *M.N.*, **88**, 729, 1928.

Actually the type of approximation one finally adopts in a special case has, of course, to be chosen from physical considerations. If, for instance, we are considering the radiative equilibrium of a tenuous atmosphere (like the chromosphere) enveloping a parent star, then, as was first pointed out by McCrea, a consistent method of approximation is always to average over $0 < \theta < \sin^{-1}(a/r)$ and $\sin^{-1}(a/r) \leqslant \theta \leqslant \pi$ (instead of $0 < \theta < \pi/2$ and $\pi/2 \leqslant \theta < \pi$), where a is the radius of the parent star. On this basis McCrea indicated the method of approximation which should replace the usual Schuster-Schwarzchild method of approximation. Here we shall also develop methods for an approximation which should replace the usual Milne-Eddington type of approximation. Of course, this method of averaging is suitable only when one has an unambiguous "a" occurring in the problem. If we are considering extended photospheres, for instance, then this method of averaging can no longer be valid. We shall then use the Milne-Eddington type of approximation which, as we have already pointed out, does not lead us to any inconsistency.

II. *The Schuster-Schwarzschild Problems*

§ 5. We shall first consider the case where the McCrea type of averaging has to be adopted.

At a point r we define θ_r to be $\sin^{-1}(a/r)$. Now, multiply the equation of transfer by $\frac{1}{2}\sin\theta d\theta$ and integrate from 0 to θ_r. On the left-hand side we have

$$\tfrac{1}{2}\int_0^{\theta_r}\left\{\cos\theta\sin\theta\frac{\partial I}{\partial r} - \frac{\sin^2\theta}{r}\frac{\partial I}{\partial\theta}\right\}d\theta,$$

which, after partially integrating the second term, transforms into

$$\tfrac{1}{2}\int_0^{\theta_r}\frac{\cos\theta\sin\theta}{r^2}\frac{\partial}{\partial r}(r^2 I)d\theta - \frac{a^2}{r^3}I(r,\theta_r),$$

which, after another partial integration, yields

$$\frac{1}{2r^2}\frac{\partial}{\partial r}r^2\int_0^{\theta_r} I\cos\theta\sin\theta d\theta - \tfrac{1}{2}I(r,\theta_r)\left[\cos\theta_r\frac{\partial\theta_r}{\partial r}\right]_r\sin\theta_r - \frac{a^2}{2r^3}I(r,\theta_r). \quad (10)$$

But

$$\sin\theta_r = a/r\,; \qquad \cos\theta_r\frac{\partial\theta_r}{\partial r} = -\frac{a}{r^2}. \quad (10')$$

The last two terms in (10) cancel and we are left with

$$\frac{1}{2r^2}\frac{d}{dr}r^2\int_0^{\theta_r} I\cos\theta\sin\theta d\theta. \quad (11)$$

Defining

$$F_1 = 2\int_0^{\theta_r} I\sin\theta\cos\theta d\theta\,; \qquad F_2 = 2\int_{\theta_r}^{\pi} I\sin\theta\cos\theta d\theta, \quad (12)$$

$$J_1 = \tfrac{1}{2}\int_0^{\theta_r} I\sin\theta d\theta\,; \qquad J_2 = \tfrac{1}{2}\int_{\theta_r}^{\pi} I\sin\theta d\theta, \quad (13)$$

we have finally
$$\frac{1}{4r^2}\frac{d}{dr}r^2F_1=\tfrac{1}{2}\kappa\rho[(1-\cos\theta_r)B-2\mathcal{J}_1]. \tag{14}$$

Similarly
$$\frac{1}{4r^2}\frac{d}{dr}r^2F_2=\tfrac{1}{2}\kappa\rho[(1+\cos\theta_r)B-2\mathcal{J}_2]. \tag{15}$$

The equation of radiative equilibrium now takes the form
$$B=\mathcal{J}_1+\mathcal{J}_2=\mathcal{J}\text{ (say)}. \tag{16}$$

§ 6. *A First Approximation.*—Equations (14), (15) and (16) are exact. Now replace I in the ranges $(0, \theta_r)$ and (θ_r, π) by their average values I_1 and I_2. We then have
$$F_1=2\int_0^{\theta_r} I\sin\theta\cos\theta d\theta=I_1\frac{a^2}{r^2}, \tag{17}$$
$$F_2=2\int_{\theta_r}^{\pi} I\sin\theta\cos\theta d\theta=-I_2\frac{a^2}{r^2}. \tag{17'}$$

Also
$$\mathcal{J}_1=\tfrac{1}{2}I_1(1-\cos\theta_r)\;;\quad \mathcal{J}_2=\tfrac{1}{2}I_2(1+\cos\theta_r). \tag{18}$$

By (17) and (17′) we have
$$F=F_1+F_2=(I_1-I_2)\frac{a^2}{r^2}. \tag{19}$$

From the flux-integral (5) we easily deduce from the above that,
$$(I_1-I_2)=F_a, \tag{20}$$
where F_a is the flux of radiation at $r=a$. The equation of radiative equilibrium now takes the form
$$B=\tfrac{1}{2}I_1(1-\cos\theta_r)+\tfrac{1}{2}I_2(1+\cos\theta_r). \tag{21}$$

Solving (20) and (21) for I_1 and I_2 we have
$$I_1=B+\tfrac{1}{2}F_a(1+\cos\theta_r), \tag{22}$$
$$I_2=B-\tfrac{1}{2}F_a(1-\cos\theta_r). \tag{22'}$$

Equations (14) and (15) in their averaged form are
$$\frac{a^2}{4r^2}\frac{dI_1}{dr}=\tfrac{1}{2}\kappa\rho(B-I_1)(1-\cos\theta_r), \tag{23}$$
$$\frac{a^2}{4r^2}\frac{dI_2}{dr}=\tfrac{1}{2}\kappa\rho(I_2-B)(1+\cos\theta_r). \tag{23'}$$

Substituting (22) and introducing the new variable τ defined by
$$d\tau=-\kappa\rho dr, \tag{24}$$
we find that
$$\frac{dI_1}{d\tau}=\frac{dI_2}{d\tau}=F_a, \tag{25}$$

or
$$I_1 = \tfrac{1}{2}F_a + B_0 + F_a\tau, \tag{26}$$
$$I_2 = -\tfrac{1}{2}F_a + B_0 + F_a\tau, \tag{26'}$$
where B_0 is a constant of integration. Let $\tau = 0$ occur at $r = R$. Then to determine B_0 we use the boundary condition that at $r = R$, $\tau = 0$ (*cf.* equation (9)),
$$B(R, 0) = \tfrac{1}{2}F(R) = \tfrac{1}{2}F_a \cdot \frac{a^2}{R^2}. \tag{27}$$
This determines B_0. We find
$$B_0 = \tfrac{1}{2}F_a\,(\cos\theta_R + \sin^2\theta_R). \tag{28}$$
From (26) and (21) we then have
$$I_1 = \tfrac{1}{2}F_a(1 + \cos\theta_R + \sin^2\theta_R + 2\tau), \tag{29}$$
$$I_2 = \tfrac{1}{2}F_a(\cos\theta_R + \sin^2\theta_R - 1 + 2\tau), \tag{29'}$$
$$B = \tfrac{1}{2}F_a\{\sin^2\theta_R + (\cos\theta_R - \cos\theta_r) + 2\tau\}. \tag{29''}$$
(29″) now corresponds to Schwarzschild's solution
$$B = \tfrac{1}{2}F(1 + 2\tau), \tag{30}$$
where F now is the *constant* net flux. If T_0 is the boundary temperature at $r = R$, then (29″) relates T_0 to the effective temperature T_e (defined by $\pi F_a = \sigma T_e^4$) by the equation
$$T_0 = \sqrt[4]{\frac{a^2}{2R^2}}\,T_e. \tag{31}$$
We may finally notice that if R is very large compared with a then we have
$$I_1 = F_a(1 + \tau)\,; \quad I_2 = F_a\tau, \tag{32}$$
$$B = \tfrac{1}{2}F_a\{(1 - \cos\theta_r) + 2\tau\}. \tag{32'}$$
(32) is identical with the corresponding solution when the curvature is neglected.

§ 7. *A Second Approximation.*—We first define the two quantities K_1 and K_2 as follows :—
$$K_1 = \tfrac{1}{2}\int_0^{\theta_r} I\sin\theta\cos^2\theta\,d\theta\,; \qquad K_2 = \tfrac{1}{2}\int_{\theta_r}^{\pi} I\sin\theta\cos^2\theta\,d\theta. \tag{33}$$
Multiply the equation of transfer by $\tfrac{1}{2}\sin\theta\cos\theta\,d\theta$ and integrate from 0 to θ_r. On the L.H.S. we have, after partially integrating the second term,
$$\tfrac{1}{2}\int_0^{\theta_r}\frac{\sin\theta\cos^2\theta}{r^2}\frac{\partial}{\partial r}(r^2I)\,d\theta - \frac{1}{2r}\int_0^{\theta_r} I\sin\theta(1 - \cos^2\theta)\,d\theta - \frac{a^2}{2r^3}\cos\theta_r I(r, \theta_r). \tag{34}$$
Again, partially integrating the first integral and using (10′), (13) and noting (33) the above reduces to
$$\frac{1}{r^2}\frac{d}{dr}r^2K_1 + \frac{1}{r}(K_1 - J_1) = \frac{dK_1}{dr} + \frac{1}{r}(3K_1 - J_1), \tag{35}$$

we have finally

$$\frac{dK_1}{dr}+\frac{1}{r}(3K_1-\mathcal{J}_1)=\frac{\kappa\rho}{4}\left(\frac{Ba^2}{r^2}-F_1\right). \tag{36}$$

Similarly

$$\frac{dK_2}{dr}+\frac{1}{r}(3K_2-\mathcal{J}_2)=\frac{\kappa\rho}{4}\left(-\frac{Ba^2}{r^2}-F_2\right). \tag{37}$$

Adding (36) and (37) and using the flux integral we have

$$\frac{d(K_1+K_2)}{dr}+\frac{1}{r}\{3(K_1+K_2)-(\mathcal{J}_1+\mathcal{J}_2)\}=-\frac{\kappa\rho}{4}F_a\cdot\frac{a^2}{r^2}. \tag{38}$$

§ 8. As before, let I_1 and I_2 be the mean intensities in the two ranges. Then we have

$$K_1=\tfrac{1}{6}I_1(1-\cos^3\theta_r)\,;\qquad K_2=\tfrac{1}{6}I_2(1+\cos^3\theta_r). \tag{39}$$

From (39), (18) and (19) we deduce that

$$3(K_1+K_2)-(\mathcal{J}_1+\mathcal{J}_2)=\frac{1}{2}\frac{a^2}{r^2}\cos\theta_r\,.\,(I_1-I_2),$$

$$=\frac{1}{2}\frac{a^2}{r^2}\cos\theta_r\,.\,F_a. \tag{40}$$

Substituting (40) in (38) we have

$$\frac{d(K_1+K_2)}{dr}=-F_a\cdot\frac{a^2}{2r^3}\cos\theta_r-F_a\cdot\frac{a^2\kappa\rho}{4r^2}. \tag{41}$$

On integration this yields

$$K_1+K_2=-\tfrac{1}{6}F_a\cos^3\theta_r+\frac{F_a}{4}\int_0^\tau\sin^2\theta_r d\tau+\tfrac{1}{3}K_0, \tag{42}$$

where K_0 is the constant of integration and $d\tau$ is defined as in (24). Hence by (16) and (40) we have

$$B=-\tfrac{1}{2}F_a\cos\theta_r+\tfrac{3}{4}F_a\int_0^\tau\sin^2\theta_r d\tau+K_0. \tag{43}$$

The constant K_0 is determined as before. We find

$$B=\tfrac{1}{2}F_a\left\{\sin^2\theta_R+(\cos\theta_R-\cos\theta_r)+\tfrac{3}{4}\int_0^\tau\sin^2\theta_r d\tau\right\}, \tag{44}$$

where at $r=R$, $\tau=0$. The above solution is analogous to the solution in the "plane problem" in the form first obtained by Milne. If R is very large compared with a, we have

$$B=\tfrac{1}{2}F_a\left\{(1-\cos\theta_r)+\tfrac{3}{4}\int_0^\tau\sin^2\theta_r\,.\,d\tau\right\}. \tag{45}$$

Relation (31) continues to be true in this order of approximation.

§ 9. We now consider the case where the type of averaging we have used in the preceding sections is not valid. As has been pointed out we now use the unmodified Milne-Eddington type of approximation. Writing

$$J = J_1 + J_2\,; \qquad F = F_0/r^2\,; \qquad K = K_1 + K_2, \tag{46}$$

we now have setting $3K = J$ (*cf.* equation (38))

$$\frac{dK}{d\tau} = -\frac{1}{4}\frac{F_0}{r^2}, \tag{47}$$

or

$$K = \tfrac{1}{4}F_0\int_0^\tau \frac{d\tau}{r^2} + \tfrac{1}{3}K_0, \tag{48}$$

where K_0 is a constant of integration. Let $\tau = 0$ occur at $r = R$. Using the usual boundary condition we derive that

$$B = \tfrac{1}{2}F_0\left\{\frac{1}{R^2} + \tfrac{3}{2}\int_0^\tau \frac{d\tau}{r^2}\right\}, \tag{49}$$

or more simply in terms of the flux πF_R at $r = R$

$$B = \tfrac{1}{2}F_R\left\{1 + \tfrac{3}{2}\int_0^\tau \frac{R^2}{r^2}d\tau\right\}. \tag{50}$$

If R is very large we have the simpler formula

$$B = \tfrac{3}{4}F_0\int_0^\tau \frac{d\tau}{r^2}. \tag{51}$$

§ 10. *Variation of Intensity.*—So far we have not attempted to determine I as a function r and θ. Having now determined B as a function of r we have to introduce this in the equation of transfer and solve it for I. The formal solution can be easily written down. If $I(\theta, p)$ refers to the intensity at a point P in a direction OA which is at a perpendicular distance p from the centre, then one easily verifies that

$$I(\theta, p) = e^{p\int_0^\theta \kappa\rho \operatorname{cosec}^2\theta d\theta}\int_\theta^\pi e^{-p\int_0^\theta \kappa\rho \operatorname{cosec}^2\theta d\theta} B(\theta)p\kappa\rho \operatorname{cosec}^2\theta d\theta. \tag{52}$$

We already know B as a function of r, and by means of the substitution $r = p \operatorname{cosec}\theta$ we can transform it into a function of θ. Thus (52) formally represents the solution to the problem. In practice, however, to perform the integrations we need to know the variation of $\kappa\rho$ with r, and for this reason it does not seem worth while to go into details over this, but we shall illustrate the general trend of the calculations corresponding to the solution (51) for B and when $\kappa\rho$ varies as some inverse power of r.

Let us then suppose that

$$\kappa\rho = cr^{-n} = cp^{-n}\sin^n\theta. \tag{53}$$

By (51) we now have

$$B = \frac{3F_0c}{4(n+1)p^{n+1}}\sin^{n+1}\theta. \tag{54}$$

Introducing this in (52) we have

$$I(\theta, p) = \frac{3F_0c^2}{4(n+1)p^{2n}} e^{cp^{-(n-1)}\int_0^\theta \sin^{n-2}\theta d\theta} \int_\theta^\pi e^{-cp^{-(n-1)}\int_0^\theta \sin^{n-2}\theta d\theta} \sin^{2n-1}\theta d\theta. \quad (55)$$

Writing

$$\alpha = cp^{-n+1} \quad (56)$$

we have

$$I(\theta, p) = \frac{3F_0\alpha}{4(n+1)p^2} e^{\alpha\int_0^\theta \sin^{n-2}\theta d\theta} \int_\theta^\pi e^{-\alpha\int_0^\theta \sin^{n-2}\theta d\theta} \alpha \sin^{2n-1}\theta d\theta. \quad (57)$$

We are interested only in the "emergent radiation," *i.e.* when $\theta = 0$. Then we have

$$I(p) = \frac{3F_0\alpha}{4(n+1)p^2} \int_0^\pi e^{-\alpha\int_0^\theta \sin^{n-2}\theta d\theta} \alpha \sin^{2n-1}\theta d\theta. \quad (58)$$

§ 11. We will consider two special cases of (58).

Case I ($n = 3$).—Now $\alpha = cp^{-2}$, and (58) simplifies to

$$I(p) = \frac{3F_0\alpha^2}{16c} \int_0^\pi e^{-\alpha(1-\cos\theta)}(1 - \cos^2\theta)^2 \alpha \sin\theta d\theta. \quad (59)$$

Introducing the new variable $\alpha \cos\theta = x$ we have

$$I(p) = \frac{3F_0\alpha^2 e^{-\alpha}}{16c} \int_{-\alpha}^{\alpha} e^x \left(1 - \frac{x^2}{\alpha^2}\right)^2 dx. \quad (60)$$

On performing the integration we find that

$$I(p) = \frac{3F_0 e^{-\alpha}}{\alpha^2 c}[(\alpha^2 + 3)\sinh\alpha - 3\alpha\cosh\alpha], \quad (61)$$

where $\alpha = c/p^2$.

The following limiting forms may be noted

$$I(p) \sim \frac{F_0c^2}{5p^6} \text{ for } p \text{ large}, \quad (62)$$

and

$$I(p) \sim \frac{3F_0}{2c}\left(1 - \frac{3p^2}{c}\right) \text{ for } p \text{ small}. \quad (63)$$

Case II ($n = 2$).—Now $\alpha = c/p$, and (58) simplifies to

$$I(p) = \frac{F_0\alpha^4}{4c^2} \int_0^\pi e^{-\alpha\theta} \sin^3\theta d\theta,$$

$$= \frac{3F_0}{2c^2} \frac{\alpha^4(e^{-\alpha\pi} + 1)}{(\alpha^2 + 9)(\alpha^2 + 1)}. \quad (64)$$

We now have

$$I(p) \sim \frac{F_0c^2}{3p^4} \text{ for } p \text{ large}, \quad (65)$$

and

$$I(p) \sim \frac{3F_0}{2c^2} \text{ for } p \text{ small.} \tag{66}$$

Comparing (65) with (62) we notice that the more rapidly $\kappa\rho$ falls off with distance the more rapidly does $I(p)$ fall off with p. This is quite a physically understandable result.

§ 12. *A Purely Scattering Atmosphere.*—When we are dealing with a purely scattering atmosphere there is no interchange of energy between the different frequencies, and we shall therefore consider each frequency separately. Let s_ν be the coefficient of scattering, σ_ν the optical thickness $\int_\infty^r s_\nu \rho dr$. Assuming the scattered radiation to be equal in all directions the equation of transfer is

$$\cos\theta \frac{\partial I_r}{\partial r} - \frac{\sin\theta}{r}\frac{\partial I_r}{\partial r} = s_\nu \rho\left(I_\nu - \tfrac{1}{2}\int_0^\pi I_\nu \sin\theta d\theta\right). \tag{67}$$

We see that formally equation (67) is identical in form with the equations (1) and (2) taken together. Hence the analysis of the preceding sections apply to this case provided we replace τ and κ wherever they occur by σ_ν and s_ν respectively, and also add a suffix ν to all the symbols.

§ 13. *The Schuster Problem.*—The problem which is specifically associated with the name of Schuster is the following * : "A layer of gas scattering monochromatic radiation of frequency ν is placed in front of a bright background radiating with a given flux πG_ν. Given the optical thickness σ_ν of the scattering material, to determine the emergent flux." In our case we have to reformulate the problem in the following terms :—

"Enveloping a bright spherical surface radiating monochromatic radiation of frequency ν with a mean intensity i_ν, is an atmosphere of gas scattering monochromatic radiation of frequency ν. Given the radial optical thickness σ_ν of the surrounding atmosphere and given also that the bright spherical surface is of radius a and that the atmosphere extends to $r=R$, to determine the mean intensity $I_\nu(0)$ of the radiation emergent through a tangent cone drawn from a point $r=R$ to the sphere $r=a$."

It is clear from this formulation that we have now to adopt the type of averaging we have used in §§ 5–8. We shall use the results of the first approximation. From (29) we now have

$$I_1 = \tfrac{1}{2}F_\nu(a)(1+\cos\theta_R+\sin^2\theta_R+2\sigma_\nu). \tag{68}$$

At $r=a$, $I=i_\nu$ and we have

$$i_\nu = \tfrac{1}{2}F_\nu(a)(1+\cos\theta_R+\sin^2\theta_R+2\sigma_\nu). \tag{69}$$

While at $r=R$, $I_1=I_\nu(0)$ and by (68)

$$I_\nu(0) = \tfrac{1}{2}F_\nu(a)(1+\cos\theta_R+\sin^2\theta_R). \tag{70}$$

Hence we have the following solution for the "Schuster's problem" :—

$$r_\nu = \frac{I_\nu(0)}{i_\nu} = \frac{1+\cos\theta_R+\sin^2\theta_R}{1+\cos\theta_R+\sin^2\theta_R+2\sigma_\nu}. \tag{71}$$

* *Cf.* E. A. Milne, *Phil. Trans. Roy. Soc.* (A), **228**, 421, 1929.

(71) replaces the usual Schuster's formula for r_ν, namely :—

$$r_\nu = \frac{1}{1+\sigma_\nu}. \tag{72}$$

As was to be expected, the value of r_ν depends not only on the optical thickness σ_ν, but also on the extent to which the scattering material is spread out. But this dependence is not very pronounced. Thus for a given σ_ν, r_ν is a *minimum* when $(1+\cos\theta_R+\sin^2\theta_R)$ is a maximum which it attains when $\theta_R=60°$. When $\theta_R=60°$, (71) takes the form

$$r_\nu = \frac{1}{1+(8/9)\sigma_\nu}. \tag{73}$$

$\theta_R=60°$ corresponds to the scattering material being spread out to a distance $2/\sqrt{3}$ ($=1{\cdot}1547$) times the radius of the original "photospheric surface."

Now it is known that in the more exact form of the Schuster's formula we have in the denominator of (72), $3/4 \cdot \sigma_\nu$ instead of simply σ_ν. For us to obtain the corresponding second approximation we really need to evaluate the integral (52) with B given by (44), for which we require the exact variation of $s_\nu\rho$ with r. As this is a rather complicated matter we shall not go into it here.

III. *The Formation of Absorption Lines* *

§ 14. In considering the formation of absorption lines it is not quite clear as to the method of approximation that has to be adopted. For the sake of simplicity we shall treat the problem on the un-modified Eddington type of approximation. The other type of averaging leads to far too complicated equations.

Let

- κ_ν be the coefficient of continuous absorption applicable to both inside and immediately outside the absorption line,
- s_ν (a function of ν) the coefficient of line absorption,
- $I_\nu'(r, \theta)$ the intensity of the radiation in the frequency range ν to $\nu+d\nu$ within the line at a point distant r from the centre and in a direction making an angle θ with the positive direction of the radius vector and $I(r, \theta)$ the corresponding intensity if there had been no scattering (*i.e.*) outside the line.

In addition to the quantities F, $\mathcal{J}$, K, we now introduce a corresponding set of quantities F_ν', $\mathcal{J}_\nu'$, K_ν' defined in exactly the same way but with the I_ν''s replacing the I's in the respective integrals.

The equation of transfer is (*cf.* Eddington, *loc. cit.*, equation (4))

$$\cos\theta\frac{\partial I_\nu'}{\partial r} - \frac{\sin\theta}{r}\frac{\partial I_\nu'}{\partial\theta} = -\rho(\kappa_\nu+s_\nu)I_\nu' + \rho(1-\epsilon)s_\nu\mathcal{J}_\nu' + \epsilon\rho s_\nu B_\nu + \rho\kappa_\nu B_\nu, \tag{74}$$

* *Cf.* E. A. Milne, *M.N.* (1), **88**, 493, 1928 ; (2), **89**, 3, 1928 ; A. S. Eddington, *M.N.*, **89**, 620, 1928.

where ϵ is the fraction of the absorbed scattered radiation which is lost in collisions of the second kind, and B_ν as usual corresponds to the Planck-function.

Multiplying (74) by $\frac{1}{2}\sin\theta d\theta$ and $\frac{1}{2}\sin\theta\cos\theta d\theta$, and integrating between 0 to π, we have (*cf.* equations (14), (15) and (38))

$$\frac{1}{4r^2}\frac{d}{dr}r^2F_\nu' = -\rho(\kappa_\nu+\epsilon s_\nu)(\mathfrak{J}_\nu'-B_\nu), \tag{75}$$

$$\frac{dK_\nu'}{dr}+\frac{1}{r}(3K_\nu'-\mathfrak{J}_\nu') = -\tfrac{1}{4}\rho(\kappa_\nu+s_\nu)F_\nu'. \tag{76}$$

Let

$$\eta = s_\nu/\kappa_\nu, \tag{77}$$

and defining the optical thickness τ_ν by

$$d\tau_\nu = -\kappa_\nu\rho dr, \tag{78}$$

we have our fundamental system of differential equations

$$\frac{1}{4r^2}\frac{d}{d\tau_\nu}r^2F_\nu' = (1+\epsilon\eta)(\mathfrak{J}_\nu'-B_\nu), \tag{79}$$

$$\frac{dK_\nu'}{d\tau_\nu} = -\frac{1}{\rho\kappa_\nu r}(\mathfrak{J}_\nu'-3K_\nu')+\tfrac{1}{4}(1+\eta)F_\nu'. \tag{80}$$

Our approximation is now to set $\mathfrak{J}_\nu'=3K_\nu'$. Hence

$$\frac{1}{4r^2}\frac{d}{d\tau_\nu}r^2F_\nu' = (1+\epsilon\eta)(\mathfrak{J}_\nu'-B_\nu), \tag{81}$$

$$\frac{d\mathfrak{J}_\nu'}{d\tau_\nu} = \tfrac{3}{4}(1+\eta)F_\nu'. \tag{82}$$

§ 15. Assume now that η and ϵ are both constants as a function of τ_ν. It is doubtful whether this approximation can be justified to the extent to which it can be justified in the case where the matter is stratified in parallel planes (*cf.* Milne (2), *loc. cit.*, p. 5). However, we shall adopt it to see the nature of the results that arise out of this assumption. From (81) and (82) we now have

$$\frac{1}{r^2}\frac{d}{d\tau_\nu}\left(r^2\frac{d\mathfrak{J}_\nu'}{d\tau_\nu}\right) = q^2(\mathfrak{J}_\nu'-B_\nu), \tag{83}$$

where

$$q^2 = 3(1+\eta)(1+\epsilon\eta). \tag{84}$$

We will now consider the case where B_ν has its "equilibrium value," *i.e.* the radiation just outside the line has its equilibrium density. We have then to adopt our earlier solution in the form (50) for B_ν:

$$B_\nu = \tfrac{1}{2}F_{R,\nu}\left\{1+\tfrac{3}{2}\int_0^{\tau_\nu}\frac{R^2}{r^2}d\tau_\nu\right\}. \tag{85}$$

For the sake of brevity we shall hereafter suppress the suffix ν to F_R, but it has to be understood that F_R always stands for the net flux of radiation in the frequency range $(\nu, \nu+d\nu)$ at the boundary $(r=R)$ of the star.

From (83) and (85) it follows that

$$\frac{1}{r^2}\frac{d}{d\tau_\nu}\left\{r^2\frac{d(\mathcal{J}_\nu'-B_\nu)}{d\tau_\nu}\right\}=q^2(\mathcal{J}_\nu'-B_\nu), \tag{86}$$

or differently as

$$\frac{d^2(\mathcal{J}_\nu'-B_\nu)}{d^2\tau_\nu}+\frac{2}{r}\frac{dr}{d\tau_\nu}\frac{d(\mathcal{J}_\nu'-B_\nu)}{d\tau_\nu}-q^2(\mathcal{J}_\nu'-B_\nu)=0. \tag{87}$$

It is seen that to solve (87) we need the variation of τ_ν with r. We shall examine the case when τ_ν varies as some inverse power m of r, *i.e.*

$$\tau_\nu=cr^{-m}\text{ (say)}. \tag{88}$$

Consistent with this assumption we should now rewrite (85) in the form

$$B_\nu=\tfrac{3}{4}F_0\int_0^{\tau_\nu}\frac{d\tau_\nu}{r^2}, \tag{85$'$}$$

where F_0 now stands for $F_{0,\nu}$. Introducing the following new quantities

$$n=\frac{m+2}{2m}, \tag{89}$$

$$\mathcal{J}_\nu'-B_\nu=Q\tau_\nu^{\,n}\,; \qquad Z=q\tau_\nu, \tag{90}$$

the equation (87) combined with the relation (88) leads to the following differential equation for Q :—

$$Z^2\frac{d^2Q}{dZ^2}+Z\frac{dQ}{dZ}-(n^2+Z^2)Q=0. \tag{91}$$

(91) is just Bessel's equation with the purely imaginary argument *in*. We need a solution of (91) which tends exponentially to zero for large values of Z, since $\mathcal{J}_\nu'$ must tend to its equilibrium value at great distances. Hence the appropriate solution we have to consider is the following :—*

$$Q(Z)=K_n(Z)=\tfrac{1}{2}\pi\{I_{-n}(Z)-I_n(Z)\}\cot n\pi. \tag{92}$$

Hence our solution is

$$\mathcal{J}_\nu'=\tfrac{3}{4}F_0\int_0^{\tau_\nu}\frac{d\tau_\nu}{r^2}+A(q\tau_\nu)^nK_n(q\tau_\nu), \tag{93}$$

where A is a constant at present undetermined. Finally, by (82), we have for the flux $\pi F_\nu'$,

$$F_\nu'=\frac{4}{3(1+\eta)}\left\{\frac{3F_0}{4r^2}-Aq(q\tau_\nu)^nK_{n-1}(q\tau_\nu)\right\}. \tag{94}$$

A is now determined from the boundary condition that at $\tau_\nu=0$, *i.e.* as $r\to\infty$, $F_\nu'-2\mathcal{J}_\nu'\to 0$. From physical considerations both must separately

* *Cf.* Whittaker and Watson, *Modern Analysis* (Cambridge, 1927), p. 373.

tend to zero as $r \to \infty$. We will not go into the details here, but one can easily prove that if $n > \frac{1}{2}$ (*i.e.* if m is finite) then $Z^n K_{n-1}(Z) \to 0$ as $Z \to 0$, while $Z^n K_n(Z)$ tends to some limiting value. Hence our boundary condition can be satisfied if, and only if, $A = 0$, in which case we simply have

$$F_\nu' = \frac{F_0 r^{-2}}{1+\eta}. \tag{95}$$

On the other hand, for the flux F_ν just outside the line,

$$F_\nu = F_0 r^{-2}. \tag{96}$$

Hence

$$\frac{F_\nu'}{F_\nu} = \frac{1}{1+\eta}. \tag{97}$$

For the case where the material is stratified in parallel planes the formula corresponding to (97) is (*cf.* Eddington, *loc. cit.*, equation (17))

$$\frac{F_\nu'}{F_\nu} = \frac{1+\frac{2}{3}q}{1+\eta+\frac{2}{3}q}. \tag{98}$$

Thus for the case we have considered the introduction of the curvature far from introducing complications essentially simplifies the problem. Though the result (97) has been obtained only for a strictly infinite atmosphere, we can expect it to be valid if only the atmosphere be sufficiently extended.

IV. *The Radiative Equilibrium of a Planetary Nebula*

§ 16. This problem has already been considered in some detail by Milne.* But he neglected the curvature of the layers. It is, however, quite easy to find the appropriate solutions, taking into account the curvature as well.

Let πS be the incident flux per cm.^{2} due to the central star on the inner boundary of the nebular shell. The equation of radiative equilibrium then takes the form

$$2\int_0^\pi (B-I)\sin\theta\, d\theta = \frac{S r_1^2}{r^2} e^{-(\tau_1-\tau)}, \tag{99}$$

where the optical thickness τ is measured from the outer boundary of the nebular shell the total radial optical thickness of which is taken to be τ_1. Further, let r_1 and r_2 be the inner and the outer radii of the planetary nebula.

The equation of transfer is the same as before (equation (1)). Multiplying the equation of transfer by $\sin\theta\, d\theta$ and integrating from 0 to π and using (99), we find that

$$\frac{d}{d\tau} r^2 F = -S r_1^2 e^{-(\tau_1-\tau)}, \tag{100}$$

which on integration yields

$$r^2 F = \text{constant} - S r_1^2 e^{-(\tau_1-\tau)}. \tag{101}$$

* E. A. Milne, *Zeits. f. Astro. Physik*, 1, 98, 1930.

At $\tau=\tau_1$, $r=r_1$ we have $F=0$. This is precisely equivalent to the boundary condition discussed by Milne (*loc. cit.*, p. 102). From this we deduce that

$$F=\frac{Sr_1^2}{r^2}(1-e^{-(\tau_1-\tau)}), \tag{102}$$

which is now our flux integral. Multiplying the equation of transfer by $\frac{1}{2}\sin\theta\cos\theta d\theta$ and integrating from 0 to π and setting $3K=J$, we have

$$\frac{dK}{d\tau}=\frac{Sr_1^2}{4r^2}(1-e^{-(\tau_1-\tau)}), \tag{103}$$

which leads to

$$J=\tfrac{3}{4}\int_0^\tau \frac{Sr_1^2}{r^2}(1-e^{-(\tau_1-\tau)})d\tau+K_0, \tag{104}$$

where K_0 is a constant of integration to be determined from our usual boundary condition that at $r=r_2$, $\tau=0$, $F=2J$. We find that

$$K_0=\tfrac{1}{2}\frac{Sr_1^2}{r_2^2}(1-e^{-\tau_1}). \tag{105}$$

We finally obtain from (99), (104) and (105) that

$$B=\tfrac{1}{2}S\left\{\frac{r_1^2}{r_2^2}(1-e^{-\tau_1})+\tfrac{1}{2}\frac{r_1^2}{r^2}e^{-(\tau_1-\tau)}+\tfrac{3}{2}\int_0^\tau\frac{r_1^2}{r^2}(1-e^{-(\tau_1-\tau)})d\tau\right\}. \tag{106}$$

If the radius of curvature be neglected, *i.e.* $r_1=r_2=r=\text{constant}$, then we easily derive from (106) that

$$B=\tfrac{1}{2}S\{1+\tfrac{1}{2}e^{-\tau}-e^{-(\tau_1-\tau)}+\tfrac{3}{2}\tau\}, \tag{107}$$

which is Milne's result (*loc. cit.*, equation (28)). The ratio of the boundary temperatures at $r=r_2$ given by (107) and (106) is seen to be $\sqrt{r_1/r_2}$, which might be appreciable for many planetaries.

Summary

In this paper the standard problems in the theory of radiative equilibrium which arise in the theory of stellar atmospheres are rediscussed without neglecting the curvature of the outer layers of the star.

The solution for the Schuster problem shows that the residual intensity depends not only on the optical thickness of the scattering material but also on the extent to which it is spread out. For a given optical thickness the minimum absorption ratio occurs when the tangent cone from a point on the boundary of the scattering atmosphere has a semivertical angle of 60°.

The formation of absorption lines has also been discussed in detail for the case when τ_ν varies as some inverse power m of r. It is shown that if m is finite then we always have

$$\frac{F_\nu'}{F_\nu}=\frac{1}{1+\eta},$$

(F_ν' being the flux of radiation in the line, while F_ν is the corresponding flux just outside the line and $\eta = s_\nu/\kappa_\nu$).

Finally, the radiative equilibrium of the planetary nebula is briefly considered. It is shown that if r_1 and r_2 are the inner and the outer radii of the nebular shell, then the temperature at the boundary is $\sqrt{r_1/r_2}$ times the value that would be predicted if we had neglected the curvature.

Trinity College, Cambridge :
1934 *March* 2.

ON THE EFFECTIVE TEMPERATURES OF EXTENDED PHOTOSPHERES

BY S. CHANDRASEKHAR, PH.D., Trinity College

[*Received* 13 May, *read* 3 June 1935]

1. For material stratified in parallel planes in local thermodynamical equilibrium we have Milne's well-known result that

$$B(\tau) = \tfrac{1}{2}F(1 + \tfrac{3}{2}\tau), \qquad (1)$$

where πF is the constant net integrated flux of radiation, τ is the optical depth and B is the "ergiebigkeit" which is related to the temperature T by the relation

$$B = \frac{\sigma}{\pi} T^4, \qquad (2)$$

σ being the Stefan-Boltzmann constant. If we define the effective temperature by the relation

$$\pi F = \sigma T_e^4, \qquad (3)$$

we have from (1)

$$T_e^4 = 0{\cdot}8 T_1^4, \qquad (4)$$

where T_1 is the temperature at $\tau = 1$. In this note we establish a similar result for *extended photospheres* where the curvature of the outer layers is properly taken into account.

2. It has already been shown by Kosirev and the author* independently that the solution corresponding to (1) for an extended photosphere is

$$B(\tau) = \tfrac{3}{4}F_0 \int_0^\tau \frac{d\tau}{r^2}, \qquad (5)$$

where r denotes the radius vector measured from the centre and F_0 is a constant such that the net flux πF at any distance r is given by

$$F = F_0 r^{-2}. \qquad (6)$$

If now $I(\theta, p)$ refers to the intensity of radiation at a point P and in a direction OA which is at a perpendicular distance p from the centre, then we have quite generally that (see Fig. 1)

$$I(\theta, p) = e^{p\int_0^\theta \kappa\rho \operatorname{cosec}^2\theta d\theta} \int_\theta^\pi e^{-p\int_0^\phi \kappa\rho \operatorname{cosec}^2\phi d\phi} B(\phi)\, p\kappa\rho \operatorname{cosec}^2\phi\, d\phi, \qquad (7)$$

where κ is the absorption coefficient and ρ is the density.

* N. A. Kosirev, *Monthly Notices Roy. Astr. Soc.*, 94 (1934), 430; S. Chandrasekhar, *ibid.* 94 (1934), 444.

We are generally interested in the emergent radiation, i.e. when $\theta = 0$. By (7),

$$I(p) = I(0, p) = \int_0^{\pi} e^{-p\int_0^{\theta}\kappa\rho \operatorname{cosec}^2\theta d\theta} B(\theta) p\kappa\rho \operatorname{cosec}^2\theta d\theta, \tag{8}$$

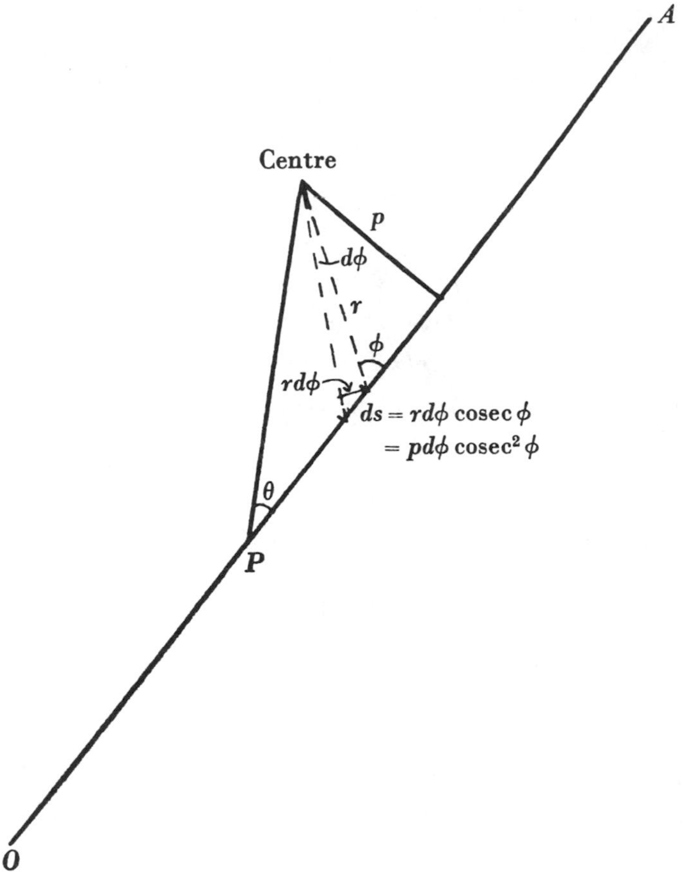

Fig. 1.

where in (8) we have replaced ϕ by θ. The integrated radiation L from all directions is therefore given by

$$L = 2\pi \int_0^{\infty} I(p) p dp. \tag{9}$$

(5), (8), (9) formally solve the problem. It is of interest, however, to obtain more precise formulae for the case when $\kappa\rho$ varies as some inverse power of r. Let

$$\kappa\rho = \frac{c}{r^n}, \tag{10}$$

where c is a constant. Then $$\tau = \frac{c}{(n-1) r^{n-1}}. \tag{10'}$$

By (5) we have
$$B=\frac{3F_0c}{4(n+1)r^{n+1}}, \tag{11}$$

or, since $r=p\operatorname{cosec}\theta$,
$$B(\theta)=\frac{3F_0c}{4(n+1)p^{n+1}}\sin^{n+1}\theta. \tag{12}$$

Substituting (12) in (8) we have

$$I(p)=\frac{3F_0\alpha^2}{4(n+1)p^2}\int_0^\pi e^{-\alpha\psi_n(\theta)}\sin^{2n-1}\theta\,d\theta, \tag{13}$$

where
$$\alpha=cp^{-n+1}, \tag{13'}$$

and
$$\psi_n(\theta)=\int_0^\theta \sin^{n-2}\theta\,d\theta. \tag{14}$$

By (9) we have for L

$$L=\frac{6\pi F_0}{4(n+1)}\int_0^\pi\int_0^\infty \frac{\alpha^2}{p}e^{-\alpha\psi_n(\theta)}\sin^{2n-1}\theta\,d\theta\,dp. \tag{15}$$

By (10) and (13′),
$$p\,d\alpha=-(n-1)\alpha\,dp. \tag{15'}$$

Changing therefore to the variable α in (15) we have

$$L=\frac{6\pi F_0}{4(n^2-1)}\int_0^\pi\int_0^\infty e^{-\alpha\psi_n(\theta)}\alpha\sin^{2n-1}\theta\,d\alpha\,d\theta. \tag{16}$$

The integration with respect to α is immediately performed. We have

$$L=\frac{6\pi F_0}{4(n^2-1)}\int_0^\pi\frac{\sin^{2n-1}\theta\,d\theta}{\psi_n^2(\theta)}. \tag{17}$$

Let $r=R_1$ and $T=T_1$ at $\tau=1$. By (10′) and (11) we have

$$(n-1)R_1^{n-1}=c;\quad B(R_1)=\frac{3F_0c}{4(n+1)R_1^{n+1}}. \tag{18}$$

Hence
$$\frac{\sigma}{\pi}T_1^4=\frac{3(n-1)F_0}{4(n+1)R_1^2}. \tag{19}$$

If we now define the "effective temperature" by the relation

$$L=\sigma R_1^2T_e^4, \tag{20}$$

we have, on comparing (17) and (19),

$$T_e^4=A_nT_1^4, \tag{21}$$

where
$$A_n=\frac{2}{(n-1)^2}\int_0^\pi\frac{\sin^{2n-1}\theta\,d\theta}{\psi_n^2(\theta)}. \tag{22}$$

A_n can be alternatively expressed as

$$A_n=\frac{2(n+1)}{(n-1)^2}\int_0^\pi\frac{\cos\theta\sin^n\theta\,d\theta}{\psi_n(\theta)}. \tag{23}$$

For $n=3, 5, 7$, A_n can be evaluated explicitly. We have

$$\left.\begin{aligned} \psi_3(\theta)&=1-\cos\theta, \\ \psi_5(\theta)&=\tfrac{1}{3}(2+\cos\theta)(1-\cos\theta)^2, \\ \psi_7(\theta)&=\tfrac{1}{15}(8+9\cos\theta+3\cos^2\theta)(1-\cos\theta)^3, \end{aligned}\right\} \tag{24}$$

and the above expressions when inserted in (23) yield elementary integrals which can be evaluated. It is found that

$$A_3=4/3,$$

$$A_5=1\cdot5\,[4-3\log_e 3]=1\cdot0563,$$

$$A_7=\frac{20}{27}\left[4-\frac{16}{\sqrt{15}}\left(\tan^{-1}\sqrt{15}-\tan^{-1}\sqrt{\frac{3}{5}}\right)\right]=0\cdot9463.$$

For $n=2$ we have, since $\psi_2(\theta)=\theta$,

$$A_2=2\int_0^\pi \frac{\sin^3\theta}{\theta^2}\,d\theta,$$

which after some transformations can be reduced to

$$A_2=\tfrac{3}{2}\,[\log_e 3+Ci\,(\pi)-Ci\,(3\pi)], \tag{25}$$

where in (25) $Ci\,(x)$ is the "cosine integral"

$$Ci\,(x)=-\int_x^\infty \frac{\cos x}{x}\,dx.$$

Numerically (25) is found to be

$$A_2=1\cdot7426.$$

Finally it is clear that $\qquad A_1=\infty;\quad A_\infty=0\cdot8.$

For other values of n numerical methods will have to be adopted. It is of interest to see that even A_7 is not very different from the limiting value 0·8.

3. A further consequence of (13) may be noted here.* As $p\to\infty$, $\alpha\to0$ and

$$I\,(p)\sim\frac{3F_0\alpha^2}{4\,(n+1)\,p^2}\int_0^\pi \sin^{2n-1}\theta\,d\theta \quad (p\to\infty), \tag{26}$$

or

$$I\,(p)\sim\sqrt{\pi}\,\frac{\Gamma\,(n)}{\Gamma\,(n+\frac{1}{2})}\,\frac{3F_0c^2}{4\,(n+1)\,p^{2n}} \quad (p\to\infty), \tag{27}$$

i.e. *the "law of darkening" is such that for large distances from the centre the intensity falls off as the square of* $\kappa\rho$.

* For the cases $n=3$ and 2, $I\,(p)$ can be evaluated, and the explicit expressions have already been given in the author's earlier paper (*loc. cit.*).

THE RADIATIVE EQUILIBRIUM OF THE OUTER LAYERS OF A STAR WITH SPECIAL REFERENCE TO THE BLANKETING EFFECT OF THE REVERSING LAYER

S. CHANDRASEKHAR, PH.D.

1. *Introduction.*—For grey material in radiative equilibrium we have Milne's well-known result that

$$B(\tau)=\tfrac{1}{2}F(1+\tfrac{3}{2}\tau), \tag{1}$$

where $B(\tau)$ is the intensity of the integrated black-body radiation for the temperature at the optical depth τ and πF is the constant net flux. From (1) we have for the specific intensity $I(\theta)$ of the emergent radiation in a direction making an angle θ with the normal, the expression

$$I(\theta)=\int_0^\infty B(\tau\cos\theta)e^{-\tau}d\tau=\tfrac{1}{2}F(1+\tfrac{3}{2}\cos\theta). \tag{2}$$

(2) corresponds to a coefficient of darkening $u=3/5$. We shall refer to (2) as the *standard darkening.*

Again (1) leads to a definite prediction concerning the intensity distribution in the continuous spectrum. Thus the intensity $I(\lambda, \theta)$ of the emergent radiation at wave-length λ and in a direction making an angle θ with the normal is given by

$$I(\lambda)=\int_0^\infty B(\lambda,\ T_{\tau\cos\theta})e^{-\tau}d\tau, \tag{3}$$

where $T_{\tau\cos\theta}$ is the temperature at optical depth $\tau\cos\theta$ and $B(\lambda, T)$ the Planck function corresponding to the temperature T. The comparison between the predicted intensity distribution in the continuous spectrum with the observed distribution in the case of the Sun has been examined by Milne * and Lindblad.† The surprising result emerged that the predictions of the theory were in much better accord with the results of observation than one had reason to expect. This fair agreement between the theory and the observations has generally been taken to mean that the overlapping of a vast number of photo-electric absorption bands from a large number of different elements in their various excited states does in fact produce an approximate " greyness " of the Sun's material.

2. However, the assumption $\kappa=$constant still leaves certain significant discordances between theory and observation. Thus a coefficient of darkening $u=0{\cdot}56$ agrees better with the results of observation. Further, as H. H. Plaskett's observations showed, the true colour temperature of the Sun at the centre of the disc is much higher than could be expected on the basis of (3). It is therefore necessary to examine whether the taking into account of the formation of the absorption lines in the higher levels of the photosphere could bring about a better agreement between the theory and observations. An obvious effect of the formation of the Fraunhofer lines in the " reversing layer " is to keep the photosphere warmer, since the energy dammed back in the Fraunhofer lines is necessarily returned to the Sun's photosphere. The reversing layer therefore effectively acts as a blanket. In his classical memoir on the subject Milne examines this blanketing effect by considering a purely scattering atmosphere placed above grey material in radiative equilibrium. Subsequently Lindblad gave a formally equivalent treatment of this problem in which he assumed that the distribution of temperature in the photosphere is unaltered, but supposed it to terminate at some fictitious optical depth $\tau=\tau_0$. At $\tau=\tau_0$ the outward flux πF_+ is found to be

$$\pi F_+=\pi F(1+\tfrac{3}{4}\tau_0). \tag{4}$$

The inward flux then is

$$\pi F_-=\pi F(1-\tfrac{3}{4}\tau_0). \tag{4'}$$

The assumption is now made that

$$F_-=\eta F_+, \tag{5}$$

where η is non-zero. (5) is usually stated in the following terms: "The reversing layer returns to the photosphere a fraction η of the radiation incident on it from the photosphere." However, the physical assumptions

* *Phil. Trans. Roy. Soc.*, A, **223**, 209, 1922.

† *Nova Acta Reg. Soc. Sc.*, Upsala, **6**, No. 1, p. 17, 1923.

involved are not quite clear, and as Professor H. H. Plaskett has kindly pointed out to the writer, Milne's earlier treatment is preferable in this respect. The following is Plaskett's version of Milne's treatment:—

For the temperature distribution assume that

$$B(\tau)=a+\tfrac{3}{4}F\tau, \tag{6}$$

where a is left arbitrary for the present. (That the coefficient of τ should always be $\tfrac{3}{4}$F is known from the theory of radiative equilibrium.*) From (6) we now find that at $\tau=0$

$$F_{+}=a+\tfrac{1}{2}F\,; \qquad F_{-}=a-\tfrac{1}{2}F. \tag{7}$$

If there had been no scattering atmosphere then $F_{+}=F$ and $F_{-}=0$. But the effect of the purely scattering atmosphere is to return to the photosphere a finite inward flux. This inward flux is mainly in the absorption lines, but it would be absorbed in the deeper photospheric layers and then converted into general radiation. We now assume that the inward flux F_{-} is a fraction η of the outward flux F_{+}. This condition determines a. We have finally

$$B(\tau)=\tfrac{1}{2}F\left[\frac{1+\eta}{1-\eta}+\tfrac{3}{2}\tau\right]. \tag{8}$$

From (8) we have the relation

$$T_0^4=\tfrac{1}{2}\cdot\frac{1+\eta}{1-\eta}T_e^4. \tag{9}$$

Again from (8) and (2) we have

$$I(\theta)=\tfrac{1}{2}F\left[\frac{1+\eta}{1-\eta}+\tfrac{3}{2}\cos\theta\right]. \tag{10}$$

(10) corresponds to a coefficient of darkening

$$u=\frac{3(1-\eta)}{5-\eta}. \tag{11}$$

As the observed coefficient of darkening is nearer 0·56, it would, according to (11), correspond to a value of $\eta=0{\cdot}083$.†

The modified intensity distribution in the continuous spectrum can be found by inserting (8) in (3). Such calculations have been made by Milne and Lindblad. Thus Lindblad finds that the detailed predictions of the theory for the darkening at each separate wave-length are fairly confirmed by observations.

In spite, however, of the formal success of the Milne-Lindblad treatment of the blanketing effect of the reversing layer, it is clear that a satisfactory theory ought really take into account the formation of the absorption lines in a much more direct way than has so far been done. In this paper an attempt in this direction is made. We shall assume in particular that *the absorption lines are uniformly distributed in the whole spectral range*, and then

* Milne, *M.N.*, **81**, 367, 1921.

† See G. F. W. Mulders, "Aequivalente Breedten van Fraunhoferlijnen in het Zonnenspectrum" (Doctor Thése, Utrecht).

calculate the altered temperature distribution and also the complete march of the integrated fluxes in the absorption lines and in the continuous spectrum. We shall also calculate the modified law of darkening and consider the circumstances under which the theory can predict a reduced centre-limb contrast in agreement with observation. We shall also estimate the true colour temperatures on the basis of our present more exact analysis of the radiative equilibrium of the outer layers. It is found that Plaskett's high value for the true colour temperature of the Sun can be accounted for theoretically as due to the blanketing effect of the reversing layer.

4. Before concluding this introductory section reference should be made to a recent paper by Woolley,* where the blanketing effect of the reversing layer is taken into account on essentially the same lines as we do. However, our analysis is much more general and Woolley's calculations appear as a very special case.

Section I

5. As stated in § 3 we shall assume that the absorption lines occur uniformly in the whole spectral range. This statement should be made more precise. Let $a_2(\nu)d\nu$ be the probability of our finding in the frequency interval $(\nu, \nu+d\nu)$ an absorption line. The probability then that the interval is free from an absorption line is

$$a_1(\nu)d\nu = \{1 - a_2(\nu)\}d\nu. \qquad (12)$$

Our assumption now is that

$$a_1(\nu) = a_1 = \text{constant}, \qquad (13$$

$$a_2(\nu) = a_2 = \text{constant}. \qquad (13')$$

It is of course clear that $a_1 + a_2 = 1$.

6. *The Equations of Transfer.*—Let $I_\nu d\nu$ be the specific intensity of the radiation in the frequency interval $(\nu, \nu+d\nu)$. Then if we are outside an absorption line the equation of transfer (on the hypothesis of local thermodynamic equilibrium) is as usual

$$\cos\theta \frac{dI_\nu}{\rho dx} = -\kappa_\nu I_\nu + \kappa_\nu B_\nu, \qquad (14)$$

where κ_ν the coefficient of absorption and $B_\nu d\nu$ the intensity of the black-body radiation at the temperature at x and in the frequency interval $(\nu, \nu+d\nu)$.

On the other hand, if we are inside an absorption line, then the appropriate equation of transfer is

$$\cos\theta \frac{dI_\nu}{\rho dx} = -(\kappa_\nu + \alpha_\nu)I_\nu + (1-\epsilon_\nu)\alpha_\nu \mathfrak{J}_\nu + (\kappa_\nu + \epsilon\alpha_\nu)B_\nu, \qquad (15)$$

where

$$\mathfrak{J}_\nu = \tfrac{1}{2}\int_0^\pi I_\nu \sin\theta d\theta, \qquad (16)$$

* R. v. d. R. Woolley, *M.N.*, **94**, 713, 1934. An error in his formula rather spoils the numerical work in this paper (see § 13 of the present paper, especially the footnote on p. 29).

and κ_ν (as in (14)) is the coefficient of continuous absorption and a_ν the *mean* coefficient of scattering in the line. Further, ϵ_ν is the probability that the radiation at $(\nu, \nu + d\nu)$ is lost by photoelectric ionizations and also by collisions of the second kind.*

We shall now assume that outside the absorption lines $\kappa_\nu = \kappa =$ constant. We shall also assume that the mean scattering coefficient a_ν in the different absorption lines and the probability coefficients ϵ_ν are all the same and are respectively a and ϵ (say).

Now integrate the equation of transfer (14) over the continuous spectrum only. By (13) then

$$\cos\theta \frac{dI_1}{\rho dx} = -\kappa(I_1 - a_1 B), \tag{17}$$

where I_1 is the intensity of the integrated radiation in the continuous spectrum and B now is the intensity of the integrated black-body radiation at x. If I_2 is the intensity of the integrated radiation in the different absorption lines, then we have

$$\cos\theta \frac{dI_2}{\rho dx} = -(\kappa + a)I_2 - (1 - \epsilon)a\mathfrak{J}_2 + a_2(\kappa + \epsilon a)B. \tag{18}$$

Notice the appearance of the probability factors a_1 and a_2 in front of B in the equations (17) and (18). ($\mathfrak{J}_2$ is defined below.)

It may be noticed here that *if the absorption lines are formed in "pure" local thermodynamic equilibrium, then* $\epsilon = 1$. On the other hand, *if the absorption lines are formed as a pure scattering phenomenon, then* $\epsilon = 0$.

7. We shall introduce the following quantities :—

$$\mathfrak{J}_1 = \tfrac{1}{2}\int_0^\pi I_1 \sin\theta d\theta\,; \qquad \mathfrak{J}_2 = \tfrac{1}{2}\int_0^\pi I_2 \sin\theta d\theta,$$

$$F_1 = 2\int_0^\pi I_1 \sin\theta \cos\theta d\theta\,; \qquad F_2 = 2\int_0^\pi I_2 \sin\theta \cos\theta d\theta,$$

$$K_1 = \tfrac{1}{2}\int_0^\pi I_1 \sin\theta \cos^2\theta d\theta\,; \qquad K_2 = 2\int_0^\pi I_2 \sin\theta \cos^2\theta d\theta.$$

On multiplying the equations of transfer successively by $\frac{1}{2}\sin\theta d\theta$ and $\frac{1}{2}\sin\theta\cos\theta d\theta$ and integrating from 0 to π we find that

$$\tfrac{1}{4}\frac{dF_1}{\rho dx} = -\kappa(\mathfrak{J}_1 - a_1 B), \tag{19}$$

$$\tfrac{1}{4}\frac{dF_2}{\rho dx} = -(\kappa + \epsilon a)(\mathfrak{J}_2 - a_2 B), \tag{19'}$$

and

$$\frac{dK_1}{\rho dx} = -\tfrac{1}{4}\kappa F_1\,; \qquad \frac{dK_2}{\rho dx} = -\tfrac{1}{4}(\kappa + a)F_2. \tag{20}$$

---

* For a precise derivation of (15) see an important paper by B. Strömgren, *Zeits. für Astrophysik*, **10**, 237, 1935.

8. *The Equation of Radiative Equilibrium and the Constancy of Net Flux.* —The equation of radiative equilibrium is obtained by equating the total absorption and the total emission in an element of volume. By an application of Kirchoff's law this is easily to be

$$\kappa \mathfrak{J}_1 + (\kappa + \epsilon\alpha)\mathfrak{J}_2 = (a_1\kappa + a_2(\kappa + \epsilon\alpha))B. \tag{21}$$

Adding equations (19) and (19′) and using (21) we have

$$\frac{dF_1}{dx} + \frac{dF_2}{dx} = 0,$$

or

$$F_1 + F_2 = \text{constant} = \mathfrak{F} \text{ (say)}, \tag{22}$$

—*i.e.* the integrated total net flux is constant.

9. We shall now assume that

$$\frac{\alpha}{\kappa} = \text{constant} = \xi \text{ (say)}. \tag{23}$$

Introducing now the optical depth τ defined by

$$d\tau = -\kappa\rho dx, \tag{24}$$

the equations (19) and (20) now reduce to

$$\tfrac{1}{4}\frac{dF_1}{d\tau} = \mathfrak{J}_1 - a_1B; \qquad \tfrac{1}{4}\frac{dF_2}{d\tau} = \lambda(\mathfrak{J}_2 - a_2B) \tag{25}$$

and

$$\frac{dK_1}{d\tau} = \tfrac{1}{4}F_1; \qquad \frac{dK_2}{d\tau} = \tfrac{1}{4}\mu F_2, \tag{26}$$

where λ and μ are defined as

$$\lambda = 1 + \epsilon\xi; \qquad \mu = 1 + \xi. \tag{27}^*$$

From the equations (26) we now have

$$\frac{dK_1}{d\tau} + \frac{1}{\mu}\frac{dK_2}{d\tau} = \tfrac{1}{4}(F_1 + F_2) = \tfrac{1}{4}\mathfrak{F} \tag{28}$$

or

$$K_1 + \frac{1}{\mu}K_2 = \tfrac{1}{4}\mathfrak{F}\tau + c, \tag{29}$$

where c is an integration constant. We shall refer to (29) as the *K-integral.*

10. We introduce now the usual approximation

$$\mathfrak{J} = 3K. \tag{30}$$

Then from the equation of radiative equilibrium

$$B = \frac{3}{a_1 + a_2\lambda}(K_1 + \lambda K_2), \tag{31}$$

* In "pure" local thermodynamical equilibrium $\epsilon = 1$ and $\lambda = \mu$. For the case where absorption lines arise as a pure scattering phenomenon $\epsilon = 0$ and $\lambda = 1$.

or using the K-integral (we have on eliminating K_1)

$$B = \frac{3}{a_1 + a_2\lambda}\left[(\tfrac{1}{4}\mathfrak{F}\tau + c) + \left(\lambda - \frac{1}{\mu}\right)K_2\right], \tag{32}$$

or alternatively eliminating K_2,

$$B = \frac{3}{a_1 + a_2\lambda}[\lambda\mu(\tfrac{1}{4}\mathfrak{F}\tau + c) - (\lambda\mu - 1)K_1]. \tag{33}$$

From (25) we now have

$$\frac{d^2K_1}{d\tau^2} = 3K_1 - a_1 B$$

or using (33)

$$\frac{d^2K_1}{d\tau^2} = \frac{3\lambda(a_1\mu + a_2)}{a_1 + a_2\lambda}K_1 - \frac{3a_1\lambda\mu}{a_1 + a_2\lambda}(\tfrac{1}{4}\mathfrak{F}\tau + c). \tag{34}$$

Similarly

$$\frac{d^2K_2}{d\tau^2} = \frac{3\lambda(a_1\mu + a_2)}{a_1 + a_2\lambda}K_2 - \frac{3a_2\lambda\mu}{a_1 + a_2\lambda}(\tfrac{1}{4}\mathfrak{F}\tau + c). \tag{35}$$

The solutions of (34) and (35) are seen to be

$$K_1 = ae^{-q\tau} + \frac{a_1\mu}{a_1\mu + a_2}(\tfrac{1}{4}\mathfrak{F}\tau + c), \tag{36}$$

$$K_2 = be^{-q\tau} + \frac{a_2\mu}{a_1\mu + a_2}(\tfrac{1}{4}\mathfrak{F}\tau + c), \tag{37}$$

where a and b are two integration constants and

$$q^2 = \frac{3\lambda(a_1\mu + a_2)}{a_1 + a_2\lambda}. \tag{38}$$

In writing (36) and (37) we have neglected a part of the solution which increases exponentially with τ. To satisfy the K-integral we must have

$$b = -a\mu. \tag{39}$$

Hence finally

$$K_1 = \frac{a_1\mu}{a_1\mu + a_2}(\tfrac{1}{4}\mathfrak{F}\tau + c) + ae^{-q\tau}, \tag{40}$$

$$K_2 = \frac{a_2\mu}{a_1\mu + a_2}(\tfrac{1}{4}\mathfrak{F}\tau + c) - a\mu e^{-q\tau}, \tag{41}$$

and

$$B = \frac{3\mu}{a_1\mu + a_2}(\tfrac{1}{4}\mathfrak{F}\tau + c) - \frac{3a(\lambda\mu - 1)}{a_1 + a_2\lambda}e^{-q\tau}. \tag{42}$$

11. From the equations (26) and (40) and (41) we obtain

$$F_1 = \frac{a_1\mu}{a_1\mu + a_2}\mathfrak{F} - 4aqe^{-q\tau}, \tag{43}$$

and
$$F_2 = \frac{a_2}{a_1\mu + a_2}\mathfrak{F} + 4aqe^{-q\tau}. \tag{44}$$

From (43) and (44) we see that F_1 and F_2 tend to constant limiting values as $\tau \to \infty$. Thus
$$\frac{F_1}{F_2} \to \frac{a_1}{a_2}\mu, \qquad (\tau \to \infty) \tag{45}$$

which is what we should have expected.

12. It now remains to determine the two constants of integrations a and c. The boundary conditions one adopts is to set
$$F(0) = 6K(0). \tag{46}$$

From (40), (41), (43) and (44) we have then
$$(6 + 4q)a + \frac{6a_1\mu}{a_1\mu + a_2}c - \frac{a_1\mu}{a_1\mu + a_2}\mathfrak{F} = 0, \tag{47}$$

$$(6\mu + 4q)a - \frac{6a_2\mu}{a_1\mu + a_2}c + \frac{a_2}{a_1\mu + a_2}\mathfrak{F} = 0. \tag{48}$$

Subtracting (48) from (47) we have the useful relation
$$\frac{\mu}{a_1\mu + a_2}c - (\mu - 1)a = \tfrac{1}{6}\mathfrak{F}. \tag{49}$$

Solving the equations (47) and (48) for a and c we find that
$$a = \frac{a_1 a_2(\mu - 1)}{(a_1\mu + a_2)[6(a_1\mu + a_2) + 4q]}\mathfrak{F}, \tag{50}$$

$$c = \frac{6(a_1\mu^2 + a_2) + 4q(a_1\mu + a_2)}{6\mu[6(a_1\mu + a_2) + 4q]}\mathfrak{F}. \tag{51}$$

Inserting the above values of a and c in (42), (43) and (44) we have finally
$$B = \frac{3\mu}{a_1\mu + a_2}\left[\tfrac{1}{4}\tau + \frac{6(a_1\mu^2 + a_2) + 4q(a_1\mu + a_2)}{6\mu(6(a_1\mu + a_2) + 4q)}\right]\mathfrak{F} - \frac{3a_1a_2(\lambda\mu - 1)(\mu - 1)}{(a_1 + a_2\lambda)(a_1\mu + a_2)(6(a_1\mu + a_2) + 4q)}\mathfrak{F}e^{-q\tau} \tag{52}$$

$$F_1 = \frac{a_1\mu}{a_1\mu + a_2}\left[1 - \frac{4a_2(1 - \mu^{-1})qe^{-q\tau}}{6(a_1\mu + a_2) + 4q}\right]\mathfrak{F}, \tag{53}$$

$$F_2 = \frac{a_2}{a_1\mu + a_2}\left[1 + \frac{4a_1(\mu - 1)qe^{-q\tau}}{6(a_1\mu + a_2) + 4q}\right]\mathfrak{F}. \tag{54}$$

13. *Discussion of the Solution.*—From (52) it now follows that as $\tau \to \infty$ we have
$$B \sim \tfrac{3}{4}\frac{\mu}{a_1\mu + a_2}\mathfrak{F}\tau. \tag{55}$$

Now the Rosseland mean absorption coefficient $\bar{\kappa}$ on our assumptions is clearly given by

$$\frac{1}{\bar{\kappa}}=\frac{a_1}{\kappa}+\frac{a_2}{\kappa+\alpha}, \tag{56}$$

or

$$\kappa=\left(a_1+\frac{a_2}{\mu}\right)\bar{\kappa}. \tag{56'}$$

Hence (55) can be rewritten as

$$B \sim \tfrac{3}{4}\mathfrak{F}\bar{\tau}, \qquad (\tau \to \infty) \tag{57}$$

where $\bar{\tau}$ is the optical depth measured in terms of the Rosseland mean absorption coefficient. (57) is an agreement with a theorem due to Milne that for material in radiative equilibrium and stratified in parallel planes, *in the far interior B increases linearly with $\bar{\tau}$ with the gradient $\tfrac{3}{4}\mathfrak{F}$.*

Again we notice that when either $a_2=0$ or $\mu=1$, (52) reduces to

$$B=\tfrac{1}{2}\mathfrak{F}(1+\tfrac{3}{2}\tau). \tag{58}$$

On the other hand, if $a_1=0$,

$$B=\tfrac{1}{2}\mathfrak{F}(1+\tfrac{3}{2}\mu\tau). \tag{59}$$

(58) and (59) are necessary for consistency. Now, on the other hand, if $\mu \to \infty$ we have

$$B=\frac{1}{2a_1}F_1(1+\tfrac{3}{2}\tau), \tag{60}$$

as in this case $F_2=0$ and $F_1=\mathfrak{F}$.*

14. *The Boundary Temperature.*—From (52) we now find that at $\tau=0$

$$B(0)=\tfrac{1}{2}\mathfrak{F}\cdot\frac{3(a_1\mu+a_2\lambda)+2q(a_1+a_2\lambda)}{(a_1+a_2\lambda)[3(a_1\mu+a_2)+2q]}. \tag{61}$$

In "pure" thermodynamic equilibrium $\epsilon=1$ and $\lambda=\mu$, in which case (61) can be expressed alternatively as

$$B(0)=\tfrac{1}{2}\mathfrak{F}\frac{2q(a_1+a_2\lambda)+3\lambda}{2q(a_1+a_2\lambda)+3\lambda+3a_1a_2(\lambda-1)^2}. \tag{62}$$

(62) shows that the boundary temperature is *less* than what it would be if there were no absorption lines. This result apparently contradicts the expectation that the effect of the formation of the absorption lines is to keep the photosphere warmer. There is, however, no real contradiction here. Our earlier result (57) shows that in the interior the temperature

* Woolley, in his paper already referred to, considers precisely the limiting case $\mu \to \infty$. But the solution he uses is seen to be

$$B=\tfrac{1}{2}F_1\left(1+\frac{3}{2a_1}\tau\right), \tag{60'}$$

which is in contradiction with our (60). One easily convinces oneself that "a_1" should occur as it does in (60) and not as in the above equation. This error in his formula, which appears to arise from the method of approximation adopted by Woolley, rather detracts from the value of his numerical work.

does increase with a larger gradient. But near the boundary it becomes relatively cooler. We shall come back to this point later.

If the absorption lines are formed as a pure scattering phenomenon, then $\epsilon = 0$ and $\lambda = 1$, and (61) now shows that

$$B(0) \equiv \tfrac{1}{2}\mathfrak{F} \tag{63}$$

—*i.e.* the formation of the absorption lines in this case has no effect on the boundary temperature.

15. *The Law of Darkening. Case* (*a*) $\lambda = \mu$.—We shall first consider the case of pure local thermodynamical equilibrium, *i.e.* $\lambda = \mu$.

Let $I_1(\theta)$ and $I_2(\theta)$ be the intensities of the integrated emergent radiations in a direction making an angle θ with the normal in the continuous spectrum and in the absorption lines respectively. Then clearly

$$I_1(\theta) = a_1 \int_0^\infty B(\tau \cos\theta) e^{-\tau} d\tau, \tag{64}$$

$$I_2(\theta) = a_2 \int_0^\infty B\left(\frac{\tau}{\mu} \cos\theta\right) e^{-\tau} d\tau. \tag{65}$$

Using the expression for B given in equation (42) we have

$$I_1(\theta) = \frac{3a_1\mu}{a_1\mu + a_2}(\tfrac{1}{4}\mathfrak{F}\cos\theta + c) - \frac{3a_1(\mu^2 - 1)}{(a_1 + a_2\mu)(1 + q\cos\theta)}a, \tag{66}$$

$$I_2(\theta) = \frac{3a_2\mu}{a_1\mu + a_2}\left(\frac{1}{4\mu}\mathfrak{F}\cos\theta + c\right) - \frac{3a_2(\mu^2 - 1)\mu}{(a_1 + a_2\mu)(\mu + q\cos\theta)}a. \tag{67}$$

Hence for the law of darkening in the general integrated radiation we have

$$I(\theta) = I_1(\theta) + I_2(\theta)$$

or

$$I(\theta) = \tfrac{3}{4}\mathfrak{F}\cos\theta + \frac{3\mu}{a_1\mu + a_2}c - \frac{3a(\mu^2 - 1)(\mu + (a_1 + a_2\mu)q\cos\theta)}{(a_1 + a_2\mu)(1 + q\cos\theta)(\mu + q\cos\theta)}. \tag{68}$$

Substituting in the above our expressions for a and c we have finally

$$I(\theta) = \mathfrak{F}\left\{\tfrac{3}{4}\cos\theta + \frac{3(a_1\mu^2 + a_2) + 2q(a_1\mu + a_2)}{2(a_1\mu + a_2)[3(a_1\mu + a_2) + 2q]} - \frac{3a_1a_2(\mu^2 - 1)(\mu - 1)[\mu + q(a_1 + a_2\mu)\cos\theta]}{2(a_1\mu + a_2)(a_1 + a_2\mu)[3(a_1\mu + a_2) + 2q](1 + q\cos\theta)(\mu + q\cos\theta)}\right\}. \tag{69}$$

The above expression reduces to the standard darkening law

$$I(\theta) = \tfrac{1}{2}\mathfrak{F}(1 + \tfrac{3}{2}\cos\theta) \tag{70}$$

for the three cases (1) $a_1 = 0$, (2) $a_2 = 0$, (3) $\mu = \infty$.

We thus see that for the models under discussion (λ and μ constants) the effect of the formation of the absorption lines is to *increase* the centre-limb contrast, and the minimum contrast corresponds to a coefficient of darkening $u=0{\cdot}6$. However, we shall show in Section III that our present analysis can be generalised in such a way that a reduced centre limb results.

[It may be noticed here that the equations (66) and (67) permit of a determination of the integration constants a and c by what is probably a more consistent procedure than the one we have adopted in § 12. For (66) and (67) lead to definite expressions for the emergent fluxes $\pi F_1(0)$ and $\pi F_2(0)$. Thus it is found that

$$F_1(0)=\frac{3a_1\mu}{a_1\mu+a_2}(\tfrac{1}{6}\mathfrak{F}+c)-\frac{6a_1(\mu^2-1)(q-\log(1+q))}{(a_1+a_2\mu)q^2}a, \tag{71}$$

$$F_2(0)=\frac{3a_2\mu}{a_1\mu+a_2}\left(\frac{1}{6\mu}\mathfrak{F}+c\right)-\frac{6a_2\mu(\mu^2-1)\left(q-\mu\log\left(1+\frac{q}{\mu}\right)\right)}{(a_1+a_2\mu)q^2}a. \tag{72}$$

Comparing the above with our expressions (43) and (44) we can determine a and c. The expressions so obtained are not so convenient to handle as our earlier ones, and, as such, for a general orientation the cruder method we have adopted is sufficient.]

16. *The Law of Darkening. The General Case.*—Let $I_1(\theta)$ and $I_2(\theta)$ have the same meanings as in § 15. Our expressions for $I_1(\theta)$ and $I_2(\theta)$ are

$$I_1(\theta)=a_1\int_0^\infty B(\tau\cos\theta)e^{-\tau}d\tau, \tag{73}$$

$$I_2(\theta)=\frac{3(1-\epsilon)(\mu-1)}{\mu}\int_0^\infty K_2\left(\frac{\tau}{\mu}\cos\theta\right)e^{-\tau}d\tau+a_2\frac{\lambda}{\mu}\int_0^\infty B\left(\frac{\tau}{\mu}\cos\theta\right)e^{-\tau}d\tau. \tag{74}$$

Using equations (40), (41) and (42) we have

$$I_1(\theta)=\frac{3a_1\mu}{a_1\mu+a_2}(\tfrac{1}{4}\mathfrak{F}\cos\theta+c)-\frac{3a_1(\lambda\mu-1)a}{(a_1+a_2\lambda)(1+q\cos\theta)}, \tag{75}$$

$$I_2(\theta)=\frac{3a_2\mu}{a_1\mu+a_2}\left(\frac{1}{4\mu}\mathfrak{F}\cos\theta+c\right)-3\left[(1-\epsilon)(\mu-1)+a_2\cdot\frac{\lambda(\lambda\mu-1)}{\mu(a_1+a_2\lambda)}\right]\frac{\mu}{\mu+q\cos\theta}a. \tag{76}$$

For the case $\lambda=\mu$, $\epsilon=1$, (75) and (76) reduce to our earlier expressions (66) and (67). When, however, the absorption lines arise as a pure scattering phenomenon, then $\epsilon=0$ and $\lambda=1$ and we have

$$I_1(\theta)=\frac{3a_1\mu}{a_1\mu+a_2}(\tfrac{1}{4}\mathfrak{F}\cos\theta+c)-\frac{3a_1(\mu-1)a}{1+q\cos\theta}, \tag{77}$$

$$I_2(\theta)=\frac{3a_2\mu}{a_1\mu+a_2}\left(\frac{1}{4\mu}\mathfrak{F}\cos\theta+c\right)-\frac{3a_2(\mu-1)(\mu+a_2)a}{\mu+q\cos\theta}. \tag{78}$$

17. *The Intensity Distribution in the Continuous Spectrum.*—To determine the intensity of the emergent radiation in a wave-length λ in the continuous spectrum at the centre of the disc we have to evaluate the integral

$$I(\lambda)=\int_0^\infty B(\lambda, T_\tau)e^{-\tau}d\tau, \tag{79}$$

where T_τ is the temperature at optical depth τ and $B(\lambda, T)$ the intensity of the black-body radiation in wave-length λ for a temperature T. The variation of T with τ is, of course, given by our solution for B. The rigorous treatment of the integral (79) is too tedious, and is perhaps not worth while considering the very crude nature of the model on which we are working. A quicker method is to adopt the following procedure first used by Eddington: * "Divide the range $e^{-\tau}$ into a large number of equal parts. Calculate the temperatures T_1, T_2, T_3, . . . at the middle of each part. Take a cu. cm. of equilibrium radiation at each of these temperatures; the simple mean gives the constitution of the emergent radiation." The above method was adopted to calculate the colour temperatures for certain special examples worked out in § 21. However, the above rough method of treating the integral was adopted only to get a general orientation to the consequences of the blanketing effect of the reversing layer. For more accurate information analytical methods should be adopted, and the equations of transfer themselves will have to be solved with greater accuracy.

We pass on now to certain numerical applications of the formulæ we have derived.

Section II

18. *The Absorption Lines formed in Pure Local Thermodynamic Equilibrium.*—To make concrete applications we need to know the value of the parameters a_1 and $\lambda(=\mu)$.

Example 1.—As a first example we shall consider the following case:—

$$a_1 = 0{\cdot}8, \quad a_2 = 0{\cdot}2, \quad \lambda = 5.$$

One finds that

$$a = 0{\cdot}003118, \quad c = 0{\cdot}1505, \quad q = 5{\cdot}916.$$

The numerical solution for B is found to be

$$B = \mathfrak{F}(0{\cdot}5374 + 0{\cdot}8929\tau - 0{\cdot}1247e^{-5{\cdot}916\tau}).$$

The relation between the boundary and effective temperatures is found to be for this case:

$$T_0^4 = 0{\cdot}4127 T_e^4 \quad \text{or} \quad T_0 = 0{\cdot}8015 T_e.$$

Values of B for different values of τ are tabulated in Table I (2 – column) (see also fig. 1). The expression for the fluxes are

$$F_1 = \mathfrak{F}(0{\cdot}9524 - 0{\cdot}07380e^{-5{\cdot}916\tau}),$$
$$F_2 = \mathfrak{F}(0{\cdot}0476 + 0{\cdot}07380e^{-5{\cdot}916\tau}).$$

The ratio of the emergent fluxes in the continuous spectrum and in the absorption lines is given by

$$\frac{F_1(0)}{F_2(0)} = 7{\cdot}24.$$

---

* A. S. Eddington, *Internal Constitution of the Stars* (Cambridge), p. 326.

For the law of darkening we have

$$I_1(\theta) = I_1(0)\left(0.3805 + 0.6322 \cos\theta - \frac{0.0883}{1 + 5.9161 \cos\theta}\right),$$

$$I(\theta) = \mathfrak{F}\left(0.5374 + 0.75 \cos\theta - \frac{0.1247(1 + 2.130 \cos\theta)}{(1 + 5.916 \cos\theta)(1 + 2.958 \cos\theta)}\right).$$

Calculations based on the above formulæ are tabulated in Table II. One notices that the centre-limb contrast is greater than for the standard darkening.

Example 2.—As a second example we shall consider the case

$$a_1 = 0.8, \quad a_2 = 0.2, \quad \lambda = 10.$$

The numerical solutions are

$$B = \mathfrak{F}(0.5547 + 0.9146\tau - 0.2149e^{-9.373\tau}),$$
$$F_1 = \mathfrak{F}(0.9756 - 0.07595e^{-9.373\tau}),$$
$$F_2 = \mathfrak{F}(0.0244 + 0.07595e^{-9.373\tau}),$$

$$B(0) = 0.3398\mathfrak{F}; \quad \frac{F_1(0)}{F_2(0)} = 8.97,$$

$$I_1(\theta) = I_1(0)\left(0.3829 + 0.6314 \cos\theta - \frac{0.1483}{1 + 9.373 \cos\theta}\right),$$

$$I(\theta) = \mathfrak{F}\left(0.5547 + 0.75 \cos\theta - \frac{0.2149(1 + 2.624 \cos\theta)}{(1 + 9.373 \cos\theta)(1 + 0.937 \cos\theta)}\right).$$

19. *Absorption Lines formed as Pure Scattering Phenomenon.*—To see the effect of the "ϵ" factor on the law of darkening the extreme case $\lambda = 1$ was considered for the examples considered in § 18.

Example 3.—

$$a_1 = 0.8, \quad a_2 = 0.2, \quad \lambda = 1, \quad \mu = 5.$$

The results of the calculations are

$$B = \mathfrak{F}(0.5464 + 0.8929\tau - 0.0464e^{-3.55\tau}),$$
$$F_1 = \mathfrak{F}(0.9524 - 0.0549e^{-3.55\tau}),$$
$$F_2 = \mathfrak{F}(0.0476 + 0.0549e^{-3.55\tau}),$$

$$B(0) = 0.5\mathfrak{F}; \quad \frac{F_1(0)}{F_2(0)} = 8.75,$$

$$I_1(\theta) = I_1(0)\left(0.3824 + 0.6248 \cos\theta - \frac{0.03249}{1 + 3.55 \cos\theta}\right).$$

Example 4.—

$$a_1 = 0.8, \quad a_2 = 0.2, \quad \lambda = 1, \quad \mu = 10.$$

$$B = \mathfrak{F}(0.5687 + 0.9146\tau - 0.0687e^{-4.96\tau}),$$
$$F_1 = \mathfrak{F}(0.9756 - 0.0505e^{-4.96\tau}),$$
$$F_2 = \mathfrak{F}(0.0244 + 0.0505e^{-4.96\tau}),$$

$$B(0) = 0.5\mathfrak{F}; \quad \frac{F_1(0)}{F_2(0)} = 12.4,$$

$$I_1(\theta) = I(0)\left(0.3864 + 0.6214 \cos\theta - \frac{0.0467}{1 + 4.96 \cos\theta}\right).$$

20. *The Discussion of the Results.*—The values of B for different values of τ for the four examples considered in §§ 18 and 19 are tabulated in Table I and the variations are graphically illustrated in fig. 1. The results for darkening are contained in Tables II and III.

An examination of Table I and fig. 1 shows that for a fixed a_1 (and a_2) as $\mu \to \infty$ the appropriate limiting solution is

$$B = \frac{1}{2a_1}\mathfrak{F}(1 + \tfrac{3}{2}\tau), \tag{80}$$

Table I

$B(\tau)$ in Units of $\mathfrak{F}$ ($a_1 = 0{\cdot}8$, $a_2 = 0{\cdot}2$).

τ	$\lambda = \mu = 5$	$\lambda = \mu = 10$	$\lambda = 1$; $\mu = 5$	$\lambda = 1$; $\mu = 5$
0	0·4127	0·3398	0·5	0·5
·02	0·4445	0·3949	0·5210	0·5248
·05	0·4893	0·4660	0·5522	0·5608
·1	0·5577	0·5620	0·6032	0·6183
·2	0·6778	0·7040	0·7022	0·7261
·3	0·7841	0·8162	0·7983	0·8276
·4	0·8829	0·9155	0·8924	0·9251
·5	0·9774	1·0100	0·9850	1·0203
·6	1·0696	1·1027	1·0766	1·1140
·8	1·2506	1·2863	1·2580	1·2991
1·0	1·4299	1·4693	1·4380	1·4828
1·2	1·609	1·652	1·617	1·666
1·4	1·787	1·835	1·796	1·849
1·6	1·966	2·018	1·975	2·032
1·8	2·145	2·201	2·153	2·215
2·0	2·323	2·384	2·332	2·398
3·0	3·216	3·299	3·225	3·313
4·0	4·109	4·213	4·118	4·227

and a certain interval on the B-axis. In the case of the absorption lines arising as a pure scattering phenomenon this interval is

$$< \frac{1}{2a_1}\mathfrak{F}, \ \tfrac{1}{2}\mathfrak{F} >, \tag{81}$$

and in the case of pure local thermodynamic equilibrium the interval is

$$< \frac{1}{2a_1}\mathfrak{F}, \ 0 >. \tag{82}$$

The apparent contradiction between (80), which predicts a boundary temperature T_0 related to the effective temperature T_e by the relation

$$T_0^4 = \frac{1}{2a_1}T_e^4, \tag{83}$$

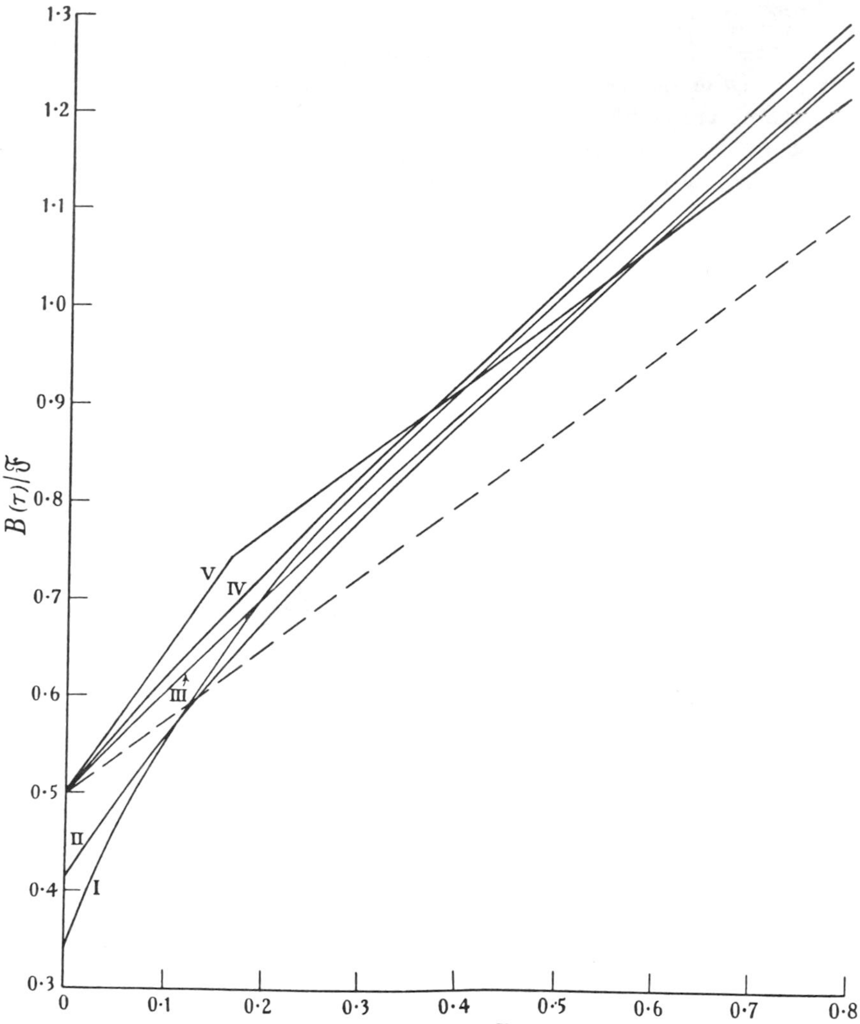

FIG. 1.—*Variation of $B(\tau)$ with τ including the blanketing effect of the reversing layer.*

I $\lambda = \mu = 10$, $a_2 = 0{\cdot}2$.
II $\lambda = \mu = 5$, $a_2 = 0{\cdot}2$.
III $\lambda = 1$, $\mu = 5$, $a_2 = 0{\cdot}2$.
IV $\lambda = 1$, $\mu = 10$, $a_2 = 0{\cdot}2$.
V $\lambda = 1$, $\mu = 10$, $\tau < 1/6$; $\lambda = \mu = 1$, $\tau \geqslant 1/6$; $a_1 = 0{\cdot}8$, $a_2 = 0{\cdot}2$.
The - - - Curve is the "standard law" : $B(\tau) = \mathfrak{F}(1 + \frac{3}{2}\tau)$.

and our earlier results in § 14 (where we saw that as $\mu \to \infty$ the boundary temperature is zero for the case $\lambda = \mu$, and $T_0 = (2)^{-4} T_e$ for the case $\lambda = 1$) is now removed.

Fig. 1 also explains why the blanketing effect of the reversing layer is to produce a larger centre-limb contrast than the standard darkening, $u = 0{\cdot}6$. (It may be pointed out explicitly here that the law of darkening *is* the standard one both for $\mu = 1$ and $\mu = \infty$.) The reason is, in the immediate

TABLE II

Law of Darkening in Pure Local Thermodynamic Equilibrium

Cos θ	$a_1=0{\cdot}8$; $a_2=0{\cdot}2$; $\lambda=\mu=5$		$a_1=0{\cdot}8$; $a_2=0{\cdot}2$; $\lambda=\mu=10$		Standard darkening $u=3/5$
	$\frac{I_1(\theta)}{I_1(0)}$	$\frac{I_1(\theta)+I_2(\theta)}{I_1(0)+I_2(0)}$	$\frac{I_1(\theta)}{I_1(0)}$	$\frac{I_1(\theta)+I_2(\theta)}{I_1(0)+I_2(0)}$	
1	1	1	1	1	1
0·9	0·9356	0·9400	0·9354	0·9386	0·94
0·8	0·8709	0·8797	0·8706	0·8768	0·88
0·6	0·7405	0·7578	0·7394	0·7514	0·76
0·4	0·6072	0·6330	0·6042	0·6163	0·64
0·2	0·4665	0·4997	0·4576	0·4808	0·52
0·1	0·3883	0·4234	0·3695	0·3963	0·44
0	0·2922	0·3241	0·2346	0·2684	0·40

TABLE III

Law of Darkening (the absorption lines formed as a pure scattering phenomenon)
($a_1=0{\cdot}8$; $a_2=0{\cdot}2$; $\lambda=1$)

Cos θ	$I_1(\theta)/I_1(0)$ ($\mu=5$)	$I_1(\theta)/I_1(0)$ ($\mu=10$)
1	1	1
0·9	0·9369	0·9372
0·8	0·8737	0·8742
0·6	0·7468	0·7475
0·4	0·6188	0·6193
0·2	0·4883	0·4873
0·1	0·4209	0·4173
0	0·3499	0·3397

outer layers the deviations from a linear increase of B with τ become pronounced, and it is precisely these regions which contribute most to the emergent radiation near the limb. One further notices that for equal values of a_1, a_2 and μ the photosphere is maintained uniformly at a higher temperature in the case where the absorption lines arise as a purely scattering phenomenon than in the case of pure local thermodynamical equilibrium.

21. *Applications to the Colour Temperature of the Sun.*—Assuming Abbot's value for the effective temperature of the Sun, namely, $T_e=5740°$, the mean intensities of the emergent radiation (at the centre of the disc) in the wave-lengths 4157 A., 6235 A. and 12470 A. for the four examples considered in §§ 18 and 19 were determined by the method sketched in § 17. The results of the calculation are summarised in Table IV. The corresponding

TABLE IV

The Effect of the Reversing Layer on the Colour Temperature

Description	Mean Intensity			$\frac{\text{Intensity}_{6235}}{\text{Intensity}_{4157}}$	Colour Temperature
	4157 A.	6235 A.	12470 A.		
Pure local thermodynamic equilibrium :					
(1) $\lambda=\mu=5$, $a_2=0{\cdot}2$	0·00408	0·02463	0·1783	6·04	6475°
(2) $\lambda=\mu=10$, $a_2=0{\cdot}2$	0·00422	0·02522	0·1807	5·98	6500°
Absorption lines formed as a pure scattering phenomenon :					
(1) $\lambda=1, \mu=5, a_2=0{\cdot}2$	0·00413	0·02496	0·1801	6·04	6475°
(2) $\lambda=1, \mu=10, a_2=0{\cdot}2$	0·00429	0·02564	0·1830	5·97	6510°
No Blanketing Effect ($a_2=0$)	0·00344	0·0221	0·168	6·43	6220°

colour temperatures (defined by the ratios of the intensities at the wave-lengths 4157 A. and 6235 A.) are given in the last column. One sees that the blanketing effect, as we have considered so far, is almost sufficient to account for Plaskett's high value for the true colour temperature of the Sun. The results in Table IV show that the colour temperature is not particularly sensitive to the "model" or to the particular values of λ and μ, provided μ is greater than about 5. It is the value of a_2 that is important. The particular value $a_2=0{\cdot}2$ adopted in our calculations is not unreasonable, especially when it is realised that a value of $\eta=1/9$ in the Milne-Lindblad treatment predicts a boundary temperature identical with that predicted by (82)—(case $\mu=\infty$)—for $a_1=0{\cdot}8$, $a_2=0{\cdot}2$.

Section III

22. In Section II we saw that on the models there discussed the theory predicts a law of darkening with a higher centre-limb contrast than the standard darkening, thus making the general agreement between the theory and the observations worse in this respect. But on the Milne-Lindblad treatment the centre-limb contrast is less than that on the standard darkening, and this suggests how we should generalise our earlier analysis in order that the theory may predict a law of darkening more in agreement with the observations.

Now in an actual stellar atmosphere both $\xi(=\alpha/\kappa)$ and ϵ will not be constants. It is, of course, clear that in the deeper layers ϵ will tend to unity (pure local thermodynamical equilibrium), while in the reversing layer ϵ will be more near zero. Further, the ratio α/κ will decrease inwards

(κ predominates in the deeper layers). Thus a better approximation to the outer layers of a star is obtained by dividing the atmosphere into two parts, ξ and ϵ taking different values in the outer and the deeper layers respectively, but constant in each of the regions separately. The analysis required for this more general problem does not present any new difficulty.

23. Let ξ and ϵ take the values ξ_1, ϵ_1 in the outer layers and the values ξ_2, ϵ_2 in the deeper layers. Let the surface of demarcation between the (ξ_1, ϵ_1) and the (ξ_2, ϵ_2) regions occur at $\tau = \tau_0$.* As before we shall now define the quantities λ_1, λ_2, μ_1, μ_2, as follows :—

$$\left.\begin{aligned}\lambda_1 &= 1 + \epsilon_1\xi_1\,; \quad \mu_1 = 1 + \xi_1,\\ \lambda_2 &= 1 + \epsilon_2\xi_2\,; \quad \mu_2 = 1 + \xi_2.\end{aligned}\right\} \tag{84}$$

Further, let q_1 and q_2 be defined in terms of (λ_1, μ_1) and (λ_2, μ_2) respectively exactly as in equation (38). Then by an analysis similar to that in §§ 9 and 10 we find that for $\tau \leqslant \tau_0$

$$\left.\begin{aligned}K_1 &= ae^{-q_1\tau} + be^{q_1\tau} + \frac{a_1\mu_1}{a_1\mu_1 + a_2}(\tfrac{1}{4}\mathfrak{F}\tau + c_1),\\ K_2 &= -\mu_1 ae^{-q_1\tau} - \mu_1 be^{q_1\tau} + \frac{a_2\mu_1}{a_1\mu_1 + a_2}(\tfrac{1}{4}\mathfrak{F}\tau + c_1),\\ F_1 &= \frac{a_1\mu_1}{a_1\mu_1 + a_2}\mathfrak{F} - 4aq_1e^{-q_1\tau} + 4bq_1e^{q_1\tau},\\ F_2 &= \frac{a_2}{a_1\mu_1 + a_2}\mathfrak{F} + 4aq_1e^{-q_1\tau} - 4bq_1e^{q_1\tau},\\ B &= \frac{3}{a_1 + a_2\lambda_1}(K_1 + \lambda_1 K_2).\end{aligned}\right\} \tag{85}$$

In the above equations a, b and c_1 are three integration constants. Similarly for $\tau \geqslant \tau_0$ we have

$$\left.\begin{aligned}K_1 &= Ae^{-q_2\tau} + \frac{a_1\mu_2}{a_1\mu_2 + a_2}(\tfrac{1}{4}\mathfrak{F}\tau + c_2),\\ K_2 &= -\mu_2 Ae^{-q_2\tau} + \frac{a_2\mu_2}{a_1\mu_2 + a_2}(\tfrac{1}{4}\mathfrak{F}\tau + c_2),\\ F_1 &= \frac{a_1\mu_2}{a_1\mu_2 + a_2}\mathfrak{F} - 4Aq_2e^{-q_2\tau},\\ F_2 &= \frac{a_2}{a_1\mu_2 + a_2}\mathfrak{F} + 4Aq_2e^{-q_2\tau},\\ B &= \frac{3}{a_1 + a_2\lambda_2}(K_1 + \lambda_2 K_2),\end{aligned}\right\} \tag{86}$$

where A and c_2 are two other integration constants.

At $\tau = 0$ we set as usual

$$6K_1(0) = F_1(0)\,; \qquad 6K_2(0) = F_2(0). \tag{87}$$

* We shall show later that the continuity of temperature at $\tau = \tau_0$ requires that $\xi_1 . \epsilon_1 = \xi_2 . \epsilon_2$ (*cf.* equation (88)).

Further, at $\tau=\tau_0$ the two sets of solutions must give the same values for K_1, K_2, F_1, F_2 and B. The continuity of K_1, K_2, F_1 and F_2 at $\tau=\tau_0$ provide *three* further conditions,* which together with the two equations provided by (87) will determine the five constants a, b, c_1, c_2 and A. On the other hand, the continuity of B requires that $\lambda_1=\lambda_2$ or

$$\epsilon_1\xi_1=\epsilon_2\xi_2. \tag{88}$$

(88) can be regarded as a statement of Kirchoff's law for our problem.

The equations which determine the integration constants can be reduced to the following set :—

$$(6+4q_1)a+(6-4q_1)b+\frac{6a_1\mu_1}{a_1\mu_1+a_2}c_1-\frac{a_1\mu_1}{a_1\mu_1+a_2}\mathfrak{F}=0,$$

$$(\mu_1-1)a+(\mu_1-1)b-\frac{\mu_1}{a_1\mu_1+a_2}c_1+\tfrac{1}{6}\mathfrak{F}=0,$$

$$(\tfrac{1}{4}\mathfrak{F}\tau_0+c_1)+e^{-q_2\tau_0}\left(\frac{\mu_2}{\mu_1}-1\right)A-\frac{\mu_2}{a_1\mu_2+a_2}\left(\frac{a_2}{\mu_1}+a_1\right)(\tfrac{1}{4}\mathfrak{F}\tau_0+c_2)=0,$$

$$\left(\frac{\mu_1}{\mu_2}-1\right)e^{-q_1\tau_0}a+\left(\frac{\mu_1}{\mu_2}-1\right)e^{q_1\tau_0}b-\frac{\mu_1}{a_1\mu_1+a_2}\left(\frac{a_2}{\mu_2}+a_1\right)(\tfrac{1}{4}\mathfrak{F}\tau_0+c_1)+(\tfrac{1}{4}\mathfrak{F}\tau_0+c_2)=0,$$

$$4q_1e^{-q_1\tau_0}a-4q_1e^{q_1\tau_0}b+\frac{a_2}{a_1\mu_1+a_2}\mathfrak{F}-4q_2e^{-q_2\tau_0}A-\frac{a_2}{a_1\mu_2+a_2}\mathfrak{F}=0.$$

The explicit solutions of the above equations are too complicated to be of any interest. It is more convenient to solve them numerically for each special case.

24. To illustrate the use of the formulæ of § 23 we shall consider the following extreme case :—

$$\left.\begin{array}{lllll}\epsilon_1=0, & \xi_1=9, & \lambda_1=1, & \mu_1=10, & (\tau\leqslant\tau_0)\\ \epsilon_2=1, & \xi_2=0, & \lambda_2=\mu_2=1, & & (\tau\geqslant\tau_0)\end{array}\right\} \tag{89}$$

—*i.e.* we are considering the case of a purely scattering atmosphere placed above material in pure local thermodynamic equilibrium and no scattering. (This case therefore corresponds to the model which underlies the cruder Milne-Lindblad treatment.) The general results are best seen from a numerical example.

Taking

$$a_1=0{\cdot}8, \qquad a_2=0{\cdot}2, \qquad \tau_0=1/6, \tag{90}$$

the numerical solutions are found to be :

$$\left.\begin{aligned}K_1&=\mathfrak{F}(0{\cdot}00310e^{-4{\cdot}96\tau}-0{\cdot}00129e^{4{\cdot}96\tau}+0{\cdot}24390\tau+0{\cdot}14631),\\ K_2&=\mathfrak{F}(-0{\cdot}03095e^{-4{\cdot}96\tau}+0{\cdot}01288e^{4{\cdot}96\tau}+0{\cdot}06095\tau+0{\cdot}03658),\\ F_1&=\mathfrak{F}(0{\cdot}9756-0{\cdot}0614e^{-4{\cdot}96\tau}-0{\cdot}0256e^{4{\cdot}96\tau}),\\ F_2&=\mathfrak{F}(0{\cdot}0244+0{\cdot}0614e^{-4{\cdot}96\tau}+0{\cdot}0256e^{4{\cdot}96\tau}),\\ B&=\mathfrak{F}(0{\cdot}5487+0{\cdot}9146\tau-0{\cdot}0836e^{-4{\cdot}96\tau}+0{\cdot}0348e^{4{\cdot}96\tau}).\end{aligned}\right\}(\tau<1/6)$$

* *Three* boundary conditions and not four, since $F_1+F_2=\mathfrak{F}$ in the two regions separately.

$$\left.\begin{aligned} K_1 &= \mathfrak{F}(-0.0174e^{-\sqrt{3}\tau}+0.2\tau+0.16508),\\ K_2 &= \mathfrak{F}(+0.0174e^{-\sqrt{3}\tau}+0.05\tau+0.04127),\\ F_1 &= \mathfrak{F}(0.8+0.1206e^{-\sqrt{3}\tau}),\\ F_2 &= \mathfrak{F}(0.2-0.1206e^{-\sqrt{3}\tau}),\\ B &= \mathfrak{F}(0.75\tau+0.61905). \end{aligned}\right\} (\tau \geqslant 1/6)$$

At $\tau = 0$

$$F_1(0) = 0.8887\mathfrak{F}; \qquad F_2(0) = 0.1113\mathfrak{F}; \qquad B(0) = \tfrac{1}{2}\mathfrak{F}.$$

At $\tau = \tau_0$ the two sets of formulæ both give the following values for the different quantities expressed in units of $\mathfrak{F}$:—

$$\begin{aligned} &F_1(\tau_0) = 0.8903; &&K_1(\tau_0) = 0.1854; &&B(\tau_0) = 0.7440,\\ &F_2(\tau_0) = 0.1097; &&K_2(\tau_0) = 0.0626. \end{aligned}$$

In fig. 1 the variation of B with τ is illustrated. To find the law of darkening we have to calculate $I_1(\theta)$. In integrating $B(\tau\cos\theta)e^{-\tau}$ we have to use the different expressions for B in the different regions. The final expression is

$$I_1(\theta) = \mathfrak{F}\left\{0.6082+0.7521\cos\theta+\frac{0.0348(e^{(4.96\cos\theta-1)/6}-1)}{4.96\cos\theta-1} - \frac{0.0836(1-e^{-(4.96\cos\theta+1)/6})}{4.96\cos\theta+1}\right\}. \quad (91)$$

Calculations based on the above formula for darkening are given in Table V and compared with a linear law with a coefficient of darkening $u = 0.56$,

TABLE V

Cos θ	$\frac{I(\theta)}{I(0)}$ according to (91)	$\frac{I(\theta)}{I(0)}$ for $u = .56$
1.0	1	1
0.9	0.9443	0.944
0.8	.8884	.888
0.6	.7768	.776
0.4	.6652	.664
0.2	.5534	.552
0.1	.4977	.496
0	.4418	.440

which value of u, according to Mulders, fits the solar observations best. By an accidental coincidence our numerical example reproduces the observed law remarkably well.

However, the prediction concerning the colour temperature is not so satisfactory. The colour temperature to be expected on the basis of our present model was also estimated by the method sketched in § 17. It is found that the resulting colour temperature is 6350°, which is rather low. The details of the calculation are given below:

Wave-length	Intensity
4137 A.	0·00386
6235 A.	0·02408
12470 A.	0·1774
$\frac{\text{Intensity at 6235 A.}}{\text{Intensity at 4137 A.}}$	6·24
T colour	6350°

The reason why our present colour temperature is much lower is of course obvious. The temperature gradient in the deeper layers is now less than on the models considered in Section II. It would, however, be possible by a more suitable adjustment of the constants λ_1, μ_1, λ_2 and μ_2 to predict a higher colour temperature than 6350° without at the same time making too great a departure from the law of darkening we have found in the above numerical example.

25. From the numerical example of § 24 it is now clear that the departures from the standard law of darkening that have been observed in the case of the Sun can be accounted for on our present analysis by considering two regions with different values for λ and μ. Clearly the formal analysis can be extended for more regions. We have solutions of the type (85) for each intermediate region, and finally a system of the type (86) for the deepest region. At each surface of demarcation there will be three boundary conditions to be satisfied, and the total number of boundary conditions provided will be just sufficient to make the problem determinate. Perhaps the best representation of the physical situation will be by a division of the outer layers into three separate regions with $\epsilon = 0{\cdot}1$ for the outermost region and $\epsilon = 1/2$ and $\epsilon = 1$ for the intermediate and the deepest regions respectively. But the calculations presented here are sufficient to illustrate the general utility of considering blanketing in the way we have done in this paper.

Summary.—In this paper an attempt has been made to include in a theory of the radiative equilibrium of the outer layers of a star the fact of the formation of the absorption lines in the reversing layer immediately above the conventional photosphere. It has been found possible to treat the problem fairly rigorously by introducing the absorption lines *uniformly* in the whole spectral range. The conclusion is reached that Plaskett's observed high value for the true colour temperature of the Sun can be accounted for. Further, it is shown that departures of the observed law of darkening from the standard darkening (with a coefficient $u = 0{\cdot}6$) can also be explained by considering a "scattering" atmosphere placed above a region of local thermodynamical equilibrium.

In conclusion, it is a pleasure to record here my best thanks to Professor H. H. Plaskett for his encouragement and advice during the progress of the work.

Trinity College, Cambridge :
1935 *November* 1.

Note added on November 29.—Mr. Greaves has kindly pointed out to the writer that the rather crude method adopted in this paper of estimating the colour temperatures is likely to lead to underestimates. Thus, in the case $a_2 = 0$ (no blanketing effect) the crude method of evaluating the integrals gives a colour temperature of 6220°, while Milne's accurate method * gave the higher value 6400°. In the same way, an accurate evaluation of the colour temperatures for the models considered in this paper might lead to colour temperatures more nearly 6700°. This higher value would definitely be in better agreement with observations. In this connection Mr. Greaves especially refers to Shajn's recent work.†

* *Handbuch der Astrophysik*, **3**/1, 153, 1930.
† *M.N.*, **94**, 7, 652, 1934.

The Radiative Equilibrium of a Planetary Nebula.

By **S. Chandrasekhar** (Cambridge).

With 6 figures. (Received November 9, 1934.)

In this paper the complete history of the ultraviolet radiation from the inner to the outer boundary a planetary nebula is traced quantitatively and the gradual transformation of the incident stellar ultraviolet radiation into that in Lyman-α is illustrated graphically. This theory forms on adequate mathematical representation of ZANSTRA's theory of nebular luminosity.

§ 1. ZANSTRA's fundamental investigations on the origin of nebular luminosity has been in recent years the basis of all the further discussions not only of nebular luminosity itself but also of the general problem of bright line emission of the Wolf - Rayet and other stars which are believed to have extensive envelopes. The general principles of ZANSTRA's theory are too well known for any special comments here but it might be recalled that it is an essential feature of his theory that it conceives of a continuous transformation of the incident stellar ultra-violet light beyond a certain limit (*eg.* Lyman limit) into radiation mainly in a particular frequency [Lyman_α ($= L_\alpha$) for instance]. A study therefore of the problem of the radiative equilibrium of a planetary nebula must incorporate in itself this feature of ZANSTRA's theory, and this was first done in an entirely satisfactory way by AMBARZUMIAN in his important work on this subject[1]).

As AMBARZUMIAN showed the best way to do this is by the following hypothesis:

After the absorption of an ultra-violet quantum there is a finite probability p of its re-emission in the same wave length and a probability $(1 - p)$ *of the re-emission of an* L_α *the* L_α *itself always sufferring only a scattering process.*

Itis easy now to write down the appropriate equations of transfer and of radiative equilibrium and solve them by the standard methods. AMBARZUMIAN in his work exclusively adopts the SCHUSTER-SCHWARZSCHILD method of approximation. But it is just as easy to obtain more accurate solutions by the MILNE-EDDINGTON[2]) type of approximation and the final results are just as, if not more, handy. In this paper we shall obtain these-

[1]) V. A. AMBARZUMIAN, Pulko. Obs. Bull. Nr. 13. Also M. N. Vol. 93, p. 50, 1931. The first paper will be referred to as l. c. — [2]) For a general discussions of the methods of approximation in such problems see the author's paper in M. N. Vol. 94, p. 444, 1934.

better approximations. The general *principles* being the same as in AMBARZUMIAN's work an apology might be needed for publishing what in effect amounts to a *formal* development in the theory. *Firstly*: In his work AMBARZUMIAN, extending some calculations of CILLIE showed that under nebular conditions $p \sim 0.5$ and he used this result exclusively in his work. In some still unpublished work of the author on the general problem of bright line emission quantities analogous to "p" appear which take all values ranging from 1 to 0 according to slight changes in the initial conditions and it was found necessary to trace *quantitatively* the complete history of the radiation from the inner to the outer boundary and to determine also the dependence of this "history" on the specified value of "p". This is done in this separate communication in as much as the general run of the various curves brings ont certain inner detailts which appear to be of some interest in itself. *Secondly*: The equations for the field of Lyman-α radiation (expecially for the case of an expanding nebula) can be put in much neater forms and the final results are immediately obvious without any discussion of the "limiting cases" in the way AMBARZUMIAN does.

I. *The field of ultra-violet radiation.*

§ 2. In solving the problem of radiative transfer we reduce it to a plane-problem but in doing so we have to be careful about the boundary conditions on the inner surface of the nebula as was first pointed out by MILNE[1]). We shall discuss these boundary conditions in due course.

The equation of transfer is as usual

$$\cos\theta \frac{d I(x,\theta)}{d x} = \varkappa \varrho \left(B(x) - I(x,\theta)\right), \tag{1}$$

where $\varkappa$ is the mean absorption coefficient for the ultra-violet radiation, ϱ the density and $4\pi\varkappa\varrho B(x)\,dx$ is the amount of energy of ultra violet quanta emitted in the layer dx per second. Instead of x we shall introduce the optical thickness τ for the ultra-violet light defined by

$$\tau = \int_x^{x_2} \varkappa p\, dx, \quad d\tau = -\varkappa\varrho\, dx, \tag{2}$$

where x_2 denotes the outer boundary of the nebula. Let x_1, denote the inner boundary of the nebula. Then the optical thickness τ_1 of the nebula for the ultra-violet light defined in an obvious way must for a successful application of ZANSTRA's theory be of magnitude not much less than unity. We

[1]) ZS. f. Astrophys. Vol. 1, p. 98, 1930.

shall however treat τ_1 as a parameter for it is one of the objects of the investigation to see how "large" τ_1 has to be. Equation (1) can be rewritten as

$$\cos\theta \frac{d I(\tau,\theta)}{d\tau} = I(\tau,\theta) - B(\tau). \tag{3}$$

We shall define the following quantities:

$$\left.\begin{aligned} J &= \tfrac{1}{2}\int_0^\pi I \sin\theta\, d\theta, \\ F &= 2\int_0^\pi I \sin\theta\cos\theta\, d\theta, \\ K &= \tfrac{1}{2}\int_0^\pi I \sin\theta\cos^2\theta\, d\theta. \end{aligned}\right\} \tag{4}$$

Multiply the equation of transfer successively $\frac{1}{2}\sin\theta\, d\theta$ and $\frac{1}{2}\sin\theta \cos\theta\, d\theta$ and integrate from 0 to π. Using the definitions (4) we obtain

$$\frac{1}{4}\frac{dF}{d\tau} = J - B(\tau). \tag{5}$$

$$\frac{dK}{d\tau} = \frac{1}{4} F. \tag{6}$$

From (5) and (6) we have

$$\frac{d^2 K}{d\tau^2} = J - B(\tau). \tag{7}$$

That is as far as we can go without the equation of radiative equilibrium. Let πS be the amount of ultra-violet energy falling on each square centimetre of the inner surface of the nebula (*ie*) at $\tau = \tau_1$. Then clearly we have

$$B(\tau) = p\left\{J(\tau) + \tfrac{1}{4} S e^{-(\tau_1 - \tau)}\right\}. \tag{8}$$

We have a factor p since for each ultra-violet quantum absorbed there is only a probability p of its re-emission in the same wavelength. From (7) and (8) we have

$$\frac{d^2 K}{d\tau^2} = (1-p) J - \frac{p}{4} S e^{-(\tau_1 - \tau)}. \tag{9}$$

§ 3. Equation (9) is exact. Our *approximation* is now to set

$$3K = J \tag{10}$$

suggested by the mean value of $\cos^2\theta$ over the sphere. From (9) then

$$\frac{d^2 J}{d\tau^2} = 3(1-p) J - \frac{3p}{4} S e^{-(\tau_1 - \tau)}. \tag{11}$$

We will consider three cases separately.

§ 4. (A) $p = 1$: This was first considered by MILNE (loc. cit.). But we repeat the analysis for the sake of completeness. (11) now can be re-written as

$$\frac{d^2 J}{d\tau^2} = -\frac{3}{4} S e^{-(\tau_1-\tau)}. \tag{12}$$

Integrating (12) we have

$$J(\tau) = 3a + 3b\tau - \tfrac{3}{4} S e^{-(\tau_1-\tau)} \tag{13}$$

or

$$K(\tau) = a + b\tau - \tfrac{1}{4} S e^{-(\tau_1-\tau)}. \tag{14}$$

In (13) and (14) a and b are two constants to be determined from appropriate boundary conditions From (6) now

$$\tfrac{1}{4} F = b - \tfrac{1}{4} S e^{-(\tau_1-\tau)}. \tag{15}$$

At $\tau = \tau_1$, $F = 0$. This vanishing of the "diffuse"—flux at the inner boundary is precisely the boundary condition that has been discussed by MILNE. This gives

$$b = \tfrac{1}{4} S \tag{16}$$

or

$$F = S\,(1 - e^{-(\tau_1-\tau)}). \tag{17}$$

To determine a we use the following condition at $\tau = 0$

$$2J(0) = F(0), \tag{18}$$

suggested by the mean value of $\cos\theta$ over the hemisphere. This determines a. We find that

$$3a = \tfrac{1}{2} S\,(1 + \tfrac{1}{2} e^{-\tau_1}). \tag{19}$$

Finally we have in a *Standard notation*, which we shall use for these functions

$$\left.\begin{aligned} J_1 &= \tfrac{1}{2} S\,(1 + \tfrac{1}{2} e^{-\tau_1} + \tfrac{3}{2}\tau - \tfrac{3}{2} e^{-(\tau_1-\tau)}), \\ B_1(\tau) &= \tfrac{1}{2} S\,(1 + \tfrac{1}{2} e^{-\tau_1} + \tfrac{3}{2}\tau - e^{-(\tau_1-\tau)}), \\ F_1 &= S\,(1 - e^{-(\tau_1-\tau)}). \end{aligned}\right\} \tag{20}$$

§ 5. As a typical case for our numerical work we shall always take $\tau_1 = 2$. The functions $J_1(\tau)$ $B_1(\tau)$, and $F_1(\tau)$ are tabulated below:

Table 1. $p = 1$, $\tau_1 = 2$; J_1, B_1 and F_1 in units of S.

τ	$J(\tau)$	$B(\tau)$	$F_1(\tau)$	$e^{-(\tau_1-\tau)}$
0	0.4323	0.4661	0.8647	0.1353
0.25	0.5910	0.6341	0.8242	0.1738
0.50	0.7415	0.7973	0.7769	0.2231
0.75	0.8815	0.9531	0.7135	0.2865
1.00	1.0079	1.0999	0.6321	0.3679
1.25	1.1171	1.2352	0.5276	0.4724
1.50	1.2039	1.3555	0.3935	0.6065
1.75	1.2622	1.4569	0.2212	0.7788
2.00	1.2838	1.5338	0	1.0000

From (20) we deduce that

$$\left.\begin{aligned} F_1(0) &= S(1 - e^{-\tau_1}), \\ J_1(\tau_1) &= 0.25\, S(3\tau_1 + e^{-\tau_1} - 1), \\ J_1(0) &= 0.5\, S(1 - e^{-\tau_1}) - 0.5\, F_1(0). \end{aligned}\right\} \tag{21}$$

For general values of τ_1 (other than 2) the value of the functions $J_1(\tau)$, $B_1(\tau)$, $F_1(\tau)$ are most often required at $\tau = 0$ and at $\tau = \tau_1$. The general variation for intermediate values is indicated by the calculations of Table 1a.

Table 1a. $p = 1$; $F_1(0)$; $J_1(\tau_1)$; $J_1(0)$.

τ_1	$J(\tau_1)$	$J(0)$	$F(0)$	τ_1	$J(\tau_1)$	$J(0)$	$F(0)$
0	0	0	0	1.25	0.759	0.357	0.7135
0.25	0.132	0.111	0.2212	1.50	0.931	0.388	0.7769
0.50	0.277	0.197	0.3935	1.75	1.106	0.413	0.8262
0.75	0.431	0.264	0.5276	2.00	1.284	0.432	0.8647
1.00	0.592	0.316	0.6321	3.00	2.013	0.475	0.9502

The above table illustrates the fact that as $\tau_1 \to \infty$ the emergent flux is almost entirely the "diffuse"-radiation. Because of a large optical thickness all trace of the initial star light is wiped out. Also we notice that $J(\tau_1) \to \infty$ as $\tau_1 \to \infty$.

§ 6. (B) $p \neq {}^2/_3$ *but* $p < 1$:

We go back to our equation (11).

Let

$$\lambda^2 = 3(1 - p); \quad (\lambda \neq 1) \tag{22}$$

[AMBARZUMIAN's definition of "λ^2" is $2(1 - p)$].

Then

$$\frac{d^2 J}{d\tau^2} = \lambda^2 J - \frac{3p}{4} S e^{-(\tau_1 - \tau)}. \tag{23}$$

The solution of (23) is easily seen to be

$$J = A e^{-\lambda\tau} + B e^{\lambda\tau} + \frac{3p}{4(2 - 3p)} S e^{-(\tau_1 - \tau)}, \tag{24}$$

where A and B are arbitrary integration constants [1]). From (6) we now deduce that

$$\tfrac{3}{4} F = -A\lambda e^{-\lambda\tau} + B\lambda e^{\lambda\tau} + \frac{3p}{4(2 - 3p)} S e^{-(\tau_1 - \tau)}. \tag{25}$$

[1]) We have abready used "B" to denote the "Ergiebigkeit" but there will no confusion as we shall *always* refer to this by "$B(\tau)$" while "B" is an integration constant.

Our boundary conditions are as before

$$\tau = \tau_1, \quad F = 0; \quad \tau = 0, \quad F = 2J. \tag{26}$$

These conditions yield

$$\left.\begin{aligned} &A\lambda e^{-\lambda\tau_1} - B\lambda e^{\lambda\tau_1} - \frac{3p}{4(2-3p)} S = 0, \\ &A(1+\tfrac{2}{3}\lambda) + B(1-\tfrac{2}{3}\lambda) + \frac{p}{4(2-3p)} S e^{-\tau_1} = 0. \end{aligned}\right\} \tag{27}$$

Solving we get

$$\left.\begin{aligned} A &= \frac{(3-2\lambda)e^{\tau_1} - \lambda e^{\lambda\tau_1}}{\lambda\{(1+\frac{2}{3}\lambda)e^{\lambda\tau_1} + (1-\frac{2}{3}\lambda)e^{-\lambda\tau_1}\}} \frac{p}{4(2-3p)} S e^{-\tau_1}, \\ B &= -\frac{(3+2\lambda)e^{\tau_1} + \lambda e^{-\lambda\tau_1}}{\lambda\{(1+\frac{2}{3}\lambda)e^{\lambda\tau_1} + (1-\frac{2}{3}\lambda)e^{-\lambda\tau_1}\}} \frac{p}{4(2-3p)} S e^{-\tau_1}. \end{aligned}\right\} \tag{28}$$

Substituting these in our equations we have finally

$$F = \left\{ e^{-(\tau_1-\tau)} - \frac{(\cosh\lambda\tau + \frac{2}{3}\lambda\sinh\lambda\tau) - \frac{\lambda}{3} e^{-\tau_1}\sinh\lambda(\tau_1-\tau)}{\cosh\lambda\tau_1 + \frac{2}{3}\lambda\sinh\lambda\tau_1} \right\} \frac{p}{(2-3p)} S, \tag{29}$$

$$J = \left\{ e^{-(\tau_1-\tau)} - \frac{(\sinh\lambda\tau + \frac{2}{3}\lambda\cosh\lambda\tau) + \frac{\lambda}{3} e^{-\tau_1}\cosh\lambda(\tau_1-\tau)}{\lambda(\cosh\lambda\tau_1 + \frac{2}{3}\lambda\sinh\lambda\tau_1)} \right\} \frac{3p}{4(2-3p)} S, \tag{30}$$

$$B(\tau) = \left\{ \frac{2}{3p} e^{-(\tau_1-\tau)} - \frac{(\sinh\lambda\tau + \frac{2}{3}\lambda\cosh\lambda\tau) + \frac{\lambda}{3} e^{-\tau_1}\cosh\lambda(\tau_1-\tau)}{\lambda(\cosh\lambda\tau_1 + \frac{2}{3}\lambda\sinh\lambda\tau_1)} \right\} \frac{3p^2}{4(2-3p)} S. \tag{31}$$

From (29) we have for the *emergent flux*

$$\pi F(0) = \pi \left\{ 1 - \frac{e^{\tau_1} - \frac{\lambda}{3}\sinh\lambda\tau_1}{\cosh\lambda\tau_1 + \frac{2}{3}\lambda\sinh\lambda\tau_1} \right\} \frac{p}{2-3p} S e^{-\tau_1}. \tag{32}$$

From the conservation of energy we easily see that flux πF_α in the Lyman-α should be given

$$\pi F_\alpha = \pi \frac{\nu_\alpha}{\nu_c} \{ S(1 - e^{-(\tau_1-\tau)}) - F(\tau) \}, \tag{33}$$

where $F(\tau)$ is given by (29). For if p were unity the first term in $\{\ \}$-brackets in (33) is simply what the flux in the ultraviolet would have been had there been no conversion of the ultraviolet radiation into Lyman-α; also $F(\tau)$ is the actual flux when p takes the specified value; from the conservation of energy theorem the difference in energy corresponding to the difference in the two fluxes must appear as energy in Lyman-α radiation. The numerical factor is clearly (ν_α/ν_c) representing the ratio of the frequency of the Lyman-α to the average frequency of the continous ultraviolet radiation beyond the LYMAN limit. (33) can be rewritten as

$$F_\alpha(\tau) = \frac{\nu_\alpha}{\nu_c}(F_1(\tau) - F_p(\tau)). \tag{33'}$$

From (32) we see that if $\lambda > 1$ (*ie.* $p < {}^2/_3$) then

$$\pi F(0) \sim \pi \frac{1+\lambda}{1+\frac{2}{3}\lambda} \frac{p}{2-3p} S e^{-\tau_1}; \; (\tau_1 \to \infty). \tag{32'}$$

For $p = 0.5$ we have asymptotically

$$\pi F(0) \sim 1.225 \, \pi \, S e^{-\tau_1}. \tag{32''}$$

(This differs naturally from the corresponding limit given by AMBARZUMIAN: cf. loc. cit. where he gives 0.7 which is about half our "accurate" value 1.225.)

On the other hand if $\lambda < 1$ (*ie.* $p > {}^2/_3$) then

$$\pi F(0) \sim \pi \frac{1}{1+\frac{2}{3}\lambda} \frac{p}{3p-2} S e^{-\lambda\tau_1}; \; (\tau_1 \to \infty). \tag{32'''}$$

Again, from (30) we see that

$$J(\tau_1) = \left\{ 1 - \frac{(\sinh \lambda\tau_1 + \frac{2}{3}\lambda \cosh \lambda\tau_1) + \frac{\lambda}{3} e^{-\tau_1}}{\lambda(\cosh \lambda\tau_1 + \frac{2}{3}\lambda \sinh \lambda\tau_1)} \right\} \frac{3p}{4(2-3p)} S, \tag{34}$$

For $\tau_1 \to \infty$ we have asymptotically

$$J(\tau_1) \sim \left(1 - \frac{1}{\lambda}\right) \frac{3p}{4(2-3p)} S. \tag{35}$$

Before proceeding to discuss further these formulae we shall treat the case $p = {}^2/_3$.

§ 7. (C) $p = {}^2/_3$:

Our equation (11) takes the form

$$\frac{d^2 J}{d\tau^2} = J - \tfrac{1}{2} S e^{-(\tau_1 - \tau)}. \tag{36}$$

Integrating (36) we have

$$J = A e^{-\tau} + B e^{\tau} - \tfrac{1}{4} S \tau e^{-(\tau_1 - \tau)}, \tag{37}$$

or

$$K = \tfrac{1}{3}(A e^{-\tau} + B e^{\tau} - \tfrac{1}{4} S \tau e^{-(\tau_1 - \tau)}). \tag{38}$$

From (6) we now have

$$\tfrac{3}{4} F = \{-A e^{-\tau} + B e^{\tau} - \tfrac{1}{4} S(\tau + 1) e^{-(\tau_1 - \tau)}\}. \tag{39}$$

Our boundary conditions yield

$$\left.\begin{aligned} 5A + B + \tfrac{1}{2} S e^{-\tau_1} &= 0, \\ -e^{-\tau_1} A + e^{\tau_1} B - \tfrac{1}{4}(\tau_1 + 1) S &= 0. \end{aligned}\right\} \tag{40}$$

Solving we have

$$\left.\begin{aligned} A &= -\frac{\tau_1 + 3}{(5 + e^{-2\tau_1})} \tfrac{1}{4} S e^{-\tau_1}, \\ B &= \frac{5(\tau_1 + 1) - 2e^{-2\tau_1}}{(5 + e^{-2\tau_1})} \tfrac{1}{4} S e^{-\tau_1}. \end{aligned}\right\} \tag{41}$$

Substituting these in our equations we have

$$F = \frac{[5(\tau_1 - \tau) - (\tau + 3) e^{-2\tau_1}] e^{-(\tau_1 - \tau)} + (\tau_1 + 3) e^{-(\tau + \tau_1)}}{3(5 + e^{-2\tau_1})} S, \tag{42}$$

$$J = \frac{[5(\tau_1 - \tau + 1) - (2 + \tau) e^{-2\tau_1}] e^{-(\tau_1 - \tau)} - (\tau_1 + 3) e^{-(\tau_1 + \tau)}}{4(5 + e^{-2\tau_1})} S, \tag{43}$$

$$B(\tau) = \frac{[5(\tau_1 - \tau + 2) - (1 + \tau) e^{-2\tau_1}] e^{-(\tau_1 - \tau)} - (\tau_1 + 3) e^{-(\tau_1 + \tau)}}{6(5 + e^{-2\tau_1})} S. \tag{44}$$

Equation (33) holds as before, but instead of (32) we now have for the emergent flux

$$\pi F(0) = \frac{1 + 2\tau_1 - e^{-2\tau_1}}{5 + e^{-2\tau_1}} S e^{-\tau_1}. \tag{45}$$ [1]

We have therefore, as $\tau_1 \to \infty$

$$\pi F(0) \sim \frac{S}{5}(1 + 2\tau_1) \cdot e^{-\tau_1}. \tag{46}$$

Again from (43) we have

$$J(\tau_1) = \frac{5 - (2\tau_1 + 5) e^{-2\tau_1}}{4(5 + e^{-2\tau_1})} S, \tag{47}$$

so that

$$J(\tau_1) \sim 0.25\, S \quad \text{as} \quad \tau_1 \to \infty. \tag{48}$$

§ 8. *Numerical Applications*: (a) The most important quantity is the *flux* which is defined by (29) ($p \neq {}^2/_3$) and (42) for $p = {}^2/_3$. Also if we denote the flux in L_a by $F_{a,p}$ then as we have abready explained

$$F_{a,p}(\tau) = \frac{\nu_a}{\nu_c} (F_1(\tau) - F_p(\tau)). \tag{49}$$

[1]) Note that when $\tau_1 = 2$. $F(0) = Se^{-\tau_1}$!

In the following tables the resultats of calculations for different values of p are tabulated. The quantities of interest are thediffuse ultraviolet flux $F_p(\tau)$, the flux in the Lyman-α $F_{\alpha,p}(\tau)$ and the *total* flux in the ultraviolet giwen by $(F_p(\tau) + e^{-(\tau_1-\tau)})$. As a typical case we have taken $\tau_1 = 2$.

Table 2. $\tau_1 = 2$. Flux-Calculations.

τ	$p = 0.9$			$p = 0.75$		
	$\frac{F(\tau)}{S}$	$\frac{(F(\tau)+e^{-(\tau_1-\tau)})}{S}$	$F_\alpha(\tau)\cdot\frac{\nu_c}{\nu_\alpha}\cdot S^{-1}$	$\frac{F(\tau)}{S}$	$\frac{(F(\tau)+e^{-(\tau_1-\tau)})}{S}$	$F_\alpha(\tau)\cdot\frac{\nu_c}{\nu_\alpha}\cdot S^{-1}$
0	0.405	0.540	0.460	0.191	0.326	0.674
0.25	0.395	0.568	0.432	0.190	0.363	0.637
0.50	0.382	0.605	0.395	0.190	0.413	0.587
0.75	0.363	0.650	0.350	0.188	0.474	0.526
1.00	0.335	0.703	0.297	0.182	0:549	0.451
1.25	0.293	0.765	0.235	0.166	0.639	0.361
1.50	0.229	0.836	0.164	0.137	0.743	0.257
1.75	0.135	0.914	0.086	0.083	0.862	0.138
2.00	0	1	0	0	1.000	0

τ	$p = 2/3$			$p = 0.5$		
	$\frac{F(\tau)}{S}$	$\frac{(F(\tau)+e^{-(\tau_1-\tau)})}{S}$	$F_\alpha(\tau)\cdot\frac{\nu_c}{\nu_\alpha}\cdot S^{-1}$	$\frac{F(\tau)}{S}$	$\frac{(F(\tau)+e^{-(\tau_1-\tau)})}{S}$	$F_\alpha(\tau)\cdot\frac{\nu_c}{\nu_\alpha}\cdot S^{-1}$
0	0.134	0.270	0.730	0.071	0.206	0.794
0.25	0.135	0.309	0.691	0.072	0.246	0.754
0.50	0.137	0.361	0.639	0.075	0.298	0.702
0.75	0.139	0.425	0.575	0.078	0.364	0.636
1.00	0.137	0.505	0.495	0.080	0.448	0.552
1.25	0.128	0.600	0.400	0.076	0.549	0.451
1.50	0.107	0.714	0.286	0.067	0.673	0.327
1.75	0.068	0.847	0.153	0.045	0.823	0.177
2.00	0	1.000	0	0	1.000	0

τ	$p = 1/3$			$p = 0$		
	$\frac{F(\tau)}{S}$	$\frac{(F(\tau)+e^{-(\tau_1-\tau)})}{S}$	$F_\alpha(\tau)\cdot\frac{\nu_c}{\nu_\alpha}\cdot S^{-1}$	$\frac{F(\tau)}{S}$	$\frac{(F(\tau)+e^{-(\tau_1-\tau)})}{S}$	$F_\alpha(\tau)\cdot\frac{\nu_c}{\nu_\alpha}\cdot S^{-1}$
0	0.036	0.171	0.829	0	0.1353	0.8647
0.25	0.037	0.211	0.789	0	0.1738	0.8262
0.50	0.039	0.263	0.737	0	0.2231	0.7769
0.75	0.042	0.329	0.691	0	0.2865	0.7135
1.00	0.044	0.412	0.588	0	0.3679	0.6321
1.25	0.044	0.516	0.484	0	0.4724	0.5276
1.50	0.039	0.645	0.355	0	0.6065	0.3935
1.75	0.026	0.805	0.195	0	0.7788	0.2212
2.00	0	1.000	0	0	1.0000	0

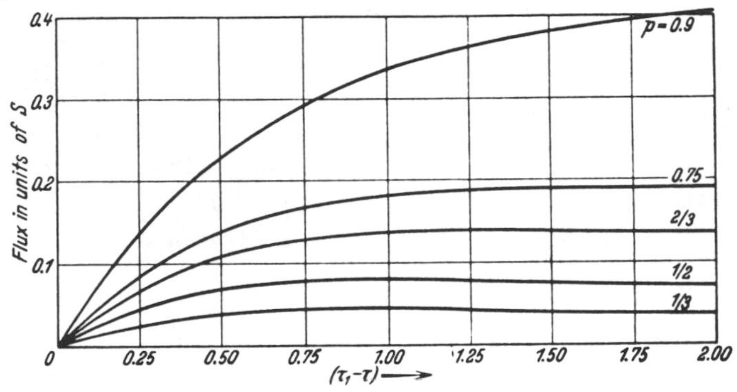

Fig. 1. "Diffuse"-ultraviolet flux in units of S for different values of p and $\tau_1 = 2$. Notice the existence of a maximum in the Curves $p \leqq 2/3$.

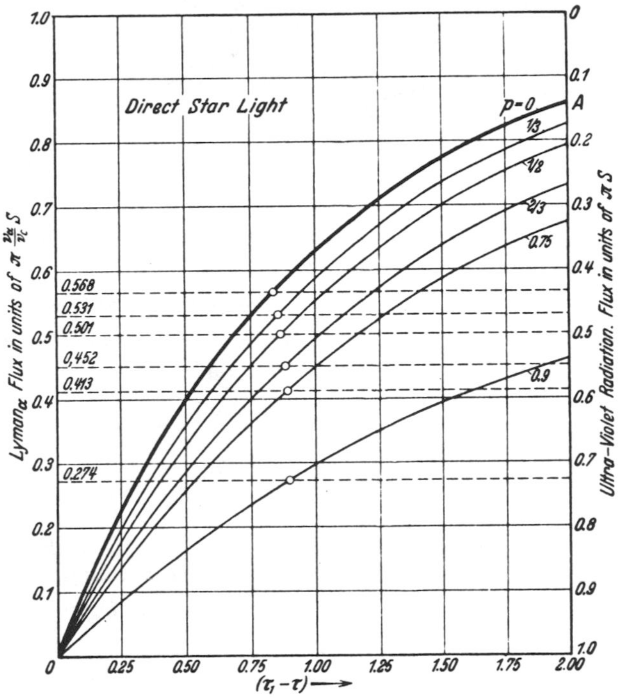

Fig. 2. Flux in the different radiations. The vertical distance between any curve and the curve OA gives the corresponding flux in the "diffuse"-ultraviolet light. If the nebula were expanding then the flux in the Lyman-α, πF_α^* differes from πF_α only by a constant. This constant in the units used above are 0.274, 0.418, 0.452, 0.501, 0.531, 0.568 for $p = 0.9$, $^3/_4$, $^2/_3$, $^1/_2$, $^1/_3$, 0, respectively.

On examining these tables the first thing we notice is that $F(\tau)$ *has a maximum in the range* $0 < \tau < \tau$, *for* $p \leqq {}^2/_3$. — (This circumstance does not appear to have been noticed before.) — Secondly the extraordinary rapidity with which the flux in the Lyman-α "builds up" as we penetrate into the outer regions of the planetary nebula. Thus when $p = 0.5$ the diffuse ultra-violet flux even at its maximum ($= 0.080\,S$) is less than a tenth of what the emergent flux ($0.865\,S$) would be had p been unity.

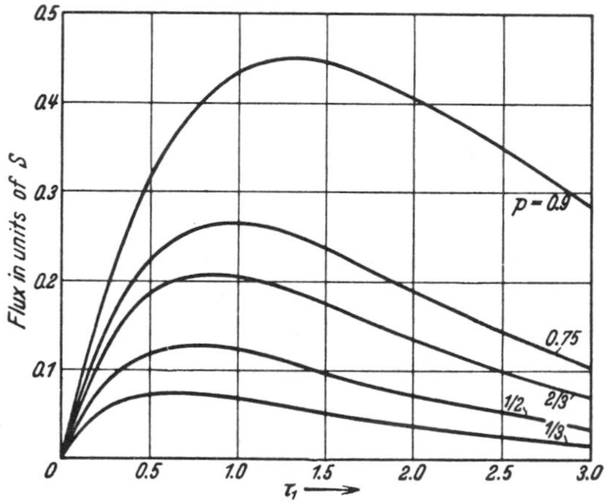

Fig. 3. Emergent diffuse ultraviolet flux as a function of the optical thickness τ_1 of the nebula.

Even for $p = 0.9$ more than 50% of the emergent diffuse flux appears as energy in the LYMAN radiation. It is precisely these facts that provide the theoretical ramifications for ZANSTRA's original ideas. For layers not immediately near the boundary we can as a crude rough approximation write

$$F_a \sim \frac{\nu_a}{\nu_c} S \left(1 - e^{-(\tau_1 - \tau)}\right). \tag{50}$$

(ZANSTRA has found this representation useful in some as yet unpublished work of his.) For $p = 0.5$ the above formula is in error by about ten per cent. The general Character of these results are illustrated in figures 1 and 2.

To examine how the energy is apportioned between, the ultraviolet and the LYMAN radiations in the emerqent radiation we have calculated $F(0)$ etc., for different values of p as a function of τ_1. The results are exemplified graphically in Figures 3 and 4.

From these tables we notice that for a given p as τ_1 increases the diffuse — flux in the ultraviolet first increases passes through a maximum and then decreases again asymptotically to zero. It is clear that for any

Table 3. Emergent-Flux.

τ_1	$p = 0.9$			$p = 0.75$		
	$\frac{F(0)}{S}$	$\frac{F(0) + e^{-\tau_1}}{S}$	$F_\alpha(0) \cdot \frac{\nu_c}{\nu_c} \Big/ S$	$\frac{F(0)}{S}$	$\frac{F(0) + e^{-\tau_1}}{S}$	$F_\alpha(0) \frac{\nu_c}{\nu_c} \Big/ S$
0	0	0	0	0	0	0
0.25	0.188	0.967	0.0328	0.146	0.924	0.0756
0.50	0.314	0.921	0.0793	0.224	0.830	0.170
0.75	0.392	0.864	0.136	0.258	0.731	0.269
1.00	0.433	0.801	0.199	0.265	0.633	0.367
1.25	0.448	0.735	0.265	0.257	0.544	0.456
1.50	0.445	0.668	0.332	0.237	0.460	0.540
1.75	0.429	0.603	0.397	0.215	0.389	0.611
2.0	0.405	0.540	0.460	0.190	0.325	0.675
3.0	0.286	0.336	0.664	0.104	0.155	0.845

τ_1	$p = {}^2/_3$			$p = 0{,}5$			$p = {}^1/_3$		
	$\frac{F(0)}{S}$	$\frac{F(0) + e^{-\tau_1}}{S}$	$F_\alpha(0) \cdot \frac{\nu_c}{\nu_\alpha} \Big/ S$	$\frac{F(0)}{S}$	$\frac{F(0) + e^{-\tau_1}}{S}$	$F_\alpha(0) \cdot \frac{\nu_c}{\nu_\alpha} \Big/ S$	$\frac{F(0)}{S}$	$\frac{F(0) + e^{-\tau_1}}{S}$	$F_\alpha(0) \cdot \frac{\nu_c}{\nu_\alpha} \Big/ S$
0	0	1.000	0	0	1.000	0	0	1.000	0
0.25	0.124	0.962	0.0971	0.0823	0.865	0.135	0.054	0.832	0.168
0.50	0.184	0.793	0.207	0.120	0.727	0.273	0.071	0.677	0.323
0.75	0.206	0.678	0.322	0.128	0.601	0.399	0.072	0.545	0.455
1.00	0.205	0.573	0.427	0.122	0.490	0.510	0.067	0.435	0.565
1.25	0.193	0.479	0.521	0.111	0.397	0.603	0.060	0.346	0.654
1.50	0.175	0.398	0.602	0.097	0.320	0.680	0.051	0.274	0.726
1.75	0.154	0.328	0.672	0.083	0.257	0.743	0.043	0.217	0.783
2.00	0.134	0.270	0.730	0.071	0.206	0.794	0.036	0.171	0.829
3.00	0.0697	0.119	0.881	0.033	0.083	0.917	0.015	0.065	0.935

specified $p < 1$, the flux in L_α tends to unity as $\tau_1 \to \infty$ but that the limit-should be so quickly be reached could not have been anticipated without these calculations. Thus for $p = 0.9$ when $\tau_1 = 3$, in the emergent

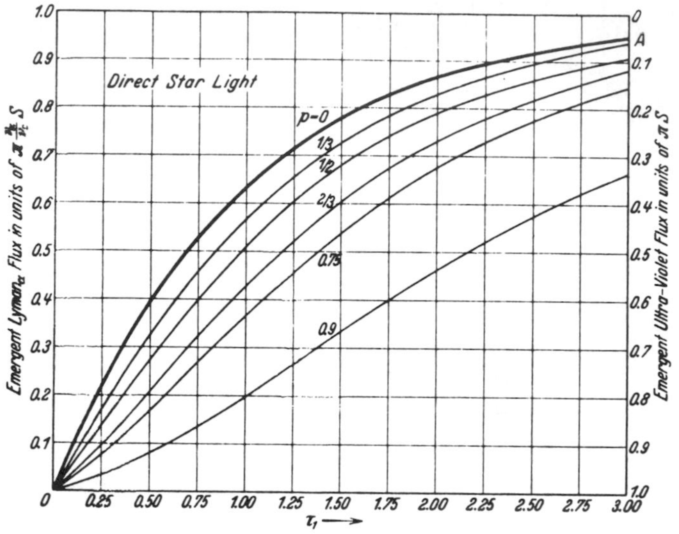

Fig. 4. Emergent flux in Lyman-α and ultraviolet radiations as a function τ_1, the optical thickness of the nebula. The vertical distance between any one curve and OA gives the amount of "diffuse"-ultraviolet-flux.

radiation about two thirds of the energy appears in L_α and only a third as ultra-violet radiation; white if $p = 0.5$ about 90% of the emergent energy in the radiation appears in Lyman-α.

(b) The quantity next in order of importance is *the density of the radiation*. The density is clearly specified by J.

First to examine the variation of J as a function of τ we again take the case $\tau_1 = 2$. The following table summarises the results of the calculations:

Table 4. $J_p(\tau)$ in units of S, $\tau_1 = 2$.

τ	$p = 1$	$p = 0.9$	$p = 0.75$	$p = {}^2/_3$	$p = {}^1/_2$	$p = {}^1/_3$
0	0.432	0.203	0.095	0.0672	0.0353	0.0179
0.25	0.591	0.278	0.130	0.0924	0.0487	0.0247
0.50	0.741	0.350	0.166	0.1180	0.0624	0.0318
0.75	0.881	0.420	0.202	0.1440	0.0767	0.0395
1.00	1.008	0.486	0.236	0.1699	0.0917	0.0476
1.25	1.117	0.545	0.269	0.1949	0.107	0.0558
1.50	1.204	0.594	0.298	0.2172	0.120	0,0637
1.75	1.262	0.629	0.317	0.2340	0.131	0.0698
2.00	1.284	0.634	0.327	0.2409	0.135	0.0726

The results are graphically illustrated in Fig. 5. We notice the quite remarkable *linear increase in $J(\tau)$ from $J(0)$ to $J(\tau_1)$.* Actually at τ_1 the differential coefficient is zero but the curve bends over too near $\tau = \tau_1$.

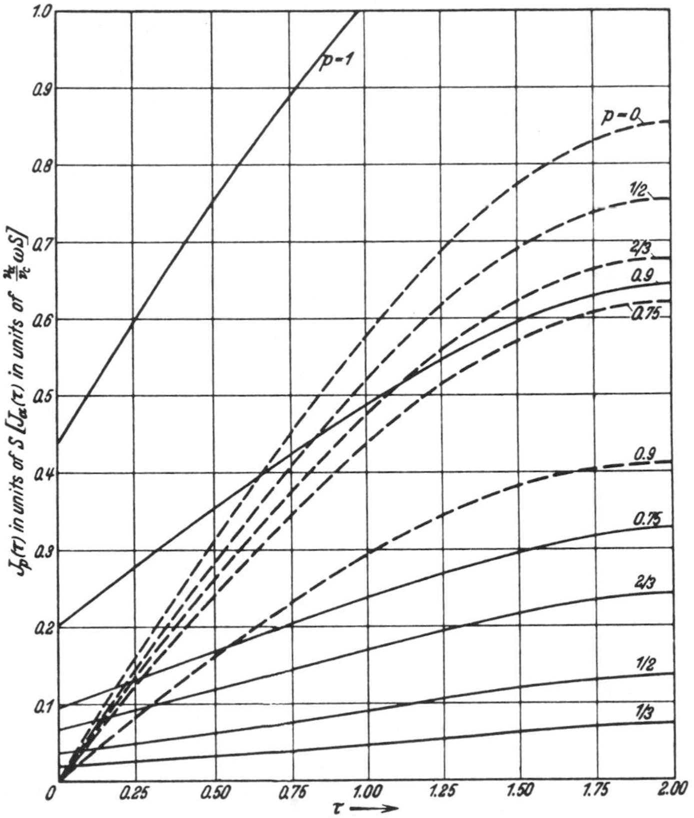

Fig. 5. $J_p(\tau)$ in units of S are represented by the full curves. Note the almost linear in crease of $J(\tau)$ for much the greater part. The dotted curves correspond to $J_\alpha(\tau)$ — density of the Lyman-α radiation — in units of $(\nu_\alpha/\nu_c)\,\omega\,S$ (see Table 6).

Hence if $J(\tau)$ were needed for other values of τ_1 then *in quite a good approximation we can assume a linear law between $J(0)$ and $J(\tau_1)$.* Actually $J(0) = 0.5\,F(0)$ and $F(0)$ has already been tabulated in Table 3. In Table 5 we give a table of values for $J(\tau_1)$. (See also fig. 6).

The quantities in the last row of Table 5 were calculated by means of equation (35). [For $p = {}^2/_3$ we have taken the result (48)]. We notice

Table 5. $J_p(\tau_1)$ in units of S.

τ_1	$p=1$	$p=0.9$	$p=0.75$	$p={}^2/_3$	$p={}^1/_2$	$p={}^1/_3$
0	0	0	0	0	0.	0
0.25	0.132	0.113	0.087	0.0742	0.0516	0.0320
0.50	0.277	0.221	0.158	0.130	0.0853	0.0503
0.75	0.431	0.321	0.213	0.170	0.1062	0.0605
1.00	0.592	0.409	0.254	0.197	0.1187	0.0662
1.25	0.759	0.485	0.283	0.216	0.1277	0.0693
1.50	0.931	0.548	0.303	0.228	0.1311	0.0711
1.75	1.106	0.600	0.317	0.236	0.1335	0.0720
2.00	1.284	0.643	0.327	0.241	0.1354	0.0726
3.00	2.013	0.740	0.340	0.249	0.1374	0.0732
∞	∞	0.796	0.348	0.250	0.1376	0.0732

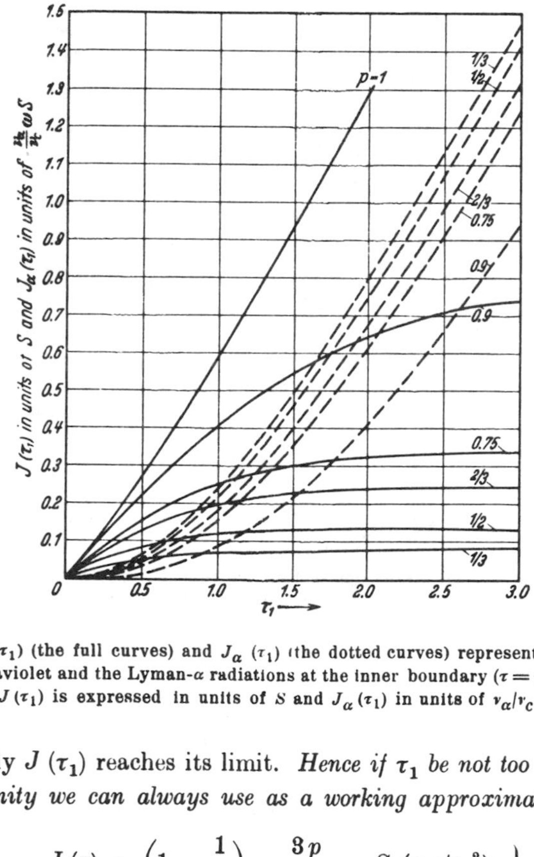

Fig. 6. $J(\tau_1)$ (the full curves) and $J_\alpha(\tau_1)$ (the dotted curves) representing the densities of the ultraviolet and the Lyman-α radiations at the inner boundary ($\tau=\tau_1$) of the nebula. $J(\tau_1)$ is expressed in units of S and $J_\alpha(\tau_1)$ in units of $\nu_\alpha/\nu_c\,\omega S$.

how quickly $J(\tau_1)$ reaches its limit. *Hence if τ_1 be not too small and p not too near unity we can always use as a working approximation*

$$\left.\begin{aligned} J(\tau) &\sim \left(1-\frac{1}{\lambda}\right)\frac{3p}{4(2-3p)}\,S \quad (p \neq \tfrac{2}{3}) \\ &= 0{,}25\,S \qquad\qquad\qquad (p = \tfrac{2}{3}) \end{aligned}\right\} \tag{51}$$

(C) We shall not pause to calculate $B(\tau)$ since it is so simply related to $J(\tau)$. Thus as we have shown already [cf. equation (8)].

$$B(\tau) = p\{J(\tau) + \tfrac{1}{4} S e^{-(\tau_1 - \tau)}\}. \tag{52}$$

In particular

$$B(\tau_1) = p\{J(\tau_1) + 0.25\, S\}. \tag{53}$$

To determine $B(\tau_1)$ we simply have to add 0.25 to the quantities tabulated in Table 5 and multiplying them by the respective p's.

This completes our discussion of the field of ultra-violet radiation.

II. The field of Lyman-α-Radiation.

§ 9. We shall introduce the coefficient of absorption $\varkappa_\alpha$ within the line L_α. The optical depth τ_α is defined by

$$\tau_\alpha = \int_x^{x_2} \varkappa_\alpha \varrho\, dx, \quad d\tau_\alpha = -\varkappa_\alpha \varrho\, dz. \tag{54}$$

In treating the problem of the radiative transfer of this Lyman radiation we shall treat τ_α/τ as a constant (ω say)

$$\frac{\tau_\alpha}{\tau} = \frac{\varkappa_\alpha}{\varkappa} = \frac{\tau_{1,\alpha}}{\tau_1} = \text{constant} = \omega. \tag{55}$$

In (55) $\tau_{1,\alpha}$ denotes the total optical thickness of the nebula.

The equation of transfer is as usual

$$\operatorname{Cos}\theta \frac{dI_\alpha}{d\tau_\alpha} = I_\alpha(\tau_\alpha) - B_\alpha(\tau_\alpha), \tag{56}$$

where all quantities with the suffix-α denote the respective quantities for the Lyman-α-radiation. Thus we have J_α, F_α, K_α, defined as our original J, F, K but with I_α replacing the I's. From (56) by our usual procedure we have

$$\frac{1}{4}\frac{dF_\alpha}{d\tau_\alpha} = J_\alpha - B_\alpha(\tau_\alpha), \tag{57}$$

$$\frac{dK_\alpha}{d\tau_\alpha} = \frac{1}{4} F_\alpha. \tag{58}$$

Combining (57) and (58) we have setting $3K_\alpha = J_\alpha$

$$\frac{d^2 J_\alpha}{d\tau_\alpha^2} = 3(J_\alpha - B_\alpha(\tau_\alpha)). \tag{59}$$

The equation of radiative equilibrium can be written down, but in doing so we have to be careful in taking account the emission in L_α due to absorption of an ultraviolet quantum. One easily verifies that (*cf.* AMBARZUMIAN loc. cit.)

$$B_\alpha(\tau_\alpha) = J_\alpha + \frac{\nu_\alpha}{\nu_c} \frac{(1-p)}{p\,\omega} B(\tau_\alpha), \tag{60}$$

where according to (8) and (25)

$$\begin{aligned} B(\tau) &= p\,\{J(\tau) + \tfrac{1}{4} S\, e^{-(\tau_1-\tau)}\}, \\ &= p\left\{A\,e^{-\lambda\tau} + B\,e^{\lambda\tau} + \frac{S}{2(2-3p)} e^{-(\tau_1-\tau)}\right\}. \end{aligned} \tag{61}$$

(61) *is true only if* $p \neq {}^2/_3$. We shall consider the case $p = {}^2/_3$ separately. We need to know $B(\tau)$ as a function of τ_α. By (55)

$$B(\tau_\alpha) = p\left\{A\,e^{-\frac{\lambda}{\omega}\tau_\alpha} + B\,e^{\frac{\lambda}{\omega}\tau_\alpha} + D\,e^{-\frac{(\tau_{1,\alpha}-\tau_\alpha)}{\omega}}\right\}, \tag{62}$$

where we have set

$$D = \frac{1}{2(2-3p)} S. \tag{63}$$

Introducing (62) in (59) we have

$$\frac{d^2 J_\alpha}{d\tau_\alpha^2} = -3\frac{\nu_\alpha}{\nu_c}\frac{1-p}{\omega}\left\{A\,e^{-\frac{\lambda}{\omega}\tau_\alpha} + B\,e^{\frac{\lambda}{\omega}\tau_\alpha} + D\,e^{-\frac{(\tau_{1,\alpha}-\tau_\alpha)}{\omega}}\right\}. \tag{64}$$

Solving (64)

$$J_\alpha = a + b\tau_\alpha - 3\frac{\nu_\alpha}{\nu_c}\frac{(1-p)\,\omega}{\lambda^2}\left\{A\,e^{-\frac{\lambda}{\omega}\tau_\alpha} + B\,e^{\frac{\lambda}{\omega}\tau_\alpha} + D\lambda^2\,e^{-\frac{(\tau_{1,\alpha}-\tau_\alpha)}{\omega}}\right\}. \tag{65}$$

Since however $\lambda^2 = 3(1-p)$ we have

$$J_\alpha = a + b\tau_\alpha - \frac{\nu_\alpha}{\nu_c}\,\omega\left\{A\,e^{-\frac{\lambda}{\omega}\tau_\alpha} + B\,e^{\frac{\lambda}{\omega}\tau_\alpha} + D\lambda^2\,e^{-\frac{(\tau_{1,\alpha}-\tau_\alpha)}{\omega}}\right\}. \tag{66}$$

From (58) now

$$\frac{3}{4}F_\alpha = b - \frac{\nu_\alpha}{\nu_c}\,\lambda\left\{-A\,e^{-\frac{\lambda}{\omega}\tau_\alpha} + B\,e^{\frac{\lambda}{\omega}\tau_\alpha} + D\lambda\,e^{-\frac{(\tau_{1,\alpha}-\tau_\alpha)}{\omega}}\right\}. \tag{67}$$

Returing to our variables τ we have finally,

$$J_\alpha = a + b\,\omega\,\tau - \frac{\nu_\alpha}{\nu_c}\,\omega\,(A\,e^{-\lambda\tau} + B\,e^{\lambda\tau} + D\lambda^2\,e^{-(\tau_1-\tau)}), \tag{68}$$

$$\frac{3}{4}F_\alpha = b - \frac{\nu_\alpha}{\nu_c}\,\lambda\,(-A\,e^{-\lambda\tau} + B\,e^{\lambda\tau} + D\lambda\,e^{-(\tau_1-\tau)}.) \tag{69}$$

§ 10. *Static-Nebula*: So far our results are applicable both to the case of a static nebula and also to one which is expanding with a uniform velocity V large compared with the mean velocity of molecular motion[1]). We shall for the present consider only the case of a static nebula. In this case our boundary conditions are as usual: at $\tau = \tau_1$, $F_\alpha = 0$ and at $\tau = 0, F_\alpha(0) = 2J_\alpha(0)$. From (69) b is determined by our first condition. We find after some minor rearrangements of the terms that

$$b = \frac{3}{4}\frac{\nu_\alpha}{\nu_c} S. \tag{70}$$

The second boundary condition yields

$$a = \frac{1}{2}\frac{\nu_\alpha}{\nu_c} S + \frac{\nu_\alpha}{\nu_c}\omega\left\{A\left(1+\frac{2}{3}\frac{\lambda}{\omega}\right) + B\left(1-\frac{2}{3}\frac{\lambda}{\omega}\right) + D\lambda^2\left(1-\frac{2}{3\omega}\right)e^{-\tau_1}\right\}. \tag{71}$$

The arbitrary constants having thus been determined (68) and (69) give us our required solution. They can be written more conveniently as follows. We had originally [cf. equations (24) and (25)]

$$\tfrac{3}{4}F = -A\lambda e^{-\lambda\tau} + B\lambda e^{\lambda\tau} + \frac{3p}{4(2-3p)} S e^{-(\tau_1-\tau)}, \tag{72}$$

and

$$J = Ae^{-\lambda\tau} + Be^{\lambda\tau} + \frac{3p}{4(2-3p)} S e^{-(\tau_1-\tau)}. \tag{73}$$

Remembering also that $\lambda^2 = 3(1-p)$ and that D has been defined in (63) we have the neater forms;

$$J_\alpha = a + b\omega\tau - \frac{\nu_\alpha}{\nu_c}\omega\left\{J(\tau) + \tfrac{3}{4} S e^{-(\tau_1-\tau)}\right\}, \tag{74}$$

$$F_\alpha = \frac{\nu_\alpha}{\nu_c}\left\{S\left(1-e^{-(\tau_1-\tau)}\right) - F(\tau)\right\}. \tag{75}$$

Of course we have already discussed in detail the result (75). (74) as such is a simple expression but it is the *exact* evaluation of a that is troublesome. We shall use a methode of approximation (originally due to AMBARZUMIAN).

§ 11. *Approximate-Solution*: The approximation consists in the following. ω is the ratio of the coefficient of line absorption in L_α to the mean coefficient of absorption for the ultraviolet. This is of the order

---

[1]) This is the case considered by AMBARZUMIAN.

of 10^4. In (71) therefore we neglect all terms cntaining ω^{-1} compared to unity. We find that

$$a = \frac{\nu_a}{\nu_c}\,\omega\,\{J(0) + \tfrac{3}{4} S\,e^{-\tau_1}\}. \tag{76}$$

Substituting this (74) we have

$$J_a = \frac{\nu_a}{\nu_c}\,\omega\,[\tfrac{3}{4} S\,(\tau - e^{-(\tau_1 - \tau)}) - (J(\tau) - J(0)) + \tfrac{3}{4} S\,e^{-\tau_1}]. \tag{77}$$

On the approximation (77), at $\tau = 0$, $J_a = 0$. This could have been anticipated. At $\tau = \tau_1$ we have

$$J_a(\tau_1) = \frac{\nu_a}{\nu_c}\,\omega\,[\tfrac{3}{4} S\,(\tau_1 + e^{-\tau_1} - 1) - (J(\tau_1) - J(0))]. \tag{78}$$

For $p \to 0$ we have

$$J_a(\tau_1) = \frac{3}{4}\,\frac{\nu_a}{\nu_c}\,S\,(\tau_1 - 1 + e^{-\tau_1})\,\omega. \tag{79}$$

If in addition $\tau \to \infty$ we have

$$J_a(\tau_1) \sim \frac{3}{4}\,\frac{\nu_a}{\nu_c}\,S\,(\tau_1 - 1)\,\omega, \tag{79'}$$

which is identical except for the factor $^3/_4$ with an earlier result of AMBARZUMIAN (loc. cit. p. 11). Before discussing (77) and (78) numerically we shall obtain the corresponding results for $p = {}^2/_3$.

§ 12. *Case* $p = {}^2/_3$: Instead of (61) we have

$$B(\tau) = \tfrac{8}{3}\,(A\,e^{-\tau} + B\,e^{\tau} + \tfrac{1}{4} S\,(1 - \tau)\,e^{-(\tau_1 - \tau)}). \tag{80}$$

On going through the algebra as before we have

$$J_a = a + b\,\omega\,\tau - \frac{\nu_a}{\nu_c}\,\omega\,\{A\,e^{-\tau} + B\,e^{\tau} + \tfrac{1}{4} S\,(3 - \tau)\,e^{-(\tau_1 - \tau)}\}, \tag{81}$$

$$\tfrac{3}{4} F_a = b - \frac{\nu_a}{\nu_c}\,\{-A\,e^{-\tau} + B\,e^{\tau} + \tfrac{1}{4} S\,(2 - \tau)\,e^{-(\tau_1 - \tau)}\}. \tag{82}$$

Our usual boundary conditions yield

$$b = \frac{3}{4}\,\frac{\nu_a}{\nu_c}\,S, \tag{83}$$

$$a = \frac{1}{2}\,\frac{\nu_a}{\nu_c}\,S + \frac{\nu_a}{\nu_c}\,\omega\Big[A\Big(1 + \frac{2}{3\,\omega}\Big) + B\Big(1 - \frac{2}{3\,\omega}\Big) + \tfrac{1}{4} S\Big(3 - \frac{4}{3\,\omega}\Big)e^{-\tau_1}\Big]. \tag{84}$$

(81) and (82) can be rewritten remembering our earlier expressions for $F(\tau)$ and $J(\tau)$ [equations (37) and (39)]

$$J_\alpha = a + b\,\omega\,\tau - \frac{\nu_\alpha}{\nu_c}\,\omega\,\{J(\tau) + \tfrac{3}{4} S\, e^{-(\tau_1 - \tau)}\}, \tag{85}$$

$$F_\alpha = \frac{\nu_\alpha}{\nu_c}\,[S(1 - e^{-(\tau_1 - \tau)}) - F(\tau)], \tag{86}$$

which are *identical in form* with our earlier results (74) and (75). One now easily verifies that (77) and (78) as they stand hold also for the case $p = {}^2/_3$.

§ 13. We shall first examine the type of variation which $J(\tau)$ follows from the inner to the outer boundary of the planetary nebula. We have in general (*ie* for $p = {}^2/_3$ as well)

$$J_\alpha(\tau) = \frac{\nu_\alpha}{\nu_c}\,\omega\,[\tfrac{3}{4} S(\tau - e^{-(\tau_1 - \tau)}) - J(\tau) + J(0) + \tfrac{3}{4} S\, e^{-\tau_1}], \tag{87}$$

(87) can be transformed into an exceedingly simple form in the following way: We have already shown that for the case $p = 1$ [cf. (20), (21)]

$$J_1(\tau) = \tfrac{1}{2} S(1 + \tfrac{1}{2} e^{-\tau_1} + \tfrac{3}{2}\tau - \tfrac{3}{2} e^{-(\tau_1 - \tau)}), \tag{88}$$

$$J_1(0) = \tfrac{1}{2} S(1 - e^{-\tau_1}). \tag{89}$$

Combining (88), (89) and (87) we have the remarkably simple result

$$J_\alpha(\tau) = \frac{\nu_\alpha}{\nu_c}\,\omega\,[\{J_1(\tau) - J(\tau)\} - \{J_1(0) - J(0)\}]. \tag{90}$$

The first thing to notice about (90) is that since there is factor ω the density of the Lyman-α radiation will be ω times [(*ie*) 10^4 times] the order of magnitude of the density of the ultra-violet radiation incident on the inner layers of a planetary nebula.

We have already evaluated numerically all the expressions occurring in (90) for our typical case ($\tau_1 = 2$). The following table provides numerical values for $J_\alpha(\tau)$ in units of $\nu_\alpha \omega S/\nu_c$. (See also the dotted curves in Fig. 5.)

In the last column we have tabulated the limiting density (*ie*) $p = 0$ or p sufficiently small. In this case we simply have

$$J_\alpha(\tau) \sim \frac{\nu_\alpha}{\nu_c}\,\omega\,(J_1(\tau) - J_1(0)). \tag{91}$$

On examining the table we notice that even for $p = {}^1/_3$ this limit is still far from having reached.

Another function of interest is $J_\alpha(\tau_1)$. By (90)

$$J_\alpha(\tau_1) = \frac{\nu_\alpha}{\nu_c}\,\omega\,[\{J_1(\tau_1) - J_1(0)\} - \{J(\tau_1) - J(0)\}]. \tag{92}$$

Table 6. $J_\alpha(\tau)$ in units of $\nu_\alpha \omega S/\nu_c$.

τ	$p = 0.9$	$p = 0.75$	$p = {}^2/_3$	$p = {}^1/_2$	$p = {}^1/_3$	$p = 0$ (limit case)
0	0	0	0	0	0	0
0.25	0.084	0.124	0.134	0.145	0.152	0.159
0.50	0.161	0.238	0.258	0.282	0.295	0.309
0.75	0.232	0.343	0.373	0.408	0.428	0.449
1.00	0.293	0.435	0.473	0.519	0.546	0.576
1.25	0.343	0.511	0.557	0.614	0.647	0.685
1.50	0.380	0.569	0.622	0.687	0.726	0.772
1.75	0.404	0.608	0.663	0.734	0.778	0.830
2.00	0.411	0.620	0.678	0.752	0.797	0.852

This function for different values of p is tabulated in Table 7. (See also the dotted curves in Fig. 6.) Now our previous calculations have shown that if τ_1 be not too small and p about a half then $J(0)$ can be neglected. On the other hand under these circumstances [cf. equation (51) we have very nearly

$$\left.\begin{aligned} J(\tau_1) &= \left(1 - \frac{1}{\lambda}\right) \frac{3p}{4(2-3p)} S \quad (p \neq \tfrac{2}{3}) \\ &= 0{,}25\, S \quad\quad (p = \tfrac{2}{3}) \end{aligned}\right\}. \tag{93}$$

We therefore have as a fairly good approximation for these cases

$$J_\alpha(\tau_1) = \frac{\nu_\alpha}{\nu_c}\, \omega \left[\tfrac{3}{4} S (\tau_1 + e^{-\tau_1} - 1) - \left(1 - \frac{1}{\lambda}\right) \frac{3p}{4(2-3p)} S\right]. \tag{94}$$

The last column in Table 6 the limiting function

$$J_\alpha(\tau_1) = \frac{\nu_\alpha}{\nu_c}\, \omega \left[\tfrac{3}{4} S (\tau_1 - 1 + e^{-\tau_1})\right], \tag{95}$$

is tabulated.

Table 7. $J_\alpha(\tau_1)$ in units of $\nu_\alpha \omega S/\nu_c$.

τ_1	$p = 0.9$	$p = 0.75$	$p = {}^2/_3$	$p = {}^1/_2$	$p = {}^1/_3$	$p = 0$
0	0	0	0	0	0	0
0.25	0.003	0.007	0.010	0.013	0.016	0.022
0.50	0.016	0.034	0.042	0.055	0.065	0.080
0.75	0.042	0.083	0.100	0.125	0.143	0.167
1.00	0.084	0.155	0.181	0,218	0.243	0.276
1 25	0.142	0.248	0.283	0,330	0.363	0.402
1.50	0.217	0.358	0.402	0.460	0.497	0.542
1.75	0.307	0.484	0.534	0.601	0.642	0.693
2.00	0.411	0.620	0.678	0.751	0.797	0.851
3.00	0.940	1.250	1.324	1.417	1.472	1.537

This completes our discussion of the field of Lyman-α radiation for the static case. We pass on to the consideration of an expanding nebula under the circumstances examined by AMBARZUMIAN.

§ 14. *Expanding-Nebula*: We shall put an asterik to all the quantities (ie. F, J, and constants of integration but not of course to such quantities as p, λ, τ_1) when they refor to the expanding nebula. Our general results (68) and (69) can now be rewritten as

$$J_a^*(\tau) = a^* + b^* \omega \tau - \frac{\nu_a}{\nu_c} \omega \{J(\tau) + \tfrac{3}{4} S e^{-(\tau_1 - \tau)}\}, \tag{96}$$

$$\tfrac{3}{4} F_a^*(\tau) = b^* - \frac{3}{4} \frac{\nu_a}{\nu_c} \{F(\tau) + S e^{-(\tau_1 - \tau)}\}. \tag{97}$$

To determine the constants a^* and b^* we have to use different boundary conditions from what we have used in the earlier sections. If the velocity of expansion be sufficiently large compared to the "molecular-motion" (this is the only case we shall consider here) then we can in a first approximation neglect the back-illumination and our appropriate boundary conditions are

$$2 J^*(\tau_1) = F^*(\tau_1); \quad 2 J^*(0) = F^*(0). \tag{98}$$

These conditions yield remembering that $F(\tau_1) = 0$

$$(a^* - \tfrac{2}{3} b^*) + b^* \omega \tau_1 = \frac{\nu_a}{\nu_c} \omega \left\{J(\tau_1) + \tfrac{3}{4} S\left(1 - \frac{2}{3\omega}\right)\right\}, \tag{99}$$

$$(a^* - \tfrac{2}{3} b^*) = \frac{\nu_a}{\nu_c} \omega \left\{J(0)\left(1 - \frac{1}{\omega}\right) + \frac{3}{4} S\left(1 - \frac{2}{3\omega}\right) e^{-\tau_1}\right\}. \tag{100}$$

Solving

$$\left.\begin{aligned} b^* &= \frac{\nu_a}{\nu_c \tau_1} \left\{J(\tau_1) - J(0)\left(1 - \frac{1}{\omega}\right) + \frac{3}{4} S\left(1 - \frac{2}{3\omega}\right)(1 - e^{-\tau_1})\right\}, \\ a^* &= \frac{\nu_a}{\nu_c} \omega \left\{J_0\left(1 - \frac{1}{\omega}\right) + \frac{3}{4} S\left(1 - \frac{2}{3\omega}\right) e^{-\tau_1}\right\} + \frac{2}{3} b^*. \end{aligned}\right\} \tag{101}$$

We do not need these complicated expressions! Neglecting terms in ω^{-1} compared to unity (which is justifiable) we have approximately

$$b^* = \frac{\nu_a}{\nu_c \tau_1} \{J(\tau_1) - J(0) + \tfrac{3}{4} S(1 - e^{-\tau_1})\}, \tag{102}$$

$$a^* = \frac{\nu_a}{\nu_c} \omega \{J(0) + \tfrac{3}{4} S e^{-\tau_1}\}. \tag{103}$$

Substituting in (96) and (97) we have

$$F_a^*(\tau) = \frac{\nu_a}{\nu_c} \left[\frac{4}{3\tau_1}(J(\tau_1) - J(0)) + \frac{1}{\tau_1} S(1 - e^{-\tau_1}) - (F(\tau) + S e^{-(\tau_1 - \tau)})\right], \tag{104}$$

$$J_a^*(\tau) = \frac{\nu_a}{\nu_c} \omega \left[\left(J(\tau_1) - J(0) + \frac{3}{4} S(1 - e^{-\tau_1})\right) \frac{\tau}{\tau_1} - \left(J(\tau) + \frac{3}{4} S e^{-(\tau_1 - \tau)}\right) + \left(J(0) + \frac{3}{4} S e^{-\tau_1}\right)\right]. \tag{105}$$

Our expression for the flux can be rewritten in terms of $F_\alpha(\tau)$ the flux in the corresponding static case as

$$F_\alpha^*(\tau) = F_\alpha(\tau) - \frac{\nu_\alpha}{\nu_c}\left[S\left\{1 - \frac{1}{\tau_1}(1 - e^{-\tau_1})\right\} - \frac{4}{3\tau_1}\{J(\tau_1) - J(0)\}\right], \quad (106)$$

or using the relations (21) we can simplity this further into

$$F_\alpha^*(\tau) = F_\alpha(\tau) - \frac{4}{3\tau_1}\frac{\nu_\alpha}{\nu_c}\{J_1(\tau_1) - J_1(0) - J(\tau_1) + J(0)\}, \quad (107)$$

or again by (92)

$$F_\alpha^*(\tau) = F_\alpha(\tau) - \frac{4}{3\omega\tau_1}J_\alpha(\tau_1), \quad (108)$$

a remarkably simple result. From (108) we see that

$$F_\alpha^*(\tau_1) = -\frac{4}{3\omega\tau_1}J_\alpha(\tau_1); \quad (109)$$

which means that the flux is *negative* in the inner regions of the planetary. This circumstance was first drawn attention to by AMBARZUMIAN who however did not notice that *the difference between the fluxes in the expanding and the corresponding static cases is just a constant* — that the inner parts of the nebula must suffer inward radiative accelerations is therefore an immediate consequence. Writing for the sake of conveniance

$$F_\alpha^*(\tau) = F_\alpha(\tau) - \frac{\nu_\alpha}{\nu_c}\beta S, \quad (110)$$

we evaluate β for different values of τ_1 and p.

The values in the above table also provide the value of the *negative flux* in units of $\nu_\alpha S/\nu_c$ at the inner boundary $\tau = \tau_1$.

Though $F_\alpha^*(\tau)$ and $F_\alpha(\tau)$ are related simply it is yet of interest to see in one special case what regions of the nebula suffer from inward radiative accebration due to line absorption. The case $\tau_1 = 2$ has been examined

Table 8. Values of β.

τ_1	$p = 0.9$	$p = 0.75$	$p = 2/3$	$p = 1/2$	$p = 1/3$	$p = 0$
0	0	0	0	0	0	0
0.25	0.016	0.039	0.051	0.070	0.087	0.115
0.50	0 042	0.090	0.112	0.146	0.174	0.213
0.75	0.074	0.148	0.178	0.222	0.254	0.297
1.00	0.111	0.207	0.242	0.291	0.324	0.368
1.25	0.151	0.265	0.302	0.352	0.387	0.429
1.50	0.193	0.318	0.357	0.409	0.442	0.482
1.75	0.234	0.368	0 407	0.458	0.489	0.528
2.00	0.274	0.413	0.452	0.501	0.531	0.568
3.00	0.418	0,556	0.588	0.630	0.654	0.683

Table 9. Flux calculations. $F_a^*(\tau)$, $\tau_1 = 2$.

τ	$p = 0.9$	$p = 0.75$	$p = {}^2/_3$	$p = {}^1/_2$	$p = {}^1/_3$	$p = 0$
0	0.186	0.261	0.279	0.293	0.298	0.298
0.25	0.161	0.224	0.239	0.253	0.258	0.259
0.50	0.121	0.174	0.188	0.201	0.206	0.209
0.75	0.076	0.113	0.123	0.135	0.140	0.146
1.00	0.023	0.038	0.043	0.051	0 057	0.065
1.25	— 0.038	- 0.052	— 0.052	0.050	— 0.047	— 0.040
1.50	— 0.110	— 0.156	— 0.166	— 0.174	— 0.176	— 0.174
1.75	— 0.188	— 0.275	— 0.299	— 0.324	— 0.336	— 0.346
2.00	— 0.274	— 0.413	— 0.452	— 0.501	— 0.531	— 0.567

and the results of the calculations summarized in Table 9. (See also fig. 2.) We notice that the regions of the planetary nebula that suffer this inward acceleration is more or less independent of p.

Finally we may note that the flux at the inner boundary according to (94) and (109) can be expressed approximately as

$$F_a^*(\tau_1) = -\frac{4\,\nu_a}{3\,\nu_c\,\tau_1}\left[\frac{3}{4}S\,(\tau_1 + e^{-\tau_1} - 1) - \left(1 - \frac{1}{\lambda}\right)\frac{3\,p}{4\,(2 - 3\,p)}\,S\right]. \quad (111)$$

This completes our discussion of the general problem of the radiative equilibrium of a planetary nebula. Applications to the general theory of emission lines mentioned in the introduction will be treated separately.

Cambridge, Trinity College, 1934 November 5.

The Nebulium Emission in Planetary Nebulae.

By **S. Chandrasekhar** (Cambridge).

Received January 24, 1935.

In this paper the radiative equilibrium of a planetary nebula in the field of the nebulium radiation is considered and it is shown that the flux and the density of radiation in the nebulium lines vary from the inner to the outer boundary of the planetary nebula in exactly the same way as in the field of the Lyman-α radiation. It is further shown that the 'mean' frequency ν_c of the ultraviolet radiation has to be obtained by averaging the frequencies beyond the head of the Lyman series weighting each frequency by the number of quanta of that frequency in the incident star light.

1. In discussions[1]) of the radiative equilibrium of a planetary nebula one has so far restricted oneself to a consideration of the fields of radiation in the Lyman-α and in the ultraviolet radiation beyond the head of the Lyman series. The question however of the radiative equilibrium in the field of the nebulium lines has not so far been studied. It is, however, quite easy to write down the equations of transfer and of radiative equilibrium (under quite general circumstances) following the general ideas of Bowen[2]) and of Zanstra[3]). In this paper we shall confine ourselves to the more elementary problem of a single nebulium line the emission in which arises according to Bowen's suggestion, namely that of electron excitation. This idealization is not so very bad since observations show that the main nebular emission (apart from the hydrogen lines) is in the N_1, N_2 nebulium lines and the separation between the two lines is sufficiently small to regard them as a single line with their combined strength. The Bowen-Zanstra theory for the emission in the N_1, N_2 lines is that they arise from transitions from a higher metastable state to the ground state the atoms being raised to the metastable state directly by the high speed electrons liberated by the primary mechanism of the photoelectric ionization of the hydrogen atoms.

The equation of transfer can be written as usual in the form

$$\cos\Theta \frac{d I_N}{d \tau_N} = I_N - B_N(\tau_N). \tag{1}$$

---

[1]) V. A. Ambarzumian, Pulkova Obs. Bulletin **13**, 3, 1933. Also S. Chandrasekhar, ZS. f. Astrophys. **9**, 266, 1935. This last paper will be referred to as *R*. — [2]) I. S. Bowen, Astrophys. Journ. **67**, 1, 1928. — [3]) H. Zanstra, Publ. Dom. Astro. Obs. **4**, No. 15.

Where the suffix N is used to denote the fact that we are dealing with nebulium-radiation; τ_N is the optical depth defined by

$$\tau_N = \int_x^{\text{outer boundary}} \alpha_N \varrho \, d x. \tag{2}$$

Further in (1) $4\pi B_N(\tau_N)\, d\tau_N$ represents the amount of energy of nebulium quanta emitted in the layer τ_N, $\tau_N + d\tau_N$ per second. We shall assume that the ratio of the absorption coefficient α_N in the nebulium lines to the absorption coefficient $\varkappa$ for the ultra-violet radiation is a constant (say w):

$$\frac{\tau_N}{\tau} = \frac{\alpha_N}{\varkappa} = w. \tag{3}$$

To obtain the equation of radiative equilibrium we argue as follows: In the layer between τ and $\tau + d\tau$ the number of ultraviolet quanta absorbed is [R equation (8)]

$$\frac{4\pi}{h\nu_c}\left\{J(\tau) + \frac{1}{4} S e^{-(\tau_1 - \tau)}\right\} d\tau, \tag{4}$$

where ν_c is the 'mean' frequency of the ultraviolet quanta. (We shall specify in the sequal how this 'mean' has to be obtained.) Now each ultraviolet quantum that is absorbed liberates a photoelectron with energy $h(\nu_c - \nu_0)$ where ν_0 represents the frequency of the head of the Lyman series. Hence the total energy of the liberated free electrons is given by

$$4\pi \frac{\nu_c - \nu_0}{\nu_c}\left\{J(\tau) + \frac{1}{4} S e^{-(\tau_1 - \tau)}\right\} d\tau. \tag{5}$$

All this energy is *not* used up to excite the nebulium spectrum, since a fraction p of the number of ultraviolet quanta specified by (4) are re-emitted as ultraviolet quanta. Hence it is only a fraction $(1 - p)$ of the energy of the liberated electrons that is used up in the excitation of the nebulium spectrum. Hence the energy emitted in the nebulium spectrum due to this process of electron-excitation is [cf. R equation (8)]

$$\frac{\nu_c - \nu_0}{\nu_c} \cdot \frac{1 - p}{p} 4\pi B(\tau)\, d\tau. \tag{6}$$

Hence our equation of radiative equilibrium is (by (3))

$$B_N(\tau_N) = J_N(\tau_N) + \frac{\nu_c - \nu_0}{\nu_c} \cdot \frac{1 - p}{w p} B(\tau_N). \tag{7}$$

(7) can be compared with the corresponding equation of radiative equilibrium for Lyman-α which is [R equation (60)]

$$B_\alpha(\tau_\alpha) = J_\alpha(\tau_\alpha) + \frac{\nu_\alpha}{\nu_c} \frac{1 - p}{\omega p} B(\tau_\alpha). \tag{8}$$

We notice the *formal* identity of the two equations with the difference that in (7) $(\nu_c - \nu_0)/\nu_c$ replaces ν_α/ν_c. From the solution for the field of the Lyman-α radiation already obtained in R we can now therefore write down the corresponding results for the field of the nebulium-radiation. Thus for the flux πF_N in the nebulium-line we have [R equation (75)]

$$F_N(\tau) = \frac{\nu_c - \nu_0}{\nu_c} \{S(1 - e^{-(\tau_1 - \tau)}) - F(\tau)\} \cdot \tag{9}$$

Also we have approximately [R equation (92)]

$$J_N(\tau) = \frac{\nu_c - \nu_0}{\nu_c} w\big(J_1(\tau) - J(\tau) + J(0) - J_1(0)\big). \tag{10}$$

From (9) and the corresponding result for L_α, namely

$$F_\alpha(\tau) = \frac{\nu_\alpha}{\nu_c} \{S(1 - e^{-(\tau_1 - \tau)}) - F(\tau)\}, \tag{11}$$

we deduce that the flux in the nebulium line increases from the inner to the outer boundary in *exactly* the same way as it does in the L_α. In fact from (9) and (11) we have

$$\frac{F_N(\tau)}{F_\alpha(\tau)} = \frac{F_N(0)}{F_\alpha(0)} = \frac{\nu_c - \nu_0}{\nu_\alpha}. \tag{12}$$

It is easy to show that (12) is in fact equivalent to a combination of ZANSTRA's equations for determining the temperature of the nuclear star of a planetary-nebula from the total energies in the Balmer series and in the nebulium lines respectively. ZANSTRA's equations are, in his well known notation

$$\int_{x_0}^{\infty} \frac{x^2\,dx}{e^x - 1} = \sum_{\text{Balmer}} \frac{x^3}{e^x - 1} A_\nu, \tag{13}$$

$$\int_{x_0}^{\infty} \frac{x^3\,dx}{e^x - 1} - x_0 \int_{x_0}^{\infty} \frac{x^2\,dx}{e^x - 1} = \sum_{\text{Neb}} \frac{x^4}{e^x - 1} A_\nu. \tag{14}$$

Remembering that equation (13) expresses merely the equality of the number of ultraviolet quanta emitted by the central star to the number of the emergent Lyman-α quanta and further that equation (14) equates the energy of the liberated photoelectrons with the energy in the nebulium lines we have

$$\frac{\mathfrak{g}_3 - x_0\,\mathfrak{g}_2}{x_\alpha\,\mathfrak{g}_2} = \frac{F_N(0)}{\sum_{\text{Balmer}} L_p \frac{\nu_\alpha}{\nu_p}} = \frac{F_N(0)}{F_\alpha(0)}, \tag{15}$$ [1]

[1]) Equation (15) is also given in R. H. Stoy, M. N. **93**, 588, 1933.

where in (15) L_p denotes the energy in a monochromatic image of the Balmer series and $\mathfrak{g}_3$ and $\mathfrak{g}_2$ are used to denote the integrals

$$\mathfrak{g}_3 = \int_{x_0}^{\infty} \frac{x^3\, d\,x}{e^x - 1}; \quad \mathfrak{g}_2 = \int_{x_0}^{\infty} \frac{x^2\, d\,x}{e^x - 1}. \tag{16}$$

From (15), (16) and (12) we have that

$$\frac{\mathfrak{g}_3 - x_0\, \mathfrak{g}_2}{x_a\, \mathfrak{g}_2} = \frac{x_c - x_0}{x_a}, \tag{17}$$

or

$$x_c = \mathfrak{g}_3\, \mathfrak{g}_2^{-1} \tag{18}$$

(18) is equivalent to

$$\nu_c = \bar{\nu} = \int_{\nu_0}^{\infty} \nu \cdot \frac{\nu^2\, d\,\nu}{e^{h\,\nu/k\,T} - 1} \Big/ \int_{\nu_0}^{\infty} \frac{\nu^2\, d\,\nu}{e^{h\,\nu/k\,T} - 1}, \tag{19}$$

(l. e.) *the mean frequency of the ultraviolet quanta has to be determined by averaging the frequencies beyond the head of the Lyman series weighting, however, each frequency by the number of ultraviolet quanta in that frequency in the incident star light.* This is of course just what we should expect as in solving for the field of ultraviolet radiation a coefficient of absorption (independent of wavelength) beyond the head of the Lyman series has been assumed.

8. Again from the formal indentity of the equations of radiative equilebrium (7) and (9) it follows that even for the case of an expanding nebula (under the circumstances examined in *R*) the result (12) continues to be true. In particular we have (cf. *R* § 14)

$$F_N^*(\tau) = F_N(\tau) - \frac{4}{3\, w\, \tau_1} J_N(\tau_1) \cdot \tag{20}$$

The result (19) is thus seen to be quite general.

Cambridge, Trinity College, 1935, January 20.

IONIZATION AND RECOMBINATION IN THE THEORY OF STELLAR ABSORPTION LINES AND NEBULAR LUMINOSITY

S. CHANDRASEKHAR

ABSTRACT

In this paper the effect of collisions and photo-ionizations on monochromatic radiative equilibrium is considered. A formalism is developed which includes the theory of stellar absorption lines (Strömgren) and the theory of nebular luminosity (Zanstra) as two limiting cases. A second approximation for the nebular problem is also sketched. The mathematical analysis is devoted to the treatment of the three-state problem including collisions. When collisions can be neglected, the approximate equations of transfer used admit of two flux integrals and two K-integrals.

1. *Introduction.*—The theory of the formation of stellar absorption lines hinges on the problem of writing down the appropriate equation of radiative transfer. If we can write down the equation of transfer, then we have, in principle, solved the problem. But in the considerations leading to the equation of transfer, we can follow one of two methods: either argue in terms of macroscopic notions, e.g., the mass absorption and emission coefficients, etc., or in terms of microscopic notions, e.g., the Einstein transition probabilities, effective cross-section for captures, etc. The two methods, if properly handled, should of course lead to identical results; but a comparison of the two methods is often useful. Thus, in the so-called "combined Schuster-Schwarzschild problem" it is assumed that we can characterize the material by a mass absorption coefficient, κ, and a mass scattering coefficient, l. The equation of transfer is then written down in the form

$$\frac{dI_\nu}{\rho ds} = -(\kappa_\nu + l_\nu)I_\nu + \kappa_\nu B_\nu + l_\nu J_\nu \,, \tag{1}$$

where I_ν is the specific intensity at frequency ν, B_ν the Planck function, and

$$J_\nu = \int I_\nu \frac{d\omega}{4\pi} = \frac{1}{2}\int I_\nu \sin\theta d\theta \,. \tag{2}$$

In writing down (1), it is assumed that the emission consists of two parts: (a) the Kirchhoff emission, $\kappa_\nu B_\nu$, and (b) the emission due to scattering, of amount $l_\nu J_\nu$. On the other hand, as E. A. Milne[1] has shown, we can derive an equation similar to (1) by an entirely different type of discussion, which does not involve an appeal to Kirchhoff's law. The assumptions, (a) a Maxwellian distribution of velocities and (b) the effect of collisions in exciting and de-exciting the atoms, lead to the equation of transfer

$$\frac{dI_{\nu_{12}}}{dt_\nu} = - I_{\nu_{12}} + \frac{J_\nu + \eta_{\nu_{12}} B_{\nu_{12}}}{1 + \eta_{\nu_{12}}} , \tag{3}$$

where $dt_{\nu_{12}}$ is the optical depth measured in the direction of s,

$$\eta_{\nu_{12}} = \frac{b_{12}}{B_{12}} \frac{1}{B_{\nu_{12}}} , \tag{4}$$

where b_{12} and B_{12} are the probabilities of an atom in state 1 being excited to state 2 by collision or by the absorption of radiation of appropriate frequency. If we write

$$\eta_{\nu 12} = \frac{\kappa_\nu}{l_\nu} \tag{5}$$

and

$$dt_\nu = (\kappa_\nu + l_\nu)\rho ds , \tag{5'}$$

we see that (1) and (3) are formally equivalent. We shall return to the discussion of the physical meaning of (5) and (5'); but it is already clear that Milne's method, if properly interpreted, will remove some of the obscurity involved in the appeal to Kirchhoff's law in the usual treatment.

However, the theory of the formation of stellar absorption lines in the frame either of (1) or of (3) encounters difficulties—among others, the matter of the central intensities.

The physical foundations have been analyzed by B. Strömgren,[2]

[1] *M.N.*, **88**, 493, 1928.

[2] *Zs. f. Ap.*, **10**, 237, 1935; *Handbuch der Ap.* **7**, 221–235, 1936.

who incorporates into the theory at an early stage the influence of electron captures and photo-ionizations. Strömgren's theory is entirely satisfactory, but at the same time it would be of some interest to examine whether Milne's method can be suitably generalized to lead to results equivalent to those obtained by Strömgren's method. We shall see that this is possible.

There is another aspect to this problem. As is well known, Zanstra's theory of nebular luminosity utilizes the processes of electron captures and ionization as a means for the conversion of the ultraviolet energy beyond the head of the Lyman series in the incident starlight into energy mainly in Lyman α. The problem of radiative transfer associated with the Zanstra theory has been studied by V. A. Ambarzumian and the writer.[3] The question now arises whether we cannot construct a scheme which will include, as limiting cases, Strömgren's equation of transfer at one end and the equations of radiative transfer for the Zanstra theory at the other end. The main object of this paper is to supply such a scheme.

2. *Definitions and fundamental equations.*—The method consists in extending Milne's procedure to include cyclical transitions between three states, one of which we eventually identify with the continuum.

First let us consider the case of three discrete levels of statistical weights, g_1, g_2, g_3. We introduce the Einstein coefficients A_{nm}, B_{nm}, B_{mn}, defined with respect to the intensity of radiation. We have the usual relations

$$\frac{A_{nm}}{B_{mn}} = \frac{2h\nu_{mn}^3}{c^2}\frac{g_m}{g_n}, \tag{6}$$

$$\frac{B_{nm}}{B_{mn}} = \frac{g_m}{g_n}. \tag{7}$$

We shall use the abbreviation

$$\sigma_{mn} = \frac{2h\nu_{mn}^3}{c^2}. \tag{8}$$

[3] S. Chandrasekhar, *Zs. f. Ap.*, **9**, 266, 1935.

If $a_{mn}(\nu)$ is the atomic absorption coefficient for the atom in the state m for radiation of frequency ν, then

$$\int a_{mn}(\nu)d\nu = \frac{B_{mn}h\nu_{mn}}{4\pi}, \tag{9}$$

the integral on the left-hand side being taken through the narrow range of frequencies about ν_{mn} within which the atom can absorb. We write (9) in the form

$$a_{mn}\Delta\nu_{mn} = \frac{B_{mn}h\nu_{mn}}{4\pi}, \tag{10}$$

where a_{mn} is the mean atomic absorption coefficient in the interval $\Delta\nu_{mn}$. We shall next introduce the probability coefficients b_{mn}, a_{nm}, to take into account the effect of collisions in exciting and de-exciting the atoms in the states m and n, $(m < n)$—more precisely, the probability that in unit time an atom in state m, (n), is excited (de-excited) by collisions of the first (second) kind is b_{mn}, (a_{nm}). If we assume that a Maxwellian distribution of the atomic velocities corresponding to a certain temperature T is maintained, then we should have

$$\frac{b_{mn}}{a_{nm}} = \frac{g_n}{g_m} e^{-h\nu_{mn}/kT}. \tag{11}$$

Now, in a steady state the number of atoms, n_1, n_2, n_3, in each of the states must remain constant. Thus, the constancy of the number of atoms in the ground state 1 yields

$$n_1(B_{12}J_{12} + b_{12} + B_{13}J_{13} + b_{13}) - n_2(A_{21} + B_{21}J_{12} + a_{21}) \\ - n_3(A_{31} + B_{31}J_{13} + a_{31}) = 0, \tag{12}$$

where the J's are used to denote

$$J_{mn} = \int I_{mn}\frac{d\omega}{4\pi} = \frac{1}{2}\int I_{mn}\sin\theta d\theta. \tag{13}$$

Using (6), (7), (8), and (11), we can re-write (12) in the form

$$\left.\begin{aligned} n_1(B_{12}J_{12} + b_{12} + B_{13}J_{13} + b_{13}) - n_2\frac{g_1}{g_2}B_{12}\left(\sigma_{12} + J_{12} + \frac{b_{12}}{B_{12}}e^{h\nu_{12}/kT}\right) \\ - n_3\frac{g_1}{g_3}B_{13}\left(\sigma_{13} + J_{13} + \frac{b_{13}}{B_{13}}e^{h\nu_{13}/kT}\right) = 0\,. \end{aligned}\right\} \quad (14)$$

Similarly, the constancy of the number of atoms in state 3 gives

$$\left.\begin{aligned} n_1(B_{13}J_{13} + b_{13}) + n_2(B_{23}J_{23} + b_{23}) - n_3\left\{\frac{g_1}{g_3}B_{13}\left(\sigma_{13} + J_{13} + \frac{b_{13}}{B_{13}}e^{h\nu_{13}/kT}\right)\right. \\ \left. + \frac{g_2}{g_3}B_{23}\left(\sigma_{23} + J_{23} + \frac{b_{23}}{B_{23}}e^{h\nu_{23}/kT}\right)\right\} = 0\,. \end{aligned}\right\} \quad (15)$$

As may easily be verified, (14) and (15) automatically provide for the constancy of the number of atoms in the second state. We introduce the abbreviations

$$\hat{\omega}_{mn} = \frac{b_{mn}}{B_{mn}}\,, \ (m < n)\,, \quad (16)$$

$$\beta_{mn} = \hat{\omega}_{mn}\, e^{h\nu_{mn}/kT}, \ (m < n)\,, \quad (17)$$

$$\xi_{mn} = \sigma_{mn} + \beta_{mn} + J_{mn}\,. \quad (18)$$

Equations (14) and (15) now take the forms

$$n_1\{B_{12}(J_{12} + \hat{\omega}_{12}) + B_{13}(J_{13} + \hat{\omega}_{13})\}$$

$$- n_2\frac{g_1}{g_2}B_{12}\xi_{12} - n_3\frac{g_1}{g_3}B_{13}\xi_{13} = 0\,, \quad (19)$$

$$n_1B_{13}(J_{13} + \hat{\omega}_{13}) + n_2B_{23}(J_{23} + \hat{\omega}_{23})$$

$$- n_3\left\{\frac{g_1}{g_3}B_{13}\xi_{13} + \frac{g_2}{g_3}B_{23}\xi_{23}\right\} = 0\,, \quad (20)$$

which are our equations of radiative equilibrium.

So far, we have restricted ourselves to three discrete states. We now wish to make state 3 the "continuum," i.e., the transitions $1 \rightarrow 3$, $2 \rightarrow 3$, now correspond to photo-ionization. As Ambarzumian[4] has shown, we can treat the continuum on the same footing as a discrete state, provided we define for the weight g_3 the expression

$$g_3 = \frac{g_+}{N_e} \frac{(2\pi m_e kT)^{3/2}}{h^3}, \tag{21}$$

where g_+ is the statistical weight of the normal state of the ionized atom and N_e is the number of free electrons per cubic centimeter. Furthermore, B_{13} and B_{23} are now no longer atomic constants but are the probability coefficients of the photoelectric transitions defined in an appropriate way.

From (19) and (20), we derive

$$\left.\begin{aligned} &\frac{n_1}{\frac{g_1}{g_2} B_{12}\xi_{12}\left\{\frac{g_1}{g_3} B_{13}\xi_{13} + \frac{g_2}{g_3} B_{23}\xi_{23}\right\} + \frac{g_1}{g_3} B_{13}\xi_{13}B_{23}(J_{23} + \hat{\omega}_{23})} \\ &= \frac{n_2}{\{B_{12}(J_{12} + \hat{\omega}_{12}) + B_{13}(J_{13} + \hat{\omega}_{13})\}\left\{\frac{g_1}{g_3} B_{13}\xi_{13} + \frac{g_2}{g_3} B_{23}\xi_{23}\right\} - \frac{g_1}{g_3} B_{13}\xi_{13}B_{13}(J_{13} + \hat{\omega}_{13})} \\ &= \frac{n_3}{B_{23}(J_{23} + \hat{\omega}_{23})\{B_{12}(J_{12} + \hat{\omega}_{12}) + B_{13}(J_{13} + \hat{\omega}_{13})\} + \frac{g_1}{g_2} B_{12}\xi_{12}B_{13}(J_{13} + \hat{\omega}_{13})} \end{aligned}\right\} \tag{22}$$

3. *The equations of transfer.*—The equations of transfer are easily seen to be (cf. Milne, *loc. cit.*)

$$\frac{dI_{12}}{\alpha_{12}ds} = -\left(n_1 - n_2\frac{g_1}{g_2}\right)I_{12} + \frac{g_1}{g_2} n_2\sigma_{12}, \tag{23}$$

$$\frac{dI_{13}}{\alpha_{13}ds} = -\left(n_1 - n_3\frac{g_1}{g_3}\right)I_{13} + \frac{g_1}{g_3} n_3\sigma_{13}, \tag{24}$$

$$\frac{dI_{23}}{\alpha_{23}ds} = -\left(n_2 - n_3\frac{g_2}{g_3}\right)I_{23} + \frac{g_2}{g_3} n_3\sigma_{23}, \tag{25}$$

[4] *M.N.*, **95**, 469, 1935.

where the a_{mn}'s are given by (10). The optical depths dt_{12}, dt_{13}, and dt_{23} can be defined by

$$dt_{12} = \alpha_{12}\left(n_1 - n_2\frac{g_1}{g_2}\right)ds = \frac{B_{12}h\nu_{12}\left(n_1 - n_2\frac{g_1}{g_2}\right)}{4\pi\Delta\nu_{12}}ds\,, \tag{26}$$

$$dt_{13} = \alpha_{13}\left(n_1 - n_3\frac{g_1}{g_3}\right)ds = \frac{B_{13}h\nu_{13}\left(n_1 - n_3\frac{g_1}{g_3}\right)}{4\pi\Delta\nu_{13}}ds\,, \tag{27}$$

$$dt_{23} = \alpha_{23}\left(n_2 - n_3\frac{g_2}{g_3}\right)ds = \frac{B_{23}h\nu_{23}\left(n_2 - n_3\frac{g_2}{g_3}\right)}{4\pi\Delta\nu_{23}}ds\,. \tag{28}$$

It should be noticed that our optical depths, dt_{mn}, are measured in the direction in which the variation of I is considered. The equations (23), (24), and (25) now take the forms:

$$\frac{dI_{12}}{dt_{12}} = -I_{12} + \frac{n_2\frac{g_1}{g_2}}{n_1 - n_2\frac{g_1}{g_2}}\sigma_{12}\,, \tag{29}$$

$$\frac{dI_{13}}{dt_{13}} = -I_{13} + \frac{n_3\frac{g_1}{g_3}}{n_1 - n_3\frac{g_1}{g_3}}\sigma_{13}\,, \tag{30}$$

$$\frac{dI_{23}}{dt_{23}} = -I_{23} + \frac{n_2\frac{g_2}{g_3}}{n_2 - n_3\frac{g_2}{g_3}}\sigma_{23}\,. \tag{31}$$

It will be noticed that (29), (30), and (31) involve the n's only through the ratios (n_1/n_2), (n_1/n_3), (n_2/n_3), respectively. The solution (22) of the equations of radiative equilibrium provides the values of the required ratios in terms of the J's; and equations (29), (30), and (31) form a system of simultaneous differential equations. To evaluate the terms involving the n's we shall adopt a method of approximation introduced by Ambarzumian. We regard J_{12}/σ_{12} and J_{23}/σ_{23} as quantities of the first order of smallness, while we regard J_{13}/σ_{13} and $J_{12}J_{23}/\sigma_{12}\sigma_{23}$ as quantities of the second order of smallness. The justification for this is found in the circumstance that, near the star, J_{12}/σ_{12} and J_{23}/σ_{13} are of the order of magnitude $e^{-h\nu_{12}/kT}$ and $e^{-h\nu_{23}/kT}$, respectively, while J_{13}/σ_{13}, for instance, is of the order of magnitude $e^{-h(\nu_{12}+\nu_{23})/kT}$. On the other hand, if we are far away from the star, the effect of the dilution factor will be to make our approximation sufficient for practical purposes. Let

$$\hat{\omega}_{mn} = \eta_{mn} B_{\nu_{mn}} , \tag{32}$$

where $B_{\nu_{mn}}$ is the Planck function:

$$B_{\nu_{mn}} = \sigma_{mn}(e^{h\nu_{mn}/kT} - 1)^{-1} . \tag{33}$$

By (18) and (32) we have

$$\xi_{mn} = \sigma_{mn} + J_{mn} + \eta_{mn}\sigma_{mn}(1 - e^{-h\nu_{mn}/kT})^{-1} . \tag{34}$$

Hence, we have

$$\xi_{12} = \sigma_{12}(1 + \eta_{12}) + \text{first-order quantities} , \tag{35}$$

$$\xi_{23} = \sigma_{23}(1 + \eta_{23}) + \text{first-order quantities} , \tag{36}$$

$$\xi_{13} = \sigma_{13}(1 + \eta_{13}) + \text{second-order quantities} . \tag{37}$$

We evaluate the expressions occurring on the right-hand sides of the equations of transfer, using (22) consistent with our method of approximation. After some rather lengthy calculations, it is found that

$$\frac{n_2 \dfrac{g_1}{g_2}}{n_1 - n_2 \dfrac{g_1}{g_2}} = \mathbf{J}_{12}\left[1 - \frac{\mu_{13}\mu_{23}}{\mu_{12}(\mu_{13} + \mu_{23})}\mathbf{J}_{23}\right] + \frac{\mu_{13}\mu_{23}}{\mu_{12}(\mu_{13} + \mu_{23})}\mathbf{J}_{13} , \tag{38}$$

$$\frac{n_3 \dfrac{g_1}{g_3}}{n_1 - n_3 \dfrac{g_1}{g_3}} = \mathbf{J}_{13}\left[1 - \frac{\mu_{23}}{\mu_{13} + \mu_{23}}\right] + \frac{\mu_{23}}{\mu_{13} + \mu_{23}}\mathbf{J}_{23}\mathbf{J}_{12} , \tag{39}$$

$$\frac{n_3 \dfrac{g_2}{g_3}}{n_2 - n_3 \dfrac{g_2}{g_3}} = \mathbf{J}_{23}\left[1 - \frac{\mu_{13}}{\mu_{13} + \mu_{23}}\right] + \frac{\mu_{13}}{\mu_{13} + \mu_{23}}\frac{\mathbf{J}_{13}}{\mathbf{J}_{12}} , \tag{40}$$

where

$$\mu_{12} = g_1 B_{12}\sigma_{12}(1 + \eta_{12}) , \tag{41}$$

$$\mu_{13} = g_1 B_{13}\sigma_{13}(1 + \eta_{13}) , \tag{42}$$

$$\mu_{23} = g_2 B_{23}\sigma_{23}(1 + \eta_{23}) . \tag{43}$$

$$\mathbf{J}_{12} = \frac{J_{12} + \eta_{12}B_{\nu_{12}}}{\sigma_{12}(1 + \eta_{12})} , \tag{44}$$

$$\mathbf{J}_{13} = \frac{J_{13} + \eta_{13}B_{\nu_{13}}}{\sigma_{13}(1 + \eta_{13})} , \tag{45}$$

$$\mathbf{J}_{23} = \frac{J_{23} + \eta_{23}B_{\nu_{23}}}{\sigma_{23}(1 + \eta_{23})} . \tag{46}$$

Let

$$p = \frac{\mu_{13}}{\mu_{13} + \mu_{23}} ; \qquad 1 - p = \frac{\mu_{23}}{\mu_{13} + \mu_{23}} . \tag{47}$$

Then

$$\frac{\mu_{13}\mu_{23}}{\mu_{12}(\mu_{13} + \mu_{23})} = (1 - p)\frac{\mu_{13}}{\mu_{12}} . \tag{47'}$$

By (46) and (47) we can re-write (38), (39), and (40) in the forms

$$\frac{n_2 \dfrac{g_1}{g_2}}{n_1 - n_2 \dfrac{g_1}{g_2}} = \mathbf{J}_{12}\left[1 - (1 - p)\frac{\mu_{13}}{\mu_{12}}\mathbf{J}_{23}\right] + (1 - p)\frac{\mu_{13}}{\mu_{12}}\mathbf{J}_{13} , \tag{48}$$

$$\frac{n_3 \dfrac{g_1}{g_3}}{n_1 - n_3 \dfrac{g_1}{g_3}} = p\mathbf{J}_{13} + (1 - p)\mathbf{J}_{23}\mathbf{J}_{12} , \tag{49}$$

$$\frac{n_3 \dfrac{g_2}{g_3}}{n_2 - n_3 \dfrac{g_2}{g_3}} = (1 - p)\mathbf{J}_{23} + p\frac{\mathbf{J}_{13}}{\mathbf{J}_{12}} . \tag{50}$$

The equations of transfer now reduce to

$$\frac{dI_{12}}{dt_{12}} = - I_{12} + \left[1 - (1 - p)\frac{\mu_{13}}{\mu_{12}}\mathbf{J}_{23}\right]\sigma_{12}\mathbf{J}_{12} + (1 - p)\sigma_{12}\frac{\mu_{13}}{\mu_{12}}\mathbf{J}_{13}\,, \tag{51}$$

$$\frac{dI_{13}}{dt_{13}} = - I_{13} + p\sigma_{13}\mathbf{J}_{13} + (1 - p)\sigma_{13}\mathbf{J}_{23}\mathbf{J}_{12}\,, \tag{52}$$

$$\frac{dI_{23}}{dt_{23}} = - I_{23} + (1 - p)\sigma_{23}\mathbf{J}_{23} + p\sigma_{23}\frac{\mathbf{J}_{13}}{\mathbf{J}_{12}}\,. \tag{53}$$

We shall now proceed to consider two limiting cases of the foregoing system of equations.

4. *The formation of absorption lines: comparison with Strömgren's results.*—According to equations (44) and (51),

$$\frac{dI_{12}}{dt_{12}} = - I_{12} + (1 - \epsilon_{12})\frac{J_{12} + \eta_{12}B_{\nu_{12}}}{1 + \eta_{12}} + \sigma_{12}\epsilon_{12}\frac{\mathbf{J}_{13}}{\mathbf{J}_{23}}\,, \tag{54}$$

where we have used ϵ_{12} to denote

$$\epsilon_{12} = (1 - p)\frac{\mu_{13}}{\mu_{12}}\mathbf{J}_{23} = \frac{\mu_{23}\mu_{13}}{\mu_{12}(\mu_{13} + \mu_{23})}\mathbf{J}_{23}\,. \tag{55}$$

We re-write (54) in the form

$$(1 + \eta_{12})\frac{dI_{12}}{dt_{12}} = - (1 + \eta_{12})I_{12} + (1 - \epsilon_{12})(J_{12} + \eta_{12}B_{\nu_{12}}) + \epsilon_{12}\sigma_{12}(1 + \eta_{12})\frac{\mathbf{J}_{13}}{\mathbf{J}_{23}}\,, \tag{56}$$

or, again, after some further rearrangement of the terms,

$$(1 + \eta_{12})\frac{dI_{12}}{dt_{12}} = - (1 + \eta_{12})I_{12} + (1 - \epsilon_{12})J_{12} + \eta_{12}B_{\nu_{12}} + \epsilon_{12}Q_{12}B_{\nu_{12}}\,, \tag{57}$$

where we have used Q_{12} to denote

$$Q_{12} = \frac{\sigma_{12}(1 + \eta_{12})\mathbf{J}_{13} - \eta_{12}B_{\nu_{12}}\mathbf{J}_{23}}{B_{\nu_{12}}\mathbf{J}_{23}}\,. \tag{58}$$

The equation of transfer (57), which we have derived, includes the effect of collisions and of the cyclical transitions (resulting from ionization and captures) on monochromatic radiative equilibrium. Consequently, it can form the basis of an analysis of the formation of absorption lines, and as such we can regard η as small, compared with unity. This simplifies our expression for Q_{12}. We can write, according to (45), (46), and (58),

$$Q_{12} \simeq \frac{\sigma_{12}}{B_{\nu_{12}}} \frac{J_{13}}{\sigma_{13}} \frac{\sigma_{23}}{J_{23}} . \tag{59}$$

Further, we have

$$\frac{J_{13}}{\sigma_{13}} \simeq e^{-h\nu_{13}/kT_c} ; \qquad \frac{J_{23}}{\sigma_{23}} \simeq e^{-h\nu_{23}/kT_c} ; \qquad \frac{B_{\nu_{12}}}{\sigma_{12}} \simeq e^{-h\nu_{12}/kT} , \tag{60}$$

where we interpret T_c as the color temperature of the radiation. On the other hand, the Planck function $B_{\nu_{12}}$ is defined with respect to the temperature of the material. Hence, we have

$$Q_{12} \simeq e^{h\nu_{12}\left(\frac{1}{kT} - \frac{1}{kT_c}\right)} . \tag{61}$$

We can also simplify our expression for ϵ_{12}. According to equations (41), (42), (43), (47), and (55), we have

$$\epsilon_{12} \simeq \frac{g_2 B_{23}\sigma_{23} \cdot g_1 B_{13}\sigma_{13}}{g_1 B_{12}\sigma_{12}(g_1 B_{13}\sigma_{13} + g_2 B_{23}\sigma_{23})} \frac{J_{23}}{\sigma_{23}} . \tag{62}$$

Now, we can further assume that $B_{23} \ll B_{13}$, since the total probability of photo-ionizations from the excited state can be neglected, compared with the probability of photo-ionization from the ground state. We can thus write

$$\epsilon_{12} \simeq \frac{g_2}{g_1 B_{12}\sigma_{12}} B_{23} J_{23} , \tag{63}$$

or, using (6), we have

$$\epsilon_{12} \simeq \frac{B_{23} J_{23}}{A_{21}} . \tag{64}$$

The term $B_{23}J_{23}$ has a simple physical meaning: it represents the probability of photo-ionization of atoms in state 2 owing to the incident radiation field. If we denote this probability by C_{2f}, we have

$$\epsilon_{12} \simeq \frac{C_{2f}}{A_{21}}. \tag{65}$$

Now, Strömgren writes the equation of transfer in the form (*loc. cit.*, Eq. [37])

$$\frac{dI}{\rho ds} = -(\kappa_\nu + l_\nu)I_\nu + (1 - \epsilon)l_\nu J_\nu + \kappa_\nu B_\nu + \epsilon Q l_\nu B_\nu. \tag{66}$$

We verify that Strömgren's ϵ and Q are identical with our ϵ_{12} and Q_{12} (cf. our equations [65] and [61] with Strömgren's [18] and the equation at the top of p. 253 of his paper). We thus see that our equation of transfer (57) will become identical with Strömgren's (66) if we write in (57)

$$dt_{12} = (\kappa_{\nu_{12}} + l_{\nu_{12}})\rho ds, \tag{67}$$

$$\eta_{12} = \frac{\kappa_{\nu_{12}}}{l_{\nu_{12}}}. \tag{68}$$

We thus see that the correspondence between our treatment and Strömgren's problem is of the same nature as the correspondence between Milne's treatment and the combined Schuster-Schwarzschild problem.

In comparing our method with Strömgren's, the following points should be noted.

1. In our method we consider (strictly speaking) only a one-constituent atmosphere, the relevant physical processes being collisions, photo-ionizations, recombinations, and radiative resonance transitions. Consequently, the only processes which give rise to general opacity in our model are collisions of the first and second kinds. It is in this sense that equation (67) has to be interpreted.

2. Our method does not take into account the contributions to the general opacity arising from factors other than collisions and, further, entirely ignores the presence of other constituents in the at-

mosphere. In Strömgren's method, however, it is assumed that *if* the general opacity, κ_ν, owing to all factors and all the constituents of the atmosphere, is known, *then* the appropriate emission is $\kappa_\nu B_\nu(T)$, where T is the temperature of the material.

3. The origin of the identity of our ϵ_{12} and Q_{12} with Strömgren's ϵ and Q is to be found in the circumstance that in both the treatments the effect of photo-ionizations on monochromatic equilibrium is considered in physically equivalent terms.

It is thus clear that the fact that in the model case we have considered—namely, the three-state problem including collisions—our method leads to an equation of transfer identical with that derived by Strömgren, can be interpreted as implying the validity of the form in which use is made of Kirchhoff's law in Strömgren's method.

5. *The equations of the planetary nebulae.*—Returning to equations (51) and (52), we see that in the case of high dilution of radiation and in the absence of collisions we can write them in the forms

$$\frac{dI_{12}}{dt_{12}} = -I_{12} + J_{12} + \sigma_{12}(1-p)\frac{g_1 B_{13}\sigma_{13}}{g_1 B_{12}\sigma_{12}}\frac{J_{13}}{\sigma_{13}}, \tag{69}$$

$$\frac{dI_{13}}{dt_{13}} = -I_{13} + pJ_{13}, \tag{70}$$

where

$$p = \frac{g_1 B_{13}\sigma_{13}}{g_1 B_{13}\sigma_{13} + g_2 B_{23}\sigma_{23}}. \tag{71}$$

Equation (69) can be further simplified by using (10):

$$\frac{dI_{12}}{dt_{12}} = -I_{12} + J_{12} + (1-p)\frac{\alpha_{13}\Delta\nu_{13}}{\alpha_{12}\Delta\nu_{12}}\frac{\nu_{12}}{\nu_{13}}J_{13}. \tag{72}$$

These are precisely the equations of transfer in the form used by Ambarzumian and the writer. The probability p introduced in that theory is identical with our present p, defined according to (71). We thus see that our formalism developed in sections 2 and 3 of this paper includes the theory of stellar absorption lines and the theory of nebular luminosity as two limiting cases.

6. *The equations of the planetary nebulae: a second approximation.* —In the absence of collisions, our equations of transfer are (cf. Eqs. [51], [52], and [53]):

$$\frac{dI_{12}}{dt_{12}} = -I_{12} + J_{12}\left[1 - (1-p)\frac{\mu_{13}}{\mu_{12}}\frac{J_{23}}{\sigma_{23}}\right] + (1-p)\frac{\mu_{13}}{\mu_{12}}\frac{\sigma_{12}}{\sigma_{13}}J_{13}\,, \tag{73}$$

$$\frac{dI_{13}}{dt_{13}} = -I_{13} + pJ_{13} + (1-p)\frac{\sigma_{13}}{\sigma_{12}\sigma_{23}}J_{12}J_{23}\,, \tag{74}$$

$$\frac{dI_{23}}{dt_{23}} = -I_{23} + (1-p)J_{23} + p\frac{\sigma_{12}\sigma_{23}}{\sigma_{13}}\frac{J_{13}}{J_{12}}\,, \tag{75}$$

where p is defined as in equation (71) and

$$\mu_{13} = g_1B_{13}\sigma_{13}\,; \qquad \mu_{12} = g_1B_{12}\sigma_{12}\,. \tag{76}$$

Hence,

$$\frac{\mu_{13}}{\mu_{12}} = \frac{B_{13}\sigma_{13}}{B_{12}\sigma_{12}}\,. \tag{77}$$

Let

$$\gamma = \frac{\sigma_{13}}{\sigma_{12}\sigma_{23}}\,. \tag{78}$$

Then

$$\frac{dI_{12}}{dt_{12}} = -I_{12} + J_{12}\left[1 - \gamma(1-p)\frac{B_{13}}{B_{12}}J_{23}\right] + (1-p)\frac{B_{13}}{B_{12}}J_{13}\,, \tag{79}$$

$$\frac{dI_{13}}{dt_{13}} = -I_{13} + pJ_{13} + \gamma(1-p)J_{12}J_{23}\,, \tag{80}$$

$$\frac{dI_{23}}{dt_{23}} = -I_{23} + (1-p)J_{23} + \frac{p}{\gamma}\frac{J_{13}}{J_{12}}\,. \tag{81}$$

It should be remembered that in the foregoing equations the optical depths are measured in the direction in which the variation of I is considered.

From our definitions of dt_{12}, dt_{13}, dt_{23} we verify that

$$\frac{dt_{13}}{dt_{23}} = \frac{B_{13}\nu_{13}\Delta\nu_{23}}{B_{23}\nu_{23}\Delta\nu_{13}} \frac{n_1 - n_3 \dfrac{g_1}{g_3}}{n_2 - n_3 \dfrac{g_2}{g_3}}, \tag{82}$$

or, using (49) and (50) and remembering that for the case under consideration $\eta = 0$, we have

$$\frac{dt_{13}}{dt_{23}} = \frac{B_{13}\nu_{13}\Delta\nu_{23}}{B_{23}\nu_{23}\Delta\nu_{13}} \frac{g_1}{g_2} \frac{\sigma_{12}}{J_{12}} = \frac{g_1}{g_2} \frac{\alpha_{13}}{\alpha_{23}} \frac{\sigma_{12}}{J_{12}}. \tag{83}$$

Hence, equation (81) can be re-written as

$$\sigma_{12} \frac{dI_{23}}{dt_{13}} = \frac{g_2\alpha_{23}}{g_1\alpha_{13}} \{- I_{23}J_{12} + (1 - p)J_{23}J_{12} + p\gamma^{-1}J_{13}\} . \tag{84}$$

We now see that equations (79), (80), and (84) form a consistent system of equations. The right-hand side of equation (79) contains terms of the first and the second order of smallness, while the right sides of equations (80) and (84) contain terms only of the second order of smallness. Consequently, it would be sufficient to find the dominant term in dt_{13}/dt_{12}. We verify that, to the order of accuracy we are working with, we have

$$\frac{dt_{13}}{dt_{12}} = \frac{B_{13}\nu_{13}\Delta\nu_{12}}{B_{12}\nu_{12}\Delta\nu_{13}} = \frac{\alpha_{13}}{\alpha_{12}}. \tag{85}$$

We introduce the abbreviations

$$\frac{\nu_{12}\Delta\nu_{13}}{\nu_{13}\Delta\nu_{12}} = \delta_1 ; \qquad \frac{\nu_{23}\Delta\nu_{13}}{\nu_{13}\Delta\nu_{23}} = \delta_2 ; \tag{86}$$

$$\frac{\alpha_{13}}{\alpha_{12}} = \frac{1}{\delta_1} \frac{B_{13}}{B_{12}} = q_1 ; \qquad \frac{g_1}{g_2} \frac{\alpha_{13}}{\alpha_{23}} = q_2 . \tag{87}$$

Considering the case of material stratified in parallel planes and denoting $d\tau$ to measure the optical depth in (1, 3) normal to the plane of stratification, the equations of transfer reduce to

$$q_1 \cos\theta \frac{dI_{12}}{d\tau} = -I_{12} + J_{12}[1 - \gamma\delta_1 q_1(1-p)J_{23}] + \delta_1 q_1(1-p)J_{13}\,, \quad (88)$$

$$\cos\theta \frac{dI_{13}}{d\tau} = -I_{13} + pJ_{13} + \gamma(1-p)J_{12}J_{23}\,, \quad (89)$$

$$q_2\sigma_{12}\cos\theta \frac{dI_{23}}{d\tau} = -I_{23}J_{12} + (1-p)J_{23}J_{12} + p\gamma^{-1}J_{13}\,. \quad (90)$$

These equations admit of three first integrals. Multiply the equations successively by $\frac{1}{2}\sin\theta d\theta$ and $\frac{1}{2}\sin\theta\cos\theta d\theta$, and integrate from 0 to π. Let

$$F_{mn} = 2\int_0^\pi I_{mn}\sin\theta\cos\theta d\theta\,, \quad (91)$$

$$K_{mn} = \frac{1}{2}\int_0^\pi I_{mn}\sin^2\theta\cos\theta d\theta\,. \quad (92)$$

Then

$$\frac{1}{4}\frac{dF_{12}}{d\tau} = \delta_1(1-p)[J_{13} - \gamma J_{23}J_{12}]\,, \quad (93)$$

$$\frac{1}{4}\frac{dF_{13}}{d\tau} = (1-p)[\gamma J_{23}J_{12} - J_{13}]\,, \quad (94)$$

$$q_2\sigma_{12}\frac{1}{4}\frac{dF_{23}}{d\tau} = p\gamma^{-1}[J_{13} - \gamma J_{23}J_{12}]\,. \quad (95)$$

Also

$$q_1\frac{dK_{12}}{d\tau} = -\frac{1}{4}F_{12}\,, \quad (96)$$

$$\frac{dK_{13}}{d\tau} = -\frac{1}{4}F_{13}\,, \quad (97)$$

$$q_2\sigma_{12}\frac{dK_{23}}{d\tau} = -\frac{1}{4}F_{23}J_{12}\,. \quad (98)$$

From (93), (94), and (95) we derive

$$\frac{dF_{12}}{d\tau} = -\delta_1 \frac{dF_{13}}{d\tau}, \tag{99}$$

$$\frac{dF_{23}}{d\tau} = -\frac{p}{\gamma q_2 \sigma_{12}(1-p)} \frac{dF_{13}}{d\tau}. \tag{100}$$

We first notice that, according to (47), (76), (78), (83), and (86), we have

$$\frac{p}{\gamma q_2 \sigma_{12}(1-p)} = \delta_2 . \tag{101}$$

Hence, (99) and (100) yield the two flux integrals

$$F_{12} + \delta_1 F_{13} = \text{Constant} = S_{12} + \delta_1 S_{13} , \tag{102}$$

$$F_{23} + \delta_2 F_{13} = \text{Constant} = S_{23} + \delta_2 S_{23} , \tag{103}$$

where we have used S_{mn} to denote the "incident" flux. From (102) and (103) we derive

$$\delta_2(F_{12} - S_{12}) = \delta_1(F_{23} - S_{23}) . \tag{104}$$

We now derive a K-integral: from (96) and (97), we have

$$q_1 \frac{dK_{12}}{d\tau} + \delta_1 \frac{dK_{13}}{d\tau} = -\tfrac{1}{4}(F_{12} + \delta_1 F_{13}) , \tag{105}$$

or, by (102),

$$\frac{d}{d\tau}\{q_1 K_{12} + \delta_1 K_{13}\} = -\tfrac{1}{4}(S_{12} + \delta_1 S_{13}) . \tag{106}$$

Hence,

$$q_1 K_{12} + \delta_1 K_{13} = -\tfrac{1}{4}(S_{12} + \delta_1 S_{13})\tau + \text{constant} . \tag{107}$$

We easily verify that (107) is equivalent to an expression which the author had derived in his paper on the radiative equilibrium of a planetary nebula (*loc. cit.*, Eq. [87]). However, we have now derived it under more general conditions.

It is possible to derive another approximate K-integral if

$$\delta_2 S_{12} = \delta_1 S_{23} \,. \tag{108}$$

Then, according to (104), $\delta_2 F_{12} = \delta_1 F_{23}$, and we can re-write (98) as

$$\delta_1 q_2 \sigma_{12} \frac{dK_{23}}{d\tau} = -\tfrac{1}{4}\delta_2 F_{12} J_{12} \,, \tag{109}$$

or, by (96),

$$\delta_1 q_2 \sigma_{12} \frac{dK_{23}}{d\tau} = \delta_2 q_1 \frac{dK_{12}}{d\tau} J_{12} \,. \tag{110}$$

If we make the approximation

$$J_{12} = 3K_{12} \,, \tag{111}$$

we have

$$\delta_1 q_2 \sigma_{12} \frac{dK_{23}}{d\tau} = \tfrac{3}{2}\, \delta_2 q_1 \frac{d}{d\tau} (K_{12}^2) \,, \tag{112}$$

or

$$\sigma_{12} K_{23} = \frac{3}{2} \frac{\delta_2 q_1}{\delta_1 q_2} [K_{12}^2 + \text{constant}] \,, \tag{113}$$

which can be written alternatively as

$$\frac{K_{23}}{\sigma_{23}} = \frac{3}{2} \frac{\delta_2 q_1}{\delta_1 q_2} \frac{\sigma_{12}}{\sigma_{23}} \left[\left(\frac{K_{12}}{\sigma_{12}} \right)^2 + \text{constant} \right] . \tag{114}$$

Since $(K_{12}/\sigma_{12})^2$ is a quantity of the second order of smallness, while (K_{23}/σ_{23}) is one of the first order of smallness, it is apparent from (114) that, if we neglect quantities of the second order, then K_{23} is a constant.

The integrals (102), (103), (107), and (113) have been derived from the equations of transfer (88), (89), and (90), which are approximations to the exact equations of the three-state problem.

However, Henyey has since shown that integrals corresponding to those we have derived can be isolated for the exact problem.

We shall next consider a little more closely the case $S_{12} = S_{23} = 0$, so that the integral (113) is valid. Let S_{13} be the incident flux in the (1, 3) radiation. We can then write

$$F_{13} = S_{13}e^{-\tau} + \mathfrak{F}_{13} , \tag{115}$$

$$K_{13} = \tfrac{1}{4}S_{13}e^{-\tau} + \mathfrak{K}_{13} , \qquad J_{13} = \tfrac{1}{4}S_{13}e^{-\tau} + \mathfrak{J}_{13} \tag{116}$$

where $\mathfrak{F}_{13}$ and $\mathfrak{K}_{13}$ are defined for the scattered radiation. Our flux integrals can be written as

$$\delta_2 F_{12} = \delta_1\delta_2[S_{13}(1 - e^{-\tau}) - \mathfrak{F}_{13}] = \delta_1 F_{23} . \tag{117}$$

Furthermore, we have

$$q_1 K_{12} + \delta_1 K_{13} = -\tfrac{1}{4}\delta_1 S_{13}(\tau + C_1) , \tag{118}$$

where C_1 is a constant. From (93) and (96) we derive

$$q_1 \frac{d^2K_{12}}{d\tau^2} = \delta_1(1 - p)[\gamma J_{23}J_{12} - J_{13}] . \tag{119}$$

We shall use the approximation

$$J_{12} = 3K_{12} ; \qquad J_{23} = 3K_{23} ; \qquad \mathfrak{J}_{13} = 3\mathfrak{K}_{13} . \tag{120}$$

Then

$$q_1 \frac{d^2K_{12}}{d\tau^2} = 3\delta_1(1 - p)[3\gamma K_{23}K_{12} - K_{13} + \tfrac{1}{6}S_{13}e^{-\tau}] . \tag{121}$$

Using the K-integral in the form (118), the foregoing equation reduces to

$$\frac{d^2K_{12}}{d\tau^2} = 3(1 - p)\left[1 + \frac{3\delta_1\gamma}{q_1}K_{23}\right]K_{12} + \frac{3(1 - p)\delta_1 S_{13}}{4q_1} \times (\tau + \tfrac{2}{3}e^{-\tau} + C_1) . \tag{122}$$

Using our second K-integral and denoting by C_2 the constant of integration occurring in (113), we finally have

$$\frac{d^2K_{12}}{d\tau^2} = 3(1 - p)\left[1 + \frac{9}{2}\frac{\delta_2}{q_2}\frac{\sigma_{13}}{\sigma_{23}}\left\{\left(\frac{K_{12}}{\sigma_{12}}\right)^2 + C_2\right\}\right]K_{12}$$

$$+ \frac{3(1 - p)\delta_1 S_{13}}{4q_1}\left(\tau + \tfrac{2}{3}e^{-\tau} + C_1\right), \qquad (123)$$

which is the fundamental differential equation of the problem. After solving (123) for K_{12}, we determine F_{12} by using (96). Once K_{12} and F_{12} are known, all the other quantities can be derived successively, using the different integrals. Concrete applications of (123) will be found in a forthcoming paper.

YERKES OBSERVATORY
February 5, 1938

Acknowledgments

The author and publisher are grateful to the following societies and publishers for their permission to reprint in this volume papers that originally appeared in print under their auspices.

The American Philosophical Society
Cambridge University Press
Editions Hermann
Kluwer Academic Publishers
Macmillan Journals Limited (*Nature*)
The Observatory
Philosophical Magazine
The Royal Astronomical Society
The Royal Society
Springer-Verlag (*Zeitschrift für Astrophysik*)